P9-AGP-398

Extracellular Matrix

Volume 2

Molecular Components and Interactions

Extracellular Matrix

Volume 2

This book, divided into two parts aims to be an in-depth profile of the extracellular matrix, the first volume looking at the tissues where it is found and the second looking at its molecular components and its interactions with the host cells.

Volume 1
Central Nervous System
The Tectorial Membrane in the Inner Ear
Dentin
Skin
Lymphatic System
Basement Membranes
Arterial Wall
Intervertebral Disc
Articular Cartilage
Tendons and Ligaments
Synovial Fluid System
The Lung
The Glomerulus
Wharton's Jelly of the Umbilical Cord

Edited by Wayne D. Comper

Extracellular Matrix

Volume 2

Molecular Components and Interactions

edited by

Wayne D. Comper

Monash University
Melbourne, Australia

harwood academic publishers
Australia • Canada • China • France • Germany • India
Japan • Luxembourg • Malaysia • The Netherlands • Russia
Singapore • Switzerland • Thailand • United Kingdom

Emmaplein 5
1075 AW Amsterdam
The Netherlands

British Library Cataloguing in Publication Data

Extracellular matrix
Vol. 2: Molecular components and interactions
1. Extracellular matrix 2. Connective tissue 3. Extracellular matrix proteins
I. Comper, Wayne D.
574.1′9245

ISBN 3-7186-5843-7 (hardback)

ISBN 3-7186-5916-6 (softback)

CONTENTS

Volume 1

Preface

Our knowledge of extracellular matrix structure and function has increased enormously over the last decade or so. It is clearly a multidisciplinary area, encompassing cell biology, biochemistry, genetics, morphology, bioengineering, biophysics, physiology and pathology. For these reasons alone there seemed to be a compelling case to bring together the structural and functional aspects of the extracellular matrix. The series is intended to not only serve the postgraduate students entering the field, but aid researchers to come to terms with connective tissue areas outside their sphere of expertise.

The question arose, in the initial planning, as how to go about presenting a relatively comprehensive treatise on the topic. I was reminded of the former Minister of Science and Education in the Australian Labour Party (well known as a television quiz show champion), Mr. Barry Jones who described scientists as being of two types, namely the lumpers and the splitters. There were pejorative overtones for both categories. The splitters, of course, were those who had somewhat tunnel vision, focused on a particular problem and did not see the bigger picture. A very common picture painted by politicians of the scientific lobby in this country in their efforts to obtain research funds from the public purse. Instead, the bemused Jones would rather deal with the rarer species of lumpers, those that take in the ramifications of their science. Yet, it was recognised that lumpers as far as the scientific community is concerned could be thought to approach their work in a superficial and glib manner.

In some aspects the balance of the reductionist view as opposed to the holistic approach in extracellular matrix-connective tissue research is similar to Jone's lumpers and splitters. For reasons which are probably based on history, expediency, simplicity, available expertise there has been far greater emphasis on the structural characterisation of extracellular matrix components rather than understanding the functional aspects of the multi-mix of components that make up the matrix.

This series attempts to redress the imbalance. The first volume of the series incorporates chapters on particular tissues whose anatomy and physiological functions are discussed in terms of their constituent extracellular matrix components. This approach entails some inevitable overlap. The more conventional second volume describes the structural characteristics of the various groups of extracellular matrix molecules and relates structure to function wherever possible. The second volume also includes a major section on the cell-extracellular matrix interaction in terms of genetic influences, physicochemical influences, cell-surface interactions, and its role in morpho-genesis.

The original aim was to make these volumes comprehensive. By and large this has been achieved although there are a couple of omissions in the tissue volume. These omissions are not for the lack of trying but invariably occur when you try and get the only expert to write yet another chapter in a time period when they are all booked up.

In editing the volumes there has been relatively strict adherence to their organisation as described above in relation to the overriding theme of structure and function. The focus was not on metabolism/regulation, disease processes or molecular genetics. These topics have only been included if they contributed to the concept of structure and function.

Wayne D. Comper

Acknowledgements

By the time this series is published it will probably be three years in the making. I certainly would like to acknowledge all the authors who contributed to this series. They are all experts in their particular areas which makes me additionally grateful that they have been able to contribute. I would also like to express my appreciation to the staff of the Gordon and Breach office in Melbourne who were instrumental in getting this project off the ground. Their advice has been invaluable through much of the preparation of this work. I would also gladly thank the staff of Harwood Academic Publishers in the U.K. for their expert and generous support in the final stages of producing the series.

Dedicated to the memory of
Joy, Rowley, Robin, Barry,
Bill and Joyce, Bill and
Nance, and John.

Contributors

John F. Bateman
Orthopaedic Molecular Biology
Research Unit
Department of Paediatrics
University of Melbourne
Royal Children's Hospital
Parkville, Victoria 3052
Australia

Mina J. Bissell
Life Sciences Division
Lawrence Berkeley Laboratory
University of California
Berkeley, CA 94720
United States of America

Thomas K. Borg
Department of Developmental Biology
and Anatomy
University of South Carolina
Columbia, SC 29208
United States of America

Nancy Boudreau
Life Sciences Division
Lawrence Berkeley Laboratory
University of California
Berkeley, CA 94720
United States of America

Michael D. Buschmann
Institute of Biomedical Engineering
Ecole Polytechnique
University of Montreal
PO Box 6079, Station Center-ville
Montreal, Quebec, H3C 3A7
Canada

Edward G. Cleary
Department of Pathology
University of Adelaide
Adelaide, SA 5005
Australia

Wayne D. Comper
Department of Biochemistry and
Molecular Biology
Monash University
Clayton, Victoria 3168
Australia

Amanda J. Fosang
Orthopaedic Molecular Biology
Research Unit
Department of Paediatrics
University of Melbourne
Royal Children's Hospital
Parkville, Victoria 3052
Australia

Eliot H. Frank
Continuum, Electromechanics Group
Department of Electrical Engineering
and Computer Science
Department of Mechanical
Engineering
Massachusetts Institute of Technology

Cambridge, MA 02139
United States of America

J. Robert E. Fraser
Department of Biochemistry and Molecular Biology
Monash University Clayton,
Victoria 3168
Australia

John T. Gallagher
CRC Department of Medical Oncology
University of Manchester
Wilmslow Road
Manchester M20 9BX
United Kingdom

Mark A. Gibson
Department of Pathology
University of Adelaide
Adelaide, S.A. 5005
Australia

Alan J. Grodinzsky
Continuum, Electromechanics Group
Department of Electrical Engineering and Computer Science
Department of Mechanical Engineering
Massachusetts Institute of Technology
Cambridge, MA 02139
United States of America

Donald Gullberg
Department of Animal Physiology
University of Uppsala
Biomedical Center
Uppsala S-75123
Sweden

Timothy E. Hardingham
Wellcome Trust Center for Cell Matrix Research
School of Biological Sciences
University of Manchester
Stopford Building, Oxford Rd
Manchester, M13PT, UK

Staffan Johansson
Medical and Physiological Chemistry Department
University of Uppsala
Biomedical Center
Uppsala S-75123
Sweden

Young-Jo Kim
Continuum, Electromechanics Group
Department of Electrical Engineering and Computer Science
Department of Mechanical Engineering
Massachusetts Institute of Technology
Cambridge, MA 02139
United States of America

Shireen R. Lamandé
Orthopaedic Molecular Biology Research Unit
Department of Paediatrics
University of Melbourne
Royal Children's Hospital
Parkville, Victoria 3052
Australia

Torvard C. Laurent
Medical and Physiological Chemistry Department
University of Uppsala
Uppsala S-75123
Sweden

Stuart A. Newman
Department Cell Biology and Anatomy
New York Medical College
Valhalla, NY 10895
United States of America

John A.M. Ramshaw
Division of Biomolecular Engineering
CSIRO

Parkville, Victoria 3052
Australia

Rolf K. Reed
Department of Physiology
University of Bergen
Bergen
Norway

Kristofer Rubin
Medical and Physiological Chemistry Department
University of Uppsala
Biomedical Center
Uppsala S-75123
Sweden

Cecilia Rydén
Medical and Physiological Chemistry Department
University of Uppsala
Biomedical Center
Uppsala S-75123
Sweden

James J. Tomasek
Department of Anatomical Sciences
University of Oklahoma Health Sciences Center
Oklahoma City, OK 73190
United States of America

Bianca Tomasini-Johannson
Medical and Physiological Chemistry Department
University of Uppsala
Biomedical Center
Uppsala S-75123
Sweden

1 Water: Dynamic Aspects

Wayne D. Comper

Department of Biochemistry and Molecular Biology, Monash University, Clayton, Victoria, Australia

INTRODUCTION

Animal connective tissues are distinguished by the types, concentrations, and organisation of the material in their extracellular matrices (ECM). The macromolecular composition of the matrix to a large extent determines the matrix hydration which, in turn, determines tissue volume, creates space for molecular transport and for dynamic organisation, and offers compressive resistance (as water is essentially incompressible).

In the ECM, water is generally extrafibrillar (in relation to collagen fibres) and its distribution is specifically determined by the concentration of dissolved and retained proteoglycans and glycoproteins. These macromolecules will be important in affecting water distribution when their concentration is ≥2 mg/ml. The hydrodynamic processes controlling the water content in the ECM will be those of osmosis, filtration, swelling and diffusion. In many cases, the macromolecular components influencing these hydrodynamic processes are nonideal. The nonideality (which yields parameters which vary nonlinearly with concentration) influences osmotic pressure through excluded volume polymer interactions and, for charged polymers, through the influence of active counterions on the ambient simple electrolyte concentration. Nonideal effects are also manifested in dynamic parameters, particularly the hydrodynamical frictional terms that describe the viscous dissipation of water over the surface of the polymer chain.

This chapter addresses the general mechanisms of these various types of hydrodynamic processes. Where appropriate, specific structure-function relationships are discussed for ECM components, particularly glycosaminoglycans, in their dynamic interaction with water.

Issues not discussed in this chapter but have been extensively treated elsewhere include water 'binding', water structure and content in ECMs (Comper and Laurent, 1978), and the influence of collagen and fibers in the ECM on water flow (Levick, 1987).

Extracellular Matrix, Volume 2, Molecular Components and Interactions, edited by Wayne D. Comper.

Simple Diffusion

The Brownian thermal motion of ECM macromolecules arises through collisions with water molecules in the tissue. The energy transferred by these collisions accounts for the various vibrational modes of the macromolecule; the most important, in terms of the water content of the ECM is the energy associated with the translational motion relative to water. This type of motion is called translational diffusion and is manifested where there are macroscopic concentration gradients. The diffusional relaxation of a concentration gradient of solute is accompanied by a corresponding relaxation in the concentration gradient or chemical activity of water. This will give rise to a net volume flow of water that is essentially an osmotic flow arising from the gradient in chemical potential of water or osmotic pressure of solute. The translational diffusional process in an isothermal system is related to the concentration gradient of solute through Fick's law such that

$$J_1 = -D_1(\text{grad } c_1) \tag{1}$$

where J_1 is the flux of component 1 per unit area and c_1 the concentration of solute component 1 (moles per unit volume). The relationship between the molecular flux of component 1 and its concentration gradient is given by the translational (or mutual) diffusion coefficient D_1. The diffusion coefficient in a binary system for a volume-fixed frame (designated by the subscript v) has been derived by Bearman (1961) for component 1 and component 2 (solvent) as

$$(D_1)_v = (D_2)_v = (1 - \phi_1)(c_1/f_{12})(\partial\mu_1/\partial c_1)_{T,P} \tag{2}$$

where ϕ_1 is the volume fraction of 1, T is the temperature, P the pressure, μ_1 the chemical potential of 1 and f_{12} is the frictional factor between 1 mole of solute and water defined as (Spiegler, 1958)

$$-(\partial\mu_1/\partial x) = f_{12}(u_1 - u_2) \tag{3}$$

where u_i is the velocity of i for unidimensional transport.

Derivation of the translational mutual diffusion coefficient corresponding to common experimental standard conditions of constant temperature and chemical potential of solvent μ_2 (Kurata, 1982; Comper *et al.*,1986) gives

$$(D_1)_v = RT(1 - \phi_1)^2(M_1/f_{12})(\partial\Pi^*/\partial C_1)_{T,\mu_2} \tag{4}$$

$$(\partial\Pi^*/\partial C_1)_{T,\mu_2} = (1/M_1) + 2A_2C_1 + 3A_3C_1^2 + \ldots. \tag{5}$$

where R is the universal gas constant, M_1 is the molecular weight of component 1, Π^* the osmotic pressure at constant T and μ_2, and A_2 and A_3 are the standard osmotic second and third virial coefficients respectively. The diffusion coefficient in equation 4 is then a composite function of an hydrodynamic parameter associated with the frictional coefficient and a thermodynamic equilibrium term associated with osmotic pressure. Therefore, the diffusional flux of solute is associated with generating a reverse flow of solvent which is driven by the osmotic pressure gradient ie., the volume flux associated with this diffusional process will be exactly balanced by an opposite volume flux of solvent (Figure 1.1) so that

$$J_1V_1 = -J_2V_2 \tag{6}$$

$$= RT(1 - \phi_1)^2 v_1(M_1/f_{12})(\partial\Pi^*/\partial x) \tag{7}$$

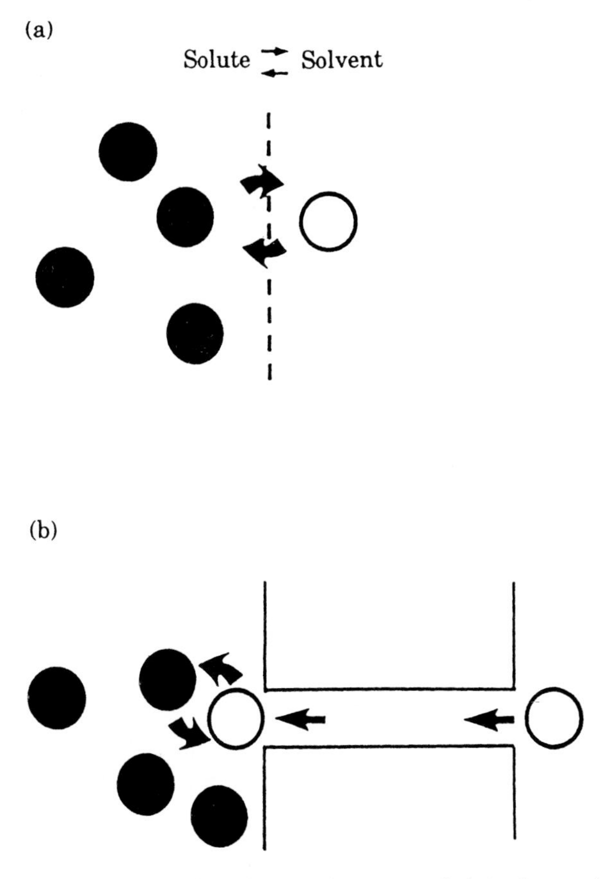

Figure 1.1 Schematic diagram of solute (filled circles)-solvent (empty circles) volume exchange a) across a free boundary and b) in a thin layer on the solution side of the semipermeable membrane.

where V_i is the partial molar volume and v_i the partial specific volume of i ($V_i = v_i M_i$). The right hand term of equation 7 could be thought of as an osmotically driven flow in free solution or the swelling dilution of component 1; the rate of which will vary as $\sqrt{(\text{time})}$.

Experimental Aspects of Diffusion

The diffusion coefficients defined by equation 1 of native Swarm rat chondrosarcoma proteoglycan aggregate (this proteoglycan is known to have similar structural and physical properties to the aggregating proteoglycans isolated from cartilage except that it does not contain keratan sulfate)(MW ~ 200×10^6), its constituent proteoglycan monomer or subunit (PGS)(MW ~ 2.6×10^6) and chondroitin sulfate (MW ~ 3.0×10^4) are markedly different in dilute solution but show molecular weight independence above concentrations of 5–10 mg/ml (Figure 1.2). The marked increase in the diffusion coefficient with concentration seen with these molecules is similar to that measured for commercial heparin and desulfated heparin. The marked increase in the diffusion coefficient with concentration is due to the predominance of the osmotic term (next section) over the hydrodynamical

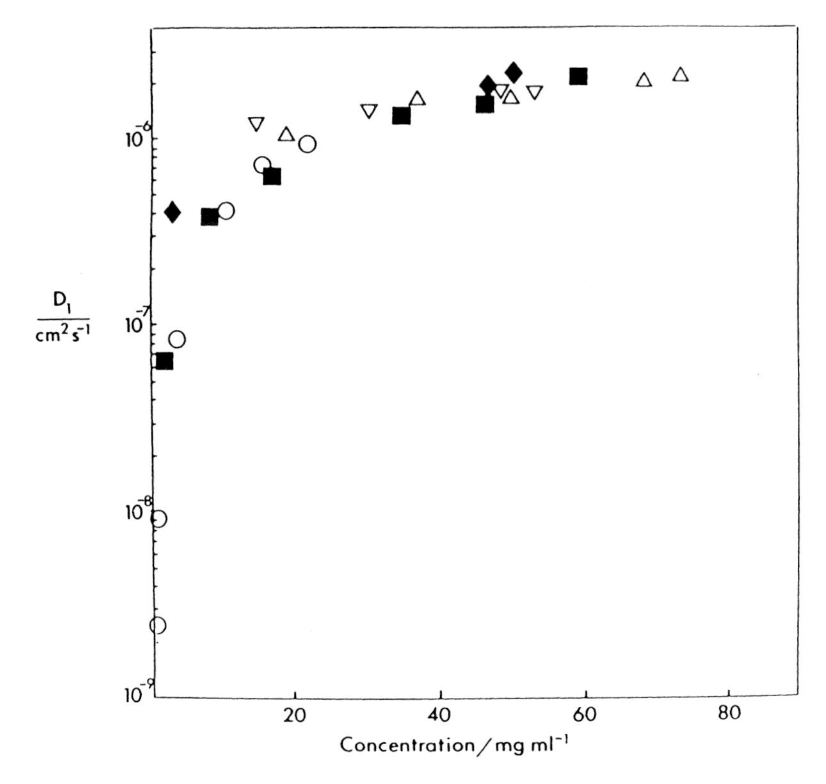

Figure 1.2 Variation of the mutual diffusion coefficient D_1, as a function of concentration for native Swarm rat chondosarcoma aggregate (○), its constituent proteoglycan monomer (■), chondroitin sulfate chains (♦), commercial heparin (△), and desulfated heparin (▽). All solutions were dialysed against phosphate buffered saline, ionic strength = 0.15, pH = 7.4 (from Comper *et al.*, 1990). Reproduced with the permission of the publishers.

frictional term. The molecular weight independence of the process demonstrates that individual molecules do not control the diffusion process but rather critical small polysaccharide segments of these molecules that have yet to be identified. It is extraordinary, though, that molecules that differ in size by a factor of 10^4 move at the same rates at higher concentrations. Further, the diffusion is independent of the degree of sulfation of the polysaccharide.

The translational diffusion of proteins is quite different. For example, the diffusion coefficient of albumin is relatively constant over a wide concentration range (Van Damme *et al.*, 1982) due, in part, to its relatively low osmotic pressure.

Osmotic Pressure

The osmotic pressure of glycosaminoglycans is very high per unit mass, because of their negative charge. We have measured osmotic pressure through sedimentation-diffusion technique in the analytical ultracentrifuge (Comper and Zamparo, 1990; Williams and Comper, 1990). Studies done in physiological saline, which is essentially 0.15 M NaCl solution which is buffered to pH 7.4, show that for the proteoglycan subunit from cartilage

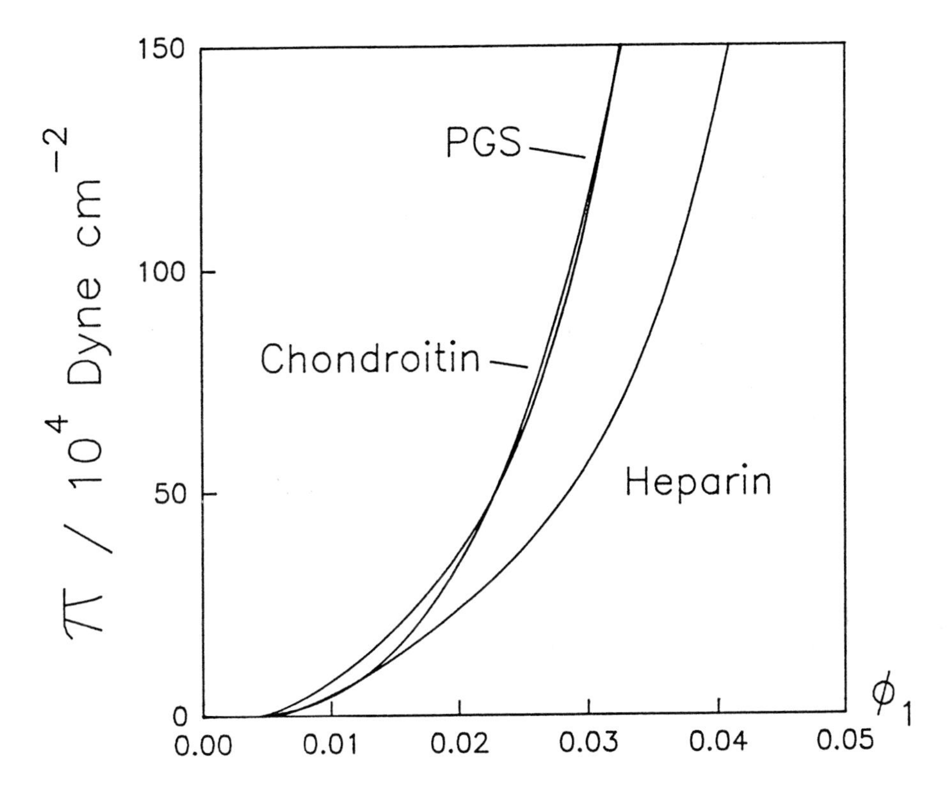

Figure 1.3 Osmotic pressure of PGS, chondroitin and heparin in thermodynamic equilibrium with phosphate-buffered saline (from Comper and Zamparo, 1990). Reproduced with the permission of the Biochemical Society & Portland Press.

the osmotic pressure increases rapidly at high concentration (Figure 1.3). The osmotic pressure of this chondroitin sulfate proteoglycan has been shown to be much higher than that of the more highly charged dextran sulfate and uncharged dextran when compared on a mass basis (Williams and Comper, 1990). Studies on hyaluronate (Preston *et al.*, 1965), an unsulfated glycosaminoglycan, show similar osmotic pressure to PGS (Williams and Comper, 1990). Consistent with this result is the finding that when the sulfate groups are chemically removed from chondroitin sulfate to form chondroitin the osmotic pressure does not change significantly (Figure 1.3). Removal of the sulfate group results in an approximately 20% reduction in the molecular weight of the disaccharide unit (mol. wt. of chondroitin sulfate disaccharide unit in the sodium form is 504 whereas that for chondroitin is 402) which may reflect, in part, the slightly higher osmotic pressure of chondroitin as seen in Figure 1.3. Yet the chondroitin and PGS have significantly higher osmotic pressures than the more highly charged heparin. The high osmotic pressure of the cartilage proteoglycan is obviously important to cartilage swelling but these results suggests that on a mass basis the sulfate group on the polysaccharide chain is not required for this swelling property. What is required is at least one anion charge per disaccharide residue and on a mass basis this is provided for most efficiently by the carboxyl group on the uronic acid residue.

Tracer Water Diffusion

Experimental studies of a variety of biological macromolecules, including albumin and proteoglycans, have demonstrated that the value of water mutual diffusion, $(D_1)_v$ according to equation 2 may vary from $10^{-9} - 10^{-6}$ cm^2 s^{-1} (Van Damme *et al.*, 1982; Comper *et al.*, 1986; Comper and Williams, 1987; Comper and Zamparo, 1990; Comper *et al.*, 1990) for semidilute solutions (volume fraction up to 0.1). This type of diffusion is quite different to the diffusion coefficients measured using water tracers such as D_2O and tritiated water. Diffusion of labelled water designated as component 2^* in the presence of component 1 may be derived (Bearman, 1961), with the condition that $c_{2^*} << c_2$, so that

$$D_{2^*} = RT/(f_{2^*2} + f_{2^*1}) \tag{8}$$

For polymer solutions with $\phi_1 \rightarrow 0.1$ the value of D_{2^*} will vary from ~2×10^{-5} to 1.6×10^{-5} cm^2 s^{-1} (Lyons and Comper, 1993). These values far exceed that obtained for mutual diffusion. This demonstrates that the exchange diffusion process between labelled water and bulk water as embodied in the f_{2^*2} term dominates the magnitude of D_{2^*}. The influence of the interaction of tracer water with the polymer in terms of its effect on D_{2^*} is small. There is also no direct theoretical or experimental relationship between the mutual diffusion coefficient of water and the tracer diffusion coefficient of water (Bearman, 1961; Lyons and Comper, 1993). As such, caution should accompany studies where tracer diffusion has erroneously been used to quantitate the mutual diffusion-osmotic process (Ussing and Anderson, 1956; Mauro, 1957) (see below).

FLOW IN SYSTEMS WITH OSMOTIC AND MECHANICAL PRESSURE GRADIENTS

Hydraulic Conductivity

The influence of mechanical pressure gradients on osmotic gradients and component fluxes is directly measured in measurements of the sedimentation of macromolecules in the ultracentrifuge. Irreversible thermodynamics provides us with an expression for the solute flux $(J_1)_v$ associated with sedimentation velocity as (Fujita, 1962; Kurata, 1982)

$$(J_1)_v = (S_1)_v c_1 \omega^2 r - (D_1)_v (dc_1/dr) \tag{9}$$

where ω is the angular speed of the rotor (radians/s), r is the distance from the centre of the rotor, and $(S_1)_v$ is the sedimentation coefficient of solute (defined as component 1) in a volume-fixed frame of reference.

At high rotor speeds, the flux J_1 is determined primarily by the sedimentation coefficient. Expressions for $(S_1)_v$ have been derived previously (Comper *et al.*, 1986) in terms of the hydrodynamic frictional coefficient,

$$(S_1)_v = (1 - \rho v_1)(1 - \phi_1) M_1 / f_{12} \tag{10}$$

where ρ is the solution density.

Mijnlieff and Jaspers (1971) theoretically established that the sedimentation process was analogous to hydraulic permeability across polymer membranes. The hydraulic permeability k (introduced by D'Arcy) through a porous plug of polymer material or polymer membrane is given by

$$k = (J_v/\Delta P)\eta_2 \qquad (11)$$

where η_2 is the solvent viscosity. This may be related to the sedimentation coefficient through the equation

$$k = \eta_2 S_1/C_1(1 - (v_1/v_2)) \qquad (12)$$

Therefore the sedimentation velocity technique can be conveniently used to measure hydraulic permeability through polymer matrices at concentrations of C_1.

Experimental Aspects of Hydraulic Conductivity

Concentration Dependence

The concentration dependence of the specific hydraulic conductivity of the proteoglycan form of chondroitin sulfate as a function of volume fraction is shown in Figure 1.4. There seems to be little difference in the concentration dependence of proteoglycan as compared to its constituent chondroitin sulfate chains. The contribution of protein to the partial specific volume and volume fraction parameters essentially normalises out the *k*-values when

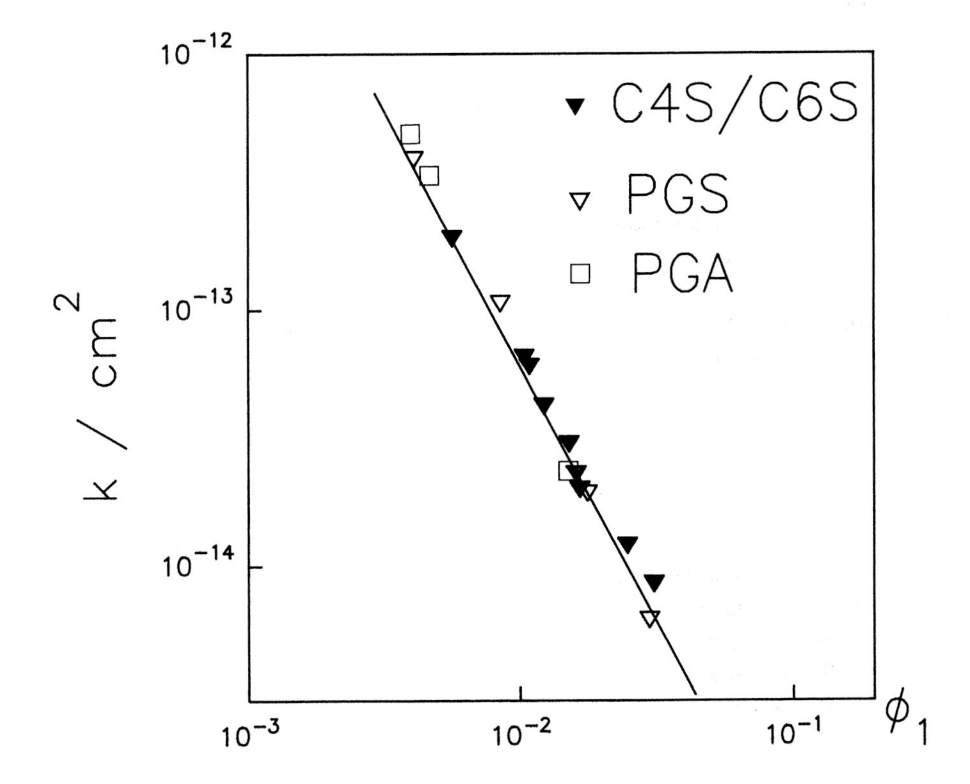

Figure 1.4 The specific hydraulic conductivity, *k*, as a function of volume fraction ϕ_1, for chondroitin sulfate (▼), Swarm rat chondrosarcoma proteoglycan subunit (monomer) (▽) and aggregate (□) in phosphate buffered saline. Lines drawn are those of best fit. Data from Williams and Comper (1987) and Comper and Zamparo (1990).

Table 1.1 Molecular Weight and Intrinsic Viscosity of Swarm Rat Chondrosarcoma Proteoglycan Preparations Measured in Phosphate-Buffered Saline (Comper and Williams, 1987)

Material	*Intrinsic Viscosity/ml g^{-1}*	*Molecular Weight/10^6*
Proteoglycan aggregate	950	211
Proteoglycan	265	2.6
Chondroitin sulfate chain	40[a]	0.02

[a] from A.Wasteson, *Biochem. J.*, 1971, *122*, 477 for chondroitin sulfate of MW = 20000

the data are expressed in terms of volume fraction. The protein core, the branched structure of the proteoglycan, the vast differences in the macroscopic viscosity and molecular weight of the different proteoglycan forms (Table 1.1) then appear to have little influence on flow resistance. Further we have previously established that there was little difference in the k-values for the chondroitin sulfate isomers namely chondroitin 4-sulfate and chondroitin 6-sulfate (Zamparo and Comper, 1989).

The molecular weight independence of k demonstrates that it is the interaction of water with segments of the polysaccharide chain that determines the magnitude of k. This is not surprising in view of the earlier data on the molecular weight independence of dextran sedimentation in semi-dilute solutions (Comper *et al.*, 1986; Ogston and Woods, 1954) and theoretical treatments involving scaling laws (de Gennes, 1979). The nature of the segment is still undefined but effective hydrodynamic units may take the form of genuine chain segments of the polymer around which flow occurs or an ensemble of segments to create a 'pore' within the transient network of the polymer solution.

Comparative Studies

Much of the stimulus for analysing the sedimentation of proteoglycans arose with the finding that chondroitin sulfate had considerably lower conductivity than other polysaccharides such as dextran and dextran sulfate, and other polymers such as albumin, poly(vinylalcohol) and poly(ethyleneglycol) (Figure 1.5). The relatively low conductivity or high resistance to flow per unit volume (or mass) for chondroitin sulfate seems to identify a specific structure-biomechanical functional relationship as its conductivity would generally account for the low conductivity found in cartilage (Zamparo and Comper, 1989). In this tissue, low conductivity is required for resistance to compression.

Variation in the glycosaminoglycan content of the proteoglycans would appear to make little difference in the magnitude of k as data in Figure 1.6 would suggest that chondroitin sulfate, hyaluronate, dermatan sulfate and keratan sulfate have similar specific hydraulic conductivities. This data clearly indicates the insensitivity of k to sulfation (as hyaluronate is not sulfated) and to the linear charge density of the glycosaminoglycan chain. Note also that while there is a 14% difference in the partial specific volume of chondroitin sulfate as compared to the unsulphated hyaluronate this does not manifest any major difference in the k-values of the two polysaccharides. Further evidence of the lack of influence of sulfation on the glycosaminoglycan chain is seen for k-data in Figure 1.7 for the comparison of desulfated chondroitin sulfate (chondroitin) to chondroitin sulfate and desulfated heparin to heparin.

The lack of influence of sulfation on the heparin chain would suggest that the range of

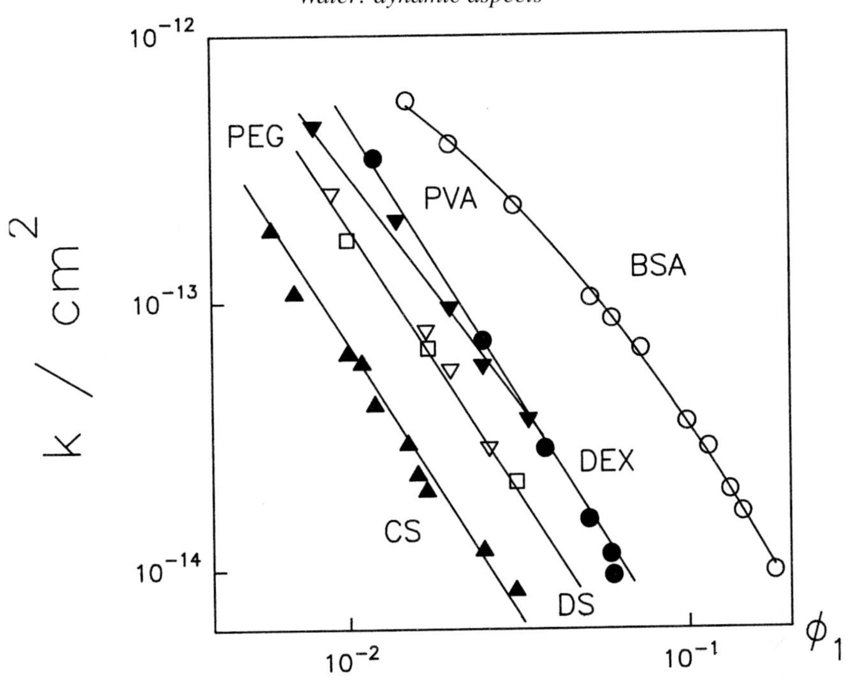

Figure 1.5 The specific hydraulic conductivity, k, as a function of volume fraction ϕ_1, for chondroitin sulfate (▲), dextran sulfate (□), poly(ethyleneglycol) (▽), dextran (●), poly(vinylalcohol) (▼) and bovine serum albumin (○) in phosphate buffered saline. Lines drawn are those of best fit. Modified from Comper and Zamparo (1989).

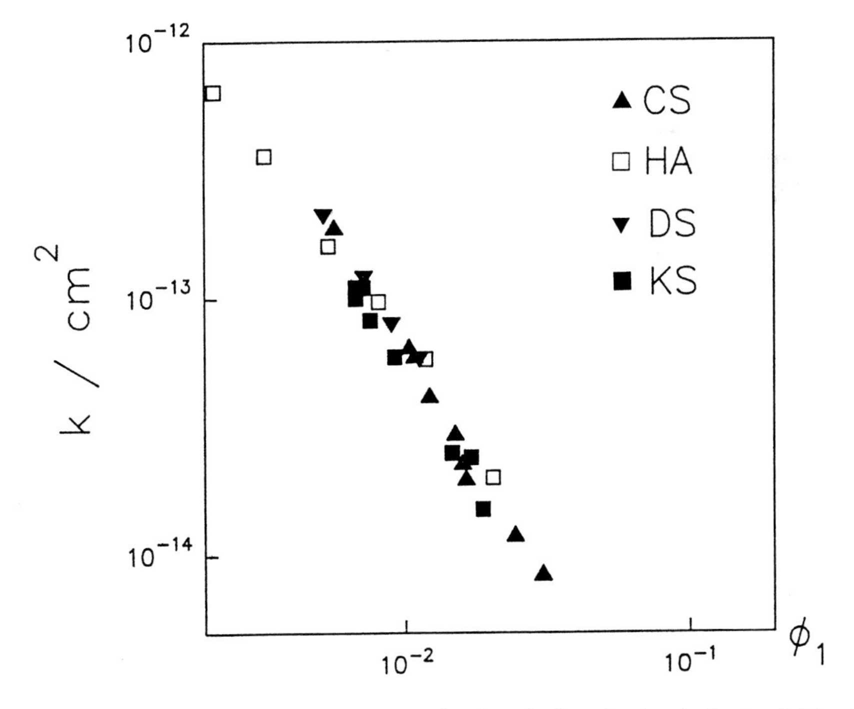

Figure 1.6 The specific hydraulic conductivity, k, as a function of volume fraction ϕ_1, for chondroitin sulfate (▲), hyaluronate (□), dermatan sulfate (▼) and keratan sulfate (■) in phosphate buffered saline. Data from Comper and Zamparo (1990).

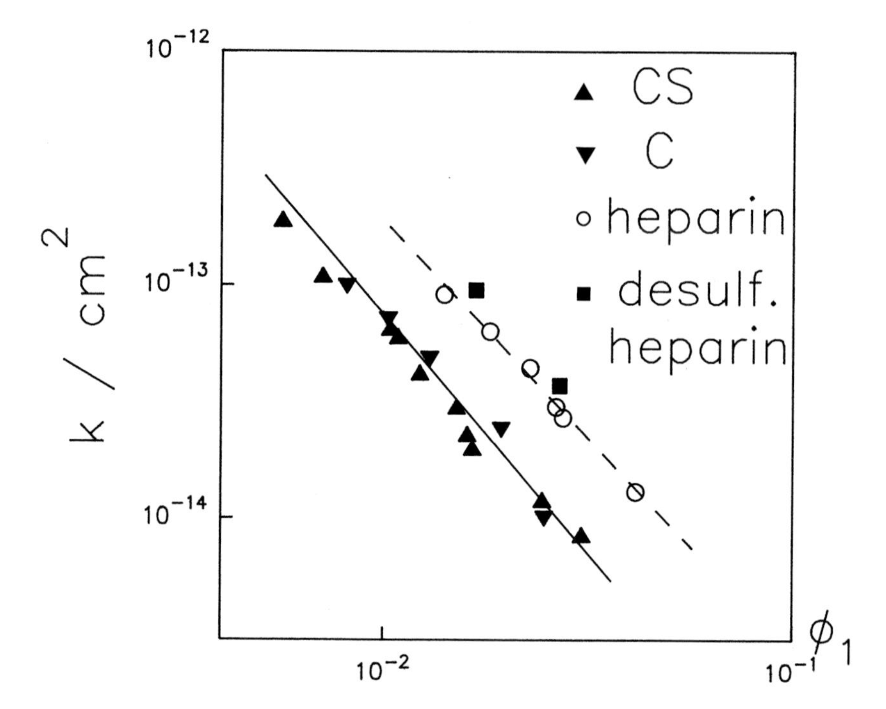

Figure 1.7 The specific hydraulic conductivity, k, as a function of volume fraction ϕ_1, for chondroitin sulfate (▲), chondroitin (desulfated chondroitin sulfate) (▼), heparin (○) and desulfated heparin (■) in phosphate buffered saline. Data from Comper and Zamparo (1990). Lines drawn are those of best fit.

heparin-like polysaccharides found in Nature most likely have similar conductivities. It is also clear from data in Figure 1.7 that the heparin-like polysaccharides have a considerably higher specific conductivity than the chondroitin sulfate/chondroitin polysaccharides. This could be viewed as consistent with biomechanical function as regions where the heparin-like polysaccharides are retained, such as the glomerular basement membrane, require high flows of water as distinct from flow resistance in cartilage which contains chondroitin sulfate proteoglycans. Recent studies by us have demonstrated that the glomerular basement membrane heparin-like polysaccharides, given that they were controlling water flow, would not be at sufficient concentration to exert a charge selective effect seen in kidney ultrafiltration (Zamparo and Comper, 1990). Rather, it turns out that the glomerular heparin-like polysaccharides exist at a low concentration (<1 mg/ml), probably do not control water flow in this system and do not exert a direct charge selective effect on kinetic charged probes as originally thought (Volume 1, Chapter 13).

Ionic Strength — pH Effects

The influence of high NaCl concentrations on hydraulic conductivity of the proteoglycan is small, with an increase in k of 1.2–1.6 fold (Figure 1.8). Even smaller effects were observed for chondroitin sulfate (similar results have also been obtained for heparin (Comper and Zamparo, 1990). Analysis of k for chondroitin sulfate at 36 mg/ml reveals also that there is only small changes in k over the NaCl concentration range of 0.035 to 1.0 mol dm^{-3}. Hydraulic conductivity of chondroitin sulfate at 23 mg/ml as a function of pH over the

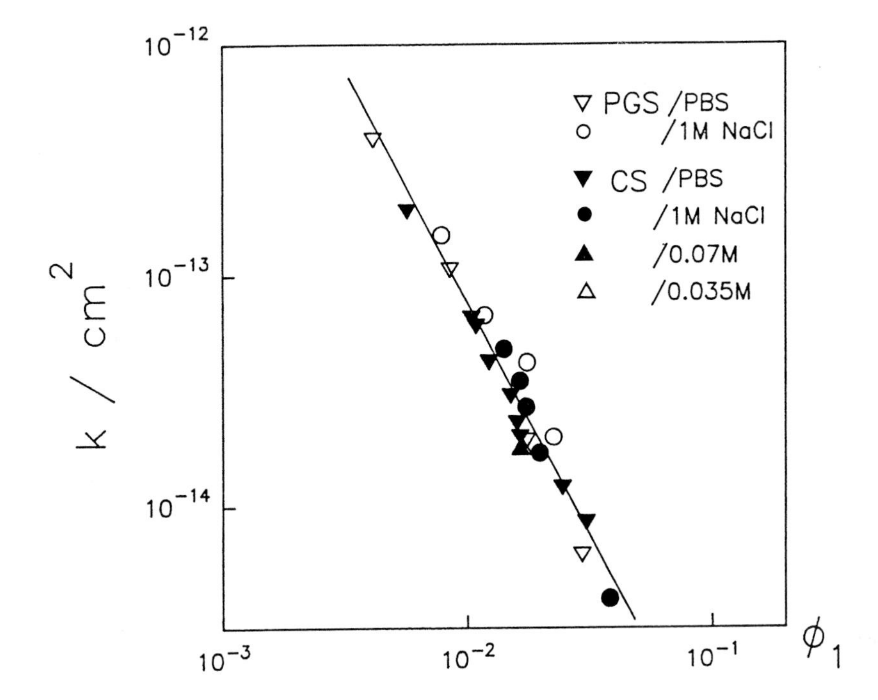

Figure 1.8 The specific hydraulic conductivity, k, as a function of volume fraction ϕ_1, for proteoglycan subunit in phosphate buffered saline (▽) and 1 mol dm^{-3} NaCl (○), and for chondroitin sulfate in phosphate buffered saline (▼), 1 mol dm^{-3} NaCl (●), 0.07 mol dm^{-3} NaCl (▲), and 0.035 mol dm^{-3} NaCl (△)(almost superimposed by the ▲ symbol). Data from Comper and Zamparo (1990) and Zamparo and Comper (1989).

range of 3.2 to 8.7 also exhibits no change (Comper and Lyons, 1993). These results are consistent with the lack of influence of the linear charge density, and particularly the degree of sulfation of the glycosaminoglycan chain on k as described previously. The low degree of electrolyte dissipation or the influence of charge separation brought about by the relative mobility of the polysaccharide chain as compared to its constituent counterions seen with glycosaminoglycan chains is due primarily to the low mobility of the polysaccharide chain and its influence on water. The studies demonstrate that the major influence of the polysaccharide is its chain conformation and flexibility that determines the low values of k.

Polysaccharide Flexibility

The difference seen for the polysaccharides represented in Figure 1.7 add to the discrimination we originally observed with chondroitin sulfate and dextran (Figure 1.5). These relative differences between various polysaccharide groups occur in spite of variations introduced through chemical modification such as the degree of sulfation.

The overriding factor associated with discriminating these polysaccharides appears to be associated with the nature of their major glycosidic linkage (a number of the polysaccharides studied have minor structural heterogeneity with respect to the type of glycosidic linkage but this issue is not raised here). This is demonstrated in Figure 1.9 where there appears to be three groups of data. The group showing the relatively lowest k values correspond to polysaccharides with alternating β-1,3 and β-1,4 linkages including

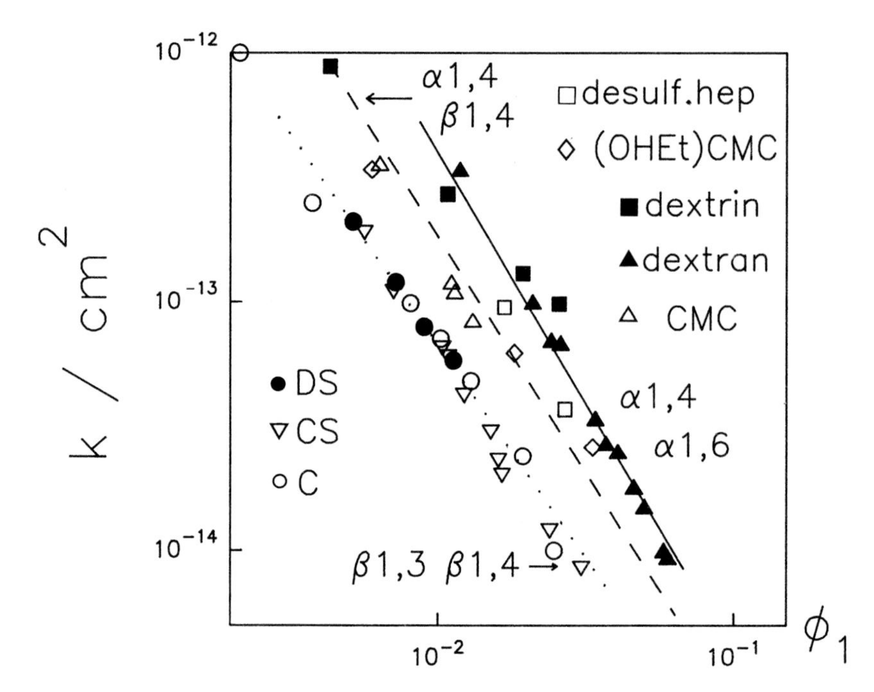

Figure 1.9 The specific hydraulic conductivity, k, as a function of volume fraction ϕ_1, for groups of polysaccharides defined by their major glycosidic linkage type. For the group with alternating β1,3 and β1,4 linkages we have dermatan sulfate (●), chondroitin sulfate (▽) and chondroitin (○). For the group with β1,4 linkages or alternating α1,4 and β1,4 (or α-L-1,4) linkages we have included desulfated heparin (□), carboxymethyl cellulose (△) and hydroxyethyl cellulose (◇) (from Sundelöf and Nyström, 1977). For the group with α1,4 linkages and/or α1,6 linkages we have included dextran (▲) and dextrin (■). Data compiled from Comper and Zamparo (1990) and Zamparo and Comper (1991) unless otherwise stated. Lines drawn are those of best fit.

dermatan sulfate, chondroitin sulfate, chondroitin and also from Figure 1.6 keratan sulfate and hyaluronate. The next group, with relatively higher values of k, correspond to polysaccharides with β-1,4 linkages or alternating α-1,4 and β-1,4 (or α-L-1,4) linkages as for the heparin-like polysaccharides (see also Figure 1.7), carboxymethyl cellulose and hydroxyethyl cellulose (Sundelöf and Nyström, 1977). The third group of data, generally exhibiting the highest value of k for any volume concentration, corresponds to polysaccharides with either α-1,6 linkages, like dextran, and α-1,4 and α-1,6 linkages for dextrin. The differentiation of these three major groups, based on the type of the glycosidic linkage, appears to be irrespective of quite major differences in the nature of the chemical groups substituted onto the native polysaccharide chain. For example, a comparison of the celluloses reveals that there is little difference between hydroxyethyl cellulose and carboxymethyl cellulose. Even more striking, in this context, is the similarity of the cellulose derivatives to the heparin-like polysaccharides, indicating the lack of influence of the iduronate group.

Details of the chain conformation of these polysaccharides, particularly with respect to conformational flexibility, unperturbed distances and water interaction are incomplete. However, some interesting data is being generated with regard to inter-residue H-bonding especially for the mammalian glycosaminoglycans. For β-1,3 and β-1,4 alternate linked glycosaminoglycans, the hydrogen-bonding groups, as well as the glycosidic oxygens, are equatorial. This allows formation of a cooperative H-bonding pattern between residues. For hyaluronate, four different hydrogen bonds within a repeat trisaccharide unit, have been identified namely glucosamine $C_{(4)}OH$ to the glucuronic ring oxygen atom, from the acetamido C=O to HO-2 of glucuronate, from the HO-3 of glucuronate to the hexosamine ring oxygen atom and from the amino NH to $O{=}C{-}O^-$ carboxyl of glucuronate (Atkins *et al.*, 1980, Heatley and Scott, 1988, Scott and Heatley, 1982, 1984). For chondroitin sulfates only three hydrogen bonds can form due to the fact that the bond from the axial galactosamine HO-4 to the glucuronate ring oxygen cannot form (Scott *et al.*, 1984). In any case, these inter-residue hydrogen bondings would clearly be expected to endow the critical hydrodynamic segments with decreased flexibility and create a stiffer chain segment. This may be a major discriminating factor in determining the correlation found between the type of polysaccharide glycosidic linkage and the magnitude of specific hydraulic conductivity. A possible exception to this interpretation is keratan sulfate as no inter-residue H-bonds have yet been identified for this glycosaminoglycan. We also know that the polysaccharides that have the relatively highest value of specific hydraulic conductivity, like α-1,6 dextrans, also have an extremely flexible type structure. For β-1,4 linked polysaccharides it has been suggested that for heparan sulfate that two possible inter-residue H-bonds may form (Scott and Heatley, 1982) yet the structure may have increased conformational flexibility through the presence of the iduronate residue (Casu *et al.*, 1988). Cellulose, too, may have inter residue H-bonds at the OH-3...O-5 and OH-2...OH-6 position (Marchessault and Sundararajan, 1983).

Therefore, the suggestion is made that a stiff-chain polysaccharide, generated through a sleeve of interresidue H-bonding yields a structure of greater surface area/volume ratio and hence offers greater resistance to flow. Efforts to disrupt the stiffness or hydrogen bonded secondary structure of chondroitin sulfate through temperature was not apparent as the variation of relative viscosity was similar to that of dextran (Comper and Lyons, 1993). The use of hydrogen bond breaking solvents were not particularly conclusive although sodium fluoride appeared to have some effect. The conclusion from these studies is that the effective stiff region was a small segment rather than extending comprehensively over the whole chain — which would have affected the viscosity measurements particularly. Deacetylation did not affect k but methylation of the carboxyl group seemed to have some effect by increasing k (Comper and Zamparo, 1990). The most striking feature result, however, of our efforts to affect the flexibility of the chain was the result obtained with periodate treatment of chondroitin sulfate. We registered a significant increase in the hydraulic conductivity (Comper and Lyons, 1993). We suggest that this was due to the increase in flexibility of critical segments in the chondroitin sulfate chain.

Combined Osmotic and Mechanical Effects

Using equations 2 and 4 in equation 9, we obtain

$$J_2 = (1 - \phi_1)(M_1/f_{12})[C_1\omega^2(1 - \rho v_1)(\partial r^2/2)/\partial r - (c_1/f_{12})(\partial\mu_1/\partial C_1)_{T,P}(dC_1/dr)] \tag{13}$$

where J_2 is the flux in mass units. This equation demonstrates the direct influence of the osmotic pressure, through the $(\partial\mu_1/\partial C_1)_{T,P}$ term, of the polymer matrix at relatively low mechanical centrifugal pressures. When $\omega \rightarrow 0$ the flux in equation 13 describes the diffusion process which is expressed as a function of the osmotic pressure gradient and f_{12} the diffusive mobility of solute. The dual nature of this frictional coefficient is readily evident in that it also describes the viscous dissipation processes occurring in sedimentation or hydraulic permeability. Furthermore, since the M_1/f_{12} term approaches molecular weight independence at high concentrations, the dissipation processes represent water interaction with relatively small critical polymer segments. Thus transient polymer networks (semi-dilute solutions) and cross-linked polymer networks (membranes) can be equivalent in terms of their water interaction.

The volume flux (J_v) of water/solvent in the sedimentation experiment can also be described by an equation similar to equation 13 since

$$J_v = J_2 v_2 = -J_1 v_1 \tag{14}$$

for a volume-fixed frame of reference. It is emphasised that the coefficient describing total solvent volume flow, namely $(1 - \phi_1)M_1/f_{12}$ will determine the rate of flow due to the osmotic pressure gradient (as in diffusion) and the rate of flow due to the pressure gradient created by the centripetal force ($C_1\omega^2(1 - \rho v_1)r^2/2$).

It is also of value to consider the implications of equation 13 to the standard equation relating volume flow across a semipermeable membrane such that (Kedem and Katchalsky, 1961; Katchalsky and Curran, 1967)

$$J_v = L_p{}^o(\Delta P - \Delta\Pi) \tag{15}$$

where $L_p{}^o$ is an hydraulic permeability coefficient and ΔP and $\Delta\Pi$ are the net pressure gradients across the membrane. The $L_p{}^o$ coefficient has been assumed to be related to the geometry of the membrane and been described as a function of the frictional interaction of solvent with the membrane. These considerations are qualitatively consistent with the coefficient in equation 13. However, the interpretation of the osmotic pressure term in equation 15 is at variance with that of equation 13. The centrifuge analogy clearly establishes that the internal osmotic pressure of the membrane should be included in the osmotic term as well as external osmotic gradients. Therefore, the expression

$$J_v = L_p{}^o[\Delta\Pi - (\Delta\Pi_s + \Pi_m)] \tag{16}$$

is more correct by including the osmotic pressure of the bathing solutions (DP_s) and an internal osmotic pressure function of the membrane (P_m)(=swelling pressure+elastic contribution). It is the latter term that is analogous to the osmotic term in equation 13 and which will be affected by all external pressure gradients on the membrane.

An example of where the correction in equation 16 could be considered is in renal transcapillary flux. This flux has normally been described by equation 15 where ΔP and $\Delta\Pi$ are the net transcapillary mechanical and osmotic pressures respectively. For renal glomerular transport equation 15 is reexpressed as

$$J_v = L_p{}^o[(P_{GC} - P_T) - (\Pi_{GC} - \Pi_T)] \tag{17}$$

where P_{GC} and Π_{GC} are the hydraulic and osmotic pressures within the glomerular capillary respectively, while P_T and Π_T are the corresponding pressures in the Bowman's space (Brenner *et al.*, 1986). It is apparent that this standard equation does not account for the internal osmotic pressure of the capillary wall which, in this case, is the glomerular basement membrane. The membrane contains osmotically active heparan sulfate which will contribute to a finite internal osmotic pressure. Experimental estimates of L_p^o using equation 17, therefore will be an overestimate because of the neglect of the Π_m term.

Osmosis

Any type of osmotic pressure, whether it is high or low, is only manifested through osmotic flow, say in ECM regions. For an isobaric ($\Delta P=0$) membrane system, the osmotic flow occurs spontaneously from the solvent compartment, across the membrane into the solution compartment. It is surprising that theoretical treatments of water flow across semi-permeable membranes (permeable to solvent but not to nonideal solute) have not incorporated terms describing the direct influence of the hydrodynamic properties of the nonideal solute. Similar concerns could be offered for water flow through compartments containing nonideal solutes. A specific example is the case of osmotic flow which would occur across a semipermeable membrane that separates a compartment containing the solvent (or permeable solutes) and nonideal solute and a compartment containing solvent alone. Water flow proceeds in the direction against the osmotic activity gradient or down the chemical potential gradient of water and is described as being directly proportional to the difference in osmotic pressure across the membrane. The proportionality constant has been generally described as a solution — independent quantity which may be related to the geometry and structure of the membrane. Yet when considering nonmembranous systems, there seems to be compelling reasons to include the influence of the dynamic properties of the osmotically active solute. A solute concentration gradient across a free liquid boundary will give rise to solute diffusion which could equally well be described by water diffusion (an osmotic volume flow) in the opposite direction. In fact, for this system the diffusion coefficient of water and that of the solute in a volume-fixed frame of reference are identical (Bearman, 1961). A more familiar process associated with osmotic flow is the swelling of gels. Gels are often composed of cross-linked nonideal macromolecules. The gels imbibe water through the difference in osmotic pressure between the gel phase and the external solvent phase. It has been demonstrated that the rate of osmotic water flow into the gel is governed by the diffusion of the gel-polymer network and that this is identical with polymer diffusion in semidilute polymer solutions (Buckley and Berger, 1962; Tanaka and Fillmore, 1979; Candau *et al.*, 1982) (semidilute polymer solutions are characterised by concentrations where there are continuous entropic molecular interactions between molecules to form a transient molecular network that manifests molecular weight independent (above a critical polymer segment size) properties).

The flow is considered to occur through two distinct regions; these are the membrane and the solution layer adjacent to the membrane, in front of the membrane pore, where solute-solvent exchange takes place. The latter region is similar to the concentration gradient in the free boundary binary system discussed above. A layer like this must exist in order for osmotic flow to occur. The driving force for solvent flow will be the tendency of the solute concentration gradient to relax in the thin layer adjacent to the membrane. It is exactly the same force that governs mutual diffusion in free solution as discussed above. In this, it was pointed out that a solute concentration gradient across a free liquid boundary will give rise to a solute diffusion which could equally well be described by water diffusion (volume flow) in the opposite direction. This water flow is driven by identical osmotic pressure gradients that drive osmotic flow across membranes. Yet it is known that in the free boundary system the diffusion coefficient of water and that of a solute in a volume fixed frame of reference are identical (Bearman, 1961) i.e. the rate of osmotic flow across a free boundary is governed by solute diffusion.

Osmotic flow across a membrane is then modelled on a series of flows consisting of i) water flow across the membrane, and ii) the diffusional exchange of osmotically active solute with solvent in the thin layer on the solution side of the membrane. This can be described as two resistances namely R_S associated with the solution layer and R_m associated with the membrane acting in series to control flow. (The R_S resistance is quite distinct from unstirred layer effects at the boundary (Massaldi and Borzi, 1982; Pedley, 1980) and will always exist in ideally stirred systems). For a given osmotic pressure Π^* the volume flow is given by

$$J_v = L_p^o \Pi^* \quad (18)$$

$$= (1/(R_s + R_m))\Pi^* \quad (19)$$

When the solute-solvent exchange resistance is zero under conditions where $C_1 = 0$ then R_m can be measured directly ($R_m = 1/L_p$) and is often related to Poiseuille's law. In experimental systems with macromolecules it becomes evident that because polymers have such concentration dependent and relatively high hydrodynamic frictional coefficients, the solution layer adjacent to the membrane may become rate limiting in governing the kinetics of the osmotic flow.

Considerable controversy has previously existed as to the mechanism of osmosis. Most interpretations recognise the flow through the membrane as the only effective kinetic process governing osmosis (Kedem and Katchalsky, 1961; Hammel and Scholander, 1976; Kiil, 1989). Many discussions have centred on the nature of the osmotic force and distribution of pressures (Hammel and Scholander, 1976; Mauro, 1981; Kiil, 1982, 1989; Ferrier, 1984) as distinct from considering factors associated with the hydrodynamics of flow. There has, however, been some investigators that have recognised the importance of the solute-solvent exchange layer adjacent to the membrane (Ray, 1960; Dainty, 1965; further endorsed by Soodak and Iberall, 1978) and that for cell membranes, at least, the rate controlling process for osmosis could be the mutual diffusion coefficient of intracellular osmotically active solutes (Dick, 1964; Dainty, 1965). While these concepts were not developed further there has been experimental evidence to suggest the influence of the concentration dependent dynamic properties of the osmotically active solute on the magnitude of L_p^o. However, these apparent deviations from ideal behaviour of water-membrane interaction have been commonly ascribed to 'unstirred boundary layer effects' (Heyer *et al.*, 1969, Massaldi and Borzi, 1982).

Using equation 7 we define R_m as

$$R_m = lf_{12}/v_1(1 - \phi_1)^2 A_{\text{eff}} M_1 \quad (20)$$

which substituted into equations 18 and 19 gives

$$L_p^o = 1/[(1/L_p) + (l/v_1(1 - \phi_1)^2 A_{\text{eff}} M_1/f_{12})] \quad (21)$$

The water-solute interaction will be included in the M_1/f_{12} term. The membrane area A_{eff} is the effective area of the pores and l represents the average distance (characteristic of the osmotically active solute) over which the osmotic solute-solvent exchange takes place in front of the pore (Figure 1.1). Any involvement of solute membrane interaction in modifying the exchange process will be included in the A_{eff}/l term.

Experimental Aspects of Osmosis

Recent experimental studies of osmotic flow has confirmed the series model (Comper and Williams, 1990; Williams and Comper, 1990). For osmotically active polymer solutions at concentration greater than 10 mg/ml the second term on the rhs of equation 21 far outweighs the first term governing water-membrane interactions for all types of commercially available membranes studied. This demonstrates that the limiting factor governing the rate of osmotic flow is the frictional term M_1/f_{12}. It was further demonstrated that the flow was independent of membrane thickness and varied with r_p^2 (where r_p is the radius of the membrane pore) (diffusion) not r_p^4 (Poiseuille). Kinetically, these studies establish osmosis as a diffusion process driven by the same forces that drive mutual diffusion. The rate limiting interaction is characterised by diffusion over a small distance (estimated to be approximately 40 nm for a 20 mg/ml solution of chondroitin sulphate proteoglycan) near the membrane or relatively large concentration gradient.

The rate limiting exchange layer also invalidates the use of the Δ terms in equation 15 because of the concentration dependent L_p^o. This could be overcome by taking some average value of the $(A_{eff}/l)M_1/f_{12}$ term or using the appropriate concentration value for solute-exchange frictional term on either side of the membrane.

The fact that osmosis was originally not thought to be a diffusion process had partly come from the fact that osmotic flow proceeds several times more rapidly than that measured for tracer diffusion of water (Mauro, 1957; Ussing and Anderson, 1956). It has been discussed earlier that there is no *a priori* relationship between mutual diffusion and tracer diffusion of water. In contrast to the membrane-osmosis, the transport of tracer water across a free liquid boundary is considerably faster than the mutual diffusion coefficient of water (or solute). For tracer water transport across the membrane, its diffusion is estimated over the whole length of the membrane so its concentration gradient is relatively lower as compared to the very sharp concentration gradient of solute (giving rise to mutual diffusion) on the solution side of the membrane that is driving flow.

Membrane Filtration

For semipermeable membranes the influence of the macromolecular solute rate-limiting layer identified for osmotic flow will be exactly the same for filtration since the same coefficient governs the kinetics of both processes as seen in equations 13 and 15. This will occur where the solution exchange layer is limiting flow in the case where well stirred macromolecular solutions are being filtered under conditions where $P >$ P or where DP = 0 for finite C_1 on both sides of the membrane.

The Special Role of Molecules Near a Pore of a Membrane

From the analysis of both osmosis and filtration of solutions containing impermeable nonideal solutes it is apparent that the flow properties of the membrane may be determined by a thin solution layer adjacent to the pores of the membrane. Molecules in this solution layer will therefore have quite distinct properties from those in the bulk solution. Molecules in the bulk solution will not determine flow processes, for example. Further significance of the solution layer-pore interaction in determining flow is discussed in the following two examples.

Osmotic Equilibration in Ternary Systems

For solute-exchange layer limiting systems, it has been established that for a given osmotic pressure the lower the frictional coefficient (or factor its effective diffusive mobility) then the greater the rate of osmotic flow. This provides an interesting situation as to the mechanism of osmotic equilibration between two solutions of different solutes with different frictional coefficients but with identical osmotic pressures. The membrane is permeable only to solvent but impermeable to both solutes. Equilibrium comes about by a long range dynamic interaction between the two impermeable solutes across the pores of the membrane (Comper, 1994) (see Appendix).

Flow in Networks Containing Physically Entrapped Nonideal Macromolecules

Flow resistance in a two phase system conferred by the interaction of a nonideal solute with a pore of a membrane is determined by the thickness of the thin solution layer in front of the pore. Clearly, the flow resistance would be enhanced with the unit of the solution layer-pore placed in series rather than in parallel as in a two dimensional membrane. The series distribution would be obtained when unidirectional flow occurs through a three dimensional fibrous network containing the nonideal solute. The flow resistance would then be a function of the number of effective pore-solution layer units that the flow would encounter. The prediction that flow resistance will depend then on the microstructure of the fibrous network and the distribution of solutes within it has yet to be investigated but it clearly has relevance to network structures that exist in cartilage and glomerular basement membrane.

Membranes with Partial Permeability to Solute

For partially permeable membranes the solute will partition itself at the membrane pore. This is a complex system particularly as it has been recognised that the dynamic coefficient describing mechanical pressure driven flow is different to that associated with osmotic flow (Hill, 1989). A relatively simple case is considered here for a solution exchange layer limiting flow. The osmotically active solute is initially on one side of the membrane only and where $\Delta P = 0$ with processes at the membrane-solution interface being rate-limiting in determining the degree of osmosis. The simplest approach is to regard the solute as an effective sphere of radius r_1 that is sterically excluded from the pore of radius r_p. This gives a partition coefficient (λ)

$$\lambda = (1 - \alpha)^2 = C_1'/C_1 \tag{22}$$

where $\alpha = r_1/r_p$ and C_1' and C_1 are the concentrations of component 1 just inside the membrane pore and in the bulk solution respectively. It will then be the corresponding difference in osmotic pressure across the mouth of the pore $\Pi(C_1) - \Pi(C_1')$ that will drive osmotic flow. This will be generated by molecules excluded by the pore and undergo diffusional exchange with solvent, with a frictional coefficient corresponding to the concentration C_1 over the characteristic distance l in front of the pore. It is assumed that all other molecules will enter the pore and do not become osmotically active in generating flows across the membrane. The volume flow generated by diffusion of solute in the pore is also assumed to be negligible. With these considerations, we can write the volume flow as

$$J_v = [\Pi(C_1) - \Pi(C_1')]/[(1/L_p(C_1') + (f_{12}(C_1)l/v_1(1 - \phi_1)^2 M_1 A_{\text{eff}})] \tag{23}$$

where $L_p(C_1')$ is the hydraulic permeability of the solution assumed to be at concentration $C_1 9$ through the membrane. Experimental studies of permeable membranes have demonstrated that equation 23 describes the data quite well (Comper and Williams, 1990).

APPENDIX

Osmotic Equilibration in Ternary Systems

For the ternary case the component nomenclature is redefined as 1 = solute on one side of the membrane 2 = solute on the other side of the membrane and 3 = solvent. The equations for ternary diffusion are

$$-J_1 = D_{11}(\partial c_1/\partial x) + D_{12}(\partial c_2/\partial x) \tag{A1}$$

$$-J_2 = D_{21}(\partial c_1/\partial x) + D_{22}(\partial c_2/\partial x) \tag{A2}$$

where D_{ij} are the ternary diffusion coefficients and $\partial c_i/\partial x_i$ are the effective solute concentration gradients on either side of the membrane. These equations can be re-expressed in terms of binary frictional coefficients and osmotic pressures of the two solution on either side of the membrane Π_1 and Π_2 such that (Miller, 1959; Comper et al., 1984)

$$J_1 = (RT/\xi)[(f_{23} + f_{21})(\partial\Pi_1/\partial x_1) + (\partial\Pi_2/\partial x_2)] \tag{A3}$$

$$J_2 = (RT/\xi)[(f_{12})(\partial\Pi_1/\partial x_1) + (f_{13} + f_{12})(\partial\Pi_2/\partial x_2)] \tag{A4}$$

where

$$\xi = -(f_{13}f_{23} + f_{13}f_{21} + f_{12}f_{23}) \tag{A5}$$

At equilibrium we have zero volume flow and equal osmotic pressures so that

$$J_1V_1 = J_2V_2$$

and

$$\Pi_1 = \Pi_2$$

When these conditions are used in combination with equations A3 and A4 we get a finite value of the cross frictional term existing across the membrane such that

$$f_{12} = -(V_1 f_{23} A_1 + V_2 f_{13} A_2)/[(V_1 A_1 + A_2)(c_1/c_2) + V_2(A_1 - A_2)] \tag{A6}$$

where $A_i = \partial\Pi_i/\partial x_i$. Assuming that $x_1 = x_2$ then

$$f_{12} = -(V_1 f_{23} + V_2 f_{13})/2(V_1(c_1/c_2) + V_2) \tag{A7}$$

This result suggests that there are thin layers on either side of the membrane, where solute-solvent exchange occurs, and where molecules will be distinct from their counterparts in the bulk solutions due to the influence of the cross membrane frictional coefficient.

This work was supported by grants from the Australian Research Council and the National Health and Medical Research Council. The major portion of the theoretical work described in this chapter has been taken from Comper (1994).

References

Atkins, E.D.T., Meader, D. and Scott, J.E. (1980) Model for hyaluronic acid incorporating 4 intramolecular hydrogen-bonds. *Int. J. Biol. Macromol.,* **2**, 318–319.

Bearman, R.J. (1961) On the molecular basis of some current theories of diffusion. *J. Phys Chem.*, **73**, 1961–1968.

Brenner, N.M., Dworkin, L.D. and Ichikawa, I. (1986) Glomerular ultrafiltration in *The Kidney* (Brenner, B.M. and Rector, F.C., eds.) Vol. 1, W.B. Saunders Company, Philadelphia, 124–144.

Buckley, D.J. and Berger, M. (1962) The swelling of polymer systems in solvents. II. Mathematics of diffusion. *J. Polym. Sci.*, **56**, 175–185.

Casu, B., Petitou, M. Provasoli, M. and Sinaij, P. (1988) Conformational flexibility: A new concept for explaining binding and biological properties of iduronic acid-containing glycosaminoglycans. *Trends Biochem. Sci.*, **13**, 221–225.

Comper, W.D. (1994) The thermodynamic and hydrodynamic properties of macromolecules that influence the hydrodynamics of porous systems. *J. Theor. Biol.*, **168**, 421–427.

Comper, W.D., Checkley, G.J. and Preston, B.N. (1984) Kinetics of multicomponent transport by structured flow in polymer solutions. 5. Ternary diffusion in the system dextran-poly(vinylpyrrolidone)-water. *J. Phys. Chem.,* **88**, 1068–1076.

Comper, W.D. and Laurent, T.C. Physiological function of connective tissue polysaccharides. *Physiol. Rev.,* **58**, 255–315.

Comper, W.D. and Lyons, K.C. (1993) Non-electrostatic factors govern the hydrodynamic properties of articular cartilage proteoglycan. *Biochem. J.,* **289**, 543–547.

Comper, W.D., Preston, B.N. and Daivis, P. (1986) The approach of dextran mutual diffusion coefficients to molecular weight independence in semidilute solutions of polydisperse dextran fractions. *J. Phys. Chem.*, **90**, 128–132.

Comper, W.D. and Williams, R.P.W. (1987) Hydrodynamics of concentrated proteoglycan solutions. *J. Biol. Chem.*, **262**, 13464–13471.

Comper, W.D. and Williams, R.P.W. (1990) Osmotic flow caused by chondroitin sulfate proteoglycan across well-defined Nuclepore membranes. *Biophys. Chem.,* **36**, 215–222.

Comper, W.D., Williams, R.P.W. and Zamparo, O. (1990) Water transport in extracellular matrices. *Connective Tissue Res.*, **25**, 89–102.

Comper, W.D. and Zamparo, O. (1990) The hydrodyanmic properties of connective tissue polysaccharides. *Biochem. J.*, **269**, 561-564.

Dainty, J. (1965) Osmotic flow. *Symp. Soc. expl. Biol.*, **19**, 75–85.

deGennes, P.G. (1979) Brownian motion of flexible polymer chains. *Nature*, **282**, 367–370.

Dick, D.A.T. (1966) *Cell Water*, Butterworths, Washington.

Ferrier, J. (1984) Osmosis and intermolecular force. *J. Theor. Biol.*, **106**, 449–453.

Fujita, H. (1962) *Mathematical Theory of Sedimentation Analysis*. Academic Press, New York.

Hammel, H. and Scholander, P. (1976) *Osmosis and Tensile Solvent*. Springer-Verlag, New York.

Heatley, F. and Scott, J.E. (1988) A water molecule participates in the secondary structure of hyaluronan. *Biochem. J.*, **254**, 489–493.

Heyer, E., Cass, A. and Mauro, A. (1969) A demonstration of the effect of permeant and impermeant solutes, and unstirred boundary layers on osmotic flow. *Yale J. Biol. Med.*, **42**, 139–153.

Hill, A. (1982) Osmosis: a bimodal theory with implications for symmetry. *Proc. R. Soc. Lond.*, **B215**, 155–174.

Hill, A. (1989) Osmotic flow equations for leaky porous membranes. *Proc. R. Soc., Lond.*, **B237**, 369–377.

Katchalsky, A. and Curran, P.F. (1967) *Nonequilibrium Thermodynamics in Biophysics.* Harvard University Press, Cambridge, Mass.

Kedem, O. and Katchalsky, A. (1961) Physical interpretation of the phenomenological coefficients of membrane permeability. *J. Gen Physiol.*, **45**, 143–179.

Kiil, F. (1982) Mechanism of osmosis. *Kidney Int.*, **21**, 303–308.

Kiil, F. (1989) Molecular mechanism of osmosis. *Am. J. Physiol.*, **256**, R801–R808.

Kurata, M. (1982) *Thermodynamics of Polymer Solutions*. Harwood Academic Publishers, London.

Levick, J.R. (1987) Flow through interstitium and other fibrous matrices. *Q. J. Exp. Physiol.*, **72**, 409–438.

Lyons, K.C. and Comper, W.D. (1993) Diffusion of tritiated water in chondroitin sulfate solutions. *Biophys. Chem.*, **47**, 61–66.

Marchessault, R.H. and Sundararajan, P.R. (1983) Cellulose in *The Polysaccharides*. Vol. 2, Chap 2 (Aspinall, G.O. ed.), Academic Press, New York, pp. 11–93.

Massaldi, H.A. and Borzi, C.H. (1982) Non-ideal phenomena in osmotic flow through selective membranes. *J. Membr. Sci.*, **12**, 87–99.

Mauro, A. (1957) Nature of solvent transfer in osmosis. *Science*, **126**, 252–253.

Mauro, A. (1981) The role of negative pressure on osmotic equilibrium and osmotic flow in water transport across epithelia: Barriers, gradients and mechanisms, Alfred Benzon Symposium 15 (Ussing, H.H., Bindslev, N., Lassen, N.A. and Sten-Knudsen, D. eds.) Munksgaard, Copenhagen, pp. 107–110.

Mijnlieff, P.F. and Jaspers, W.J.M. (1971) Solvent permeability of dissolved polymer material. The direct determination from sedimentation measurements. *Trans. Far. Soc.*, **67**, 1837–1854.

Miller, D.G. (1959) Ternary isothermal diffusion and the validity of the Onsager reciprocal relations. *J. Phys. Chem.*, **63**, 570–578.

Ogston, A.G. and Woods, E.F. (1954) The sedimentation of some fractions of degraded dextran. *Trans. Far. Soc.*, **50**, 635–643.

Pedley, T.J. (1980) The interaction between stirring and osmosis. Part 1, *J. Fluid Mech.*, **101**, 843–861.

Preston, B.N., Davies, M. and Ogston, A.G. (1965) The composition and physicochemical properties of hyaluronic acids prepared from ox synovial fluid and from a case of mesothelomia. *Biochem. J.*, **96**, 449–474.

Ray, P.M. (1960) On the theory of osmotic water movement. *Pl. Physiol.*, **35**, 783–795.

Scott, J.E. and Heatley, F. (1982) Detection of secondary structure in glycosaminoglycans via the H-1-NMR signal of the acetamido NH group. *Biochem. J.*, **207**, 139–144.

Scott, J.E., Heatley, F. and Hull, W.E. (1984) Secondary structure of hyaluronate in solution. *Biochem. J.*, **220**, 197–205.

Soodak, H. and Iberall, A. (1978) Osmosis, diffusion, convection. *Am. J. Physiol.*, **235**, R3–R17.

Spiegler, K.S. (1958) Transport processes in ionic membranes. *Trans. Far. Soc.*, **54**, 1408–1428.

Sundelöf, L-O. and Nyström, B. (1977) Sedimentation behaviour and particle shape in concentrated macromolecular solutions. *J. Polym. Sci., Polym. Lett. Ed.*, **15**, 377–384.

Tanaka, T. and Fillmore, D.J. (1979) Kinetics of swelling of gels. *J. Chem. Phys.*, **70**, 1214–1218.

Ussing, H.H. and Andersen, B. (1956) The relation between solvent drag and active transport of ions in *Proceedings of Third International Congress of Biochemistry* (Liebecq, C., ed.), Academic Press, New York, pp. 434–440.

Van Damme, M-P., Comper, W.D. and Preston, B.N. (1982) Experimental measurements of polymer unidirectional fluxes in polymer and solvent systems with non-zero chemical-potential gradients. *J. Chem. Soc. Soc. Far Trans. I.*, **78**, 3357–3367.

Williams, R.P.W. and Comper, W.D. (1987) Osmotic flow caused by nonideal macromolecular solutes. *J. Phys. Chem.*, **91**, 3443–3448.

Williams, R.P.W. and Comper, W.D. (1990) Osmotic flow caused by polyelectrolytes. *Biophys. Chem.*, **36**, 223–234.

Zamparo, O. and Comper, W.D. (1989) Hydraulic conductivity of chondroitin sulfate proteoglycan solutions. *Arch Biochem. Biophys.*, **274**, 259–269.

Zamparo, O. and Comper, W.D. (1990) Model anionic polysaccharide matrices exhibit lower charge selectivity than is normally associated with kidney ultrafiltration. *Biophys. Chem.*, **38**, 167–178.

2 Collagen Superfamily

John F. Bateman[1], Shireen R. Lamandé[1] and John A.M. Ramshaw[2]

[1]*Orthopaedic Molecular Biology Research Unit, Department of Paediatrics, University of Melbourne, Royal Children's Hospital, Parkville, Victoria, Australia*

[2]*Division of Biomolecular Engineering, CSIRO, Parkville, Victoria, Australia*

INTRODUCTION

The extracellular matrix (ECM) of connective tissues is a complex composite material composed of insoluble fibres, microfibrils and a wide range of soluble proteins and glycoproteins. The primary role of the ECM is to endow tissues with their specific mechanical and physiochemical properties and provide a scaffolding for cell attachment and migration. While these physiological functions may be obvious, the mechanisms of how cells regulate the synthesis and turnover of the numerous protein and glycoprotein components of the matrix, and control the organisation of the individual "building blocks" into an architecturally-precise network of interconnecting fibrils and microfibrils remains poorly understood. Thus, although ECM macromolecules themselves contain much of the information necessary to allow assembly and promote the formation of the ECM network, the interaction of specific cell ECM receptors and cell binding sequences on several matrix proteins (Bissell and Barcellos-Hoff, 1987; Adams and Watt, 1993; Lin and Bissell, 1993; Venstrom and Reichardt, 1993) suggests that the cell plays a role in orchestrating this assembly *in vivo*. The ECM, via its composition and structure, in turn, also exerts a regulatory role in promoting or maintaining cellular differentiation and phenotypic expression (see other chapters in this volume under the sub heading Cell-Matrix Interactions).

ECM composition varies dramatically between tissues due to different functional requirements, reflected in patterns of gene expression, altered post-translational processing of matrix proteins and because of microheterogeneity in protein structure due to differential mRNA splicing. The regulation of the composition of the matrix thus has the potential to provide specific localised information to cells. The physical structure of the matrix can in itself also impart regulatory information by altering cell morphology and thus phenotypic expression (Bissell and Barcellos-Hoff, 1987). The binding of growth factors to specific

Extracellular Matrix, Volume 2, Molecular Components and Interactions, edited by Wayne D. Comper. Published in The Netherlands by Harwood Academic Publishers GmbH.

matrix components (Yamaguchi *et al.*, 1990) may also provide a mechanism for the local control of cellular behaviour.

Extracellular Matrix Composition

The most abundant proteins of the ECM, and indeed of the vertebrate body, are members of the collagen family of proteins and these form the major structural elements of connective tissues. The most commonly occurring collagens are types I, II and III which form the long-recognised characteristic fibre bundles seen in many tissues (Figure 2.1). Connective tissues are far from uniform in composition and tissue-specific expression of many other protein and glycoprotein components results in the unique functional and biological characteristics of these tissues. These other important constituents include members of the proteoglycan family, hyaluronan, elastin and microfibrillar-associated components, fibronectin, thrombospondin, and a plethora of other specialised matrix components such as

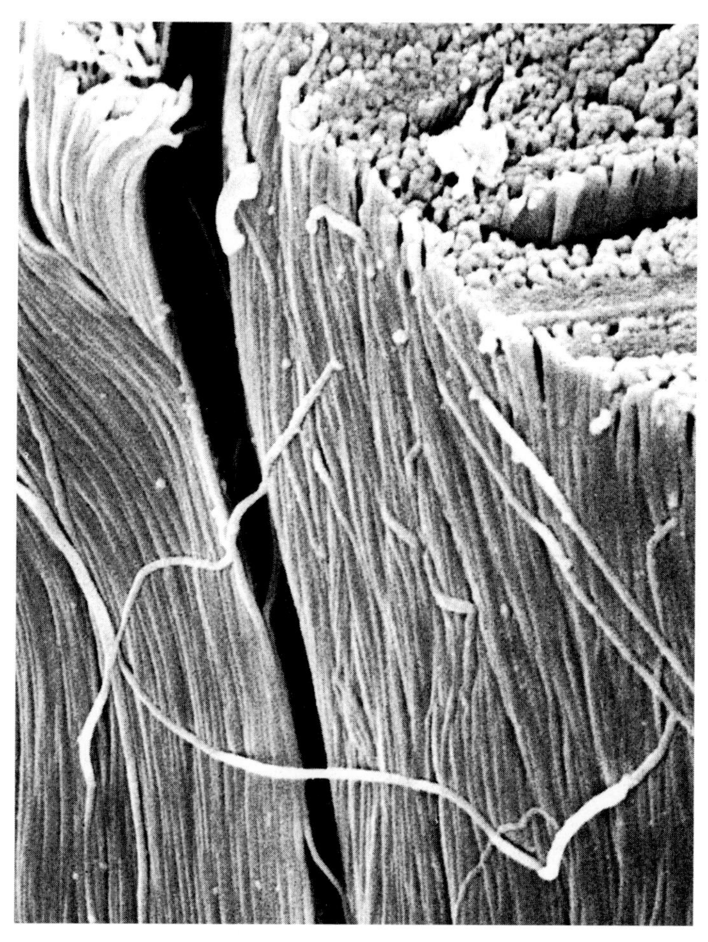

Figure 2.1 Scanning electron microscopy of dermis showing collagen fibre bundles.

laminin, cartilage oligomeric protein (COMP), cartilage matrix protein (CMP), bone sialoprotein (BSP) and tenascin. The biochemistry of many of these is discussed elsewhere in this volume.

Collagen Structure

The name "collagen" is used as a generic term to cover a wide range of protein molecules which form supramolecular extracellular matrix structures and share the basic structural motif of three polypeptide chains wound in a characteristic triple helical configuration. Some 19 different collagen types, comprised of at least 33 individual genetically distinct polypeptide chains have been identified (Table 2.1), although not all have been well characterised.

The basic conformation of the triple-helix has been deduced from high angle x-ray diffraction studies on collagen in tendon (Rich and Crick, 1961; Ramachandran, 1967; Fraser and MacRae, 1973). Each of the three polypeptide chains in the molecule forms an

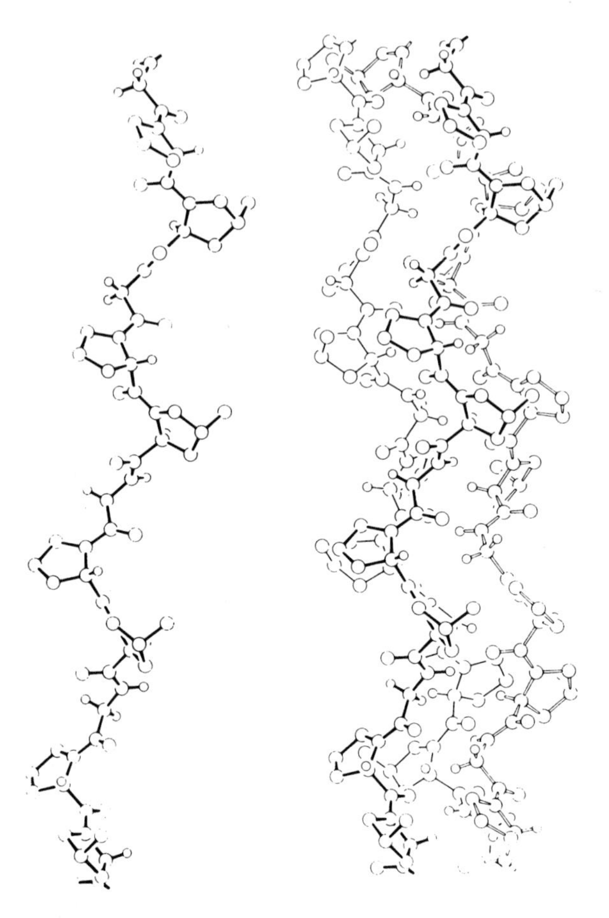

Figure 2.2 A model for the structure of the collagen molecule. This was generated using the linked-atom least squares refinement procedure. For clarity, hydrogen atoms not attached to the main chain have been omitted from the drawing. From Fraser *et al.* (1979) with permission from Academic Press Ltd., London.

Table 2.1 Genetically Distinct Collagen Types

Type	*Chain(s)*	*Molecular Forms*	*Main Distribution*	*Key Features*
I	α1(I) α2(I)	$[\alpha1(I)]_2\,\alpha2(I)$	Very widespread; dermis, bone, ligament, tendon, etc	Most abundant collagen type. Main constituent of major fibre bundles that give strength to connective tissues.
		$[\alpha1(I)]_3$	Dermis, dentin	Apparently a minor form.
II	α1(II)	$[\alpha1(II)]_3$	Cartilage, intervertebral disk	Main collagen of cartilage. Forms the main fibrils of this tissue.
III	α1(III)	$[\alpha1(III)]_3$	Blood vessels, dermis, intenstine, etc	Fibrillar, frequently associated with type I, in extensible tissues. More abundant in foetal tissues.
IV	α1(IV) α2(IV) α3(IV) α4(IV) α5(IV) α6(IV)	$\alpha1[(IV)]_2\,\alpha2(IV)$ $\alpha3[(IV)]_2\,\alpha4(IV)$ (?) $\alpha5[(IV)]_2\,\alpha6(IV)$ (?) (?)	Basement membranes	A non-fibril forming collagen. Forms a two-dimensional network.
V	α1(V) α2(V) α3(V)	$[\alpha1(V)]_3$ $[\alpha1(V)]_2\,\alpha2(V)$ α1(V)α2(V)α3(V)	Widespread in low quantity; appears associated with collagen I fibrils	Forms fibrils; may form fibril core with collagen I.
VI	α1(VI) α2(VI) α3(VI)	α1(VI)α2(VI)α3(VI)	Widespread	Forms beaded filaments. Alternative spliced forms
VII	α1(VII)	$[\alpha1(VII)]_3$	Skin, oral mucosa, cervix	Forms anchoring structure linking epithelial basement membrane to underlying tissue.
VIII	α1(VIII) α2(VIII)	(?)	Associated with endothelial cell layers, e.g. Descemet's membrane	May form a hexagonal lattice in some tissues. Short chain length.
IX	α1(IX) α2(IX) α3(IX)	α1(IX)α2(IX)α3(IX)	Cartilage, vitreous body	A collagen II fibril-associated collagen with an interrupted triple helix; can contain a glycosaminoglycan chain.
X	α1(X)	$[\alpha1(X)]_3$	Hypertrophic mineralising cartilage	A short-chain collagen; has similar structure to collagen VIII.
XI	α1(XI) α2(XI) α3(XI)	α(XI)α2(XI)α3(XI)	Cartilage, intervertebral disc	Forms fibrils which are associated with collagen II.
XII	α1(XII)	$[\alpha1(XII)]_3$	Ligament, tendon	A collagen I fibril-associated collagen with an interrupted triple helix.
XIII	α1(XIII)	(?)	Widespread in low quantity	Shows a complex pattern of alternative splicing.
XIV	α1(XIV)	$[\alpha1(XIV)]_3$	Skin, tendon	A fibril-associated collagen with an interrupted triple helix.
XV	α1(XV)	(?)	Expressed in fibroblasts, smooth muscle cells	Triple helix is interrupted in several places.
XVI	α1(XVI)	(?)	Expressed in fibroblasts, keratinocytes	Triple helix is interrupted in several places.
XVII	α1(XVII)	(?)	Bullous pemphigoid antigen, expressed at dermal-epidermal junction	Triple helix is interrupted in several places. Potential hydrophobic transmembrane domain.

Table 2.1 Genetically Distinct Collagen Types (*continued from p. 25*)

Type	*Chain(s)*	*Molecular Forms*	*Main Distribution*	*Key Features*
XVIII	α1(XVIII)	(?)	Expressed in highly vascularised tissues	Triple helix is interrupted in several places.
XIX	α1(XIX)	(?)	Expressed in very small amounts by cultured skin fibroblasts and tumour cells	Five triple-helical domains.

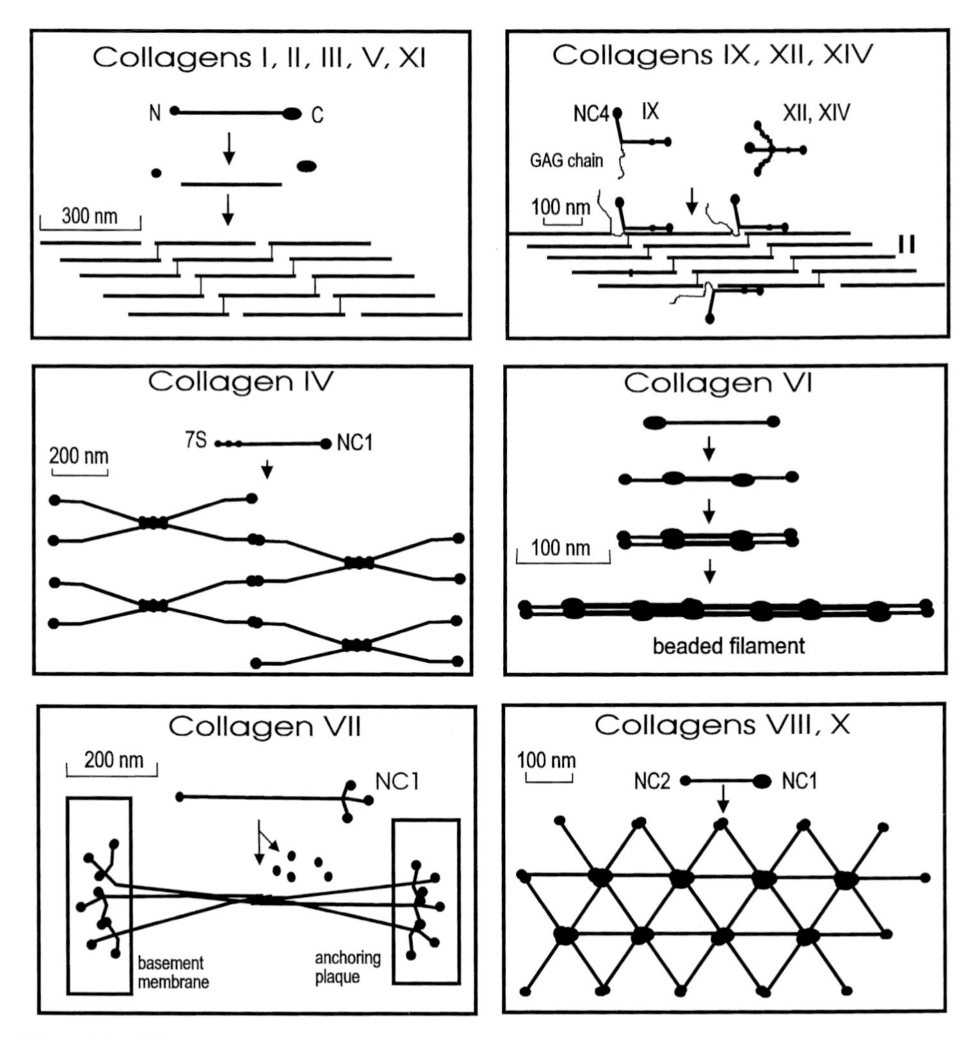

Figure 2.3 Collagen molecular assemblies. These structures are based on rotary shadowing electron microscopy data and are the most likely structures formed by the collagen monomers. However, it should be noted that there is some evidence for alternative structures for several collagen types and *in vivo* specific collagens may have the ability to assume different supramolecular structures.

extended left-handed polyproline II-type helix, which is stabilised by the high imino acid content. The three chains, staggered by one residue relative to each other, are supercoiled about a common axis in a right-handed manner to form the triple-helix. This helical structure has an absolute requirement that every third amino acid is a glycine residue, the smallest amino acid, which allows the close packing along the central axis and hydrogen bonding between the three chains (Figure 2.2). Thus collagens have the general structure of $(Gly\text{-}X\text{-}Y)_n$ where the other residues in the sequence, X and Y, are frequently the imino acids proline and hydroxyproline. The number of repeats, n, varies between collagen types. In the case of the fibrillar collagens, this helix-forming (Gly-X-Y) repeat is continuous over the approximately 1000 amino acid length of the three α-chain subunits of the molecule, while the non-fibrillar collagens commonly contain imperfections and non-collagenous interruptions in the helical sequence.

A second characteristic of collagens is the post-translational enzymatic hydroxylation of specific proline and lysine residues and the further modification by glycosylation of some of the hydroxylysines. The formation of hydroxyproline is important for stabilising the structure of the collagen triple helix by the formation of extra hydrogen bonds.

The distinctive functional feature of collagens is their ability to form highly-organised supramolecular assemblies in the extracellular space. The most commonly occurring collagens, collagens I, II and III, form fibrils in which adjacent molecules are staggered with respect to each other and stabilised by covalent intermolecular crosslinks. In addition, other collagen types form diverse and characteristically-organised assemblies (Figure 2.3).

In this Chapter we attempt to provide an overview of the current understanding of collagen biochemistry in health and disease. We have not attempted to provide a comprehensive review of the complexities of collagen structure and biochemistry, the reader is referred to numerous excellent specialist review articles for this information.

THE COLLAGEN PROTEIN FAMILY

All the different collagen types show characteristic tissue distributions (Table 2.1), and while some may be present in very low quantities overall, in the specialised tissues or tissue elements where they are found they may exist as the predominant collagen species. Detailed reviews are available describing the structure and distribution of different collagen types (Weiss and Ayad, 1982; Mayne and Burgeson, 1987; Burgeson, 1988; Vuorio and de Crombrugghe, 1990; van der Rest *et al.*, 1991; van der Rest and Garrone, 1991; Hulmes, 1992; Kielty *et al.*, 1993).

Fibril-Forming Collagens: Collagens I, II, III,V and XI

Collagen I is the most abundant collagen, representing the principle fibrillar component of many tissues. Largely because of this abundance, collagen I has been the most intensively studied biochemically, and has become the archetypical collagen molecule. Much of our discussion on collagen structure and biosynthesis will relate to collagen I and the other classical fibrillar collagens, II and III, and it should be borne in mind that while the general structure and biochemistry of other collagen types may be similar, significant differences in biosynthesis and assembly of these collagens may occur.

Fibrillar Architecture

Collagen fibrils formed from collagens I, II and III can be visualised by transmission electron microscopy (Chapman and Hulmes, 1984) and with negative staining show a characteristic banding pattern along the fibre axis of about 65–67 nm, called the D-period. This D-period results from the packing of individual collagen molecules, which are about 300 nm or 4.4D long, into a staggered overlap arrangement. The molecules are packed side by side with adjacent molecules staggered along their axis by 67 nm or 1.0D, resulting in a gap of about 0.6D between the end of one molecule and the beginning of the next (Figure 2.3).

X-ray diffraction studies of tendon, where native samples are examined and potential preparation artefacts are avoided, confirm the 67 nm D-period (Brodsky and Eikenberry, 1982). In addition to tendon the 67 nm D-period is also observed in bone, dura mater and cartilage (Brodsky and Eikenberry, 1982; Brodsky *et al.*, 1988; Brodsky *et al.*, 1982), and has been frequently cited as the characteristic collagen fibril periodicity. However, in other tissues, including skin (Brodsky *et al.*, 1980; Stinson and Sweeny, 1980; Gathercole *et al.*, 1987a), periodontal ligament, gingival tissue (Gathercole *et al.*, 1987b), heart valves (James *et al.*, 1991) and granuloma tissue (Brodsky and Ramshaw, 1994) a shorter D-period of about 65–66 nm has been observed using X-ray diffraction.

Thus, there appears to be at least two different sets of intermolecular relationships possible for collagen molecules within fibrils. In one arrangement, the molecules are in a pseudo-hexagonal array, staggered by multiples of D = 67 nm in all three planes (Hulmes and Miller, 1979; Fraser *et al.*, 1983). In the other arrangement, adjacent molecules are staggered by a multiple of D = 65–66 nm except for one plane where they are not staggered, but are tilted by about 12–16° in that plane (Katz and David, 1992).

A common feature of all these tissues which exhibit a shorter D-period is that they are more elastic and flexible and contain a significant amount of collagen III. Although the presence of collagen III in these tissues suggests a possible role for this collagen in the alternative packing arrangement of collagen I, the packing difference may also reflect the effect of tissue-specific matrix proteins interacting with the mature fibril, or influencing the formation of the fibril. Thus while skin shows a shorter D-period, reconstituted fibrils of purified skin collagen I or III both give the longer 67 nm D-period (Brodsky and Eikenberry, 1982). These data suggest that while collagen molecules have the ability to self-associate *in vitro*, there are more complex mechanisms regulating fibril formation *in vivo*. Strong evidence is accumulating that *in vivo*, collagen fibrils are comprised of several collagen types (see later), and this co-polymerisation may play an important role in regulating fibril architecture.

Molecular Structure

Collagen I is a heterotrimer of two identical α1(I) chains and one α2(I) chain, $[\alpha 1(I)]_2\alpha 2(I)$, and is the major fibrillar component of a wide variety of tissues such as skin, bone, tendon, and ligament. As well as the heterotrimeric form, collagen I can exist as a homotrimer of $[\alpha 1(I)]_3$ (Jimenez *et al.*, 1977; Uitto, 1979). It is not clear if this minor molecular form of collagen I plays a specific functional role.

Collagen III is a homotrimer of α1(III)-chains, and is distributed widely in many collagen-I-containing tissues. In many tissues, for example skin, the amount of collagen III is substantially greater during the foetal stage of development, and gradually decreases in proportion with age (Epstein, 1974). The enrichment of collagen III in tissues such as blood vessels, suggests that it may play a role in tissues with elastic characteristics.

Collagen II is also a homotrimer, consisting of three $\alpha 1$(II)-chains. It is the predominant collagen of cartilage and vitreous humor and is transiently expressed in many tissues during embryonic development (Cheah *et al.*, 1991).

Collagens V and XI are closely related structurally. Collagen V is a minor component of virtually all tissues which contain collagens I and III while collagen XI is co-expressed with collagen II (Sandberg *et al.*, 1993). These two collagens are thought to be buried in the fibrils since their helical domains are immunologically masked in tissues (Birk *et al.*, 1988; Petit *et al.*, 1993) and it has been suggested that collagens I and III may initially copolymerise around a collagen V core and collagen II around a core of collagen XI. Collagen V exists in several molecular forms, $[\alpha 1(V)]_3$, $[\alpha 1(V)]_2\alpha 2(V)$, and $\alpha 1(V)\alpha 2(V)\alpha 3(V)$ (Haralson *et al.*, 1980; Rhodes and Miller, 1981; Niyibizi *et al.*, 1984), while collagen XI has been described as an $\alpha 1(XI)\alpha 2(XI)\alpha 3(XI)$ heterotrimer. The $\alpha 3$(XI) chain is probably a more highly glycosylated a1(II) chain (Burgeson and Hollister, 1979). However, collagens V and XI may not be entirely distinct as the copolymerisation of $\alpha 1$(XI) chains with collagen V chains has been described (Niyibizi and Eyre, 1989).

The component α-chains of the fibrillar collagens are synthesised as large precursor pro α-chains with N- and C-terminal globular extensions called propeptides (Figure 2.4). Following secretion, the N- and C-propeptide domains are proteolytically removed to produce the collagen monomers consisting of the triple helix and short non-helical telopeptide sequences at both the C- and N-termini.

The C-propeptide globular domains of the fibrillar procollagens are highly homologous, both in primary amino acid sequence (Figure 2.5) and predicted secondary structure (Bernard *et al.*, 1983; de Wet *et al.*, 1987; Dion and Myers, 1987; Bernard *et al.*, 1988; Kimura *et al.*, 1989; Janeczko and Ramirez, 1989; Su *et al.*, 1989; Takahara *et al.*, 1991).

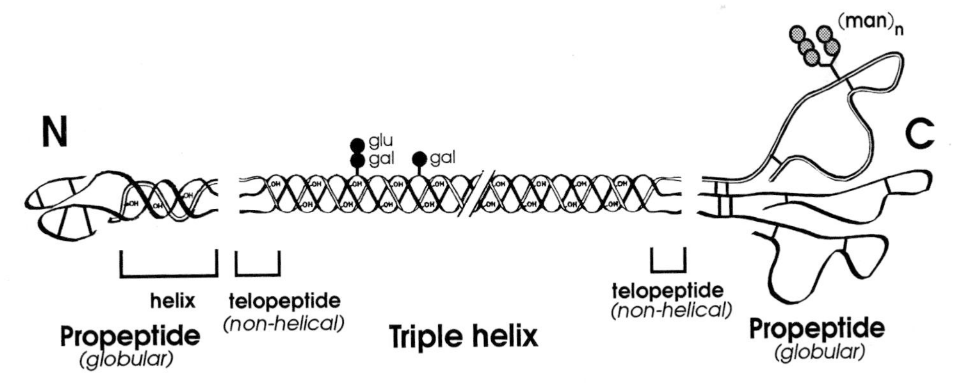

Figure 2.4 The structure of a type I procollagen molecule. Diagrammatic representation of procollagen I showing the triple-helical domain, the short non-helical N- and C-terminal telopeptides and the largely globular N- and C-terminal propeptides contributed by two pro-$\alpha 1$(I) chains (*solid ribbons*) and one pro-$\alpha 2$(I) chain (*open ribbon*). Intrachain disulphide bridges (*lines*) are present within both N- and C-terminal propeptides but interchain disulphide bonds are present between the C-propeptides only. One high-mannose N-linked oligosaccharide ($(\text{man})_n$) is attached to each C-propeptide. Specific proline and lysine residues are post-translationally hydroxylated (OH) and some hydroxylysine residues are further modified by the addition of galactose (gal) or galactose and glucose (glu).

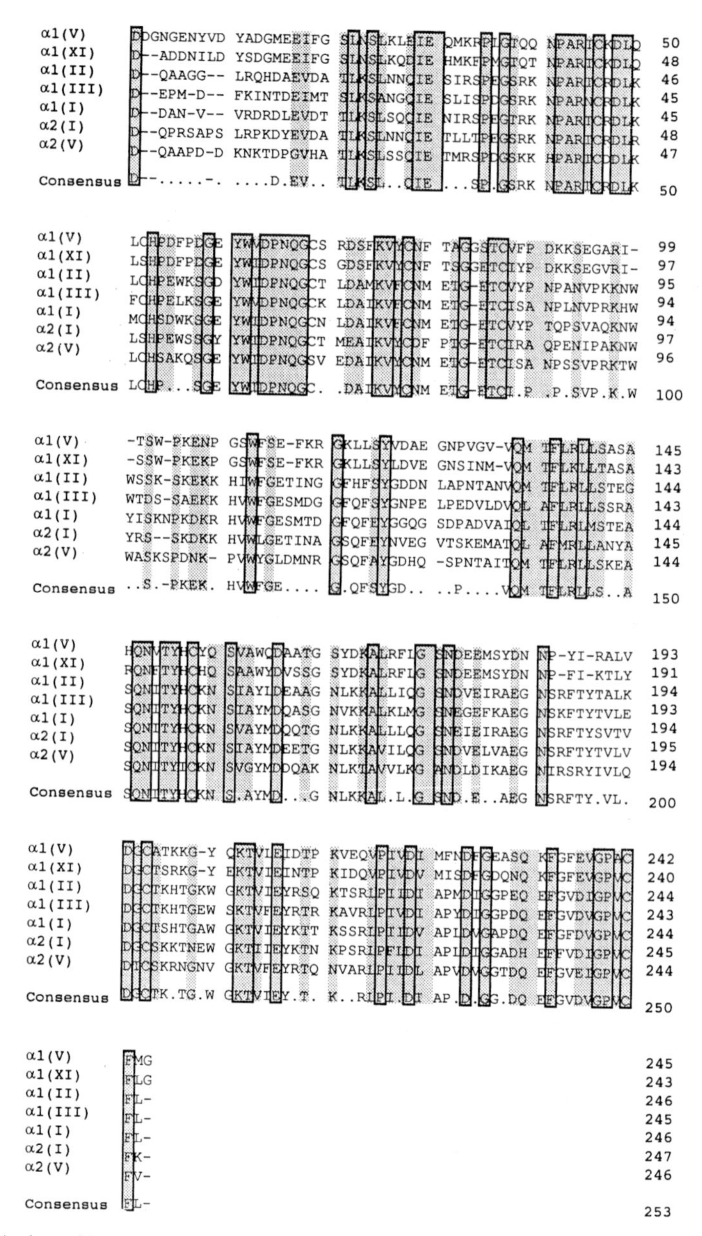

Figure 2.5 Amino acid sequence comparison of the fibrillar collagen C-propeptides. The amino acid sequences of seven C-propeptides were aligned using the Geneworks software (Intelligenetics). The chains are identified on the left and amino acid sequence numbers on the right. *Dashes* indicate gaps inserted to achieve optimal alignment. A consensus sequence was generated when the sequence was conserved between four or more of the chains and is shown below. Completely conserved amino acids are *boxed*, residues conserved between 4–6 of the chains are *shaded*, and *periods* indicate positions where the sequence was conserved between fewer than four of the chains.

This conservation of structure probably reflects their similar, essential role in initiating molecular assembly and helix formation. The more variable regions of the C-propeptides may encode information which directs the selective assembly of procollagen chains when different collagen types are co-expressed in the same cell.

A highly conserved feature of the C-propeptides is the number and location of cysteine residues. The four most C-terminal cysteines seem to be invariant and form intrachain disulphide linkages; the remaining three or four cysteines are involved in intermolecular disulphide bonds (Doege and Fessler, 1986). A second highly conserved feature is the tripeptide acceptor sequence for N-linked oligosaccharide attachment (Asn-X-Ser/Thr) found at amino acid 153–155 of the C-propeptide consensus sequence (Figure 2.5) in all the chains except α2(XI) (Kimura *et al.*, 1989).

The N-propeptide procollagen domains show a much higher degree of structural divergence. The α1(I) and α1(III) N-propeptides are similar, comprising a globular cysteine-rich domain, a short triple helical domain and, adjacent to the main helix, a small globular region which contains the N-proteinase cleavage site (Tromp *et al.*, 1988; Janeczko and Ramirez, 1989). The α2(I) N-propeptide lacks the larger cysteine-rich globular domain (de Wet *et al.*, 1987). Alternative splicing of the collagen II pre-mRNA results in two forms of the α1(II) chain, one including and one lacking the cysteine-rich globular domain (Ryan and Sandell, 1990). The functional significance of these alternative protein forms is not known, but in general, α1(II) transcripts in non-skeletal tissues contain this globular domain (collagen IIA) while the majority of the α1(II) mRNAs in cartilage do not (collagen IIB) (Sandberg *et al.*, 1993). These experiments suggest that this globular cysteine-rich domain may play a developmental role and the production of collagen IIB may be an important marker of cartilage maturation. The α1(V), α2(V) and α1(XI) N-propeptides are larger than those of the major fibril forming collagens and their collagenous domains contain two interruptions to the Gly-X-Y sequence (Woodbury *et al.*, 1989; Takahara *et al.*, 1991; Yoshioka and Ramirez, 1990).

The N- and C-telopeptides are short non-triple helical domains which remain at the ends of the collagen molecule following proteolytic removal of the propeptides. The sequence of both domains varies between chain types and between species (Ramirez, 1989). The C- and N-telopeptide lysine residues are important cross-linking sites in collagens I, II, and III (Eyre *et al.*, 1984) suggesting that the main function of the telopeptides is in stabilising the extracellular collagen fibrils.

Fibril-Associated Collagens: Collagens IX, XII, XIV and XVI

The Fibril-Associated Collagens with Interrupted Triple helices, or FACIT collagens, have small triple helical domains interrupted by short non-helical domains. They do not form fibrils alone but interact with fibrils containing collagens I or II and project their amino-terminal globular domains out into the matrix. These collagens may link fibrils to each other or modulate interactions with other matrix components (Olsen, 1989; Shaw and Olsen, 1991; van der Rest *et al.*, 1991; Kielty *et al.*, 1993).

Collagen IX is a heterotrimer with the molecular formula α1(IX)α2(IX)α3(IX). Molecules of collagen IX are located periodically along the surface of collagen II-containing fibrils and this interaction is stabilised by covalent lysine-derived crosslinking to the N-telopeptide of collagen II (Eyre *et al.*, 1987; Van der Rest and Mayne 1988; Wu *et al.*,

1992). Tissue-specific forms of the molecule are generated by the use of an alternative α1(IX) transcription start site. In cornea the majority of α1(IX) chains have a smaller N-terminal globular domain than the cartilage form (Nishimura *et al.*, 1989). The α2(IX) chain contains a site for the covalent attachment of a glycosaminoglycan chain, the size of which may also vary between tissues (Yada *et al.*, 1990).

Collagens XII and XIV are homotrimeric molecules, indistinguishable by rotary shadowing, which are present mainly in tissues that contain collagen I as the major fibril forming type, although both collagens have also recently been identified in developing cartilage (Watt *et al.*, 1992). They have a much larger N-terminal globular domain than collagen IX, and in collagen XII this domain is variable in size. Whether the synthesis of these variants is due to the use of alternative transcription start sites or alternative splicing has not been established (Gerecke *et al.*, 1993). Like collagen IX, some forms of both collagens XII and XIV can be substituted with chondroitin sulphate glycosaminoglycan side-chains (Watt *et al.*, 1992).

Non-Fibrillar Collagens

Short-chain Collagens: Collagen VIII and X

Collagens VIII and X are structurally related short-chain collagens. Collagen VIII is probably a heterotrimer composed of α1(VIII) and α2(VIII) chains (Yamaguchi *et al.*, 1991; Muragaki *et al.*, 1991). It is the major structural protein of Descemet's membrane, a specialised basement membrane synthesised by corneal endothelial cells, and has also been observed immunologically in a wide range of tissues (Muragaki *et al.*, 1992). In contrast, collagen X $[\alpha 1(X)]_3$ has a very restricted pattern of expression, being confined to terminally-differentiated hypertrophic chondrocytes. The supramolecular structures formed by these collagens are not yet fully resolved. In the Descemet's membrane, collagen VIII forms hexagonal lattices (Sawada *et al.*, 1990), although in other tissues the packing has not been defined. Similiar hexagonal structures were seen by rotary shadowing of *in vitro* aggregated type X (Kwan *et al.*, 1991). However in cartilage, type X has been localised in fine mat-like filaments and in association with type II fibrils (Schmid and Linsenmayer, 1990).

Basement Membrane Collagen IV

Collagen IV occurs exclusively in basement membranes and associated structures. Basement membranes are thin, sheet-like structures associated with epithelial and endothelial cells and surrounding muscle, nerve and fat cells. In addition to providing a mechanical support for cells, the basement membrane has a molecular sieving role in the kidney and regulates cell migration, growth and differentation (Yurchenco and Schittny, 1990).

Collagen IV is composed of three domains; a central triple-helix containing numerous short interruptions to the Gly-X-Y sequence, the N-terminal (7S) domain and the C-terminal globular (NC1) domain. Individual collagen IV molecules assemble to form a sheet-like network in the matrix (Yurchenco and Schittny, 1990). The NC1 domain is linked by disulphide bonds with the NC1 domain of another molecule, and four 7S domains associate and form cross-links. In addition to these globular domain interactions, the flexible interrupted triple-helical domains intertwine with NC1 domains and form supercoiled structures

(Hudson *et al.*, 1993). Collagen IV exists predominately as an $[\alpha 1(IV)]_2\alpha 2(IV)$ heterotrimer but six individual collagen IV chains ($\alpha 1 - \alpha 6$), some with restricted tissue distribution, have now been described.

Anchoring Fibril Collagen VII

Anchoring fibrils extend from the basement membrane lamina densa to anchoring plaques in the dermis. Collagen VII is the major component of anchoring fibrils (Burgeson *et al.*, 1990). It is a large homotrimeric molecule, consisting of a small C-terminal globular domain (NC2), a central interrupted triple-helical region and a very large N-terminal globular (NC1) domain. Secreted collagen VII assembles into antiparallel dimers, in which the C-terminal regions overlap and form intermolecular disulphide bonds. The NC2 domain is then proteolytically removed. Dimers aggregate laterally to form fibres (Morris *et al.*, 1986).

Microfibrillar Collagen VI

Collagen VI forms an extensive network of microfilaments in the extracellular matrix of virtually all connective tissues. It is commonly composed of three genetically distinct subunits but alternative assemblies have also been suggested (Kielty *et al.*, 1990). Collagen VI has a short triple helical domain and very large terminal globular domains which account for more than two-thirds of the molecule (Kielty *et al.*, 1993; Timpl and Chu, 1994). Collagen VI is not secreted as a triple-helical monomer but assembles intracellularly into antiparallel, overlapping dimers which then align in a parallel manner to form tetramers (Bonaldo *et al.*, 1990). The secreted tetramers aggregate to form filaments. The $\alpha 3(VI)$ chain is much larger than the homologous $\alpha 1(VI)$ and $\alpha 2(VI)$ chains with an extended N-terminal globular domain. Multiple alternatively spliced forms of the $\alpha 3(VI)$ and $\alpha 2(VI)$ mRNAs give rise to protein variants which have been proposed to modulate the organisation of the microfilaments and interactions with cells and other matrix molecules (Chu *et al.*, 1989; Saitta *et al.*, 1990; Stokes *et al.*, 1991; Zanussi *et al.*, 1992). However, precise structure/function relationships of the alternative protein forms have not yet been defined.

Bullous Pemphigoid Antigen: Collagen XVII

It has been shown that a hemidesmosomal protein initially characterised as an autoantigen in the blistering skin disease bullous pemphigoid has identity with collagen XVII (Giudice *et al.*, 1991, 1992). It is not known yet if the molecule is a homotrimer or heterotrimer. Its sequence predicts a membrane-associated domain suggesting that collagen XVII may be a transmembrane protein with an intracellular N-terminal globular domain and the collagenous domain extending into the matrix where it may serve as an attachment site to other components of the basement membrane (Li *et al.*, 1993).

Other Collagen Types: Collagen XIII, XV, XVI, XVIII, and XIX

Collagens XIII, XV, XVI, XVIII and XIX are unique collagen chains known only from cDNA and genomic sequencing. Collagens XVI and XIX show features characteristic of the FACIT group and so may also associate with major collagen fibrils and modulate matrix interactions (Pan *et al.*, 1992; Inoguchi *et al.*, 1995; Myers *et al.*, 1994). Collagen XIII is expressed in a wide range of tissues (Sandberg *et al.*, 1989). Its mRNA transcripts undergo complex alternative splicing of both non-collagenous and triple helical encoding exons. At least nine exons can be alternatively spliced (Juvonen and Pihlajaniemi, 1992). The physiological function, macromolecular structure and significance of alternatively spliced forms of this collagens is not known. Recently, two novel homologous collagens, XV and XVIII, have been identified with multiple triple helical "cassettes" flanked and separated by non-collagenous sequences (Myers *et al.*, 1992; Kivirikko *et al.*, 1994; Muragaki *et al.*, 1994; Oh *et al.*, 1994). The name "multiplexin" has been proposed for this subclass of the collagen superfamily (Muragaki *et al.*, 1994). Collagen XV mRNA is expressed predominantly in internal organs such as the adrenal gland, kidney and pancreas (Muragaki *et al.*, 1994) and collagen XVIII is present in many organs, with highest mRNA levels in liver, kidney and placenta (Oh *et al.*, 1994).

Collagen Sequences in Other Proteins

The "collagen cassette" is also used by a number of non-collagenous proteins to achieve a specific protein structure. The collagen motif is found in complement component C1q (Reid, 1979; Acton *et al.*, 1993), serum mannan binding protein (Drickamer *et al.*, 1986), and in integral membrane components acetylcholinesterase (Mays and Rosenberry, 1981), the macrophage scavenger receptor (Matsumoto *et al.*, 1990; Acton *et al.*, 1993), pulmonary surfactant proteins (White *et al.*, 1985), and conglutinin (Davis and Lachmann, 1984). The two characteristics of the collagen helix, the ability to fold into a triple helix and to form a "semi-rigid" elongated structure are used in these proteins to promote subunit oligomerisation and to provide a rigid spacing element. In some of these proteins the collagen sequence element may, in addition to being a structural element, play a distinct functional role such as platelet adhesion or fibronectin binding.

COLLAGEN GENE ORGANISATION

While the 9bp unit coding for the (Gly-X-Y) repetitive amino acid sequence motif results in a general homology between all collagen types within the gene domains coding for the helix, important differences in the gene structure of the collagens exist, reflecting differences in functionally-important protein regions. The allocation of collagen types into two subgroups, fibril-forming and non-fibril-forming, also serves to differentiate the general organisation of the collagen genes. While the fibrillar collagens have conserved gene structures, the non-fibrillar collagen genes are less conserved, reflecting the protein structural heterogeneity of this group. In this chapter we will present only a brief discussion of the major points of fibrillar collagen gene structure and the reader is referred to several recent review articles for further detailed information on the organisation of fibrillar and non-fibrillar collagen genes (Ramirez, 1989; Olsen *et al.*, 1989; Vuorio and de Crombrugghe, 1990; Chu and Prockop, 1993).

The fibrillar collagen chains are encoded by large and complex genes comprising 51–53 exons. While the exons are homologous in these genes and have been given the same numerical designation, intron sizes vary widely resulting in total gene sizes of 18kb for

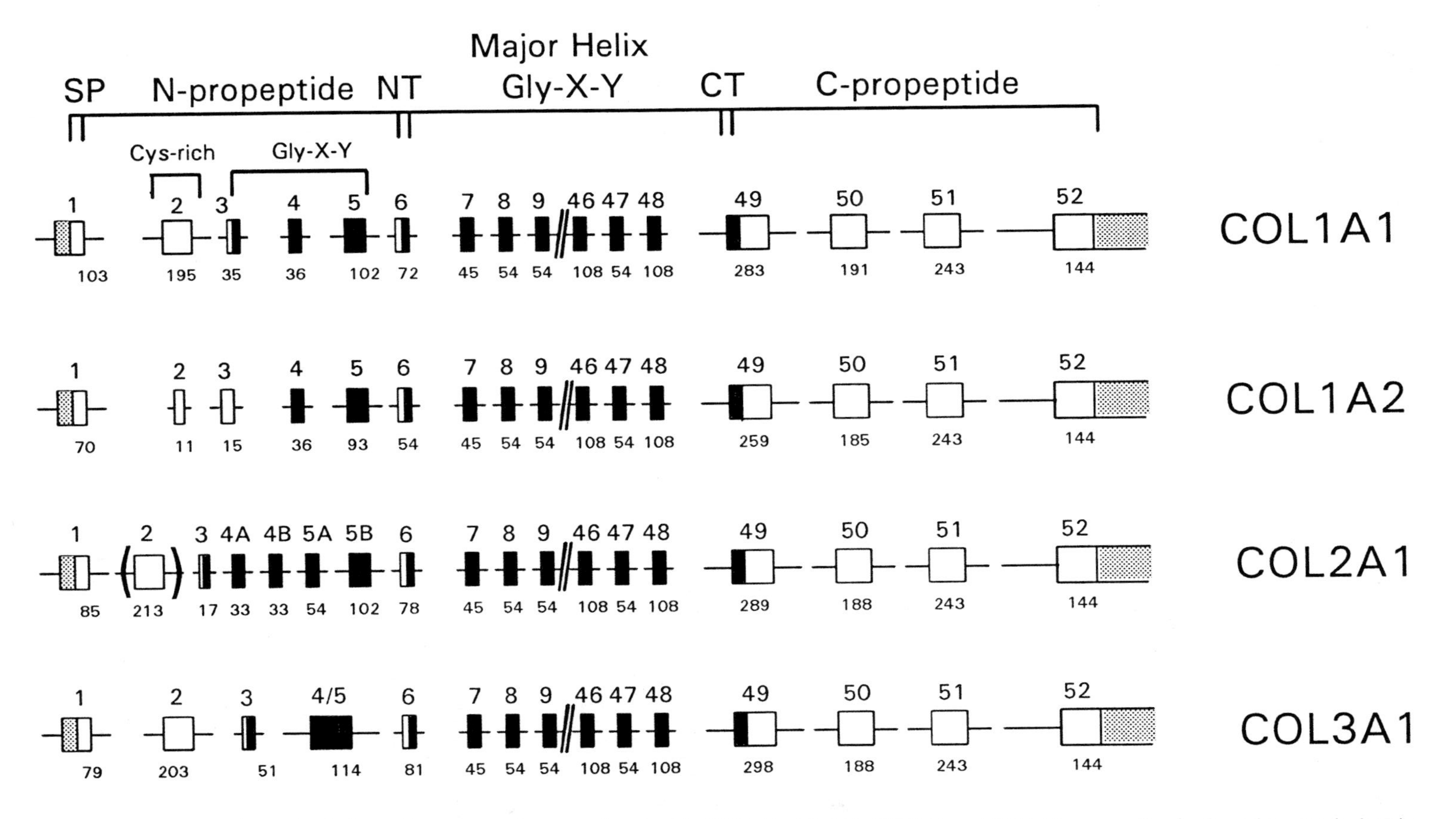

Figure 2.6 Organisation of the fibrillar collagen genes. The exons (*boxes*) of the COL1A1, COL1A2, COL2A1, and COL3A1 genes coding for the various protein domains are shown schematically. SP, signal peptide; NT, N-terminal telopeptide; CT, C-terminal telopeptide. Exon numbers are indicated above, and the size of the exons in base pairs, below the boxes. Regions coding for triple-helical Gly-X-Y sequences are indicated by *filled* boxes, and globular protein encoding sequences by the *open* boxes. *Shaded* regions indicate the sequences coding for the 5′ and 3′ non-translated regions of the mRNA. The alternatively spliced exon 2 of the COL2A1 gene is depicted in parentheses.

the human α1(I) gene and about 40 kb for the α2(I) (Ramirez, 1989; Chu and Prockop, 1993). The structure of these complex genes can be separated into three domains coding for the collagen helix, the C-propeptide protein region and the N-propeptide domain (Figure 2.6). The collagen triple helical domain is encoded by 44 exons. These exons are commonly 54 bp, coding for 6 Gly-X-Y repeats, and less frequently 45 bp, 99 bp, 108 bp and 162 bp. The finding that the majority of the exons are 54 bp suggested that the collagen progenitor gene arose through duplication of a 54 bp ancestral exon cassette (Ramirez, 1989). Each of the exons coding for the triple helix sequence (exons 7 to 48) starts with a complete glycine codon and ends with a complete Y amino acid codon.

The N- and C-terminal ends of the triple helix are coded for by junctional exons. These exons (exon 6 at the N-terminal end and exon 49 at the C-terminal end of the helix) code for the sequence spanning the triple helical end, the short telopeptide sequences, and some propeptide regions. The exons coding for the C-propeptide, part of exon 49 and exons 50–52, are much larger than the exons coding for the collagen helix sequence (Figure 2.6) and are highly conserved between the fibrillar collagens, probably reflecting the importance of this domain in chain selection and trimer assembly. Exon 52 codes for the C-terminal portion of the propeptide and also for the 3′ untranslated sequence of the mRNA. Several polyadenylation sites are used in the gene transcripts for the fibrillar collagens, leading to the production of a range of distinct transcript sizes.

The exons encoding the N-propeptide are more variable, both in number and size, reflecting the different protein structures of this domain in the fibrillar collagens (Figure 2.6). The globular cysteine-rich domain is encoded by exon 2 in α1(I) and α1(III). This domain is not present in the α2(I), being replaced with a short globular domain coded by two exons of 11 and 15 bp. The cysteine-rich domain is encoded by exon 1B in α1(II) and is alternatively spliced (Ryan and Sandell, 1990). The genes for the major fibril-forming collagens have a dispersed human chromosomal localisation, with COL1A1 (α1(I)) on chromosome 17, COL1A2 (a2(I)) on 7, COL3A1 (α1(III)) on 2 and COL2A1 (α1(II)) on chromosome 12.

Only limited gene sequence information is available for the minor fibril-forming collagens, collagen V and XI (Chu and Prockop, 1993). However, this information suggests that these genes conform to the overall conserved gene structures of the major fibril-forming collagens. The gene for α2(V), COL5A2, has been localised to chromosome 2, COL11A1 (α1(XI)) to chromosome 1 and COL11A2 (α2(XI)) to chromosome 6.

COLLAGEN BIOSYNTHESIS

The pathway of collagen biosynthesis from gene transcription to secretion and aggregation of collagen monomers into functional fibrils is complex (Figure 2.7). While the individual steps follow the pattern of assembly and secretion of other multi-subunit proteins, the complex mRNA processing and extensive and diverse post-translational modifications, make collagen biosynthesis an exquisite problem of co-ordinating a very large number of biochemical events temporally and spatially. The biosynthesis pathway has been most carefully studied in the fibrillar collagens, and the discussion here relates mainly to collagen I. While many of the pathways are probably common to all collagens, highly specialised collagens may require additional components or steps for assembly, secretion and

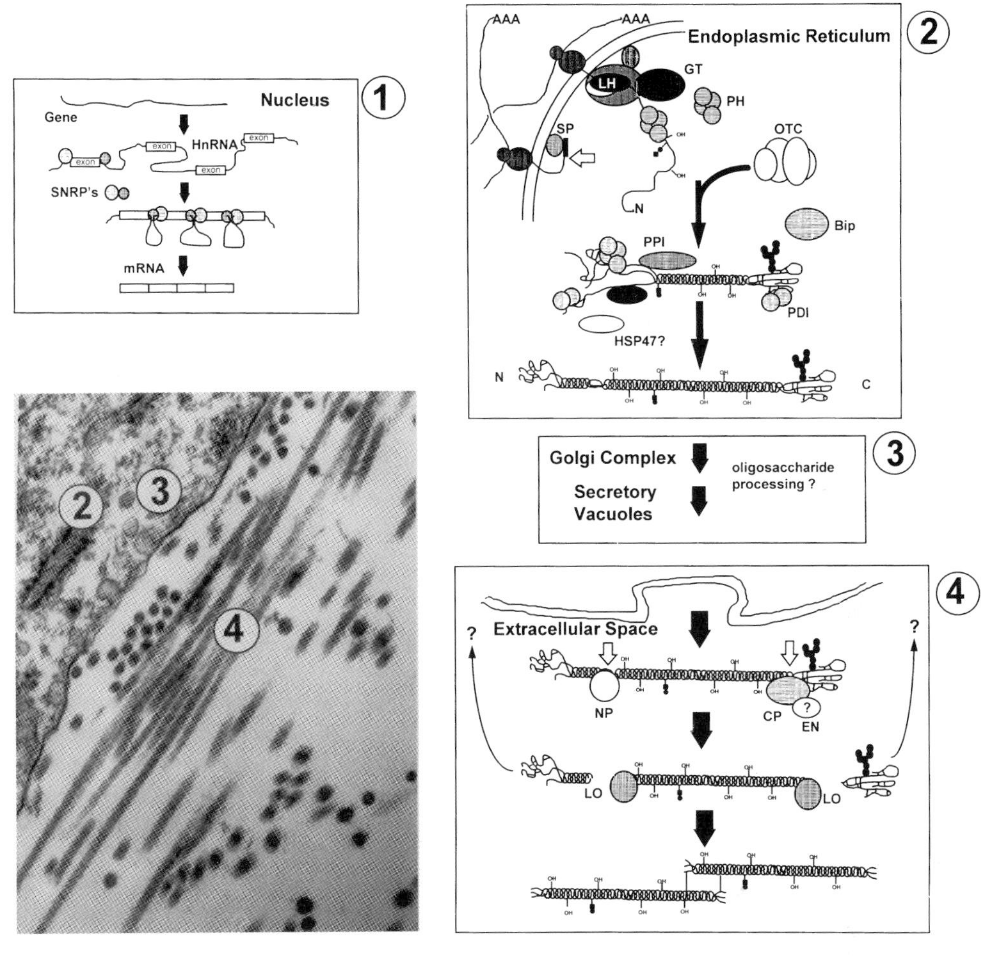

Figure 2.7 The collagen biosynthetic pathway.

1. Collagen gene transcription and mRNA processing within the nucleus follow the general pathway described for other eukaryotic genes. The entire gene, including non-coding intronic sequences, is transcribed (HnRNA), and the transcripts modified at the 5′ end by capping, and by polyadenylation at the 3′ end. Intron sequences are spliced out in reactions performed by 'splicosome' complexes — small nuclear ribonucleoproteins (SNRP's). The mature mRNA is transported to the cytoplasm.

2. Collagen translation products are directed into the endoplasmic reticulum by the N-terminal signal peptide which is then cleaved from the polypeptide chain (*open arrow*) by signal peptidase (SP). Individual procollagen chains undergo complex enzyme-catalysed post-translational modifications prior to the completion of chain assembly and the folding of the triple helix. Proline residues in the Y position of the Gly-X-Y triplet sequence are hydroxylated (–OH) by prolyl 4-hydroxylase (PH) (the dimeric β-subunits of this enzyme are protein disulphide isomerase), and some lysine residues are hydroxylated by lysyl hydroxylase (LH). Hydroxylysine residues can be further modified by the addition of galactose or galactose and glucose (■●). These reactions are catalysed by the enzymes hydroxylysyl galactosyltransferase and galactosylhydroxylysyl glucosyltransferase respectively (GT). An *N*-linked carbohydrate group is added to each C-propeptide by the oligosaccharyl transferase complex (OTC). Folding of the procollagen chains is catalysed by further enzymes and probably assisted by several molecular chaperones. The formation of disulphide bonds is catalysed by protein disulphide isomerase (PDI) and the *cis-trans* isomerisation of prolyl peptide bonds by peptidyl-prolyl *cis-trans* isomerase (PPI). The polypeptide chain binding proteins BiP, and HSP47 have been shown to associate intracellularly with procollagen chains and although their precise role in collagen biosynthesis has not been established they

may assist and regulate oligomeric assembly (see text). The actions of these endoplasmic reticulum-resident enzymes and molecular chaperones are likely to be coupled and cooperative with the proteins forming large complexes. The C-propeptides of three procollagen chains associate and the triple-helix folds towards the amino-terminal end.

3. Triple-helical molecules are transported through the Golgi complex, where *N*-linked oligosaccharide processing may occur, and packaged into secretory vacuoles which release the procollagen into the extracellular space by exocytosis.

4. Following secretion the N- and C-propeptides are cleaved by specific endopeptidases, a procollagen N-proteinase (NP) and the procollagen C-proteinase (CP). The activity of the C-proteinase may be regulated by a glycoprotein known as the C-proteinase enhancer (EN). Cleavage of the C-propeptide is a prerequisite for fibril formation and N-propeptide cleavage is essential for regular fibril morphology. Cleaved N- and C-propeptides may have a role in feedback regulation of procollagen biosynthesis. The processed collagen molecules assemble into fibrils which are stabilised by lysyl oxidase (LO) catalysed cross-links. The electron micrograph (×44,000) shows collagen fibrils **(4)** in the extracellular matrix deposited by 3T6 cells when grown for an extended period in the continuous presence of ascorbic acid . The endoplasmic reticulum **(2)** and location of the Golgi **(3)** are indicated.

extracellular organisation. Collagen IX represents one example of this, where addition of glycosaminoglycan side chains to at least a proportion of the molecules also occurs intracellularly.

Transcription and mRNA Processing

Three major kinds of nuclear RNA processing events have been identified and are common to the vast majority of protein-encoding RNA molecules transcribed from eukaryotic genes. Transcripts are modified by capping at their 5′ end in a reaction resulting in the addition of a terminal G which is then methylated. Polyadenylation occurs at the 3′ end of the transcript after the poly(A) additional signal AATAAA. Finally, the non-coding intronic sequences are spliced out in reactions performed by a 'spliceosome' complex containing protein and small nuclear RNAs. Correct processing depends on the presence of consensus splice donor and acceptor sequences at the intron/exon boundaries and a branch consensus site located towards the 3′ end of each intron (Hames and Glover, 1988; Lamond, 1993). The fibrillar collagen genes contain more than 50 exons and so the large RNA transcripts must undergo multiple exon definition and splicing events prior to their transport to the cytoplasm.

mRNA Translation

Like other secreted proteins, collagen I translation products are initially directed into the endoplasmic reticulum (ER). As the hydrophobic N-terminal signal peptide is translated it is bound to a subunit of the signal recognition particle (SRP) (Krieg *et al.*, 1986; Walter and Lingappa, 1986; Bernstein *et al.*, 1989) and further translation is then delayed until the SRP has bound to its ER-integral membrane receptor or docking protein. Translocation begins, the SRP is released and translation of the mRNA continues (Rothman, 1989a). It is not clear how polypeptides are translocated across the ER membrane but the process involves at least 5 proteins found in the ER membrane and may also be facilitated by binding of ER luminal proteins to translocated portions of the protein (Rapoport, 1992; Sanders and Schekman, 1992). Collagen mRNAs may also be translated while associated into "supramolecular assemblies" which may facilitate interactions and chain registration (Kirk *et al.*, 1987; Veis and Kirk, 1989).

The signal, or pre-peptide, of preprocollagen is cleaved from the nascent polypeptide chain, probably as it emerges into the lumen of the ER, by signal peptidase (Palmiter *et al.*, 1979; Graves *et al.*, 1981). Signal peptidase is an integral membrane protein, located on the inner face and present in a complex of five polypeptides (Evans *et al.*, 1986).

Post-Translational Modifications

Hydroxyproline and hydroxylysine are found almost exclusively in collagens and other proteins with collagen-like domains. The formation of these amino acids is catalysed post-translationally by three separate enzymes; prolyl 4-hydroxylase, prolyl 3-hydroxylase and lysyl hydroxylase. All three enzymes require Fe^{2+}, 2-oxoglutarate, O_2 and ascorbate for activity (Kivirikko and Myllylä, 1982). Prolyl 4-hydroxylase reacts with the minimum sequence X-Pro-Gly, and lysyl hydroxylase with X-Lys-Gly but both enzymes are able to hydroxylate some other triplets. Prolyl 3-hydroxylase requires a Pro-4-Hyp-Gly triplet and probably does not hydroxylate the sequence Pro-Pro-Gly (Kivirikko and Myllylä, 1982). The function of 3-hydroxyproline is not known. The presence of 4-hydroxyproline is essential for the stability of the collagen triple helix at physiological temperatures and almost complete hydroxylation of proline in the Y position of Gly-X-Y sequences is necessary for a molecule to be stable at 37°C (Rosenbloom and Harsch, 1973; Berg and Prockop, 1973a,b). The extent of prolyl hydroxylation is species dependent, with those living at lower environmental temperatures showing a lower level of hydroxylation (Mathews, 1975). For species, such as fish, which can adapt to different environmental temperatures the extent of hydroxylation changes to meet the needs for collagen stability (Cohen-Solal *et al.*, 1986). In fibril-forming collagens approximately 50% of the proline residues are 4-hydroxylated. The extent of lysine hydroxylation varies between tissues and collagen types (Kivirikko and Myllylä, 1979) and is generally lower than the maximum level achievable in *in vitro* assays (Kivirikko *et al.*, 1973). Certain hydroxylysine residues are essential for the stability of intermolecular collagen crosslinks and are the site of attachment for carbohydrate units.

Although the hydroxylases are catalytically similar (Kivirikko and Myllylä, 1987) there is no significant homology between lysyl hydroxylase and prolyl 4-hydroxylase either in their primary sequence or subunit structure (Myllylä *et al.*, 1991). Active prolyl 4-hydroxylase is a tetramer composed of two α- and two β-subunits (Kivirikko *et al.*, 1989), both of which are required for activitiy. However, the β-subunit is a multifunctional protein, and has separate activities which are independent of the α-subunit. As a homodimer it is the enzyme protein disulphide isomerase (PDI), and functions as a glycosylation site binding protein in the oligosaccharide transferase enzyme complex (Pihlajaniemi *et al.*, 1987; Freedman, 1989; LaMantia *et al.*, 1991). The α-subunit, containing the substrate binding site of the enzyme (de Waal and de Jong, 1988; Kivirikko *et al.*, 1989), associates with pre-existing β-subunits (PDI) to form the active tetramers (Berg *et al.*, 1980a). Prolyl 4-hydroxylase is resident in the ER and maintained there by the presence of the amino acid sequence Lys-Asp-Glu-Leu (KDEL) at the C-terminus of the β-subunits but not the α-subunits (Helaakoski *et al.*, 1989; Munro and Pelham, 1987). This sequence is recognised by the KDEL receptor, an intrinsic membrane protein which prevents secretion by recycling bound protein back to the ER from the salvage compartment located between the ER and the Golgi (Vaux *et al.*, 1990).

In contrast, lysyl hydroxylase is a dimer made up of only one type of subunit. The mechanism of ER retention of lysyl hydroxylase is not presently clear since it doesn't contain the KDEL retention signal or the retention signal of ER transmembrane proteins (Myllylä *et al.*, 1991; Jackson *et al.*, 1990). As the enzyme appears to be membrane bound (Kivirikko and Myllylä, 1982), retention may involve undefined interactions with transmembrane proteins. It has been suggested that since purified lysyl hydroxylase does not hydroxylate the telopeptide lysyl residues, a second lysyl hydroxylase may exist (Royce and Barnes, 1985). However, no direct eveidence for such an additional enzyme is available.

Two specific enzymes catalyse the formation of hydroxylysine-linked carbohydrate groups on procollagen chains. Hydroxylysyl galactosyltransferase catalyses the addition of galactose to hydroxylysine residues and galactosylhydroxylysyl glucosyltransferase the addition of glucose to some galactosylhydroxylysine residues. Carbohydrate is present as both the monosaccharide galactose and the disaccharide glucosylgalactose. For a detailed review of hydroxylysine glycosylation see Kivirikko and Myllylä (1979). Although the precise function of the hydroxylysine-linked carbohydrates is not clear, they may influence the biological properties of the collagen by altering the lateral packing of individual molecules within fibrils.

All the collagen specific hydroxylases and glycosidases involved in intracellular modifications have similar substrate requirements. Longer polypeptide chains interact more efficiently with the enzymes and, importantly, folding of the chains into the triple helix prevents any further synthesis of modified amino acids (Berg and Prockop, 1973a; Kivirikko *et al.*, 1973; Menashi *et al.*, 1976; Risteli *et al.*, 1976, 1977). These modifications thus occur *in vivo* on nascent chains before the triple helix has formed.

A conserved structural feature of the C-propeptides of the fibril-forming collagens is the presence of an Asn-X-Ser/Thr consensus sequence for N-linked oligosaccharide addition (Dion and Myers, 1987). Asparagine-linked oligosaccharides are first synthesised as a lipid-linked precursor, dolichol pyrophosphate oligosaccharide (Kornfeld and Kornfeld, 1985). Transfer of the preformed oligosaccharide precursor to the acceptor asparagine residue occurs co-translationally in the lumen of the ER and is catalysed by the membrane bound enzyme oligosaccharyl transferase (Kaplan *et al.*, 1987; Hirschberg and Snider, 1987). Following transfer, oligosaccharides undergo a series of processing events, including both trimming and further addition reactions which eventually produce the wide variety of high-mannose, complex and hybrid oligosaccharide structures which have been identified (Kornfeld and Kornfeld, 1985; Rademacher *et al.*, 1988; Goochee *et al.*, 1991). This N-linked oligosaccharide processing begins in the ER and continues as the protein moves through the Golgi compartments.

The C-propeptides of chick tendon collagen I each carry a high-mannose N-linked oligosaccharide consisting of 9–13 mannose and 2 N-acetylglucosamine residues (Olsen *et al.*, 1977; Clarke, 1979; Pesciotta *et al.*, 1981). The conservation of this motif has led to the proposal that it has functional importance but a role in collagen function or biosynthesis has not been unambiguously defined (Duksin and Bornstein, 1977; Tanzer *et al.*, 1977; Duksin *et al.*, 1978; Housley *et al.*, 1980; Peltonen *et al.*, 1980; Duksin *et al.*, 1983). Recent data using site-directed mutagenesis to delete this attachment site in mouse $\alpha 1(I)$, followed by cell transfection and expression of the mutant chain, directly demonstrated that the absence of this mannose does not significantly affect collagen assembly, secretion, processing or deposition of the collagen into the collagenous extracellular matrix (Lamandé and

Bateman, 1995). A possible role of the mannose oligosaccharide *in vivo* is clearance of the cleaved C-propeptide from circulation by liver endothelial cells via specific mannose-receptor-mediated endocytosis (Smedsrød *et al.*, 1990).

Procollagen Assembly and Secretion

The simultaneous synthesis by a single cell of several fibrillar collagen types introduces a first critical step in the assembly process, appropriate chain recognition and selection. It has been suggested that chain selection occurs in the very early stages of mRNA translation. The model proposes that either mRNA or partially translated sequences are recognised by a component on the ER surface and this leads to an organised arrangement and syncronised attachment of collagen synthesing ribosomes (Kirk *et al.*, 1987; Veis and Kirk, 1989). Procollagen chains would thus be aligned at the completion of translation, their C-propeptides in close proximity. Alternatively, chain selection may occur during folding in the lumen of the ER. Specific recognition by additional polypeptide chain binding proteins or molecular chaperones are likely to play an important role in regulating these folding and assembly pathways.

The initial interactions leading to collagen triple-helix formation occur between the C-propeptides, thus a crucial function of this domain is to align the chains and facilitate the nucleation event of triple-helix folding. The recognition events have not been characterised, but are presumed to occur via specific protein-protein interactions between the propeptides involving sequence motifs within the propeptide and/or conformational determinants on the folded individual propeptides. The folded propeptide structures are stabilised by two intra-chain disulfide bonds (Doege and Fessler, 1986) which form before trimeric association. Thus procollagen assembly probably occurs via the folding of individual subunit C-propeptide domains, followed by association and alignment of the chains and finally the formation of interchain disulphide links stabilising the final propeptide assembly.

It is now clear that intracellular protein folding is not a spontaneous, uncatalysed process but that interactions with enzymes and polypeptide chain binding proteins or molecular chaperones, catalyse slow folding steps, assist in oligomer assembly and prevent aggregation. These molecular chaperones bind transiently to newly synthesised polypeptides, but are not part of the fully folded and assembled products, and can remain associated with mutant or denatured proteins (Pelham, 1986; Rothman, 1989b; Fischer and Schmid, 1990; Ellis and van der Vies, 1991; Gething and Sambrook, 1992; Seckler and Jaenicke, 1992; Hendrick and Hartl, 1993). Recent data suggests that several molecular chaperones may be involved in procollagen association and folding.

Formation of native disulphide bonds is catalysed *in vivo* by the enzyme protein disulphide isomerase (PDI), the β subunit of prolyl-4-hydroxylase, which promotes the rapid formation of the correct disulphide-linkages (Bassuk and Berg, 1989; Freedman, 1989; Noiva and Lennarz, 1992). Experiments *in vitro* have shown that PDI accelerates the formation of native interchain disulphide bonds between the C-propeptides of both procollagens I and II (Koivu and Myllylä, 1987; Forster and Freedman, 1984).

An important rate-limiting step in protein folding is the *cis-trans* isomerisation of prolyl peptide bonds. This process has been shown to be accelerated by enzymes with peptidyl-prolyl *cis-trans* isomerase (PPI) activity (Lang *et al.*, 1987). PPI activity is inhibited *in vitro* by the drug cyclosporin A (Shieh *et al.*, 1989). Cyclosporin A slowed the folding of procollagen I and III in cultured human fibroblasts providing indirect evidence for a role for PPI in procollagen folding and assembly *in vivo* (Steinmann *et al.*, 1991). The triple helix of procollagens I and III folds from the C- to the N-terminus at a rate which is consistent with the *cis-trans* isomerisation of proline bonds being the rate limiting step in helix formation (Bruckner *et al.*, 1981; Bächinger *et al.*, 1980). The efficiency of protein disulphide isomerase is enhanced in the presence of PPI (Schönbrunner and Schmid, 1992) suggesting that these two activities cooperate in the catalysis of protein folding.

A further function of the post-translational modifying enzyme prolyl 4-hydroxylase may be to ensure that only triple-helical collagen is efficiently secreted (Bassuk and Berg, 1989). Non-helical collagen synthesised in the presence of α, α'-dipyridyl or *cis*-hydroxyproline continues to bind prolyl 4-hydroxylase, suggesting that this stable enzyme substrate interaction may be an important quality control mechanism acting to prevent secretion of incorrectly or incompletely folded molecules.

A 78 kD ER resident protein, BiP (for binding protein), was first described as the immunoglobulin heavy chain binding protein (Haas and Wabl, 1983) but is now thought to to play a general role in protein folding and oligomeric assembly in the ER. BiP is synthesised constitutively but its synthesis can be further induced by the presence in the ER of mutant proteins or by stress conditions that cause incorrectly folded proteins to accumulate (Gething *et al.*, 1986; Kassenbrock *et al.*, 1988; Kozutsumi *et al.*, 1988; Hurtley and Helenius, 1989). It is likely that BiP is also involved in procollagen folding and subunit assembly. BiP synthesis was increased in three cell lines from OI patients in which mutations in the proα1(I) C-propeptide slowed procollagen assembly, and in these cases, BiP was stably associated with the mutant chains (Chessler and Byers, 1993). Induction and binding of BiP seems to be specific to mutations which slow chain association, implying that its role may be restricted to assisting the folding of the C-propeptides. Mutations which do not affect association but disturb the folding of the triple helix do not result in BiP binding or increased synthesis (Chessler and Byers, 1992; Chessler and Byers, 1993).

HSP47 (also known as colligin, gp46, and cb48) is a 47 kD, stress-inducible, collagen binding glycoprotein which is localised in the lumen of the ER and may be a collagen-specific molecular chaperone (Clarke *et al.*, 1991; Vaillancourt and Cates, 1991; Takechi *et al.*, 1992). Similarly, the binding of another ER-resident stress-induced protein GRP94, to collagen has been demonstrated (Nakai *et al.*, 1992). The precise role of these potential molecular chaperones in collagen biosynthesis has yet to be elucidated.

The biosynthesis and assembly of procollagen molecules thus involves post-translational modification by at least nine ER-resident enzymes and the action of a number of molecular chaperones (Table 2.2). How the actions of the modifying enzymes and chaperones are regulated temporally and spatially has not yet been addressed. However, following chemical crosslinking, procollagen chains can be coprecipitated with the ER resident proteins HSP47, BiP, PDI, and GRP94 (Nakai *et al.*, 1992), suggesting that these proteins may form large complexes or close associations in the ER and that their actions are likely to be coupled or cooperative.

The secretion pathway of triple helical procollagen is similar to that of other extracellular proteins. The molecules pass through the Golgi and are then packaged into secretory vacuoles which move to the cell surface and release the procollagen by exocytosis. A number of workers have observed SLS (segment-long-spacing)-like procollagen aggre-

Table 2.2 Enzymes and protein chaperones involved in post-translational events of collagen I biosynthesis.

Location	*Enzyme/Chaperone*	*Event Catalysed*
endoplasmic reticulum	signal peptidase	removal of signal peptide
	prolyl 4-hydroxylase	4-hydroxylation of proline
	prolyl 3-hydroxylase	3-hydroxylation of proline
	lysyl hydroxylase	hydroxylation of lysine
	hydroxylysyl galactosyl-transferase	O-glycosylation of hydroxylysine
	hydroxylysyl glucosyl-transferase	O-glycosylation of galactosyl hydroxylysine
	oligosaccharyl transferase	N-glycosylation of Asn-X-Ser/Thr
	protein disulphide isomerase	native disulphide bond formation
	prolyl-peptidyl cis/trans isomerase	interconversion of the cis and trans form of proline peptide bonds
	BiP	oligomeric assembly
	HSP47	?
	GRP94	?
endoplasmic reticulum/golgi	various exoglycosidases and transferases	processing of N-linked oligosaccharides
extracellular	procollagen N-proteinase	removal of N-propeptides
	procollagen C-proteinase	removal of C-propeptides
	C-proteinase enhancer	modulate C-propeptide processing?
	lysyl oxidase	cross-link formation

gates within the secretory vacuoles suggesting that procollagen is probably secreted in an aggregated form and not as monomeric molecules (Trelstad and Hayashi, 1979; Bruns *et al.*, 1979; Hulmes *et al.*, 1983). The ends of the intracellular aggregates form an electron-dense interface with the vacuole membrane (Trelstad and Hayashi, 1979) implying that alignment and packing of the molecules could be regulated by interactions between the propeptide domains and the membrane.

Extracellular Processing

During the secretion and matrix assembly process, the N- and C-propeptides are cleaved by specific endopeptidases, a procollagen N-proteinase and the procollagen C-proteinase. Procollagen I N-proteinase cleaves the N-propeptides of procollagens I and II. It is a multisubunit cell-layer-associated proteinase which is Ca^{2+}-dependent, inhibited by metal chelators, and requires a triple-helical procollagen substrate (Tuderman and Prockop, 1982; Tanzawa *et al.*, 1985; Hojima *et al.*, 1989). A separate enzyme, procollagen III N-proteinase, cleaves the N-propeptide of collagen III. Like collagen I N-proteinase, it is a neutral, Ca^{2+} requiring proteinase and correct conformation of the cleavage site is critical for its action (Halila and Peltonen, 1984).

Procollagen C-proteinase specifically cleaves the C-propeptides of procollagens I, II and III. In contrast to the native substrate requirement of the N-proteinases, the C-proteinase cleaves heat denatured procollagen chains at a similiar rate to native procollagen (Njieha *et al.*, 1982; Hojima *et al.*, 1985). The activity of the C-proteinase is enhanced, and may be regulated, by a 55 kDa connective tissue glycoprotein known as the C-proteinase enhancer (Kessler and Adar, 1989; Kessler *et al.*, 1990). The enhancer protein specifically binds to the C-propeptides of collagen I and so may function to ensure the correct conformation of the cleavage site or to optimise interactions between the enzyme and the procollagen substrate (Kessler and Adar, 1989).

Although it is clear that these N- and C-proteinases are the specific and main enzymes processing procollagen in the tissues, other enzymes in the matrix are capable of producing some cleavage. For example, skin from patients with dermatosparaxis (an absence of functional procollagen I N-proteinase) contains a small amount of fully processed collagen I (Smith *et al.*, 1992). In fibroblast cultues cell-layer-associated proteolytic activities can remove both N- and C-terminal telopeptides (Bateman *et al.*, 1987), and proα2(I) chains containing site-directed mutations which block cleavage by the C-proteinase can be processed by a different, cell-associated protease activity (Lee *et al.*, 1990). The importance of these alternative processing activities is not known.

Collagen Fibrillogenesis

The precise biochemical mechanisms of collagen fibril assembly and growth are less well defined than the biosynthetic pathway and our current understanding is based largely on *in vitro* experimentation (see review by Veis and Payne, 1988). While these approaches have been useful in exploring the mechanisms of self-assembly, the *in vivo* process may be far more complex, involving modulation by the extent of propeptide processing, the ratio of coassembled collagen types and by the presence of other matrix molecules. Many of these findings are supported by ultrastructural and biochemical studies of tissues.

Individual processed collagen I molecules will spontaneously 'self-assemble' into ordered fibrillar structures *in vitro* (Wood, 1960; Veis and Payne, 1988). A critical feature of this *in vitro* fibrillogenesis is that collagens I, II, III, V (and possibly collagen XI) all share the tendency to aggregate into ordered filamentous assemblages which show by electron microscopy (Chapman and Hulmes, 1984) and X-ray diffraction (Brodsky *et al.*, 1982) the features of morphology that are found with *in vivo* fibrils. This fibril-forming ability is encoded in the structure of the collagens, implying that precise interactions between collagen domains are involved in directing axial organisation of the fibrillar aggregates. Both hydrophobic and electrostatic interactions between adjacent chains have been proposed as the mechanism (Veis and Payne, 1988). The extra-helical telopeptide sequences also appear to play a key role in the regulation of collagen I *in vitro* fibrillogenesis (Gross and Kirk, 1958; Comper and Veis, 1977; Helseth *et al.*, 1979; Helseth and Veis, 1981) and may have to assume specific conformations as an integral part of the process of fibrillogenesis (Weiss, 1976; Veis and Payne, 1988).

Various models have been proposed for initial nucleation and fibril elongation structures, but the evidence to support these various models is limited. These data have been reviewed (Veis and Payne, 1988) and will only be discussed briefly here. *In vivo*, the earliest structures observed are arrangements analogous to the organised SLS structures in vacuoles within various cells. These structures where the procollagen molecules are aligned without stagger maybe advantageous for intracellular transport as fibril formation is prevented, and may provide a favourable orientation for subsequent cleavage of the propeptides allowing fibrils to then form. The question of the precise molecular packing of collagen I into fibrils is not fully resolved, even in simplified *in vitro* systems. Several models have also been developed to describe the building of collagen into fibrils, including the formation of a five-stranded microfibril. Recently, a model has been proposed (Silver *et al.*, 1992) which includes two specific binding steps defined by 3.4 D-period and 0.4 D-period overlaps.

Collagen I fibril formation proceeds after the proteolytic removal of the procollagen C- and N-propeptides. Recent studies using site-directed mutagenesis directly demonstrated that C-propeptide removal is obligatory (Fenton and Bateman, unpublished data). In these studies a mutation was introduced into the α1(I) mouse collagen gene which deleted the C-proteinase cleavage site and the engineered gene was expressed in transfected fibroblasts. The procollagen produced by the mutant gene retained the C-propeptide and was unable to be incorporated into the substantial fibrillar cross-linked collagen matrix that formed in long-term cell culture. These studies confirmed previous *in vitro* fibrillogenesis studies which indicated that the removal of the large C-propeptide domain is a prerequisite for fibril formation *in vitro* (Kadler *et al.*, 1987).

Cleavage of the collagen I N-propeptide is essential for regular fibril morphology. The presence of the N-propeptide (PN) markedly alters the circularity of collagen I fibrils formed *in vitro*. Unlike the circular fibrils formed from fully processed collagen I, pN collagen alone forms sheet-like polymers, and copolymers containing pN-collagen I and collagen I are ribbon-like (Hulmes *et al.*, 1989; Romanic *et al.*, 1992). Similar twisted-ribbon collagen structures with hieroglyphic cross-section profiles are seen in the skin of patients with dermatosparaxis (Nusgens *et al.*, 1992; Smith *et al.*, 1992). However, cleavage of the α1(I) N-propeptide but not the α2(I) N-propeptide, as occurs in Ehlers-Danlos syndrome type VIIB (where mutations in the α2(I) chain prevent N-proteinase cleavage), changes the conformation of the still covalently associated N-propeptides and allows the formation of nearly cylindrical fibrils (Holmes *et al.*, 1993).

pNα1(III) is a normal component of tissues where it appears to coat the type I collagen fibrils thus masking type I antigenic sites (Fessler *et al.*, 1981; Fleischmajer *et al.*, 1981, 1990). These observations have led to the idea that the type III N-propeptide may limit fibril diameter by steric hindrance. Indeed, *in vitro* copolymerisation of pN-collagen III and collagen I generates thinner fibrils than those formed from type I collagen alone and increasing the pN-collagen III concentration progressively decreases the fibril diameter (Romanic *et al.*, 1991). While it may be tempting to speculate that the type III N-propeptide is responsible for the observed modulation of fibril size this study did not investigate the effect of fully processed type III collagen on copolymer fibril diameter.

The type V collagen N-terminal domain, a globular region which remains after processing, has also been proposed to regulate type I fibril diameter. This is supported by two lines of evidence. *In vitro* heterotypic fibrillogenesis experiments have shown that increasing the proportion of type V collagen progressively decreases the diameter of the copolymers. This effect resides largely within the N-terminal domain of the type V molecule (Birk *et al.*, 1990a). In addition, while the type V collagen helical domain is immunologically masked, antibodies against the N-terminal domain react with mature fibrils of the corneal stroma indicating that at least some of this domain is exposed on the surface of the fibril (Linsenmayer *et al.*, 1993). The fact that the N-terminal domain is not removed during processing means that it may be able to project out from the centre of the fibril and maintain a constant fibril diameter over time (Chapman, 1989).

Other extracellular matrix molecules are known to interact with collagen fibrils in tissues and it is likely that at least some of them are also involved in regulating the formation of

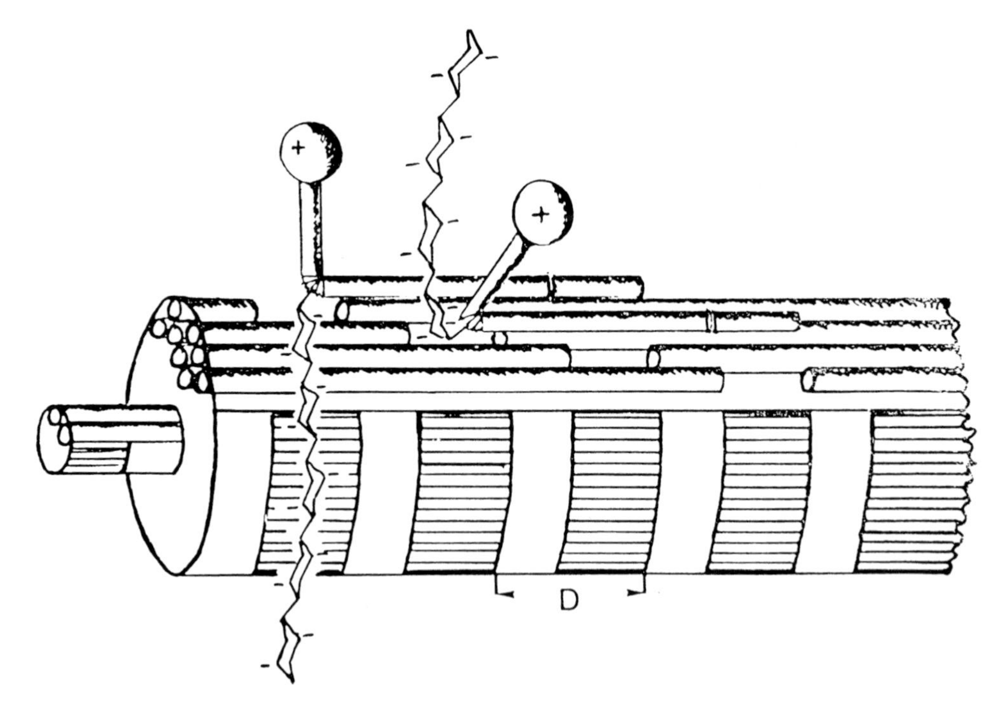

Figure 2.8 Proposed arrangment of collagen in cartilage fibrils. The core filament is formed by collagen XI and surrounded by collagen II. Collagen IX with its chondroitin sulfate chain is at the surface. The cationic nature of the N-terminal globular domain (NC4) is indicated by (+) and the negative charges on chondroitin sulfate by (–). Taken from Eikenberry *et al.* (1992) with permission from Raven Press, Ltd., New York.

fibrillar collagen structures. Since it is likely that fibril formation *in vivo* occurs in close association with the cell surface (Trelstad and Hayashi, 1979; Birk *et al.*, 1990b), the cell may play a fundamental role in finely regulating fibril architecture and function by controlling the availability of interacting macromolecular components. For example, type IX collagen reduces type II fibril diameter in proportion to its relative concentration in *in vitro* fibrillogenesis experiments (Wotton *et al.*, 1988) (see Figure 2.8). The small proteoglycan decorin, a widely distributed matrix molecule, inhibits *in vitro* fibrillogenesis of both type I and type II collagen and this property resides in the core protein, rather than the glycosaminoglycan component of the molecule (Vogel *et al.*, 1984). Decorin core protein interacts with collagen fibrils in cartilage, skin and bone and is localised periodically along the fibrils where it may be in a position to prevent further lateral growth (Scott, 1988; Fleischmajer *et al.*, 1991). Another structurally-related keratan sulphate proteoglycan, fibromodulin, has also been shown to bind to collagen via core protein interactions and affect *in vitro* fibrillogenesis (Heinegård and Oldberg, 1989).

The preceeding discussion has concentrated on the periodic fibrillar assemblies generated by the fibril-forming collagens (I,II, III) and the probable role of their fibril-associated partners in this process. However other collagens can also associate into highly organised supramolecular structures, but these may be very different in structure to the classic interstitial collagen fibril, reflecting both differences in interacting sequence domains, and molecular assembly mechanisms (Figure 2.3).

Cross-Linking

The molecular arrangements within collagen fibrils are rapidly stabilised in the tissues by the formation of covalent cross-links. The cross-links confer physical and mechanical properties to the fibrils and provide the tensile strength fundamental to the structural role of collagen fibrils in connective tissues. Detailed reviews of collagen cross-linking are available (Eyre *et al.*, 1984; Yamauchi and Mechanic, 1988).

Lysyl oxidase is the only enzyme known to be required for cross-link formation. It is a copper-dependent enzyme and is most active against native collagen fibrils. The copper-

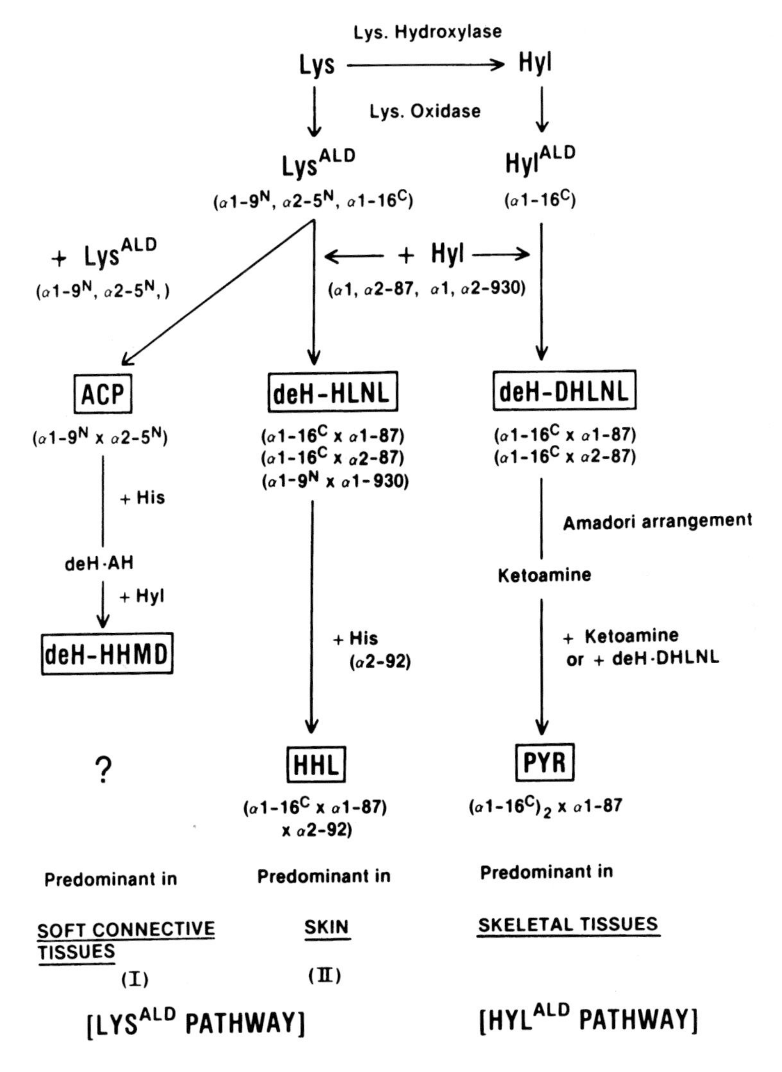

Figure 2.9 Major crosslink pathways in collagen I. Numbers in parenthesis coresspond to the α-chain amino acid residue numbers that participate in the crosslink. Superscripts C and N represent the C-terminal and N-terminal non-helical telopeptide amino acid sequence numbers. Reprinted with permission from Yamauchi and Mechanic (1988). Copyright CRC Press, Boca Raton, Florida.

binding, catalytic domain is located in the C-terminal portion of the molecule (Hämäläinen *et al.*, 1991; Mariani *et al.*, 1992; Svinarich *et al.*, 1992). Cloning of the rat and human lysyl oxidase genes revealed extensive homology to the murine *ras* recision gene product, a protein that counteracts the transforming activity of the *ras* oncogene, and suggested that they may be the same gene product (Kenyon *et al.*, 1991). It is therefore possible that lysyl oxidase mutations may result in certain malignancies as well as structural connective tissue disorders.

Lysyl oxidase initiates cross-linking by catalysing the formation of aldehydes from the telopeptide lysine- and hydroxylysine residues. Further spontaneous reactions result initially in the formation of various di-functional crosslinks, either though formation of aldol condensation products or through the formation of aldimines or ketoimines involving amino groups from specific helical residues (Figure 2.9). These crosslinks, which can be reduced using borohydride, decline in quantity as a tissue increases in age, being replaced by other, multivalent crosslinks which provide the strength of tissues through formation of a network of intermolecular cross-links. Despite the importance of these multivalent crosslinks, the complexity of their analysis has meant that they are yet to be fully characterised. Some, such as histidinohydroxylysinonorleucine and pyridinoline are now well defined, and the pathway for their formation elucidated (Yamauchi and Mechanic, 1988) (Figure 2.9). For others, such as those including the Ehrlich Chromogen (EC) the sequences of the cross-linked collagen chains have been determined (Kuypers *et al.*, 1992) but the pathway of their formation is less clear. In some cases, the sequence observed for the EC cross-link is the same found for a pyridinoline cross-link. The number and proportions of the various cross-links show tissue specificity and it is likely that this is regulated by the steric relationship between molecules, the type of collagens copolymerised, and glycosylation and hydroxylation of the participating amino-acid residues (Reiser *et al.*, 1992). Although most work on cross-links has focused on collagen I, the key residues are also found in the other fibrillar collagens so that similar intra-type and inter-type cross-links may form (Henkel and Glanville, 1982). In addition to these aldehyde derived cross-links, as tissues age they also accumulate further crosslinks due to non-enzymatic glycosylation of collagen.

REGULATION OF COLLAGEN BIOSYNTHESIS

Changes in collagen gene expression occur during development, between tissues and in a number of pathological conditions. Indeed, precise temporal, spatial and quantitative regulation is likely to be critical for normal morphogenesis and differentation. Collagen expression is known to be sensitive to modulation by growth factors, cytokines, glucocorticoids, and viral and chemical transformation (Bornstein and Sage, 1989; Raghow and Thompson, 1989; Slack *et al.*,, 1993).

Regulation of Transcription

Gene transcription is modulated by multiple *cis*-acting elements — promoters, enhancers, and silencers. Promoters are located 5′ to the transcription start site and are required to initiate transcription. The level of transcription is up or down regulated by enhancers and silencers which may be close to the promoter or located further away, often in introns. These DNA sequences are recognised by *trans*-acting factors which bind to specific sites. Much recent work has focused on defining the *cis*-regulatory elements in collagen genes and the transcription factors that bind to, and interact with these regions, and it is already clear that the level of collagen gene expression and tissue specificity involves complex interactions between multiple, sometimes distant DNA sequences that are mediated by the availability and binding of multiple transcription factors.

Regulatory sequences have been identified in the 5′-flanking region as well as within the first intron of the α1(I), α2(I), α1(II), and α1(III) genes. In addition intron 2 of the α2(I) gene contains a second, cartilage specific promoter which directs transcription of an alternative, smaller α2(I) mRNA in this tissue (Bennett and Adams, 1990).

An enhancer within the first intron of the collagen II gene and two silencers in the promoter, active in chondrocytes and fibroblasts respectively, contribute to tissue specific expression of this gene (Horton *et al.*, 1987; Savagner *et al.*, 1990). The first intron of the mouse α2(I) gene has been shown to enhance expression in NIH3T3 cells while the human α2(I) first intron inhibits expression in chick tendon fibroblasts (Rossi and de Crombrugghe, 1987b; Sherwood *et al.*, 1990). Experiments using transient transfection assays and reporter gene constructs must therefore be interpreted with caution. Removing regulatory elements from the correct chromosomal location, species and cell type may have profound effects on gene expression.

The role of the α1(I) first intron has been more extensively studied. Multiple positive and negative regulatory elements have been described and their effect varies depending on the size of the promoter and intronic segments used, their location and orientation within the expression construct, and the cell type (Rossouw *et al.*, 1987; Bornstein *et al.*, 1987; Bornstein and McKay, 1988; Sherwood and Bornstein, 1990; Boast *et al.*, 1990; Olsen *et al.*, 1991). Despite finding different effects, the importance of the first intron in collagen gene expression is not disputed and experiments *in vivo* also confirm a role for these sequences. Transgenes containing a 2.3 kb α1(I) promoter show tissue specificity whether or not they contain the first intron (Slack *et al.*, 1991; Sokolov *et al.*, 1993). However, when a 440 bp α1(I) promoter was used, constructs without the first intron lacked tissue specificity (Slack *et al.*, 1991).

The insertion of a Moloney murine leukaemia virus in the first intron of the mouse α1(I) collagen gene completely blocks transcription of the gene in mesodermal cells and causes recessive embryonic lethality (Schnieke *et al.*, 1983; Hartung *et al.*, 1986). Odontoblasts and osteoblasts, however, transcribe the gene normally (Schwarz *et al.*, 1990; Kratochwil *et al.*, 1989) suggesting that rather than causing a general block in transcription the retrovirus insertion in the first intron interferes with the cell-type specific regulation of the gene.

Regulation of Translation

A highly conserved nucleotide sequence surrounding the translation initiation site in the α1(I), α2(I), and α1(III) mRNAs has been proposed to modulate translational efficiency (Yamada *et al.*, 1983). This sequence of about 50 bp, contains an inverted repeat which could form a stable stem-loop structure in the RNA. The start site of translation would be within the stem of such a structure. Pre-incubation of α2(I)-chloramphenicol acetyltransferase chimeric mRNA with a ribosomal salt extract caused the formation of an mRNA dimer in which the conserved sequences apparently bound to each other, and inhibited cell-free translation of the mRNA. RNAs with deletions in the inverted repeat did not form dimers and their translation was not inhibited by the ribosomal extract (Rossi and de Crombrugghe, 1987a). However, Bornstein *et al.* (1988) found no changes in translational efficiency using a deleted α2(I)-bovine growth hormone fusion gene construct in either stable or transient transfection assays. These divergent findings may have been due to the different constructs or assay systems that were used and so the functional role of the inverted repeat sequences remains unclear.

The 5′-untranslated region of these mRNAs also contain two or more initiation (AUG) triplets preceeding the one that is used to translate the preprocollagen. The upstream AUGs are followed by short open-reading-frames then stop codons. The significance of these sequences in the collagen mRNAs has not been explored but four upstream AUG codons have been shown to mediate translational control of GCN4 mRNA in yeast (Mueller and Hinnebusch, 1986).

Feedback Regulation by the N- and C-propeptides

The idea that the N-propeptide may play a role in modulating collagen synthesis came initially from studies on fibroblasts from patients with Ehlers-Danlos syndrome type VII. These cells showed a significantly higher rate of collagen synthesis which was proposed to be due to a reduction in the amount of free N-propeptide in the medium (Lichtenstein *et al.*, 1973). When added to the medium of cultured cells the free N-propeptides of type I and III collagens specifically inhibit collagen synthesis by fibroblasts but not chondrocytes (Wiestner *et al.*, 1979; Paglia *et al.*, 1981). A role in translational control was suggested by experiments showing that in cell-free translation systems the intact type I N-propeptide selectively inhibits procollagen mRNA translation by inhibiting polypeptide chain elongation. However, the unfolded peptide and some of its proteolytic fragments exert non-specific inhibition of cell-free translation by inhibiting chain initiation (Hörlein *et al.*, 1981). The possibility of a direct regulatory effect of the N-propeptide on collagen biosynthesis was further supported by the demonstration that the type I N-propeptide specifically bound to the fibroblast cell membrane and was internalised via the endocytic pathway (Schlumberger *et al.*, 1988). Expression of a metallothionein-α1(I) N-propeptide minigene in fibroblasts directed the synthesis of a chimeric protein which lacked a signal sequence and was localised in the cytoplasm of transfected cells. This protein specifically inhibited the synthesis of type I procollagen demonstrating that the N-propeptide could directly modulate procollagen synthesis in intact cells (Fouser *et al.*, 1991).

The collagen I C-propeptides have also been implicated in feedback regulation of procollagen synthesis. The propeptide can be internalised by fibroblasts and become associated with the nucleus (Wu *et al.*, 1991). This suggested a direct effect on transcription which was confirmed by nuclear run-off assays demonstrating decreased collagen transcription in the presence of the propeptide. However, studies using smaller fragments of the collagen I C-propeptides have shown the peptides to have both positive and negative regulatory effects on cultured cells which are mediated post-transcriptionally (Aycock *et al.*, 1986; Katayama *et al.*, 1991; Katayama *et al.*, 1993). Thus, the question of their feedback role *in vivo* remains unresolved.

Intracellular Collagen Degradation

Intracellular degradation of newly synthesised secretory proteins is common and may represent an important post-translational process for regulating the functional levels of extracellular proteins (Bienkowski, 1983). There is now ample evidence that newly synthesised collagens are also subjected to intracellular degradation with approximately 15% of collagen synthesised by fibroblasts degraded prior to secretion (Bienkowski, 1983). This process has been termed "basal degradation" and is dramatically increased if the collagen is structurally abnormal, either due to abnormal helix formation and reduced secretion resulting from the inclusion of proline analogues *in vitro* (Bienkowski *et al.*, 1978; Berg *et al.*, 1980b), or due to structural mutations in inherited diseases such as osteogenesis imperfecta (Bateman *et al.*, 1984). In these cicumstances, this degradation is thought to represent an important "quality control" mechanism, minimising the release into the extracellular space of structurally abnormal, and thus potentially deleterious, collagens.

The exact mechanism of this degradation process is still unclear, but recent evidence suggests that basal degradation is not lysosomal and occurs primarily after the collagen has exited the endoplasmic reticulum, possibly in the trans-Golgi network (Berg *et al.*, 1980b; Barile *et al.*, 1990; Ripley *et al.*, 1993). There is, however, evidence using specific proteinase inhibitors, that the increased degradation above basal levels of structurally-abnormal collagen is a lysosomal process (Berg *et al.*, 1984).

The role of this degradation in controlling normal collagen synthesis is unclear, but in pathological states it may represent a critical regulatory step. This rapid intracellular degradation may represent a mechanism whereby cells can finely regulate collagen secretion levels on a faster time-scale than other transcriptional or translational regulatory mechanisms, but this putative regulatory role has yet to be demonstrated *in vivo*.

MOLECULAR PATHOLOGY OF COLLAGEN

The complex collagen biosynthetic pathway presents numerous opportunities for the appearance of molecular defects. However to date, the vast majority of mutations affecting collagen biochemistry which have been defined are structural mutations in the procollagen genes themselves which result in distinct heritable connective tissue disorders. The clinical manifestations of the known collagen mutations reflect the tissue distribution of the molecules and highlight their structural and functional roles in the matrix (Table 2.3). Thus, mutations in type I collagen result in the fragile bone disease osteogenesis imperfecta (OI), and Ehlers-Danlos syndrome (EDS) type VII where the major tissues affected are skin and ligaments, while type II collagen mutations result in the cartilage diseases spondyloepiphyseal dysplasia (SED), achondrogenesis, and Stickler syndrome.

Unlike many other genetic diseases such as cystic fibrosis and the globin disorders where the severe disease is seen only in homozygotes, heterozygous collagen mutations are commonly dominant. If a mutant collagen I $\alpha 1$(I)-chain is able to be incorporated into trimeric molecules, three-quarters of the molecules produced will contain a mutant chain and be structurally abnormal, with a severe disease usually resulting. The effect is further amplified because mutant collagen molecules are not normally able to form fully functional

Table 2.3 Collagen Diseases

Collagen Type		*Disease*
Collagen I	α1(I)	Osteogenesis Imperfecta Ehler's Danlos Syndrome VII Osteoporosis
	α2(I)	Osteogenesis Imperfecta Ehler's Danlos Syndrome VII Osteoporosis
Collagen II	α1(II)	Spondyloepiphyseal dysplasia Achondrogenesis — Hypochondrogenesis Stickler syndrome Osteoathritis
Collagen III	α1(III)	Ehler's Danlos Syndrome IV Familial aneurysm
Collagen IV	α5(IV)	Alport's syndrome
	α6(IV)	Alport's syndrome / leimyomatosis
Collagen V	α2(V)	Spinal deformaties, severe skin fragility*
Collagen VII	α1(VII)	Epidermolysis bullosa
Collagen IX	α1(IX)	Osteoarthritis*
Collagen X	α1(X)	Metaphyseal Chondrodysplasia Schmid
Collagen XI	α2(XI)	Osteochondrodysplasias

* Demonstrated in transgenic mice — no human patients defined.

extracellular assemblies; the presence of even a small number of structurally abnormal molecules can disturb the entire matrix architecture. When a mutant chain is fully excluded from intracellular trimeric assembly or a reduced amount of collagen is synthesised, the phenotypic consequences are less severe. This trans-dominant effect of collagen mutations has been successfully exploited in the production of transgenic mice expressing proα1(I), proα1(II), proα1(IX) and proα1(X) chains with structurally abnormal triple-helical domains (Stacey *et al.*, 1988; Khillan *et al.*, 1991; Garofalo *et al.*, 1991; Rintala *et al.*, 1993; Nakata *et al.*, 1993; Jacenko *et al.*, 1993). Transgenic mice synthesising mutant proα1(I) and proα1(II) chains displayed phenotypes which closely resembled the human diseases osteogenesis imperfecta and spondyloepiphyseal dysplasia respectively.

There have been numerous excellent recent reviews on the molecular basis of collagen diseases (Byers, 1989a,b; Kuivaniemi *et al.*, 1991; Briggaman, 1992; Hudson *et al.*, 1992; Shapiro and Chipman, 1992; Wenstrup, 1992; Byers, 1993; Kivirikko, 1993; Steinmann *et al.*, 1993) and the reader is referred to these for detailed information. In this chapter we focus on one well-studied disease of collagen I, osteogenesis imperfecta (OI) as a collagen disease model. Biochemical characterisation of the defects causing this syndrome have provided valuable insights into collagen biosynthesis, and collagen gene and protein structure. Many of the types of mutations that disturb the helical structure of collagen I in OI also occur in other collagen types resulting in a range of inherited connective tissues diseases according to the tissue distribution and physiological function of the affected collagens. For these reasons the biochemical lessons learnt in studies on OI have a wide significance in defining the molecular basis of many collagen diseases.

Osteogenesis Imperfecta — The Archetypical Collagen Disease

Osteogenesis imperfecta is a heterogeneous group of genetically-determined disorders whose main clinical feature is bone fragility. Four main catagories of OI are currently recognised based on the clinical presentation (Sillence *et al.*, 1979). These range in clinical severity from type I OI, the mildest form, through to types IV and III with progressively

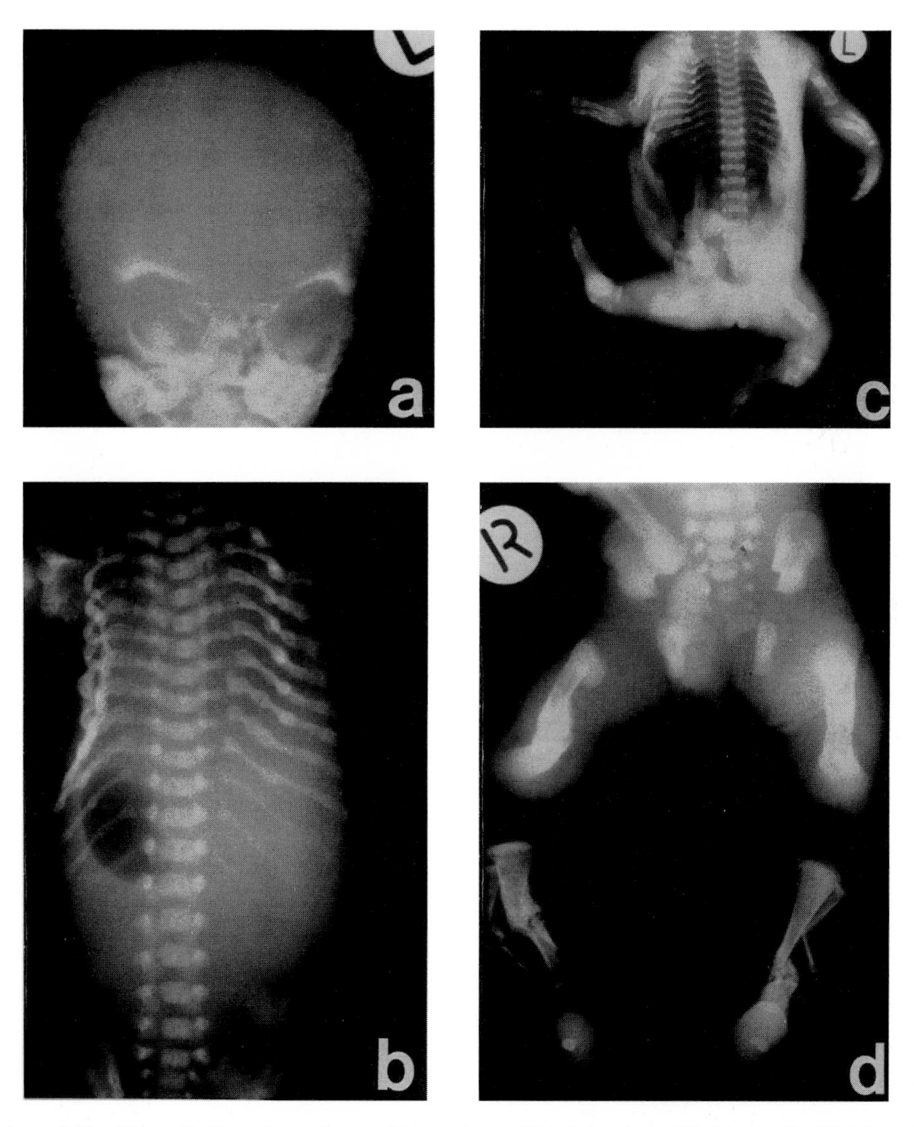

Figure 2.10 OI type II. The radiographs are of two patients with lethal perinatal OI (OI type II). OI26 (panels a,b,d) has a frameshift mutation in the pro-α1(I) C-propeptide domain, and in OI59 (panel c) α1(I) glycine 973 is substituted by a valine. **(a)**, Radiograph of the skull of OI26 showing many hundreds of wormian bones. The base of the skull, orbits, face, and jaw are better ossified than the calvarium. Radiographs of the torso of OI26 **(b)** and OI59 **(c)** show discontinuous beading of the ribs with distortion of the chest wall. The vertebrae and pelvis are well formed. (**d**), Radiograph of the pelvis and legs of OI26. The femora are broad with multiple healing fractures. The tibial and fibular metaphyses are broad and fractured. Their shafts are overmodelled and fractured.

more severe bone fragility, to the lethal perinatal form, type II OI (Byers, 1989b), which is usually fatal within a few hours of birth. These babies have short, bowed limbs with severe osteopenia, abnormal bone modelling and multiple fractures. The ribs are generally beaded (Figure 2.10).

The vast majority of the more than 100 structural mutations that have now been characterised in the proα1(I) and proα2(I) chains are within the triple helical domain. Most commonly these are point mutations leading to the substitution of glycine residues. Heterozygous glycine substitutions have been identified along the entire triple helical domain; the predominance of mutations towards the C-teminus of both chains most probably reflects bias introduced by the methods used to detect mutations. Single exon deletions, usually the result of point mutations altering the RNA splicing signals, and genomic deletions and insertions are less common.

The dramatic consequences of the disruption of the triple-helical domain by glycine substitutions, deletions and insertions have been explained in terms of the proposed mechanisms of helix assembly and folding, and indeed the analysis of the mutations has contributed greatly to our understanding of this process. The model of procollagen folding

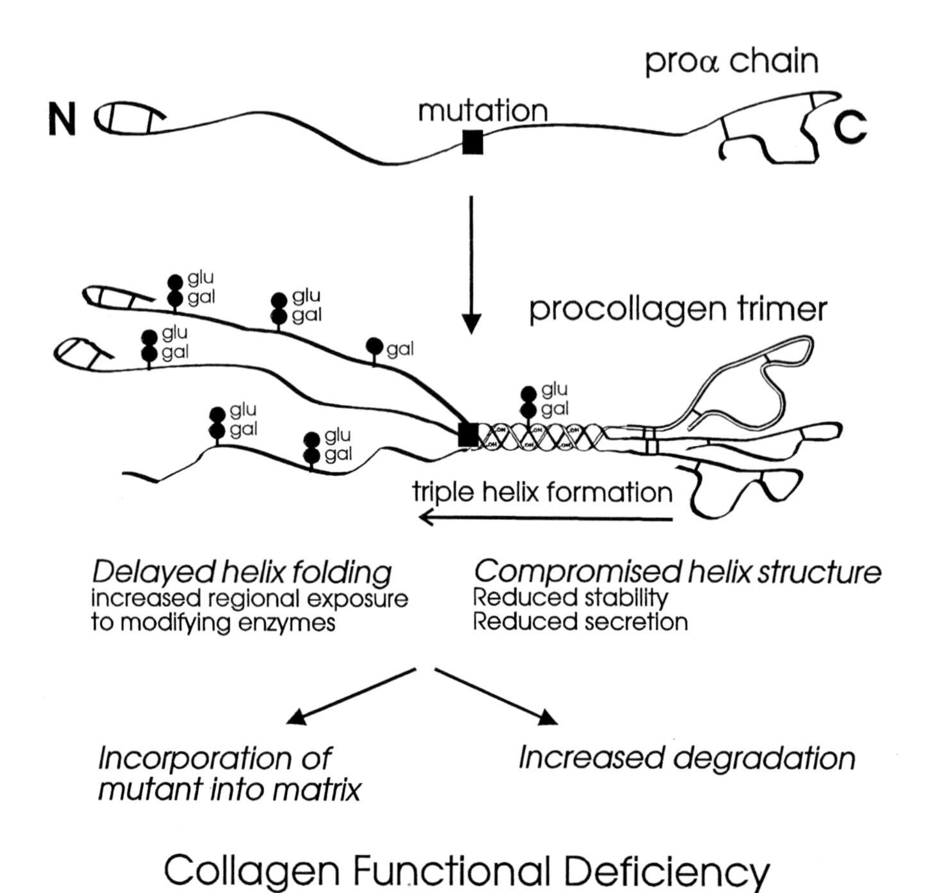

Figure 2.11 The effect of structural mutations on collagen helix folding.

predicts that the C-propeptides of the component α-chains come together and disulphide bond, and the helix then folds from the C- to the N-terminus. The finding of collagen chains that migrated slowly on electrophoresis because of excess post-translational lysine hydroxylation and glycosylation led to the idea that mutations disrupting the Gly-X-Y sequence delayed the folding of the triple-helix and exposed the unfolded chains on the N-terminal side of the mutation to the modifying enzymes for a longer period (Bateman *et al.*, 1984; Bonadio and Byers, 1985; Bateman *et al.*, 1986). The consequences of this abnormal folding are reduced collagen secretion, increased intracellular breakdown and compromised helical stability (Figure 2.11). The validity of this model has been directly demonstrated by the production of transgenic mice with lethal OI due a glycine to a cysteine subsitution in the proα1(I) chain (Stacey *et al.*, 1988) and later, a deletion of some α1(I) triple helical sequences (Khillan *et al.*, 1991). These transgenic animals faithfully reproduced the collagen biochemical abnormalities seen in patients with similar mutations.

The clinical severity of the mutation depends on the chain involved, the type of substitution, its location within the α-chain and the surrounding sequence. While there are many exceptions, a gradient of severity from the C- to N-terminus is often noted and is most apparent with glycine to cysteine substitutions. However, similar mutations produce OI with varying clinical severity and one of the current challenges is to understand the link between the mutation and the phenotype.

Although structural mutations in collagen I have been shown to be a cause of mild type I OI, fibroblasts from patients with this form of the disease often produce about half the normal amount of collagen I and show no evidence of structurally abnormal collagen in their tissues (Byers, 1993). Many of these cases have reduced α1(I) to α2(I) mRNA ratios (Willing *et al.*, 1993; Bateman *et al.*, 1988) suggesting that the underlying cause of the disease may be mutations which decrease the steady state levels of α1(I) mRNA. These could be promoter or enhancer mutations preventing normal levels of transcription, or frameshift and point mutations which result in the introduction of premature stop codons and interfere with mRNA stability, translocation to the cytoplasm or translatability (McIntosh *et al.*, 1993).

CONCLUSIONS

The collagens were once considered to be a small arcane group of inert structural proteins with a well-defined biochemical pathway, which played a relatively passive structural role as an extracellular scaffolding. However, over the last 20 years our knowledge has expanded dramatically, and now it is apparent that the collagens are members of an ever-expanding protein superfamily of extracellular matrix structural macromolecules which form diverse supramolecular assemblies performing an equally diverse variety of functional roles. Many of the details of collagen biosynthesis remain poorly defined and studies on the mechanisms of intracellular collagen subunit assembly, protein folding and transport are likely to describe important mechanisms involving molecular chaperones and other specific protein-protein interactions, fundamental to our understanding of protein synthesis in general.

The extracellular pathways of collagen fibril assembly and the interaction of collagens and other matrix components also remain poorly understood. This information is clearly critical if we are to unravel the complexities of how the cells orchestrate the tissue-specific assembly of the extracellular matrix, and how the extracellular matrix interacts with cells to maintain, or modulate, phenotypic expression in a tissue-specific manner during growth, development and during the many pathological processes that affect connective tissues.

The bewildering array of "new" collagens that are being defined share the common triple-helical motif with the "old" collagens, and the rules learnt from extensive biochemical and structural studies on these "old" collagens provide valuable signposts for the further characterisation of these newer family members. The helical motif may, however, be the only thing these collagens have in common with the the fibril-forming collagen family, since they form very different supramolecular structures. As we have discussed, many of these newer collagens have large non-collagenous domains which may have important roles in interacting with with other matrix components. The specialised distribution of many of these collagens implies an important functional role and new molecular approaches using dominant negative, or "gene knockout", transgenic mice promise to provide critical information on their role *in vivo*.

References

Acton, S., Resnick, D., Freeman, M., Ekkel, Y., Ashkenas, J. and Krieger, M. (1993) The collagenous domains of macrophage scavenger receptors and complement component C1q mediate their similar, but not identical, binding specificities for polyanionic ligands. *J. Biol. Chem.*, **268**, 3530–3537.

Adams, J.C. and Watt, F.M. (1993) Regulation of development and differentiation by the extracellular matrix. *Development*, **117**, 1183–1198.

Aycock, R.S., Raghow, R., Stricklin, G.P., Seyer, J.M. and Kang, A.H. (1986) Post-transcriptional inhibition of collagen and fibronectin synthesis by a synthetic homolog of a portion of the carboxyl-terminal propeptide of human type I collagen. *J. Biol. Chem.*, **261**, 14355–14360.

Bächinger, H.P., Bruckner, P., Timpl, R., Prockop, D.J. and Engel, J. (1980) Folding mechanism of the triple helix in type-III collagen and type-III pN-collagen. Role of disulphide bridges and peptide bond isomerization. *Eur. J. Biochem.*, **106**, 619–632.

Barile, F.A., Guzowski, D.E., Ripley, C., Siddiqi, Z. and Bienkowski, R.S. (1990) Ammonium chloride inhibits basal degradation of newly synthesized collagen in human fetal lung fibroblasts. *Arch. Biochem. Biophys.*, **276**, 125–131.

Bassuk, J.A. and Berg, R.A. (1989) Protein disulphide isomerase, a multifunctional endoplasmic reticulum protein. *Matrix*, **9**, 244–258.

Bateman, J.F., Mascara, T., Chan, D. and Cole, W.G. (1984) Abnormal type I collagen metabolism by cultured fibroblasts in lethal perinatal osteogenesis imperfecta. *Biochem. J.*, **217**, 103–115.

Bateman, J.F., Chan, D., Mascara, T., Rogers, J.G. and Cole, W.G. (1986) Collagen defects in lethal perinatal osteogenesis imperfecta. *Biochem. J.*, **240**, 699–708.

Bateman, J.F., Pillow, J.J., Mascara, T., Medvedec, S., Ramshaw, J.A.M. and Cole, W.G. (1987) Cell-layer-associated proteolytic cleavage of the telopeptides of type I collagen in fibroblast culture. *Biochem. J.*, **245**, 677–682.

Bateman, J.F., Chan, D., Lamande, S., Mascara, T. and Cole, W.G. (1988) Biochemical heterogeneity of type I collagen mutations in osteogenesis imperfecta. *Ann. N.Y. Acad. Sci.*, **543**, 95–105.

Bennett, V.D. and Adams, S.L. (1990) Identification of a cartilage-specific promoter within intron 2 of the chick α2(I) collagen gene. *J. Biol. Chem.265*, 2223–2230.

Berg, R.A., Kao, W.W.-Y. and Kedersha, N.L. (1980a) The assembly of tetrameric prolyl hydroxylase in tendon fibroblasts from newly synthesized α-subunits and from preformed cross-reacting protein. *Biochem. J.*, **189**, 491–499.

Berg, R.A., Schwartz, M.L. and Crystal, R.G. (1980b) Regulation of the production of secretory proteins: Intracellular degradation of newly synthesized "defective" collagen. *Proc. Natl. Acad. Sci. U.S.A.*, **77**, 4746–4750.

Berg, R.A., Schwartz, M.L., Rome, L.H. and Crystal, R.G. (1984) Lysosomal function in the degradation of defective collagen in cultured lung fibroblasts. *Biochemistry*, **23**, 2134–2138.

Berg, R.A. and Prockop, D.J. (1973a) The thermal transition of a non-hydroxylated form of collagen. Evidence for a role for hydroxyproline in stabilizing the triple-helix of collagen. *Biochem. Biophys. Res. Commun.*, **52**, 115–120.

Berg, R.A. and Prockop, D.J. (1973b) Purification of [^{14}C]protocollagen and its hydroxylation by prolyl-hydroxylase. *Biochem.*, **12**, 3395–3401.
Bernard, M., Yoshioka, H., Rodriguez, E., van der Rest, M., Kimura, R., Ninomiya, Y., Olsen, B.R. and Ramirez, F. (1988) Cloning and sequencing of proα1(XI) cDNA demonstrates that type XI belongs to the fibrillar class of collagens and reveals that the expression of the gene is not restricted to cartilagenous tissue. *J. Biol. Chem.*, **263**, 17159–17166.
Bernard, M.P., Chu, M-L., Myers, J.C., Ramirez, F., Eikenberry, E.F. and Prockop, D.J. (1983) Nucleotide sequences of complementary deoxyribonucleic acids for the proα1 chain of human type I procollagen. Statistical evaluation of structures that are conserved during evolution. *Biochemistry*, **22**, 5213–5223.
Bernstein, H.D., Rapoport, T.A. and Walter, P. (1989) Cytosolic protein translocation fatcors. Is SRP still unique? *Cell*, **58**, 1017–1019.
Bienkowski, R.S., Baum, B.J. and Crystal, R.G. (1978). Fibroblasts degrade newly synthesized collagen within the cells before secretion. *Nature*, **276**, 413–416.
Bienkowski, R.S. (1983) Intracellular degradation of newly synthesized secretory proteins. *Biochem. J.*, **214**, 1–10.
Birk, D.E., Fitch, J.M., Babiarz, J.P. and Linsenmayer, T.F. (1988) Collagen type I and V are present in the same fibril in the avian corneal stroma. *J. Cell Biol.*, **106**, 999–1008.
Birk, D.E., Fitch, J.M., Babiarz, J.P., Doane, K.J. and Linsenmayer, T.F. (1990a) Collagen fibrillogenesis *in vitro*: Interaction of types I and V collagen regulates fibril diameter. *J. Cell Sci.*, **95**, 649–657
Birk, D.E., Zycband, E.I., Winkelmann, D.A. and Trelstad, R.L. (1990b) Collagen fibrillogenesis *in situ*. Discontinuous segmental assembly in extracellular compartments. *Ann. N.Y. Acad. Sci.*, **580**, 176–194.
Bissell, M.J. and Barcellos-Hoff, M.H. (1987) The influence of extracellular matrix on gene expression: Is structure the message? *J. Cell Sci. Suppl.*, **8**, 327–343.
Boast, S., Su, M-W., Ramirez, F., Sanchez, M. and Avvedimento, E.V. (1990) Functional analysis of *cis*-acting DNA sequences controlling transcription of the human type I collagen genes. *J. Biol. Chem.*, **265**, 13351–13356.
Bonadio, J. and Byers, P.H. (1985) Subtle structural alterations in the chains of type I procollagen produce osteogenesis imperfecta type II. *Nature*, **316**, 363–366.
Bonaldo, P., Russo, V., Bucciotti, F., Doliana, R. and Colombatti, A. (1990) Structural and functional features of the α3 chain indicate a bridging role for chicken collagen VI in connective tissue. *Biochemistry*, **29**, 1245–1254.
Bornstein, P., McKay, J., Morishima, J.K., Devarayalu, S. and Gelinas, R.E. (1987) Regulatory elements in the first intron contribute to transcriptional control of the human α1(I) collagen gene. *Proc. Natl. Acad. Sci. U.S.A.*, **84**, 8869–8873.
Bornstein, P., McKay, J., Devarayala, S. and Cook, S.C. (1988). A highly conserved 5′ untranslated inverted repeat sequence is ineffective in translational control of the α1(I) collagen gene. *Nucleic Acids Res.*, **16**, 9721–9736.
Bornstein, P. and McKay, J. (1988)The first intron of the α1(I) collagen gene contains several transcriptional regulatory elements. *J. Biol. Chem.*, **263**, 1603–1606.
Bornstein, P. and Sage, H. (1989) Regulation of collagen gene expression. *Prog. Nucleic Acids Res. Mol. Biol.*, **37**, 67–106.
Briggaman, R.A. (1992) Pathobiology of type VII collagen: A role in the pathogenesis of epidermolysis bullosa. In *Collagen. Pathobiochemistry*, Kang, A.H. and Nimni, M.E. (eds.) pp.1–16.Boca Raton: CRC Press.
Brodsky, B., Eikenberry, E.F. and Cassidy, K. (1980) An unusual collagen periodicity in skin. *Biochim. Biophys. Acta*, **621**, 162–166.
Brodsky, B., Eikenberry, E.F., BelBruno, K.C. and Sterling, K. (1982)Variations in collagen fibril structure in tendons. *Biopolymers*, **21**, 935–951.
Brodsky, B., Tanaka, S. and Eikenberry, E.F. (1988) X-ray diffraction as a tool for studying collagen structure. In *Collagen: Volume I, Biochemistry*, Nimni, M. (ed.) pp. 95–112. Boca Raton: CRC Press.
Brodsky, B. and Eikenberry, E.F. (1982) Characterization of fibrous forms of collagen. *Methods in Enzymology*, **82**, 127–174.
Brodsky, B. and Ramshaw, J.A.M. (1994) Collagen orientation in a oriented fibrous capsule. *Int. J. Biol. Macromol.*, **16**, 27–30.
Bruckner, P., Eikenberry, E.F. and Prockop, D.J. (1981) Formation of the triple helix of type I procollagen *in cellulo*. A kinetic model based on *cis-trans* isomerization of peptide bonds. *Eur. J. Biochem*, **118**, 607–613.
Bruns, R.R., Hulmes, D.J.S., Therrien, S.F. and Gross, J. (1979) Procollagen segment-long-spacing crystallites: Their role in collagen fibrillogenesis. *Proc. Natl. Acad. Sci. U.S.A.*, **76**, 313–317.
Burgeson, R.E. (1988) New collagens, new concepts. *Ann. Rev. Cell Biol.*, **4**, 551–577.
Burgeson, R.E., Lunstrum, G.P., Rokosova, B., Rimberg, C.S., Rosenbaum, L.M. and Keene, D.R. (1990) The structure and function of type VII collagen. *Ann. N.Y. Acad. Sci.*, **580**, 32–43.
Burgeson, R.E. and Hollister, D.W. (1979) Collagen heterogeneity in human cartilage: Identification of several new collagen chains. *Biochem.Biophys.Res.Commun.*, **87**, 1124–1131.

Byers, P.H. (1989a) Inherited disorders of collagen gene structure and expression. *Am. J. Med. Genet.*, **34**, 72–80.

Byers, P.H. (1989b) Disorders of collagen biosynthesis and structure. In *The Metabolic Basis of Inherited Disease. II*, Scriver, C.R., Beaudet, A.L., Sly, W.S., Valle, D. (eds.) New York: McGraw-Hill, Inc..

Byers, P.H. (1993) Osteogenesis imperfecta. In *Connective Tissue and its Heritable Disorders. Molecular, Genetic, and Medical Aspects*, Royce, P.M. and Steinmann, B. (eds.) pp. 317–350. New York: Wiley-Liss, Inc.

Chapman, J.A. (1989) The regulation of size and form in the assembly of collagen fibrils *in vivo*. *Biopolymers*, **28**, 1367–1382.

Chapman, J.A. and Hulmes, D.J.S. (1984) Electron microscopy of the collagen fibril. In *Ultrastructure of the Connective Tissue Matrix*, Ruggeri, A. and Motta, P.M. (eds.) pp.1–33.Boston: Martinus Nijhoff Publishers.

Cheah, K.S.E., Lau, E.T., Au, P.K.C. and Tam, P.P.L. (1991). Expression of the mouse α1(II) collagen gene is not restricted to cartilage during development. *Development*, **111**, 945–953.

Chessler, S.D. and Byers, P.H. (1992) Defective folding and stable association with protein disulphide isomerase/ prolyl hydroxylase of type I procollagen with a deletion in the proα2(I) chain that preserves the Gly-X-Y repeat pattern. *J. Biol. Chem.*, **267**, 7751–7757.

Chessler, S.D. and Byers, P.H. (1993)BiP binds type I procollagen proα-chains with mutations in the carboxyl-terminal propeptide synthesized by cells from patients with osteogenesis imperfecta. *J. Biol. Chem.*, **268**, 18226–18233.

Chu, M-L., Pan, T., Conway, D., Kuo, H-J., Glanville, R.W., Timpl, R., Mann, K. and Deutzmann, R. (1989) Sequence analysis of α1(VI) and α2(VI) chains of human type VI collagen reveals internal triplication of globular domains similar to the A domains of van Willebrand factor and two α2(VI) chain variants that differ in the carboxy terminus. *EMBO J.*, **8**, 1939–1946.

Chu, M-L. and Prockop, D.J. (1993) Collagen: Gene structure. In *Connective Tssue and its Heritable Disorders. Molecular, Genetic, and Medical Aspects*, Royce, P.M. and Steinmann, B. (eds.) pp. 149–165. New York: Wily-Liss, Inc..

Clarke, C.C. (1979) The distribution and initial characterization of oligosaccharide units on the COOH-terminal propeptide extensions of the pro-α1 and proα2 chains of type I procollagen. *J. Biol. Chem.*, **254**, 10798–10802.

Clarke, E.P., Cates, G.A., Ball, E.H. and Sanwal, B.D. (1991) A collagen-binding protein in the endoplasmic reticulum of myoblasts exhibits relationship with serine protease inhibitors. *J. Biol. Chem.*, **266**, 17230–17235.

Cohen-Solal, L., Castanet, J., Meunier, F.J. and Glimcher, M.J. (1986) Proline hydroxylation of collagens synthesized at different temperatures *in vivo* by two poikilothermic species. *Comp Biochem Physiol [B]*, **83**, 483–486.

Comper, W.D. and Veis, A. (1977) Characterization of nuclei in *in-vitro* collagen fibril formation. *Biopolymers*, **16**, 2133–2142.

Davis, A.E. and Lachmann, P.J (1984) Bovine conglutinin is a collagen-like protein. *Biochem.*, **23**, 2139–2144.

de Waal, A. and de Jong, L. (1988) Processive action of the two peptide binding sites of prolyl 4-hydroxylase in the hydroxylation of procollagen. *Biochem.*, **27**, 150–155.

de Wet, W., Bernard, M., Benson-Chanda, V., Chu, M-L., Dickson, L., Weil, D. and Ramirez, F. (1987) Organization of the human proα2(I) collagen gene. *J. Biol. Chem.*, **262**, 16032–16036.

Dion, A.S. and Myers, J.C. (1987) COOH-terminal propeptides of the major human procollagens. Structural, functional and genetic comparisons. *J. Mol. Biol.*, **193**, 127–143.

Doege, K.J. and Fessler, J.H. (1986) Folding of carboxyl domain and assembly of procollagen I. *J. Biol. Chem.*, **261**, 8924–8935.

Drickamer, K., Dordal, M.S. and Reynolds, L. (1986) Mannose-binding proteins isolated from rat liver contain carbohydrate-recognition domains linked to collagenous tails. Complete primary structures and homology with pulmonary surfactant apoprotein. *J. Biol. Chem.*, **261**, 6878–6887.

Duksin, D., Davidson, J.M. and Bornstein, P. (1978) The role of glycosylation in the enzymatic conversion of procollagen to collagen: Studies using tunicamycin and concanavalin A. *Arch. Biochem. Biophys.*, **185**, 326–332.

Duksin, D., Gal, A. and Goodman, R.M. (1983) Altered protein glycosylation and procollagen to collagen conversion in human fibroblasts. *Lab. Invest.*, **49**, 346–352.

Duksin, D. and Bornstein, P. (1977) Impaired conversion of procollagen to collagen by fibroblasts and bone treated with tunicamycin, an inhibitor of protein glycosylation. *J. Biol. Chem.*, **252**, 955–962.

Eikenberry, E.F., Mendler, M., Bürgin, R., Winterhalter, K.H. and Bruckner, P. (1992) Fibrillar organization in cartilage. In *Articular Cartilage and Arthritis*, Kuettner, K.E., Schleyerbach, R., Peyron, J.G. and Hascall, V.C. (eds.) pp.133–149. New York: Raven Press.

Ellis, R.J. and van der Vies, S.M. (1991) Molecular chaperones. *Annu. Rev. Biochem.*, **60**, 321–347.

Epstein, E.H. (1974)[α1(III)]$_3$ human skin collagen. Release by pepsin digestion and preponderance in fetal life. *J Biol Chem*, **249**, 3225–3231.

Evans, E.A., Gilmore, R. and Blobel, G. (1986) Purification of microsomal signal peptidase as a complex. *Proc. Natl. Acad. Sci. U.S.A.*, **83**, 581–585.

Eyre, D.R., Paz, M.A. and Gallop, P.M. (1984) Cross-linking in collagen and elastin. *Ann. Rev. Biochem.*, **53**, 717–748.

Eyre, D.R., Apone, S., Wu, J-J., Ericsson, L.H. and Walsh, K.A. (1987) Collagen type IX: evidence for covalent linkages to type II collagen in cartilage. *FEBS lett.*, **220**, 337–341.

Fessler, L.I., Timpl, R. and Fessler, J.H. (1981)Assembly and processing of procollagen type III in chick embryo blood vessels. *J.Biol.Chem.*, **256**, 2531–2537.

Fischer, G. and Schmid, F.X. (1990)The mechanism of protein folding. Implications of *in vitro* refolding models for *de novo* protein folding and translocation in the cell. *Biochemistry*, **29**, 2205–2212.

Fleischmajer, R., Timpl, R., Tuderman, L., Raisher, L., Wiestner, M., Perlish, J.S. and Graves, P.N. (1981) Ultrastructural identification of extension aminopropeptides of type I and III collagens in human skin. *Proc. Natl.Acad. Sci. U.S.A.*, **78**, 7360–7364.

Fleischmajer, R., Perlish, J.S., Burgeson, R.E., Shaikh-Bahai, F. and Timpl, R. (1990) Type I and type III collagen interactions during fibrillogenesis. *Ann. N.Y. Acad. Sci.*, **580**, 161–175.

Fleischmajer, R., Fisher, L.W., MacDonald, E.D., Jacobs, L., Perlish, J.S. and Termine, J.D. (1991) Decorin interacts with fibrillar collagen of embryonic and adult skin. *J. Struct. Biol.*, **106**, 82–90.

Forster, S.J. and Freedman, R.B. (1984) Catalysis by protein disulphide-isomerase of the assembly of trimeric procollagen from procollagen polypeptide chains. *Bioscience Reports*, **4**, 223–229.

Fouser, L., Sage, E.H., Clark, J and Bornstein, P. (1991) Feedback regulation of collagen gene expression: A Trojan horse approach. *Proc.Natl.Acad.Sci.U.S.A.*, **88**, 10158–10162.

Fraser, R.B.D. and MacRae, T.P. (1973). *Conformation in Fibrous Proteins*, New York:Academic Press,

Fraser, R.B.D., MacRae, T.P., and Suzuki, E. (1979). Chain conformation in the collagen molecule. *J. Mol. Biol.*, **129**, 463–481.

Fraser, R.B.D., MacRae, T.P., Miller, A. and Suzuki, E. (1983). Molecular conformation and packing in collagen fibrils. *J. Mol. Biol.*, **167**, 497–521.

Freedman, R.B. (1989)Protein disulfide isomerase: multiple roles in the modification of nascent secretory proteins. *Cell*, **57**, 1069–1072.

Garofalo, S., Vuorio, E., Metsaranta, M., Rosati, R., Toman, D., Vaughan, J., Lozano, G., Mayne, R., Ellard, J., Horton, W. and de Crombrugghe, B. (1991) Reduced amounts of cartilage collagen fibrils and growth plate anomalies in transgenic mice harboring a glycine-to-cysteine mutation in the mouse type II procollagen 1-chain gene. *Proc. Natl. Acad. Sci. U.S.A.*, **88**, 9648–9652.

Gathercole, L.J., Porter, S. and Scully, C. (1987a) Axial periodicity in periodontal collagens. Human periodontal ligament and gingival connective tissue collagen fibers possess a dermis-like D-period. *J. Periodont. Res.*, **22**, 408–411.

Gathercole, L.J., Shah, J.S. and Nave, C. (1987b) Skin-tendon differences in collagen D-period are not geometric or stretch-related artefacts. *Int. J. Biol. Macromol.*, **9**, 181–183.

Gerecke, D.R., Foley, J.W., Castagnola, P., Gennari, M., Dublet, B., Cancedda, R., Linsenmayer, T.F., van der Rest, M., Olsen, B.R. and Gordon, M.K. (1993) Type XIV collagen is encoded by alternative transcripts with distinct 5' regions and is a multidomain protein with homologies to von Willebrand's factor, fibronectin, and other matrix proteins. *J. Biol. Chem.*, **268**, 12177–12184.

Gething, M-J., McCammon, K. and Sambrook, J. (1986) Expression of wild-type and mutant forms of influenza haemagglutinin: the role of folding in intracellular transport. *Cell*, **46**, 939–950.

Gething, M-J. and Sambrook, J. (1992) Protein folding in the cell. *Nature*, **355**, 33–45.

Giudice, G.J., Squiquera, H.L., Elias, P.M. and Diaz, L.A. (1991) Identification of two collagen domains within the bullous pemphigoid autoantigen, BP180. *J. Clin. Invest.*, **87**, 734–738.

Giudice, G.J., Emery, D.J. and Diaz, L.A. (1992) Cloning and primary structural analysis of the bullous pemphigoid autoantigen BP180. *J. Invest. Dermatol.*, **99**, 243–250.

Goochee, C.F., Gramer, M.J., Anderson, D.C., Bahr, J.B. and Rasmussen, J.R. (1991) The oligosaccharides of glycoproteins: bioprocess factors affecting oligosaccharide structure and their effect on glycoprotein proterties. *Bio/Technology*, **9**, 1347–1355.

Graves, P.N., Olsen, B.R., Fietzek, P.P., Prockop, D.J. and Monson, J.M. (1981) Comparison of the NH_2-terminal sequences of chick type I preprocollagen chains synthesized in an mRNA-dependent reticulocyte lysate. *Eur. J. Biochem.*, **118**, 363–369.

Gross, J. and Kirk, D. (1958). The heat precipitation of collagen from neutral salt solutions: Some rate-limiting factors. *J. Biol. Chem.*, **233**, 355–360.

Haas, I.G and Wabl, M. (1983) Immunoglobulin heavy chain binding protein. *Nature*, **306**, 387–389.

Halila, R. and Peltonen, L. (1984) Neutral protease cleaving the N-terminal propeptide of type III procollagen: Partial purification and characterization of the enzyme from smooth muscle cells of bovine aorta. *Biochemistry*, **23**, 1251–1256.

Hämäläinen, E-R., Jones, T.A., Sheer, D., Taskinen, K., Pihlajaniemi, T. and Kivirikko, K.I. (1991)Molecular cloning of human lysyl oxidase and assignment of the gene to chromosome 5q23.3-31.2. *Genomics*, **11**, 508–516.

Hames, B.D., Glover, D.M. (Eds.) (1988) *Transcription and Splicing*. Oxford. IRL Press Ltd.

Haralson, M.A., Mitchell, W.M., Rhodes, R.K., Kresina, T.F., Gay, R. and Miller, E.J. (1980) Chinese hamster lung cells synthesize and confine to the cellular domain a collagen composed solely of B chains. *Proc. Natl. Acad. Sci. U.S.A.*, **77**, 5206–5210.

Hartung, S., Jaenisch, R. and Breindl, M. (1986) Retrovirus insertion inactivates mouse α1(I) collagen gene by blocking initiation of transcription. *Nature*, **320**, 365–367.

Heinegård, D. and Oldberg, A. (1989). Structure and biology of cartilage and bone matrix noncollagenous macromolecules. *FASEB J.*, **3**, 2042–2051.

Helaakoski, T., Vuori, K., Myllylä, R., Kivirikko, K.I. and Pihlajaniemi, T. (1989)Molecular cloning of the α-subunit of human prolyl 4-hydroxylase: The complete cDNA-derived amino acid sequence and evidence for alternative splicing of RNA transcripts. *Proc. Natl. Acad. Sci. U.S.A.*, **86**, 4392–4396.

Helseth, D.L., Lechner, J.H. and Veis, A. (1979) Role of the amino-terminal extrahelical region of type I collagen in directing the 4D-overlap in fibrillogenesis. *Biopolymers*, **18**, 3005–3014.

Helseth, D.L. and Veis, A. (1981) Collagen self-assembly *in vitro*. Differentiating specific telopeptide-dependent interactions using selective enzyme modification and the addition of the free amino telopeptide. *J.Biol.Chem.*, **256**, 7118–7128.

Hendrick, J.P. and Hartl, F-U. (1993) Molecular chaperone functions of heat-shock proteins. *Annu. Rev. Biochem.*, **62**, 349–384.

Henkel, W. and Glanville, R.W. (1982) Covalent crosslinking between molecules of type I and type III collagen. The involvement of the N-terminal, nonhelical regions of the α1 (I) and α1 (III) chains in the formation of intermolecular crosslinks. *Eur. J. Biochem.*, **122**, 205–213.

Hirschberg, C.B. and Snider, M.D. (1987) Topography of glycosylation in the rough endoplasmic reticulum and Golgi apparatus. *Ann. Rev. Biochem.*, **56**, 63–87.

Hojima, Y., van der Rest, M. and Prockop, D.J. (1985) Type I procollagen carboxyl-terminal proteinase from chick embryo tendons. Purification and characterization. *J. Biol. Chem.*, **260**, 15996–16003.

Hojima, Y., McKenzie, J., van der Rest, M. and Prockop, D.J. (1989) Type I procollagen N-proteinase from chick embryo tendons. Purification of a new 500-kDa form of the enzyme and identification of the catalytically active polypeptides. *J. Biol. Chem.*, **264**, 11336–11345.

Holmes, D.F., Watson, R.B., Steinmann, B. and Kadler, K.E. (1993) Ehlers-Danlos syndrome type VIIB. Morphology of type I collagen fibrils formed *in vivo* and *in vitro* is determined by the conformation of the retained N-propeptide. *J. Biol. Chem.*, **268**, 15758–15765.

Hörlein, D., McPherson, J., Goh, S.H. and Bornstein, P. (1981) Regulation of protein synthesis: Translational control by procollagen-derived fragments. *Proc. Natl. Acad. Sci. U.S.A.*, **78**, 6163–6167.

Horton, W., Miyashita, T., Kohno, K., Hassell, J.R. and Yamada, Y. (1987) Identification of a phenotype-specific enhancer in the first intron of the rat collagen II gene. *Proc. Natl. Acad. Sci. U.S.A.*, **84**, 8864–8868.

Housley, T.J., Rowland, F.N., Ledger, P.W., Kaplan, J. and Tanzer, M.L. (1980) Effects of tunicamycin on the biosynthesis of procollagen by human fibroblasts. *J. Biol. Chem.*, **255**, 121–128.

Hudson, B.G., Wisdom, B.J., Gunwar, S. and Noelken, M.E. (1992) Collagen IV: Role in Goodpasture syndrome, Alport-type familial nephritis, and diabetic nephropathy. In *Collagen. Pathobiochemistry*, Kang, A.H. and Nimni, M.E. (eds.) pp.17–30. Boca Raton: CRC Press.

Hudson, B.G., Reeders, S.T. and Tryggvason, K. (1993) Type IV collagen: Structure, gene organisation, and role in human diseases. *J. Biol. Chem.*, **268**, 26033–26036

Hulmes, D.J.S. and Miller, A. (1979) Quasi-hexagonal molecular packing in collagen fibrils. *Nature*, **282**, 878–880.

Hulmes, D.J.S., Bruns, R.R. and Gross, J. (1983) On the state of aggregation of newly secreted procollagen. *Proc. Natl. Acad. Sci. U.S.A.*, **80**, 388–392.

Hulmes, D.J.S., Kadler, K.E., Mould, A.P., Hojima, Y., Holmes, D.F., Cummings, C., Chapman, J. and Prockop, D.J. (1989)Pleomorphism in type I collagen fibrils produced by persistence of the procollagen N-propeptide. *J. Mol. Biol.*, **210**, 337–345.

Hulmes, D.J.S. (1992) The collagen superfamily – diverse structures and assemblies. *Essays in Biochemistry*, **27**, 49–67.

Hurtley, S.M. and Helenius, A. (1989) Protein oligomerization in the endoplasmic reticulum. *Annu. Rev. Cell Biol.*, **5**, 277–307.

Inoguchi, K., Yoshioka, H., Khaleduzzaman, M. and Ninomiya, Y. (1995) The mRNA for α1(XIX) collagen chain, a new member of FACITs, contains a long unusual 3´ untranslated region and displays many unique splicing variants. *J. Biochem.* **117**, 137–146.

Jacenko, O., LuValle, P.A. and Olsen, B.R. (1993) Spondylometaphyseal dysplasia in mice carrying a dominant negative mutation in a matrix protein specific for cartilage-to-bone transition. *Nature*, **365**, 56–61.

Jackson, M.R., Nilsson, T. and Peterson, P.A. (1990) Identification of a consensus motif for retention of transmembrane proteins in the endoplasmic reticulum. *EMBO J.*, **9**, 3153–3162.

James, V.J., McConnell, J.F. and Capel, M. (1991) The d-spacing of collagen from mitral heart valves changes with ageing, but not with collagen type III content. *Biochim. Biophys. Acta*, **1078**, 19–22.

Janeczko, R.A. and Ramirez, F. (1989) Nucleotide and amino acid sequences of the entire human α1(III) collagen. *Nucl. Acids Res.*, **17**, 6742.
Jimenez, S.A., Bashey, R.I., Benditt, M. and Yankowski, R. (1977) Identification of collagen α1(I) trimer in embryonic chick tendons and calvaria. *Biochem. Biophys. Res. Commun.*, **78**, 1354–1361.
Juvonen, M. and Pihlajaniemi, T. (1992) Characterization of the spectrum of alternative splicing of α1(XIII) collagen transcripts in HT-1080 cells and calvarial tissue resulted in identification of two previously unidentified alternatively spliced sequences, one previously unidentified exon, and nine new mRNA variants. *J. Biol. Chem.*, **267**, 24693–24699.
Kadler, K.E., Hojima, Y. and Prockop, D.J. (1987) Assembly of collagen fibrils *de novo* by cleavage of the type I pC-collagen with procollagen C-proteinase. Assay of critical concentration demonstrates that collagen self-assembly is a classical example of an entropy-driven process. *J. Biol. Chem.*, **260**, 15696–15701.
Kaplan, H.A., Welply, J.K. and Lennarz, W.J. (1987) Oligosaccharyl transferase: the central enzyme in the pathway of glycoprotein assembly. *Biochim. Biophys. Acta*, **906**, 161–173.
Kassenbrock, C.K., Garcia, P.D., Walter, P. and Kelly, R.B. (1988) Heavy-chain binding protein recognizes aberrant polypeptides translocated *in vitro*. *Nature*, **333**, 90–93.
Katayama, K., Seyer, J.M., Raghow.R., and Kang, A.H. (1991) Regulation of extracellular matrix production by chemically synthesized subfragments of type I collagen carboxy propeptide. *Biochemistry*, **30**, 7097–7104.
Katayama, K., Armendariz-Borunda, J., Raghow, R., Kang, A.H. and Seyer, J.M. (1993)A pentapeptide from type I procollagen promotes extracellular matrix production. *J. Biol. Chem.*, **268**, 9941–9944.
Katz, E.P. and David, C.W. (1992) Unique side-chain conformation encoding chirality and azimuthal orientation in the molecular packing of skin. *J. Mol. Biol.*, **228**, 963–969.
Kenyon, K., Contente, S., Trackman, P.C., Tang, J., Kagan, H.M. and Friedman, R.M. (1991) Lysyl oxidase and rrg messenger RNA. *Science*, **253**, 802.
Kessler, E., Mould, A.P. and Hulmes, D.J.S. (1990) Procollagen type I C-proteinase enhancer is a naturally occurring connective tissue glycoprotein. *Biochem. Biophys. Res. Commun.*, **173**, 81–86.
Kessler, E. and Adar, R. (1989) Type I procollagen C-proteinase from mouse fibroblasts. Purification and demonstration of a 55-kDa enhancer protein. *Eur. J. Biochem.*, **186**, 115–121.
Khillan, J.S., Olsen, A.S., Kontusaari, S., Sokolov, B. and Prockop, D.J. (1991) Transgenic mice that express a mini-gene version of the human gene for type I procollagen (COL1A1) develop a phenotype resembling a lethal form of osteogenesis imperfecta. *J. Biol. Chem.*, **266**, 23373–23379.
Kielty, C.M., Boot-Handford, R.P., Ayad, S., Shuttleworth, C.A. and Grant, M.E. (1990) Molecular composition of type VI collagen. Evidence for chain heterogeneity in mammalian tissues and cultured cells. *Biochem. J.*, **272**, 787–795.
Kielty, C.M., Hopkinson, I. and Grant, M.E. (1993) Collagen. The collagen family: structure, assembly, and organization in the extracellular matrix. In *Connective Tissue and its Heritable Disorders. Molecular, Genetic, and Medical Aspects*, Royce, P.M. and Steinmann, B. (eds.) pp.103–147. New York: Wiley-Liss, Inc..
Kimura, T., Cheah, K.S.E., Chan, S.D.H., Lui, V.D.C., Mattei, M-G., Van der Rest, M., Ono, K., Solomon, E., Ninomiya, Y. and Olsen, B.R. (1989)The human α2(XI) collagen (COL11A2) chain. *J. Biol. Chem.*, **264**, 13910–13916.
Kirk, T.Z., Evans, J.S. and Veis, A. (1987) Biosynthesis of type I procollagen. Characterization of the distribution of chain sizes and extent of hydroxylation of polysome-associated proa-chains. *J. Biol. Chem.*, **262**, 5540–5545.
Kivirikko, K.I., Ryhänen, L., Anttinen, H., Bornstein, P. and Prockop, D.J. (1973) Further hydroxylation of lysyl residues in collagen by protocollagen lysyl hydroxylase *in vitro*. *Biochem.*, **12**, 4966–4971.
Kivirikko, K.I., Myllylä, R. and Pihlajaniemi, T. (1989) Protein hydroxylation: prolyl 4-hydroxylase, an enzyme with four cosubstrates and a multifunctional subunit. *FASEB J.*, **3**, 1609–1617.
Kivirikko, K.I. (1993) Collagens and their abnormalities in a wide spectrum of diseases. *Ann. Med.*, **25**, 113–126.
Kivirikko, K.I., Heinamaki, P., Rehn, M., Honkanen, N., Myers, J.C. and Pihlajaniemi, T. (1994) Primary structure of the a1 chain of human type XV collagen and exon intron organization in the 3′ region of the corresponding gene. *J. Biol. Chem.*, **269**, 4773–4779.
Kivirikko, K.I. and Myllylä, R. (1979) Collagen glycosyltransferases. *Int. Rev. Connect. Tiss. Res.*, **8**, 23–72.
Kivirikko, K.I. and Myllylä, R. (1982) Posttranslational enzymes in the biosynthesis of collagen: Intracellular enzymes. *Methods Enzymol.*, **82**, 245–304.
Kivirikko, K.I. and Myllylä, R. (1987) Recent developments in posttranslational modification: Intracellular processing. *Methods Enzymol.*, **144**, 96–114.
Koivu, J. and Myllylä, R. (1987) Interchain disulphide bond formation in types I and II procollagen. *J. Biol. Chem.*, 6159–6164.
Kornfeld, R. and Kornfeld, S. (1985) Assembly of asparagine-linked oligosaccharides. *Ann. Rev. Biochem.*, **54**, 631–664.
Kozutsumi, Y., Segal, M., Normington, K., Gething, M-J. and Sambrook, J. (1988) The presence of malfolded proteins in the endoplasmic reticulum signals the induction of glucose-regulated proteins. *Nature*, **332**, 462–464.

Kratochwil, K., von der Mark, K., Kollar, E.J., Jaenisch, R., Mooslehner, K., Schwarz, M., Haase, K., Gmachi, I. and Harbers, K. (1989) Retrovirus-induced insertional mutation in Mov13 mice affects collagen I expression in a tissue-specific manner. *Cell*, **57**, 807–816.

Krieg, U.C., Walter, P. and Johnson, A.E. (1986) Photocrosslinking of the signal sequence of nascent preprolactin to the 54-kilodalton polypeptide of the signal recognition particle. *Proc. Natl. Acad. Sci. U.S.A.*, **83**, 8604–8608.

Kuivaniemi, H., Tromp, G. and Prockop, D.J. (1991)Mutations in collagen genes: causes of rare and some common diseases in humans. *FASEB J.*, **5**, 2052–2060.

Kuypers, R., Tyler, M., Kurth, L.B, Jenkins, I.D. and Horgan, D.J. (1992) Identification of the loci of the collagen-associated Ehrlich chromogen in type I collagen confirms its role as a trivalent cross-link. *Biochem. J.*, **283**, 129–136.

Kwan, A.P.L., Cummings, C.E., Chapman, J.A. and Grant, M.E. (1991) Macromolecular organization of chicken type X collagen in vitro. *J. Cell Biol.*, **114**, 597–604.

Lamandé, S.R. and Bateman, J.F. (1995) The type I collagen proα1(I) COOH-terminal propeptide *N*-linked oligosaccharide: Functional analysis by site-directed mutagenesis. *J. Biol. Chem.* **270**, 17858–17865.

LaMantia, M., Miura, R., Tachikawa, H., Kaplan, H.A., Lennarz, W.J. and Mizunaga, T. (1991) Glycosylation site binding protein and protein disulfide isomerase are identical and essential for cell viability in yeast. *Proc. Natl. Acad. Sci. U.S.A.*, **88**, 4453–4457.

Lamond, A.I. (1993)The spliceosome. *Bio Essays*, **15**, 595–603.

Lang, K., Schmid, F.X. and Fischer, G. (1987) Catalysis of protein folding by prolyl isomerase. *Nature*, **329**, 268–270.

Lee, S-T., Kessler, E. and Greenspan, D.S. (1990) Analysis of site-directed mutations in human proα2(I) collagen which block cleavage by the C-proteinase. *J. Biol. Chem.*, **265**, 21992–21996.

Li, K., Tamai, K., Tan, E.M.L. and Uitto, J. (1993) Cloning of type XVII collagen. Complementary and genomic DNA sequences of mouse 180-kilodalton bullous pemphigoid antigen (BPAG2) predict an interrupted collagenous domain, a transmembrane segment, and unusual features in the 5′-end of the gene and the 3′-untranslated region of the mRNA. *J. Biol. Chem.*, **268**, 8825–8834.

Lichtenstein, J.R., Martin, G.R., Kohn, L.D., Byers, P.H. and McKusick, V.A. (1973). Defect in the conversion of procollagen to collagen in a form of Ehler's-Danlos syndrome. *Science*, **182**, 298–300.

Lin, C.Q. and Bissell, M.J. (1993) Multi-faceted regulation of cell differentiation by extracellular matrix. *FASEB J.*, **7**, 737–743.

Linsenmayer, T.F., Gibney, E., Igoe, F., Gordon, M.K., Fitch, J.M., Fessler, L.I. and Birk, D.E. (1993) Type V collagen: Molecular structure and fibrillar organization of the chicken α1(V) NH_2-terminal domain, a putative regulator of corneal fibrillogenesis. *J. Cell Biol.*, **121**, 1181–1189.

Mariani, T.J., Trackman, P.C., Kagan, H.M., Eddy, R.L., Shows, T.B., Boyd, C.D. and Deak, S.B. (1992) The complete derived amino acid sequence of human lysyl oxidase and assignment of the gene to chromosome 5. Extensive sequence homology with the murine RAS recision gene. *Matrix*, **12**, 242–248.

Mathews, M.B. (1975). *Connective Tissue: Macromolecular Structure and Function*, New York: Springer-Verlag,

Matsumoto, A., Naito, M., Itakura, H., Ikemoto, S., Asaoka, H., Hayaakawa, I., Kanamori, H., Aburatani, H., Takaku, F., Suzuki, H., Kobari, Y., Miyai, T., Takahashi, K., Cohen, E.H., Wydro, R., Housman, D.E. and Kodama, T. (1990) Human macrophage scavenger receptors: Primary structure, expression, and localisation in atherosclerotic lesions. *Proc. Natl. Acad. Sci. U.S.A.*, **87**, 9133–9137.

Mayne, R. and Burgeson, R.E. (Eds.) (1987) *Structure and Function of Collagen Types*. pp. 1–317, Orlando, Academic Press.

Mays, C. and Rosenberry, T.L. (1981) Characterisation of pepsin-resistant collagen-like tail subunit fragments of 18S and 14S acetylcholinesterase from *Electrophorus electricus*. *Biochemistry*, **20**, 2810–2817.

McIntosh, I., Hamosh, A. and Dietz, H.C. (1993) Nonsense mutations and diminished mRNA levels. *Nature Genet.*, **4**, 219.

Menashi, S., Harwood, R. and Grant, M.E. (1976) Native collagen is not a substrate for the collagen glucosyltransferase of platelets. *Nature*, **264**, 670–672.

Morris, N.P., Keene, D.R., Glanville, R.W., Bentz, H. and Burgeson, R.E. (1986) The tissue form of type VII is an antiparallel dimer. *J. Biol. Chem.*, **261**, 5638–5644.

Mueller, P.P. and Hinnebusch, A.G. (1986) Multiple upstream AUG codons mediate translational control of GCN4. *Cell*, **45**, 201–207.

Munro, S. and Pelham, H.R.B. (1987) A C-terminal signal prevents secretion of luminal ER proteins. *Cell*, **48**, 899–907.

Muragaki, Y., Jacenko, O., Apte, S., Mattei, M.G., Ninomiya, Y. and Olsen, B.R. (1991) The α2(VIII) collagen gene. A novel member of the short chain collagen family located on the human chromosome 1. *J. Biol. Chem.*, **266**, 7721–7727.

Muragaki, Y., Shiota, C., Inoue, M., Ooshima, A., Olsen, B.R. and Ninomiya, Y. (1992) α1(VIII)-collagen gene transcripts encode a short-chain collagen polypeptide and are expressed by various epithelial, endothelial and mesenchymal cells in newborn mouse tissues. *Eur. J. Biochem.*, **207**, 895–902.

Muragaki, Y., Abe, N., Ninomiya, Y., Olsen, B.R. and Ooshima, A. (1994) The human α1(XV) collagen chain contains a large amino terminal non triple helical domain with a tandem repeat structure and homology to α1(XVIII) collagen. *J. Biol. Chem.*, **269**, 4042–4046.

Myers, J.C., Kivirikko, S., Gordon, M.K., Dion, A.S. and Pilajaniemi, T. (1992) Identification of a previously unknown human collagen chain, α1(XV), characterized by extensive interruptions in the triple-helical region. *Proc. Natl. Acad. Sci. U.S.A.*, **89**, 10144–10148.

Myers, J.C., Yang, H., D'Ippolito, J.A., Miller, M.K. and Dion, A.S. (1994) The triple-helical region of human type XIX collagen consists of multiple collagenous subdomains and exhibits limited sequence homology to α1(XVI). *J. Biol. Chem.*, **269**, 18549–18557.

Myllylä, R., Pihlajaniemi, T., Pajunen, L., Turpeenniemi-Hujanen, T. and Kivirikko, K.I. (1991) Molecular cloning of chick lysyl hydroxylase. Little homology in primary structure to the two types of subunit of prolyl 4-hydroxylase. *J. Biol. Chem.*, **266**, 2805–2801.

Nakai, A., Satoh, M., Hirayoshi, K. and Nagata, K. (1992) Involvement of the stress protein HSP47 in procollagen processing in the endoplasmic reticulum. *J. Cell Biol.*, **117**, 903–914.

Nakata, K., Ono, K., Miyazaki, J., Olsen, B.R., Muragaki, Y., Adachi, E., Yamamura, K-I. and Kimura, T. (1993) Osteoarthritis associated with mild chondrodysplasia in transgenic mice expressing 1(IX) collagen chains with a central deletion. *Proc. Natl. Acad. Sci. U.S.A.*, **90**, 2870–2874.

Nishimura, I., Muragake, Y. and Olsen, B.R. (1989) Tissue-specific forms of type IX collagen-proteoglycan arise from the use of two widely separated promoters. *J. Biol. Chem.*, **264**, 20033–20041.

Niyibizi, C. and Eyre, D. (1989) Identification of the cartilage α1(XI) chain in type V collagen from bovine bone. *FEBS Lett.*, **242**, 314–318.

Niyibizi, C., Fietzek, P.P. and van der Rest, M. (1984) Human placenta type V collagens. Evidence for the existence of an α1(V)α2(V)α3(V) collagen molecule. *J. Biol. Chem.*, **259**, 14170–14174.

Njieha, F.K., Morikawa, T., Tuderman, L. and Prockop, D.J. (1982)Partial purification of a procollagen C-proteinase. Inhibition by synthetic peptides and sequential cleavage of type I procollagen. *Biochemistry*, **21**, 757–764.

Noiva, R. and Lennarz, W.J. (1992) Protein disulfide isomerase. A multifunctional protein resident in the lumen of the endoplasmic reticulum. *J. Biol. Chem.*, **267**, 3553–3556.

Nusgens, B.V., Verellen-Dumoulin, C., Hermanns-LW, T., De-Paepe, A., Nuytinck, L., Pierard, G.E. and Lapiere, C.M. (1992) Evidence for a relationship between Ehlers-Danlos type VII C in humans and bovine dermatosparaxis. *Nature Genetics.*, **1**, 214–217.

Oh, S.P., Warman, M.L., Seldin, M.F., Sou-De, C., Knoll, J.H.M., Timmons, S. and Olsen, B.R. (1994) Cloning of cDNA and genomic DNA encoding human type XVIII collagen and localization of the α1(XVIII) collagen gene to mouse chromosome 10 and human chromosome 21. *Genomics*, **19**, 494–499.

Olsen, A.S., Geddis, A.E. and Prockop, D.J. (1991) High levels of expression of a minigene version of the human proα1(I) collagen gene in stably transfected mouse fibroblasts. *J. Biol. Chem.*, **266**, 1117–1121.

Olsen, B.J., Guzman, N.A., Engel, J., Condit, C. and Aase, S. (1977) Purification and characterization of a peptide from the carboxy-terminal region of chick tendon procollagen type I. *Biochemistry*, **16**, 3030–3036.

Olsen, B.R. (1989)The next frontier. Molecular biology of extracellular matrix. *Connect. Tiss. Res.*, **23**, 115–121.

Olsen, B.R., Gerecke, D., Gordon, M., Green, G., Kimura, T., Konomi, H., Muragaki, Y., Ninomiya, Y., Nishimura, I. and Sugrue, S. (1989) A new dimension in the extracellular matrix. In *Collagen. Molecular Biology*, Olsen, B.R. and Nimni, M.E. (eds.) pp.1–19. Boca Raton: CRC Press.

Paglia, L.M., Wiestner, M., Duchene, M., Ouellette, L.A., Hörlein, D., Martin, G.R. and Müller, P.K.. (1981) Effects of procollagen peptides on the translation of type II collagen messenger ribonucleic acid and on collagen biosynthesis in chondrocytes. *Biochemistry*, **20**, 3523–3527.

Palmiter, R.D., Davidson, J.M., Gagnon, J., Rowe, D.W. and Bornstein, P. (1979) NH_2-terminal sequence of the chick-proα1(I) chain synthesized in the reticulocyte lysate system. *J. Biol. Chem.*, **254**, 1433–1436.

Pan, T-C., Zhang, R-Z., Mattei, M-G., Timpl, R. and Chu, M-L. (1992) Cloning and chromosomal location of human (XVI) collagen. *Proc. Natl. Acad. Sci. U.S.A.*, **89**, 6565–6569.

Pelham, H.R.B. (1986) Speculations on the functions of the major heat shock and glucose-regulated proteins. *Cell*, **46**, 959–961.

Peltonen, L., Palotie, A. and Prockop, D.J. (1980) A defect in the structure of type I procollagen in a patient who had osteogenesis imperfecta: Excess mannose in the COOH-terminal propeptide. *Proc. Natl. Acad. Sci. U.S.A.*, **77**, 6179–6183.

Pesciotta, D.M., Dickson, L.A., Showalter, A.M., Eikenberry, E.F., de Crombrugghe, B., Fietzek, P.P. and Olsen, B.R. (1981) Primary structure of the carbohydrate-containing regions of the carboxyl propeptides of type I procollagen. *FEBS Lett.*, **125**, 170–174.

Petit, B., Ronziére, M.C., Hartmann, D.J. and Herbage, D. (1993) Ultrastructural organisation of type XI collagen in bovine epiphyseal cartilage. *Histochem.*, **100**, 231–239.

Pihlajaniemi, T., Helaakoski, T., Tasanen, K., Myllylä, R., Huhtala, M-L, Koivu, J. and Kivirikko, K.I. (1987) Molecular cloning of the α-subunit of human prolyl 4-hydroxylase. This subunit and protein disulphide isomerase are products of the same gene. *EMBO J.*, **6**, 643–649.

Rademacher, T.W., Parekh, R.B. and Dwek, R.A. (1988) Glycobiology. *Ann. Rev. Biochem.*, **57**, 785–838.
Raghow, R. and Thompson, J.P. (1989) Molecular mechanisms of collagen gene expression. *Mol. Cell. Biochem.*, **86**, 5–18.
Ramachandran, G.N. (1967) Structure of collagen at the molecular level. In *Treatise on Collagen*, Ramachandran, G.N. (ed.) pp.103–183. New York: Academic Press.
Ramirez, F. (1989) Organization and evolution of the fibrillar collagen genes. In *Collagen. Molecular Biology*, Olsen, B.R. and Nimni, M.E. (eds.) pp. 21–30. Boca Raton.: CRC Press, Inc.
Rapoport, T.A. (1992) Transport of proteins across the endoplasmic reticulum membrane. *Science*, **258**, 931–936.
Reid, K.B.M. (1979) Complete amino acid sequences of the three collagen-like regions present in subcomponent C1q of the first component of human complement. *Biochem. J.*, **179**, 367–371.
Reiser, K., McCormick, R.J. and Rucker, R.B. (1992) Enzymatic and nonenzymatic cross-linking of collagen and elastin. *FASEB J.*, **6**, 2439–2449.
Rhodes, R.K. and Miller, E.J. (1981) Evidence for the existence of an α1(V)α2(V)α3(V) collagen molecule in human placental tissue. *Collagen Relat. Res.*, **1**, 337–343.
Rich, A. and Crick, F.H.C. (1961). The molecular structure of collagen *J. Mol. Biol.*, **3**, 483–506.
Rintala, M., Metsaranta, M., Garofalo, S., de Crombrugghe, B., Vuorio, E. and Ronning, O. (1993) Abnormal craniofacial morphology and cartilage structure in transgenic mice harboring a Gly → Cys mutation in the cartilage-specific type II collagen gene. *J. Craniofac. Genet. Dev. Biol.*, **13**, 137–146.
Ripley, C.R., Fant, J. and Bienkowski, R.S. (1993) Brefeldin A inhibits degradation as well as production and secretion of collagen in human lung fibroblasts. *J. Biol. Chem.*, **268**, 3677–3682.
Risteli, J., Tryggvason, K. and Kivirikko, K.I. (1977) Prolyl 3-hydroxylase: Partial characterization of the enzyme from rat kidney cortex. *Eur. J. Biochem.*, **73**, 485–492.
Risteli, L., Myllylä, R. and Kivirikko, K.I. (1976) Partial purification and characterization of collagen galactosyltransferase from chick embryos. *Biochem. J.*, **155**, 145–153.
Romanic, A.M., Adachi, E., Kadler, K.E., Hojima, Y. and Prockop, D.J. (1991) Copolymerization of pNcollagen III and collagen I. pNcollagen III decreases the rate of incorporation of collagen I into fibrils, the amount of collagen I incorporated, and the diameter of the fibrils formed. *J. Biol. Chem.*, **266**, 12703–12709.
Romanic, A.M., Adachi, E., Hojima, Y., Engel, J. and Prockop, D.J. (1992) Polymerisation of pNcollagen I and copolymerisation of pNcollagen I with collagen I. A kinetic, thermodynamic, and morphologic study. *J. Biol. Chem.*, **267**, 22265–22271.
Rosenbloom, J. and Harsch, M. (1973) Hydroxyproline content determines the denaturation temperature of chick tendon collagen. Formation of a type I collagen RNA dimer by intermolecular base-pairing of a conserved sequence around the translation initiation site. *Arch. Biochem. Biophys.*, **158**, 478–484.
Rossi, P. and de Crombrugghe, B. (1987a) Formation of a type I collagen RNA dimer by intermolecular base-pairing of a conserved sequence around the translation initiation site. *Nucl. Acids Res.*, **15**, 8935–8956
Rossi, P. and de Crombrugghe, B. (1987b) Identification of a cell-specific transcriptional enhancer in the first intron of the mouse α2 (type I) collagen gene. *Proc. Natl. Acad. Sci. U.S.A.*, **84**, 5590–5594.
Rossouw, C.M.S., Vergeer, W.P., du Plooy, S.J., Bernard, M.P., Ramirez, F. and de Wet, W.J. (1987) DNA sequences in the first intron of the human pro-α1(I) collagen gene enhance transcription. *J. Biol. Chem.*, **262**, 15151–15157.
Rothman, J.E. (1989a) Signal peptide recognition. GTP and methionine bristles. *Nature*, **340**, 433–434.
Rothman, J.E. (1989b) Polypeptide chain binding proteins: Catalysts of protein folding and related processes in cells. *Cell*, **59**, 591–601.
Royce, P.M. and Barnes, M.J. (1985) Failure of highly purified lysyl hydroxylase to hydroxylate lysyl residues in the non-helical regions of collagen. *Biochem. J.*, **230**, 475–480.
Ryan, M.C. and Sandell, L.J. (1990) Differential expression of a cysteine-rich domain in the amino-terminal propeptide of type II (cartilage) procollagen by alternative splicing of mRNA. *J. Biol. Chem.*, **265**, 10334–10339.
Saitta, B., Stokes, D.G., Vissing, H., Timpl, R. and Chu, M-L. (1990) Alternative splicing of the human α2(VI) collagen gene generates multiple mRNA transcripts which predict three protein variants with distinct carboxyl termini. *J. Biol. Chem.*, **265**, 6473–6480.
Sandberg, M., Tamminen, M., Hirvonen, H., Vuorio, E. and Pihlajaniemi, T. (1989) Expression of mRNAs coding for the α1 chain of type XIII collagen in human fetal tissues: Comparison with expression of mRNAs for collagen types I, II, and III. *J. Cell Biol.*, **109**, 1371–1379.
Sandberg, M.M., Hirvonen, H.E., Elima, K.J.M. and Vuorio, E.I. (1993) Co-expression of collagens II and XI and alternative splicing of exon 2 of collagen II in several developing human tissues. *Biochem. J.*, **294**, 595–602.
Sanders, S.L. and Schekman, R. (1992) Polypeptide translocation across the endoplasmic reticulum membrane. *J. Biol. Chem.*, **267**, 13791–13794.
Savagner, P., Miyashita, T. and Yamada, Y. (1990) Two silencers regulate the tissue-specific expression of the collagen II gene. *J. Biol. Chem.*, **265**, 6669–6674.
Sawada, H., Konomi, H. and Hirosawa, K. (1990) Characterization of the collagen in the hexagonal lattice of

Descemet's membrane: Its relation to type VIII collagen. *J. Cell Biol.*, **110**, 219–227.

Schlumberger, W., Thie, M., Volmer, H., Rauterberg, J. and Robenek, H. (1988) Binding and uptake of col 1(I), a peptide capable of inhibiting collagen synthesis in fibroblasts. *Eur. J. Cell Biol.*, **46**, 244–252.

Schmid, T.M. and Linsenmayer, T.F. (1990) Immunoelectron microscopy of type X collagen: Supramolecular forms within embryonic chick cartilage. *Develop. Biol.*, **138**, 53–62.

Schnieke, A., Harbers, K. and Jaenisch, R. (1983) Embryonic lethal mutation in mice induced by retrovirus insertion into the $\alpha 1$(I) collagen gene. *Nature*, **304**, 315–320.

Schönbrunner, E.R. and Schmid, F.X. (1992) Peptidyl-prolyl *cis-trans* isomerase improves the efficiency of protein disulphide isomerase as a catalyst of protein folding. *Proc. Natl. Acad. Sci. U.S.A.*, **89**, 4510–4513.

Schwarz, M., Harbers, K. and Kratochwil, K. (1990) Transcription of a mutant collagen I gene is a cell type and stage-specific marker for odontoblast and osteoblast differentation. *Development*, **108**, 717–726.

Scott, J.E. (1988) Proteoglycan-fibrillar collagen interactions. *Biochem. J.* **252**, 313–323.

Seckler, R. and Jaenicke, R. (1992) Protein folding and protein refolding. *FASEB J.*, **6**, 2545–2552.

Shapiro, J.R. and Chipman, S.D. (1992) Osteogenesis imperfecta. In *Collagen. Pathobiochemistry*, Kang, A.H. and Nimni, M.E. (eds.) pp. 49–86. Boca Raton: CRC Press.

Shaw, L.M. and Olsen, B.R. (1991) FACIT collagens. Diverse molecular bridges in extracellular matrices. *TIBS*, **16**, 191–194.

Sherwood, A.L., Bottenus, R.E., Martzen, M.R. and Bornstein, P. (1990) Structural and functional analysis of the first intron of the human $\alpha 2$(I) collagen-encoding gene. *Gene*, **89**, 239–244.

Sherwood, A.L. and Bornstein, P. (1990) Transcriptional control of the $\alpha 1$(I) collagen gene involves orientation- and position-specific intronic sequences. *Biochem. J.*, **265**, 895–897.

Shieh, B-H., Stamnes, M.A., Seavello, S., Harris, G.L. and Zuker, C.S. (1989). The ninA gene required for visual transduction in *Drosophila* encodes a homologue of the cyclosporin A-binding protein. *Nature*, **338**, 67–70.

Sillence, D.O.A, Senn, A. and Danks, D.M. (1979) Genetic heterogeneity in osteogenesis imperfecta. *J. Med. Genet.*, **16**, 101–116.

Silver,D., Miller, J., Harrison, R. and Prockop, D.J. (1992) helical model of nucleation and propagation to account for the growth of type I collagen fibrils from assymetrical pointed tips: A special example of self-assembly of rod-like monomers. *J. Biol. Chem.* **89**, 9860–9864.

Slack, J.L., Liska, D.J. and Bornstein, P. (1991) An upstream regulatory region mediates high-level, tissue-specific expression of the human $\alpha 1$(I) collagen gene in transgenic mice. *Mol. Cell. Biol.*, **11**, 2066–2074.

Slack, J.L., Liska, D.J. and Bornstein, P. (1993) Regulation of expression of the type I collagen genes. *Am. J. Med Genet.*, **45**, 140–151.

Smedsrød, B., Melkko, J., Risteli, L. and Risteli, J. (1990) Circulating C-terminal propeptide of type I procollagen is cleared mainly via the mannose receptor in liver endothelial cells. *Biochem. J.*, **271**, 345–350.

Smith, L.T., Wertelecki, W., Milstone, L.M., Petty, E.M., Seashore, M.R., Braverman, I.M., Jenkins, T.G. and Byers, P.H. (1992) Human dermatosparaxis: A form of Ehlers-Danlos syndrome that results from failure to remove the amino-terminal propeptide of type I procollagen. *Am. J. Hum. Genet.*, **51**, 235–244.

Sokolov, B.P., Mays, P.K., Khillan, J.S. and Prockop, D.J. (1993) Tissue- and development-specific expression in transgenic mice of a type I procollagen (COL1A1) minigene construct with 2.3 kb of the promoter region and 2 kb of the 3′-flanking region. Specificity is independent of the putative regulatory sequences in the first intron. *Biochemistry*, **32**, 9242–9249.

Stacey, A., Bateman, J., Choi, T., Mascara, T., Cole, W. and Jaenisch, R. (1988) Perinatal lethal osteogenesis imperfecta in transgenic mice bearing an engineered mutant pro-$\alpha 1$(I) collagen gene. *Nature*, **332**, 131–136.

Steinmann, B., Bruckner, P. and Superti-Furga, A. (1991) Cyclosporin A slows collagen triple-helix formation *in vivo*: Indirect evidence for a physiologic role of peptidyl-prolyl *cis-trans*-isomerase. *J. Biol. Chem.*, **266**, 1299–1303.

Steinmann, B., Royce, P.M. and Superti-Furga, A. (1993) The Ehlers-Danlos syndrome. In *Connective Tissue and Its Heritable Disorders*, Royce, P.M. and Steinmann, B. (eds.) pp. 351–407. New York: Wiley-Liss, Inc..

Stinson, R.H. and Sweeny, P.R. (1980) Skin collagen has an unusual d-spacing. *Biochim. Biophys. Acta*, **621**, 158–161.

Stokes, D.G., Saitta, B., Timpl, R. and Chu, M-L. (1991) Human $\alpha 3$(VI) collagen gene. Characterization of exons coding for the amino-terminal globular domain and alternative splicing in normal and tumor cells. *J. Biol. Chem.*, **266**, 8626–8633.

Su, M.-W., Lee, B., Ramirez, F., Machado, M. and Horton, W. (1989) Nucleotide sequence of the full length cDNA encoding human type II procollagen. *Nucleic Acids Res.* 17, 9473

Svinarich, D.M., Twomey, T.A., Macauley, S.P., Krebs, C.J., Yang, T.P. and Krawetz, S.A. (1992) Characterization of the human lysyl oxidase gene locus. *J. Biol. Chem.*, **267**, 14382–14387.

Takahara, K., Sato, Y., Okazawa, K., Okamoto, N., Noda, A., Yaoi, Y. and Kato, I. (1991) Complete primary structure of human collagen $\alpha 1$(V) chain. *J. Biol. Chem.*, **266**, 13124–13129.

Takechi, H., Hirayoshi, K., Nakai, A., Kudo, H., Saga, S. and Nagata, K. (1992) Molecular cloning of a mouse 47-kDa heat-shock protein (HSP47), a collagen-binding stress protein, and its expression during the differentiation of F9 teratocarcinoma cells. *Eur. J. Biochem.*, **206**, 323–329.

Tanzawa, K., Berger, J. and Prockop, D.J. (1985)Type I procollagen N-proteinase from whole chick embryos. Cleavage of a homotrimer of pro-α1(I) chains and the requirement for procollagen with a triple-helical conformation. *J. Biol. Chem.*, **260**, 1120–1126.
Tanzer, M.L., Rowland, F.N., Murray, L.W. and Kaplan, J. (1977) Inhibitory effects of tunicamycin on procollagen biosynthesis and secretion. *Biochim. Biophys. Acta*, **500**, 187–196.
Timpl, R. and Chu, M-L. (1994) Microfibrillar collagen type VI. In *Extracellular Matrix Assembly and Structure*, Yurchenco, P.D., Birk, D. and Mecham, R.P. (eds.) In press. Orlando: Academic Press.
Trelstad, R.L. and Hayashi, K. (1979) Tendon collagen fibrillogenesis: Intracellular subassemblies and cell surface changes associated with fibril growth. *Develop. Biol.*, **71**, 228–242.
Tromp, G., Kuivaniemi, H., Stacey, A., *et al.*, (1988) Structure of a full-length cDNA clone for the prepro-α1(I) chain of human type I procollagen. *Biochem. J.*, **253**, 919–922.
Tuderman, L and Prockop, D.J. (1982) Procollagen N-proteinase. Properties of the enzyme purified from chick embryo tendons. *Eur. J. Biochem.*, **125**, 545–549.
Uitto, J. (1979)Collagen polymorphism. Isolation and partial characterization of α1(I) trimer molecules in normal human skin. *Arch. Biochem. Biophys.*, **192**, 371–379.
Vaillancourt, J.P. and Cates, G.A. (1991) Purification and reconstitution of a collagen-binding heat-shock glycoprotein from L6 myoblasts. *Biochem. J.*, **274**, 793–798.
Van der Rest, M., Aubert-Foucher, E., Dublet, B., Eichenberger, D., Font, B. and Goldschmidt, D. (1991) Structure and function of the fibril-associated collagens. *Biochem. Soc. Trans.*, **19**, 820–824.
Van der Rest, M. and mayne, R. (1988) Type IX collagen proteoglycan is covalently crosslinked to type II collagen. *J. Biol. Chem.* **263**, 1615–1618.
Van der Rest, M. and Garrone, R. (1991) Collagen family of proteins. *FASEB J.*, **5**, 2814–2823.
Vaux, D., Tooze, J. and Fuller, S. (1990) Identification by anti-idiotype antibodies of an intracellular membrane protein that recognizes a mammalian endoplasmic reticulum retention signal. *Nature*, **345**, 495–502.
Veis, A. and Kirk, T.Z. (1989) The coordinate synthesis and cotranslational assembly of type I procollagen. *J. Biol. Chem.*, **264**, 3884–3889.
Veis, A. and Payne, K. (1988) Collagen fibrillogenesis. In *Collagen. Biochemistry*, Nimni, M.E. (ed.) pp. 113–137. Boca Raton: CRC Press.
Venstrom, K.A. and Reichardt, L.F. (1993) Extracellular matrix 2: Role of extracellular matrix molecules and their receptors in the nervous system. *FASEB J.*, **7**, 996–1003.
Vogel, K.G., Paulsson, M. and Heinegård, D. (1984) Specific inhibition of type I and type II collagen fibrillogenesis by the small proteoglycan of tendon. *Biochem. J.*, **223**, 587–597.
Vuorio, E. and de Crombrugghe, B. (1990) The family of collagen genes. *Ann. Rev. Biochem.*, **59**, 837–872.
Walter, P. and Lingappa, V.R. (1986). Mechanism of protein translocation across the endoplasmic reticulum membrane. *Ann. Rev. Cell Biol.*, **2**, 499–516.
Watt, S.L., Lunstrum, G.P., McDonough, M., Keene, D.R., Burgeson, R.E. and Morris, N.P. (1992) Characterization of collagen types XII and XIV from fetal bovine cartilage. *J. Biol. Chem.*, **267**, 20093–20099.
Weiss, J.B. (1976) Enzymic degradation of collagen. *Int. Rev. Connect. Tissue. Res.*, **7**, 101–157.
Weiss, J.B. and Ayad, S. (1982) An introduction to collagen. In *Collagen in Health and Disease*, Weiss, J.B. and Jayson, M.I.V. (eds.) pp.1–27. Edinburgh: Churchill Livingstone.
Wenstrup, R.J. (1992) Collagen abnormalities in the Ehlers-Danlos syndromes and the Marfan syndrome. In *Collagen. Pathobiochemistry*, Kang, A.H. and Nimni, M.E. (eds.) pp. 31–48. Boca Raton.: CRC Press.
White, R.T., Damm, D., Miller, J., Spratt, K., Schilling, J., Hawgood, S., Benson, B. and Cordell, B. (1985) Isolation and characterisation of the human pulmonary surfactant apoprotein gene. *Nature*, **317**, 361–363.
Wiestner, M., Krieg, T., Hörlein, D., Glanville, R.W., Fietzek, P. and Müller, P.K. (1979) Inhibiting effect of procollagen peptides on collagen biosynthesis in fibroblast cultures. *J. Biol. Chem.*, **254**, 7016–7023.
Willing, M.C., Pruchno, C.J. and Byers, P.H. (1993) Molecular heterogeneity in osteogenesis imperfecta type I. *Am. J. Med. Genet.*, **45**, 223–227.
Wood, G.C. (1960). The formation of fibrils from collagen solutions. 2. A mechanism of collagen fibril formation. *Biochem. J.*, **75**, 598–605.
Woodbury, D., Benson-Chanda, V. and Ramirez, F. (1989) Amino-terminal propeptide of human pro-α2(V) collagen conforms to the structural criteria of a fibrillar procollagen molecule. *J. Biol. Chem.*, **264**, 2735–2738.
Wotton, S.F., Duance, V.C. and Fryer, P.R. (1988) Type IX collagen: a possible function in articular cartilage. FEBS Lett., **234**, 79–82.
Wu, C.H., Walton, C.M. and Wu, G.Y. (1991)Propeptide-mediated regulation of procollagen synthesis in IMR-90 human lung fibroblast cell cultures. *J. Biol. Chem.*, **266**, 2983–2987.
Wu, J.-J., Woods, P.E. and Eyre, D.R. (1992) Identification of cross-linking sites in bovine cartilage type IX collagen reveals an antiparallel type II-type IX molecular relationship and type IX to type IX bonding. *J. Biol. Chem.*, **267**, 23007–23014.
Yada, T., Suzuki, S., Kobayashi, K., Kobayashi, M., Hoshino, T., Horie, K. and Kimata, K. (1990)Occurence in chick embryo vitreous humor of a type IX collagen proteoglycan with an extraordinarily large chondroitin

sulphate chain and short α1 polypeptide. *J. Biol. Chem.*, **265**, 6992-6999.

Yamada, Y., Mudryj, M. and de Crombrugghe, B. (1983) A uniquely conserved regulatory signal is found around the translation initiation site in three different collagen genes. *J. Biol. Chem.*, **258**, 14914–14919.

Yamaguchi, N., Mayne, R. and Ninomiya, Y. (1991) The α1(VIII) collagen gene is homologous to the α1(X) collagen gene and contains a large exon encoding the entire triple helical and carboxyl-terminal non-triple helical domains of the α1(VIII) polypeptide. *J. Biol. Chem.*, **266**, 4508–4513.

Yamaguchi, Y., Mann, K. and Ruoslahti, E. (1990) Negative regulation of TGF-B by the proteoglycan decorin. *Nature*, **346**, 281–283.

Yamauchi, M. and Mechanic, G. (1988) Cross-linking of collagen. In *Collagen: Biochemistry*, Nimni, M. (ed.) pp. 157–172. Boca Raton: CRC Press.

Yoshioka, H. and Ramirez, F. (1990) Proα1(XI) collagen. Structure of the amino-terminal propeptide and expression of the gene in tumour cell lines. *J. Biol. Chem.*, **265**, 6423–6426.

Yurchenco, P.D. and Schittny, J.C. (1990) Molecular architecture of basement membranes. *FASEB J.*, **4**, 1577–1590.

Zanussi, S., Doliana, R., Sega, D., Bonaldo, P. and Colombatti, A. (1992)The human type VI collagen gene. mRNA and protein variants of the α3 chain generated by alternative splicing of an additional 5'-end exon. *J. Biol. Chem.*, **267**, 24082–24089.

3 Non-Collagenous Matrix Proteins

Staffan Johansson

Department of Medical and Physiological Chemistry, Biomedical Center, Uppsala, Sweden

INTRODUCTION

The components of the extracellular matrix are often classified as collagens, proteoglycans and glycoproteins. Since essentially all extracellular proteins, including collagens and the core proteins of proteoglycans, are glycosylated, this nomenclature is somewhat incorrect. Historically, the reason for categorizing all matrix proteins that were not collagens or proteoglycans in one pool was that their identity and functions were unknown, partly because of the difficulty in isolating them from tissues without the use of proteases or strongly denaturing agents. In the 70s, the proteins fibronectin and laminin were identified and found to mediate adhesion of cells to matrices or culture dishes. Since then, a growing number of cell adhesion proteins, as well as proteins with other functions, have been isolated from extracellular matrix and characterized. Thus, anti-adhesive proteins, growth factors, proteases, protease inhibitors, and transglutaminase have been shown to be constituents of various matrices. This progress has made clear that the matrix not only has supportive functions, but it is also an important source of information affecting cell behaviour.

A characteristic feature of several matrix proteins is that they are multifunctional, which is made possible by their composition of repeated, discrete structural units of different amino acid sequence motifs. The units are often coded for by separate exons, and are then named protein modules (Baron *et al.*, 1991) One or several units together may form a domain within the protein which exerts its function independently of other domains. The genes of these so called **mosaic proteins** have apparently evolved by duplication, mutation, and shuffling of ancestor exons. The known mosaic proteins are evolutionary "young" and, with few exceptions, unique to eucaryotic cells. Since they are mainly located extracellularly in matrix, blood plasma, or on cell surfaces, it has been suggested that the mechanisms for generation of such genes developed during evolution in parallel with the appearance of multicellular organisms (Patthy, 1991).

The number of identified matrix proteins has now grown too big to easily be covered in one context. Therefore, this chapter will be restricted to those which have a function in cell adhesive events. The best known proteins of this group are briefly presented in Table 3.1, and a few examples of relatively well characterized cell adhesion proteins will be described in more detail. Although collagens and proteoglycans structurally and functionally overlap with the "glycoproteins", they are not included in this chapter, but are discussed separately in this book.

FIBRONECTINS

Fibronectins[1] are major components of extracellular matrices and are also present in body fluids. More than twenty years ago they were the first cell adhesion proteins to become identified, and they have since then in many respects served as a prototype for the investigation of other cell adhesion proteins. Fibronectins are a group of closely similar proteins (twenty possible variants in humans) generated by alternative splicing of pre-mRNA which is transcribed from a single gene (Tamkun *et al.*, 1984; Kornblihtt *et al.*, 1985). They are important for several basic processes, including cell migration, cell differentiation, embryonic development, wound healing, blood coagulation, tumor formation, and metastasis of tumor cells. All aspects of fibronectin research until 1990 have been thoroughly reviewed by Hynes in a highly recommendable book (Hynes, 1990). The most important progress that has occurred in this area since then concerns the tertiary structure of isolated parts of the protein, the process of fibronectin fibril assembly, and the generation of fibronectin deficient mouse embryos.

Fibronectins have Widespread Distribution in Tissues

All vertebrates appear to express fibronectins, and several invertebrates have also been reported to possess fibronectin-like proteins (Hynes, 1990). Fibronectins exist in two major forms – as soluble dimers and as insoluble fibrillar networks. Different splice variants of the protein may have different solubility properties, but they are all synthesized as soluble dimers and can polymerize into fibrils by a reaction controlled by cells.

The major source of soluble fibronectin is blood plasma. It contains 300–400 μg/ml of one type of fibronectin ("plasma fibronectin") (Mosesson and Umfleet, 1970), which is produced by hepatocytes (Tamkun and Hynes, 1983). Upon coagulation of plasma, varying amounts of the protein will bind to the fibrin clot depending on the time and the temperature of the reaction (Mosesson and Umfleet, 1970; Ruoslahti and Vaheri, 1975; Mosher, 1976). Typically, serum contains 150–250 μg/ml of fibronectin which is partially degraded due to exposure to the proteolytic enzymes of the coagulation cascade. Seminal plasma (Lilja *et al.*, 1987) and joint fluid (Carsons *et al.*, 1989) are also rich in fibronectin, while it is

[1] The name indicates that these proteins link to several filamentous networks e.g. collagens, fibrin and the cytoskeleton (Vaheri *et al.*, 1977).

Table 3.1 Adhesive glycoproteins of extracellular matrices

Protein	*Structure*	*Main location*	*Function*
Bone sialoprotein (Heinegård and Oldberg, 1993)	Monomeric, 57 kDa. ~40% carbohydrate	Bone	Mediats adhesion of osteoblasts to mineralized matrix via integrins.
Fibrinogen (Ruggeri, 1993)	Heterohexamer of α, β, γ chains of 46.5–66.5 kDa, encoded by three related genes.	Blood plasma	Causes thrombus formation by binding to integrin $\alpha IIB\beta 3$ of activated platelets. After conversion to fibrin by thrombin, it forms a temporary matrix through self assembly.
Fibronectins	Dimers of 250 kDa chains, encoded by one gene. Several splice variants.	Blood plasma, mesenchymal matrix	Mediat adhesion of cells to matrices. Bind to integrins, proteoglycans, collagens, and fibrin. Form fibrillar networks.
Laminins (LN1-7) (Paulsson, this book)	Heterotrimers of α, β, γ chains of 140–400 kDa, enchoded by at least 8 related genes.	Basement membranes	Mediate adhesion of cells to basement membranes. Bind to integrins, proteoglycans and nidogen.
Osteopontin (Heinegård and Oldberg, 1993)	Monomeric, 57 kDa ~40% carbohydrate	Bone, kidney, placenta	Mediats adhesion of osteoclasts to mineralized matrix via integrins.
Tenascins (TN-C, -R, -X)	Hexamers and trimers of 160–400 kDa chains, encoded by three related genes. Several splice variants.	Around migrating or proliferating cells during embryogenesis, repair, and oncogenesis.	Functions largely unknown. Mediat adhesion of some cell types, anti-adhesive for others.
Thrombospondins (TSP1-4, COMP) (Adams and Lawler, 1993)	TSP1-4 are trimers, COMP is a pentamer. The subunits are 100–180 kDa and encoded by five related genes.	TSP1: Matrix of growing tissues, platelet granules. TSP2: Many connectiv tissue matrices. TSP3: Lung, skinn. TSP4: heart and skeletal muscle. COMP: Cartilage.	Adhesive for some cell types, reduce adhesion of others. Modulate cell shape and growth. Inhibit angiogenesis. Interact with several matrix components and cellular receptors, including collagen, proteoglycans, integrins.
Vitronectin	Monomeric, 75 kDa.	Blood plasma, subendothelial matrix	Regulates extracellular proteolysis, coagulation, and complement factors. Binds to integrins.
von Willenbrand factor (Ruggeri, 1993)	Multimer of 220 kDa chains.	Blood plasma, platelets, subendothelial matrix	Binds platelets to subendithelial collagen during vascular injury. In complex with factor VIII in the blood.

present in much lower concentration in most other fluids, e.g. in saliva (<1 μg/ml) (Linde *et al.*, 1982) and in cerebrospinal fluids (1–3 μg/ml) (Kuusela *et al.*, 1978). However, all body fluids investigated have been found to contain fibronectin. Often the protein is synthesized locally rather than being derived from the blood, as evident by analysis of its splicing pattern (Hynes, 1990).

The insoluble matrix form of fibronectin is abundant in most embryonic and adult mesenchymal tissues, particularly in loose connective tissue (Stenman and Vaheri, 1978; Hynes, 1990). It is a prominent component in the walls of arteries and veins, and in stromal connective tissue of kidney, spleen, lung etc. Many cell types, including fibroblasts, chondrocytes, myoblasts, smooth muscle cells, endothelial cells, and epithelial cells, secrete and assemble fibronectins into fibrils at their surface (Hynes, 1990). Macrophages (Alitalo *et al.*, 1980), activated neutrophils (La *et al.*, 1987), and activated T-lymphocytes (Hauzenberger *et al.*, 1995) are examples of cells that synthesize and secrete fibronectin but which do not assemble fibrils.

Fibronectins are Composed of Three Repeated Structural Motifs

The fibronectin molecule is a dimer of two chains of Mw ~250 000 Da, which are covalently linked in an antiparallel manner by two interchain disulfide bridges close to the very C-terminal end (Figure 3.1) (Skorstengaard *et al.*, 1986; Kar *et al.*, 1993). Each chain consists of a number of independently folded domains, which are arranged like pearls on a string. By electron microscopy the subunits are often shown to be elongated and flexible. However, depending on the composition of the surface on which the molecules are sprayed out, different conformations are observed (Erickson and Carrell, 1983; Tooney *et al.*, 1983; Hynes, 1990). Biophysical data indicate that the chains of the native soluble protein actually are partially folded, possibly due to ionic interactions between alternating basic and acidic domains (Figure 3.2) (Hormann, 1982; Erickson and Carrell, 1983; Rocco *et al.*, 1987).

The amino acid sequence reveal that the polypeptide is almost entirely composed of three

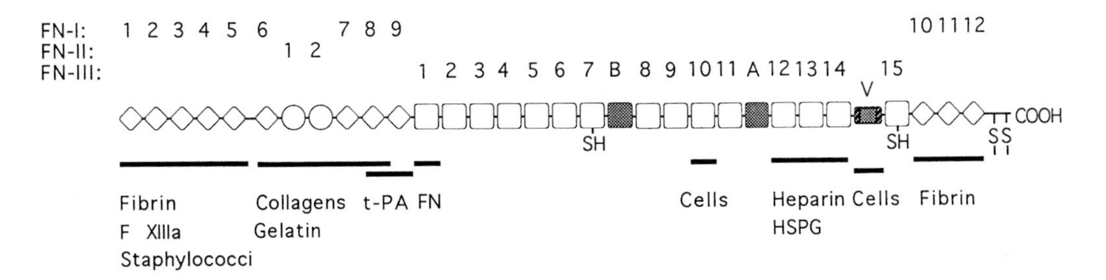

Figure 3.1 Schematic model of the primary sequence of the fibronectin subunit variants. The protein is a dimer of two identical or slightly different subunits liked together at the C-terminus by disulfide bridges. The subunits consist mainly of three types of repeating structural units called I, II, and III. The dotted or hatched segments represent sequences that can be included or excluded during the splicing of the pre-mRNA. At the V region the RNA can be processed in 5 different ways (in human) to become completely missing, included as segments of 64, 89, or 95 amino acids, or completely included (120 amino acids). It is not homologous to any known protein motif. The two unpaired cysteins in the type III_7 and III_{15} units may be involved in the formation insoluble polymers. The bars indicate the location of identified functional domains in each subunit interacting with the compoments listed below.

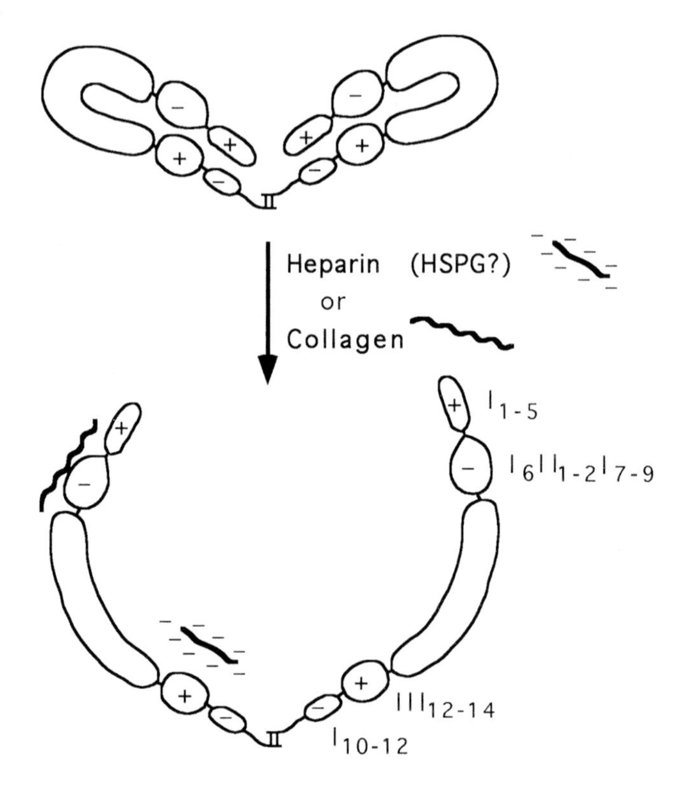

Figure 3.2 Hypothetical model for the conformational transition of fibronectin from a folded state in solution to the extended form induced by interactions with heparin-like polysaccharides or collagens (here represented as short fragments). The folded confomation may be stabilized by opposit charge of the indicated domains. The conformational transition appears to promote polymerization of the protein.

types of repeated structures (Kornblihtt *et al.*, 1985; Skorstengaard *et al.*, 1986; Patel *et al.*, 1987). The so called type I, II, and III repeats contain about 45, 60, and 90 amino acid residues, respectively. Each of the type I and II repeats are encoded by a separate exon (except FN-I_{12}), thus representing true modules (Patel *et al.*, 1987). Most of the type III repeats are encoded by two exons (Schwarzbauer *et al.*, 1987), but nevertheless, this structural motif has been frequently used during evolution as building block of many proteins. Alternative splicing of the pre-mRNA have been found to occur at three positions: the exons III_A and III_B can either be entirely present or missing in the polypeptide (Kornblihtt *et al.*, 1985; Schwarzbauer *et al.*, 1987), while the exon coding for the region called V ("variable") or CS ("connecting segment") can be spliced in five different ways in humans (three in rat) (Figure 3.1) (Tamkun *et al.*, 1984; Vibe *et al.*, 1986). The tertiary structure has been determined by NMR spectroscopy or by crystallography for selected repeats of all three types found in fibronectin.

Type I Modules

The structure of four type I modules have been reported, all of which fit into a consensus structure where the peptide chain is folded into two antiparallel β-sheets (Williams *et al.*, 1994). The sheets are stabilized by two disulfide bridges and are stacked on top of each other. If generalized from studies of one recombinant protein containing a pair of type I modules (modules 4 and 5 of fibronectin), neighbouring type I modules are linked through a common β-strand with the β-sheets stacked into an elongated rod of fixed structure and little flexibility (Williams *et al.*, 1994). Type I modules are present in twelve copies in fibronectin, and occur also in tissue plasminogen activator (tPA) and the in blood coagulation factor XII. Several of these modules participate in binding to fibrin.

Type II Modules

The two copies of type II modules present in fibronectin are homologous to collagen binding modules present in gelatinases A, B, and C, and in seminal plasma proteins BSP-A3 and PDC-109 (Constantine *et al.*, 1991; Banyai *et al.*, 1994). Type II modules occur also in a few other proteins which do not bind to collagen. Like in the type I modules, a general feature of this structural unit is a pair of disulfide bridges. The type II modules of fibronectin have not yet been subjected to closer structural analysis, but the folding is probably similar to that reported for a type II domain in PCD-109: several loops forming a tight global structure with a surface-exposed hydrophobic depression likely representing a ligand-binding site (Constantine *et al.*, 1991).

Type III Units

The type III unit is a very common structural element of animal proteins. It has been estimated to occur in about 2% of all animal proteins and is thus at least as frequent as the epidermal growth factor (EGF) repeat (Bork and Doolittle, 1992). Type III units have been found also in several bacterial proteins. Analysis of the evolutionary relationship between these proteins clearly indicate that the bacterial type III units have originally been acquired from an animal gene and then been spread between related bacteria (Bork and Doolittle, 1992). Type III units are often found as multiple copies (15–18 in fibronectins) in a repeated array, which appears as a relatively straight strand when studied by electron microscopy (Rocco *et al.*, 1987; Leahy *et al.*, 1992). The tenth type III unit of fibronectin (FN-III_{10}) is of particular interest and has been studied in detail. FN-III_{10} contains the amino acid sequence RGD, which was identified as the smallest active peptide of the major cell binding site in fibronectin (Pierschbacher and Ruoslahti, 1984). Later it was found to be a widely used cell binding motif in various proteins (Ruoslahti and Pierschbacher, 1986) FN-III_{10} is composed of seven β-strands forming two layers of β-sheets. The unit is held together mainly by a hydrophobic core; no disulphides are present in type III units. The RGD sequence is well exposed at the apex of a loop. This loop is flexible, allowing the cell recognition sequence to adopt more than one conformation (Main *et al.*, 1992; Dickinson *et al.*, 1994). This may be relevant for the ability of fibronectin to interact with several different receptors.

Posttranslational Modifications of Fibronectin

Secondary modifications of the protein contribute to the heterogeneity of fibronectins. The degree of glycosylation, as well as of phosphorylation and sulfation, varies depending on the cellular source. The main phosphorylation site is at the serine three amino acids from the C-terminus (Skorstengaard *et al.*, 1982), while sulfation appears to occur on a tyrosine in the V region (Paul and Hynes, 1984; Hynes, 1990). The function of these substituents are poorly understood.

Fibronectins Interact with many Matrix Components and Cellular Receptors

Interactions of the N-Terminal Domain

At the N-terminus a distinct domain is located which can be isolated as a ~30 kDa fragment after cleavage by plasmin or several other proteases (Hynes, 1990). It consists of FN-I_{1-5} and has binding sites for several components (Figure 3.1). A major function of this domain is to bind fibrin (Sekiguchi *et al.*, 1981), an interaction which can become stabilized by covalent crosslinks. The crosslinking reaction is catalyzed by factor XIIIa and connects amino acid residue Gln3 in fibronectin with lysine residues in fibrin (McDonagh *et al.*, 1981). The incorporation of fibronectin into fibrin clots during coagulation is important for the immigration of fibroblasts and other cells to the wounded area to allow healing and remodelling (Knox *et al.*, 1986; Hynes, 1990). In fibrinolytic situations, the cleavage of fibronectin by plasmin to release the N-terminal domain will contribute to the dissolution of the fibrin-fibronectin network. In addition, the N-terminal fibronectin domain may mediate binding of fibrin to macrophages, possibly to promote internalization and degradation of fibrin (Blystone and Kaplan, 1993). This domain has also been shown to be involved in the formation of fibronectin fibrils by binding to surface components of fibroblasts and myocytes (McKeown and Mosher, 1985; Woods *et al.*, 1988; Fogerty and Mosher, 1990; Schwarzbauer, 1991; Dzamba *et al.*, 1994).

A quite different role for this part of fibronectin concerns its binding to specific proteins on the surface of Staphylococcus aureus and several other bacterial strains (Hook *et al.*, 1989; McGavin *et al.*, 1993). This interaction appears to be of benefit for the bacteria during colonization of fibronectin-rich tissues.

The Collagen-Binding Domain

The N-terminal domain is followed by a collagen-binding region which can be isolated as protease resistant fragments of 30–45 kDa (Skorstengaard *et al.*, 1984). This domain is separated from the N-terminal domain by the connecting strand between FN-I_5 and FN-I_6, which is somewhat longer than between other type I modules in fibronectin. This protease sensitive site appears to represent a hinge region in the protein. The FN-II modules as well as FN-I_{7-9} have been shown to have affinity for collagen and both parts may contribute to a common binding site in fibronectin (Owens and Baralle, 1986; Ingham *et al.*, 1989; Banyai *et al.*, 1990). In cell cultures and tissues fibronectins are often found associated with collagen fibers (Vaheri *et al.*, 1978; Fleischmajer and Timpl, 1984), and the ability to bind triple helical collagens has been found to be a general feature of all fibronectins (Hynes, 1990). However, the interaction occurs only at sites in the collagen molecules where the triple helix is less tight, primarily at the single collagenase sensitive site of collagens I, II, and III (Kleinman *et al.*, 1981; Dzamba *et al.*, 1993). The binding of fibronectins to denatured collagens (gelatin) is more efficient, because several binding sites in each collagen chain become available after unfolding (Guidry *et al.*, 1990; our unpublished data). Affinity chromatography on immobilized gelatin offers an efficient method for isolation of fibronectins. However, it is important to notice that other gelatin-binding molecules, such as widely occurring gelatinases, may be co-purified by this procedure (Johansson and Smedsrod, 1986; Smilenov *et al.*, 1992).

The Self-Associating Site

The function of the central part of fibronectin, encompassing the type III units, is less well understood although several important activities have been identified in this region. $FN\text{-}III_1$ contains a cryptic site involved in fibronectin-fibronectin interactions (Chernousov *et al.*, 1991). A 14 kDa fragment of $FN\text{-}III_1$ retains this ability to bind to fibronectin at a yet unidentified site (Morla and Ruoslahti, 1992). Most interestingly, the interaction induces formation of large disulfide-bonded fibronectin multimers (Morla *et al.*, 1994). Apparently, a chain reaction is initiated where binding of the 14 kDa fragment triggers a conformational change in fibronectin, which has two effects: 1) exposure of the fibronectin binding structure in $FN\text{-}III_1$ of the interacting fibronectin molecule, which then can bind to and actvate another fibronectin molecule, 2) disulfide exchange between associated fibronectin subunits. This is probably one step in the polymerization reaction of fibronectins (see below).

The Integrin-Binding Domains

The cell adhesion promoting activity of fibronectins are mediated mainly by two different domains in the type III region. The RGD site in $FN\text{-}III_{10}$ is recognized by several cell surface receptors of the integrin family, including integrins $\alpha_3\beta_1$, $\alpha_5\beta_1$, $\alpha_v\beta_1$, $\alpha_v\beta_3$, $\alpha_v\beta_6$, and $\alpha_{IIB}\beta_3$ (Hynes, 1992). This site is part of a domain which can be isolated as fragments of 85–120 kDa after proteolytic digestion of the protein (Ruoslahti *et al.*, 1981; Johansson, 1985). Within this domain so called synergistic sites have been localized to $FN\text{-}III_8$ and $FN\text{-}III_9$, which contribute to the RGD dependent binding by increasing the affinity for the receptors (Aota *et al.*, 1991). Neighbouring structures of the RGD sequence, possibly the synergy sites, affect also the specificity for the receptors, since for example integrin $\alpha_5\beta_1$ binds only to fibronectin but not to other RGD containing proteins (Hynes, 1992).

The second integrin-binding region is located in the V region and contains two sites recognized by integrin $\alpha_4\beta_1$ (Humphries *et al.*, 1987; Mould *et al.*, 1991). Since this part is subjected to alternative splicing, it is available for binding cells only in some fibronectin variants. However, integrin $\alpha_4\beta_1$ can also bind to a site in the neighbouring $FN\text{-}III_{14}$ (Mould and Humphries, 1991), and may even be activated to recognize the RGD site (Sanchez *et al.*, 1994).

The Heparin-Binding Domain

$FN\text{-}III_{12\text{-}14}$ form a domain having high affinity for heparin-like polysaccharides (Ingham *et al.*, 1990; Barkalow and Schwarzbauer, 1991). Heparan sulfate proteoglycans at the cell surface appear to be physiologically relevant ligands for this domain (Bernfield *et al.*, 1992). In some cells, binding of fibronectin to both integrins and such a proteoglycan, possibly syndecan 4, is needed for formation of stable adhesive contacts ("focal adhesion sites") at which actin bundles end (Woods *et al.*, 1986; LeBaron *et al.*, 1988; Woods and Couchman, 1994). The heparin binding domain may also associate with proteoglycans present in extracellular matrices and thereby contribute to the supramolecular organization of the matrix (Hedman *et al.*, 1982).

The Alternatively Spliced Segments

Little is known about the functions of the III_A and III_B domains, but their distinct expression during e.g. embryogenesis and wound healing suggests that the differently spliced forms of fibronectin have specific physiological roles (Ffrench and Hynes, 1989; Ffrench *et al.*, 1989). This is supported by the fact that III_B is the most conserved region in fibronectin between human, rat and chicken. The inclusion of III_B in fibronectin has been shown to affect the conformation of the molecule and to unmask a cryptic epitope (Carnemolla *et al.*, 1992). However, no specific interactions have yet been ascribed to the isolated III_B or III_A, or to fibronectins containing these units. In contrast, it is obvious that splicing of the V segment is functionally important as this domain carries cell binding sites. In addition to the cell binding activity, the V region contains a still uncharacterized signal required for secretion of fibronectin. At least one chain of a dimer has to contain the V segment (or a part of it) in order to enter the secretory pathway and to avoid intracellular degradation (Schwarzbauer *et al.*, 1989). However, the significance of the complexity of the splicing events in this region and the funtion of the different splice segments within the V region remains to be clarified. Furthermore, the knowledge of the distribution of each of the V splice variants in the body is still rudimental.

Interactions of the C-terminal Domain

At the C-terminus of the chain three additional type I modules are located. They constitute a second fibrin binding domain in fibronectin (Sekiguchi *et al.*, 1981).

Polymerization of Fibronectin is Controlled by Cells

The mechanism by which fibronectins polymerize into matrix fibrils is a major unsolved problem regarding this protein. It is an important issue since the reaction, which is a cell-mediated process, does not occur on transformed cells. The lack of matrix assembly is thought to contribute to the invasive behaviour of malignant cells (Vaheri *et al.*, 1989).

Assembly of fibronectin fibrils is a multistep process which involves several domains in fibronectin and probably several cell surface components (Fogerty and Mosher, 1990). Integrins containing the β_1-subunit are important in this process, as demonstrated both by use of blocking antibodies directed against the β_1-subunit (McDonald *et al.*, 1987), and by restoration of fibril formation through expression of β_1-cDNA in β_1-deficient cells (Wennerberg *et al.*, 1995). A possible role for β_1-integrins could be to capture soluble fibronectin dimers to the cell surface. Integrin $\alpha_5\beta_1$ can serve this function, but fibronectin fibrils are formed also in integrin α_5-deficient mouse embryos (Yang *et al.*, 1993). While the fibronectin-binding integrins $\alpha_4\beta_1$ and $\alpha_v\beta_1$ are unable to induce fibronectin polymerization, at least in CHO cells (Wu *et al.*, 1995; Zhang *et al.*, 1993), $\alpha_v\beta_3$ was unexpectedly found to have this ability (Wennerberg *et al.*, 1995). Further, the N-terminal domain of fibronectins (FN-$I_{1\text{-}5}$) is required at an early step by binding to another cellular receptor (McKeown and Mosher, 1985; Sottile *et al.*, 1991). A membrane protein of 66 kDa has been implicated as such a receptor (Moon *et al.*, 1994). Subsequently, alignment of one fibronectin molecule to the next has to occur, an interaction in which the self-association site in FN-III_1 may participate (see above). Finally, formation of interchain disulfides stabilizes the fibril. Conformational changes in the fibronectin molecules appear to be required for some of these reactions, including binding of soluble fibronectins to integrin $\alpha_5\beta_1$, self-association of fibronectins, and disulfide exchange (Johansson *et al.*, 1985; Morla and Ruoslahti, 1992; Morla *et al.*, 1994). The physiologic stimuli for such conformational rearregements are not known. *In vitro*, binding of heparin or short collagen fragments to fibronectin has by biophysical measurements been found to cause a partial unfolding of the soluble molecule (Figure 3.2) (Williams *et al.*, 1982; Homandberg, 1987), as well as to stimulate the interaction with integrin $\alpha_5\beta_1$ (Johansson, 1985). Thus, interactions of fibronectin with collagens or heparan sulfate proteoglycans in the matrix or at the cell surface may promote fibronectin fibril formation (Dzamba *et al.*, 1993).

Fibronectin Deficiency is Lethal

Mouse embryos lacking fibronectins die early. This has been demonstrated by inactivation of the fibronectin gene by homologous recombination ("gene knockout"). The fibronectin deficient embryos implant normally and initiate gastrulation. However, severe defects such as abscence of somites and notochord, abnormal heart and vasculature, and deformed neural tubes are soon displayed (George *et al.*, 1993). These defects can be explained by abnormal cell proliferation, differentiation or migration in the abscence of fibronectins, particularly among mesodermally derived cells.

TENASCINS

Tenascins are a family of large and complex extracellular matrix proteins (for reviews see Chiquet, 1991; Erickson, 1993). Unlike fibronectins they have a restricted distribution in the body, and are only weakly adhesive or even anti-adhesive for different cells. The physiological functions of tenascins are yet unclear, but the expression pattern of the proteins suggests that they are involved in processes of tissue formation and remodelling. Three different tenascins — C, R, and X — encoded by separate genes have been identified[2]. They occur in a number of splice variants, each of them presumably having specific functions.

Tenascins have Unique Distributions in Tissues

Distribution of Tenascin-C

Much of the interest in tenascins originate from their specific and transient occurrence in tissues. The distribution of tenascin-C has been extensively studied and the general conclusions are that a) the protein is a major extracellular protein of several tissues during embryonal development, although with a much more restricted occurrence than

[2] The name tenascin was created from the words tendon and nascent, since the first discovered protein of the family was found to be abundant in myotendenous junctions and growing tissues (Chiquet *et al.*, 1986). According to the nomenclature of Bristow *et al.*, (1993) the letter C refers to cytotactin, which was one of several other names previously used for this protein. R refers to the protein originally called restrictin in chicken and J1-160/180 in rat, and tenascin-X is the protein encoded by the human gene X.

fibronectin, b) it is absent from most adult tissues but is strongly re-expressed during wound healing and in tumors.

During embryogenesis tenascin-C is found at several interesting locations. It appears early around the neural tube and subsequently continues to be a prominent component of the extracellular matrix in the developing central nervous system (Bourdon *et al.*, 1983; Kruse *et al.*, 1985; Crossin *et al.*, 1986). Neural crest cells, the highly migratory progenitors of several different cell types in the body, migrate along pathways in the embryo which precisely coincide with depositions of tenascin-C (Tan *et al.*, 1987; Mackie *et al.*, 1988; Tucker and McKay, 1991). A third intriguing location of this protein is at sites of epithelial-mesenchymal interactions during morphogenic events, e.g. during formation of mammary gland, teeth, kidney, and lung (Aufderheide and Ekblom, 1988; Young *et al.*, 1994). In addition to these locations, the protein is prominent during develpoment in mesenchymal tissues such as muscle, tendon, and cartilage (Chiquet and Fambrough, 1984; Mackie *et al.*, 1987).

Tenascin-C is expressed very sparsely in normal adult tissues, but it has been found in specific regions of adult brain (Grumet *et al.*, 1985), cartilage (Vaughan *et al.*, 1987), skin (Mackie *et al.*, 1987), bone marrow (Ekblom *et al.*, 1993; Klein *et al.*, 1993) and thymus (Ocklind *et al.*, 1993). The restricted distribution of the protein may partly be caused by downregulation of the mRNA synthesis by glucocorticoids (Ekblom *et al.*, 1993). In case of injury, tenascin-C rapidly accumulates in the wounded area due to induced expression of the protein in both surrounding epithelial cells and in mesenchymal cells (Mackie *et al.*, 1988). Following healing of the wound tenascin-C disappears from the area, indicating that efficent means for removal of the protein from the matrix exist. Also damage of tissue by disease may result in strong expression of tenascin-C, as shown in fibrotic liver (Van *et al.*, 1992; Yamada *et al.*, 1992).

The large amount of tenascin-C found in many different types of tumors, particularly in the most undifferentiated and malignant forms (Chiquet *et al.*, 1986; Lynch *et al.*, 1987; Borsi *et al.*, 1992), makes it an interesting candidate as clinical diagnostic marker. Tenascin-C has even been considered as a potential target for directing drugs to the vicinity of the malignant cells by specific antibodies.

Distribution of Tenascin-R

Tenascin-R is found exclusively in the central nervous system at very restricted locations, why it was originally named restrictin (Rathjen *et al.*, 1991; Fuss *et al.*, 1993). It is present in certain areas of the retina, the brain and the spinal cord during embryonic and postnatal periods which coincide in time with myelination of axons. Some neurons continue to synthesize the protein at adult stages (Fuss *et al.*, 1993).

Distribution of Tenascin-X

Tenascin-X appears to be more widely expressed than its relatives. The mRNA was detected in most human fetal tissues, with by far the highest levels found in skeletal, cardiac and smooth muscle, and in testis (Bristow *et al.*, 1993). At the protein level tenascin-X was also identified in several embyonic and adult mouse tissues (Matsumoto *et al.*, 1994). The protein is particularly abundant along skeletal and heart muscles at all stages. The matrix around blood vessels generally contains the protein, which probably explains the ubiquitous occurrence of tenascin-X mRNA in tissue extracts. In the skin and the gut of mouse embryos the distribution of tenascin-X was nearly complimentary to, and clearly distinct from, that of tenascin-C. Noteworthy, tenascin-X was not detected in the brain at any stage, in contrast to tenascin-C and -R (Saga *et al.*, 1991; Bristow *et al.*, 1993; Matsumoto *et al.*, 1994).

Tenascins are Large and Occur as Several Variants

Tenascins are oligomers of elongated subunits. The molecular mass of the subunits are 190–230 kDa for tenascin-C, 160-180 kDa for tenascin-R, and >400 kDa for tenascin-X. All known subunits are composed of four types of amino acid sequence motifs (Pearson *et al.*, 1988; Jones *et al.*, 1989; Nies *et al.*, 1991; Weller *et al.*, 1991; Norenberg *et al.*, 1992; Bristow *et al.*, 1993; Fuss *et al.*, 1993; Matsumoto *et al.*, 1994) (Figure 3.3). In the amino-terminal part a strech of four repeating "heptad sequences"[3] allows three protein chains to assemble into an α-helical coiled coil. The trimers are stabilized by interchain disulfide bonds, probably at both sides of the triple helical region. Tenascin-C and -R have in addition an extra cysteine in each subunit which could link two trimers into hexamers. This is the typical form of tenascin-C, while tenascin-R mostly has been isolated as trimers from tissues. Tenascin-X is thought to a be a trimer since it lacks the extra cysteine.

Next to the assembly domain several cysteine-rich protein units are located which are

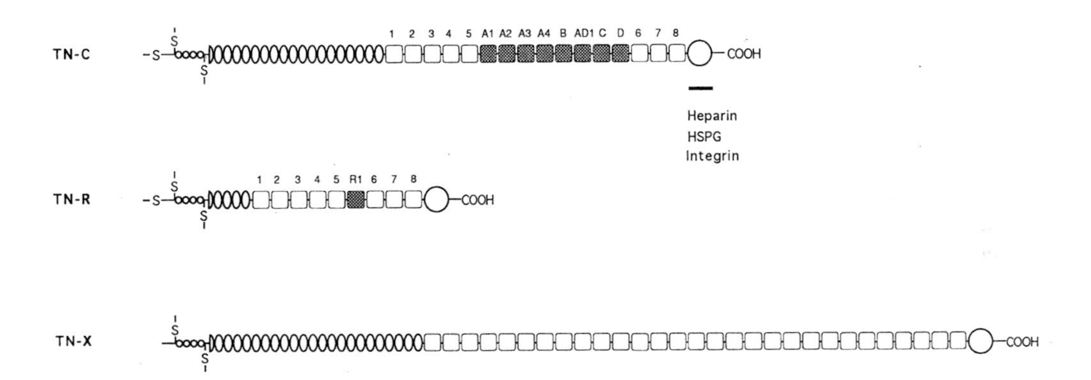

Figure 3.3 Schematic model of the primary structure of tenascin-C, -R, and -X subunits. While tenascin-C is a hexameric protein it is unclear whether intact tenascin-R is a trimer or a hexamer. Tenascin-X is assumed to be trimeric. The dotted segments represent sequences that can be included or excluded during the splicing of the pre-mRNAs. All subunits contain four types of structural units homologous to motifs found in other proteins: heptad sequences (small circles), EGF repeats (ovals), fibronectin type III repeats (squeres), and a fibrinogen-like domain (big circle). The bar indicates the location of the domain in tenascin-C shown to interact with the components listed below.

[3] Heptad sequences are typically found in two- or three-stranded α-helical coiled coil domains. When the amino acid sequence of heptad repeats is written in groups of seven, hydrophobic residues are often found in positions 1 and 4 (Beck *et al.*, 1991) This structure is found in many proteins, including laminins, thrombospondins, and fibrinogen.

homologous to EGF modules. Tenascin-C, -R, and -X has 14.5, 4.5, and 18.5, EGF modules, respectively. These are immediately followed by a variable number of fibronectin-like type III repeats. Some of the type III units are always present (named by numbers), while others (named by letters) are subject to alternative splicing (Figure 3.3). So far, eight splice variants of the human tenascin-C, and two of tenascin-R, have been demonstrated (Norenberg *et al.*, 1992; Fuss *et al.*, 1993; Sriramarao and Bourdon, 1993) . Different forms of tenascin-C have been shown to be expressed in different contexts, i.e. by specific cells, at certain developmental stages, and in response to growth factors (Weller *et al.*, 1991; Borsi *et al.*, 1992; Sriramarao and Bourdon, 1993; Tucker *et al.*, 1993; Tucker *et al.*, 1994). The physiological roles for these variants remain to be identified, although some functional differences have been suggested (see sections below). The third type III unit (TN-III_3) in human and chick tenascin-C contains an RGD sequence at the corresponding location as in FN-III_{10} but in a shorter loop (Leahy *et al.*, 1992). This sequence is not conserved in several other species (Onda *et al.*, 1991; Saga *et al.*, 1991; Weller *et al.*, 1991), and it appears not to be a physiologic cell binding site in tenascin-C (Joshi *et al.*, 1993) (see below).

The fourth type of structural unit in tenascins is the carboxy-terminal domain, which forms a globule at the end of each arm in the oligomeres. It is homologous to the carboxy-terminal domain of the β- and γ- chains of fibrinogen.

Interactions of Tenascins with Extracellular Matrix are Poorly Understood

There is no evidence for self-binding of tenascins; while tenascin-X has not been studied in this respect, tenascin-C and -R appear to be deposited in the matrix as single oligomers. This contrasts to the fibrillar polymers of fibronectin. Furthermore, very few interactions of tenascins with other extracellular macromolecules have been observed. The best documented case is the interaction between tenascin-C and chondroitin sulfate proteoglycan (CSPG). CSPGs have been isolated in complex with tenascin-C from brain, cartilage, and cultured muscle cells (Chiquet and Fambrough, 1984; Vaughan *et al.*, 1987; Hoffman *et al.*, 1988). The core proteins of these CSPGs have not been identified, but they probably represent different members of the versican or aggrecan families of large hyaluronan-binding proteoglycans (Yamagata *et al.*, 1993; Yamada *et al.*, 1994).

Whether tenascin-C binds to fibronectin has been a matter of controversy. However, in careful studies a weak interaction with fibronectin was demonstrated for the larger splice variants of tenascin-C (Lightner and Erickson, 1990), and small splice variants of the protein was found to have higher affinity for fibronectin than large forms (Chiquet *et al.*, 1991). The presence of tenascin-C in many preparations of fibronectin from cell cultures may reflect a physiological interaction between certain forms of the two proteins. The hyaluronan-binding activity previously reported for preparations of "cell fibronectin" (Yamada *et al.*, 1980; Laterra and Culp, 1982) was actually due to a contamination of tenascin-C in complex with a hyaluronan-binding CSPG (Yamagata *et al.*, 1986).

Interactions of Tenascins with Cells may be Adhesive or Anti-adhesive

Adhesive as well as anti-adhesive activities have been reported for tenascin-C and -R. Both activites are yet poorly understood and contradictory results have been reported. The anti-adhesive effect could potentially be important for creating borders that would direct cell migration or neurite outgrowth. The CSPGs associated with tenascin-C have also been implicated to participate in such a function (Perris and Johansson, 1987; Tan *et al.*, 1987; Perris and Johansson, 1990).

Anti-Adhesive Activities of Tenascins

Different anti-adhesive mechanisms have been suggested for tenascins. Due to its large size, tenascin-C can sterically interfere with cell binding to adhesion proteins. For example, if tenascin-C is added to a fibronectin-coated culture dish, some tenascin-C molecules will bind to remaining sites on the plastic surface and like an umbrella prevent the cells from reaching neighbouring fibronectin molecules (Lightner and Erickson, 1990). A similar situation may occur *in vivo*. In addition to this passive effect, active anti-adhesive signals may be triggered by binding of tenascins to cell surface receptors. One example is the ability of soluble tenascin-C to dissolve focal adhesions and cause rounding of well spread endothelial cells. This activity was located to the alternatively spliced domain A–D (Murphy *et al.*, 1991). The proposed receptor for this domain and the signals generated are unknown. Other less characterized anti-adhesive domains in tenascin-C have also been suggested.

Both tenascin-C and -R bind specifically to F3/F11/contactin (here called contactin), a neuronal cell adhesion molecule anchored in the plasma membrane by a glyco-phosphoinositol moiety (Rathjen *et al.*, 1991; Zisch *et al.*, 1992; Pesheva *et al.*, 1993). Contactin has been reported to mediate the initial adhesion of neurons to tenascin-R. However, the result of the interaction appears to be anti-adhesive, since inhibition of neurite outgrowth and detachment of the cells was induced within a few hours (Pesheva *et al.*, 1993). The function of the interaction of contactin with tenascin-C is unknown, but it is probably similar to that of tenascin-R. Interestingly, contactin binds preferentially to small splice variants of tenascin-C, while the binding site is largely disrupted by inclusion of the alternatively spliced type III repeats in the protein (Zisch *et al.*, 1992).

Adhesive Activities of Tenascins

Cell adhesive activity of tenascin-C has been located mainly to $TN\text{-}III_3$ and the fibrinogen-like carboxy-terminal domain (Aukhil *et al.*, 1993; Joshi *et al.*, 1993; Prieto *et al.*, 1993; Sriramarao *et al.*, 1993). The cell binding site in the isolated human $TN\text{-}III_3$ fragment involves the RGD sequence and is recognized by α_v-containing integrins. However, this site is cryptic in the intact protein, due to steric blocking by $TN\text{-}III_2$ (Joshi *et al.*, 1993). Instead the major native cell binding site appears to reside in the fibrinogen-like domain. It promotes attachment, but not flattening of several types of cells. Cell surface heparan sulfate proteoglycans have been implicated as receptors for this domain on some cell types (Aukhil *et al.*, 1993), while other cells bind via unidentified integrins (Joshi *et al.*, 1993; Prieto *et al.*, 1993).

Obviously, an important unresolved question is how the various adhesive and anti-adhesive activities of tenascins work together in physiological situations.

Tenascin-C Deficient Mice Appear Normal

Recently, a mutant mouse strain was generated in which the tenascin-C gene had been "knocked out" by homologous recombination (Saga *et al.*, 1992). Although the protein was believed to be important for embryonic development, no obvious abnormalities were detected in the tenascin-C defective animals. Since the protein has been strongly conserved during vertebrate evolution, it is unlikely that it would be unnecessary. The possibility that functionally similar proteins could compensate for each other is probably not valid in this case, as the known tenascins are expressed at distinct stages and locations. Rather, the lack of tenascin-C may have effects which are difficult to observe, e.g. reduced tolerance toward various types of stress, less efficient brain functions, etc.

VITRONECTIN

Vitronectin is, together with fibronectin, the major cell adhesion protein in blood plasma and serum[4]. In addition, it is an important regulator of inflammatory reactions in blood and tissues, taking part in the control of extracellular proteolysis as well as of cytolysis by complement. Vitronectin is a well studied example of a matrix protein with inducible activities, regulated through conformational alterations (for reviews see Preissner, 1991; Tomasini and Mosher, 1991).

Vitronectin is mainly found in Blood

Vitronectin is synthesized by hepatocytes and secreted to the blood (Barnes and Reing, 1985; Seiffert *et al.*, 1991). The concentration of vitronectin in plasma is 200–400 mg/l, a value not significantly altered by clotting to serum (Shaffer *et al.*, 1984). Vitronectin synthesis has been demonstrated also in macrophages (Hetland *et al.*, 1989), megakaryocytes (Preissner, 1991), and neural crest cells (Delannet *et al.*, 1994) but it is not expressed by most other cells *in vivo* or cell lines *in vitro*. Until recently, vitronectin was thought to be widely distributed in the matrix of several organs (Hayman *et al.*, 1983; Preissner, 1991). However, the monoclonal antibody used in those localization studies mainly detected a crossreacting 30 kDa microfibril associated protein (Tomasini *et al.*, 1993). While it is clear that vitronectin is much less abundant in normal tissues than previously thought, detailed studies of the distribution with specific reagents have not yet been performed for most organs. So far, vitronectin has been found associated with the surface of several types of embryonic cells, in particular with neural crest cells (Delannet *et al.*, 1994). Further, it has been detected in the matrix of vessel walls (Guettier *et al.*, 1989) and in elastic fibers of skin from adults (Dahlback *et al.*, 1989). Possibly, this vitronectin is derived from the blood (Volker *et al.*, 1993). Inflammatory tissues may contain increased amounts of vitronectin (Tomasini-Johansson, 1993), which could be due to both increased permeability of the vessel wall and to local production by macrophages.

Vitronectin has a Conformationally Labile Structure

The 75 kD glycoprotein is composed of four domains (Figure 3.4) and contains at least two structural motifs found in other proteins (Jenne and Stanley, 1985; Suzuki *et al.*, 1985). The

[4] As the name indicates, vitronectin binds strongly to glass surfaces (*vitro* = glass) (Hayman *et al.*, 1985)

N-terminal sequence of 44 amino acids is identical to somatomedin B, a peptide of unknown function present in human serum which possibly is derived from vitronectin by proteolysis. It is a compact, disulfide rich part of the molecule encoded by one exon. Somatomedin B-like domains have been found also in the membrane protein PC-1 on B-lymphocytes (Patthy, 1988). The next domain in vitronectin, “the connecting segment” (amino acids 45–130), carries several important functions. It starts with an RGD sequence which is responsible for the cell adhesive activity of the protein (Cherny *et al.*, 1993; Zhao and Sane, 1993). Close to the RGD site is a cluster of negatively charged amino acids (residues 53–64). In addition, the two sulfated tyrosines in vitronectin are probably located here (Jenne *et al.*, 1989). This very acidic region is believed to interact with a highly basic region in the C-terminal part of the molecule and thereby stabilize one of its conformations. Finally, a cryptic collagen binding site has tentatively been mapped to the connecting segment (Gebb *et al.*, 1986; Izumi *et al.*, 1988).

The remaining part of vitronectin (amino acids 131–489) consists of two similar domains which are homologous to structures present in hemopexin and several metalloproteases. Vitronectin and hemopexin both undergo a major conformational change upon binding of certain ligands (Figure 3.5), which at least in the latter case involves intramolecular movements around a hinge between two such domains (Smith *et al.*, 1988). In the conformation of native plasma vitronectin, a basic region (residues 343–379) in the second hemopexin-like domain of vitronectin has been suggested to bind to the tyrosine sulfate-containing region in the connecting segment (Jenne *et al.*, 1989). The basic region constitutes the binding site for heparin-like polysaccharides, which becomes accesible after conformational rearrangements in the protein (Suzuki *et al.*, 1984; Preissner and Muller, 1987; Yatohgo *et al.*, 1988). This transition probably involves disruption of the interaction between the basic and the acidic regions of vitronectin, accompanied by exposure of new antigenic epitopes and reactive sulfhydryl groups (Tomasini *et al.*, 1989; Hogasen *et al.*, 1992; Stockmann *et al.*, 1993). The exposed basic region can also interact with another vitronectin molecule, and thereby induce it to adopt the conformation with exposed basic region. This chain reaction results in formation of disulfide-stabilized vitronectin multimers consting of up to 16 monomers in a global arrangement (Stockmann *et al.*, 1993). The process appears to resemble the polymerization of fibronectin induced by exposure of a region in the FN-III_1 unit.

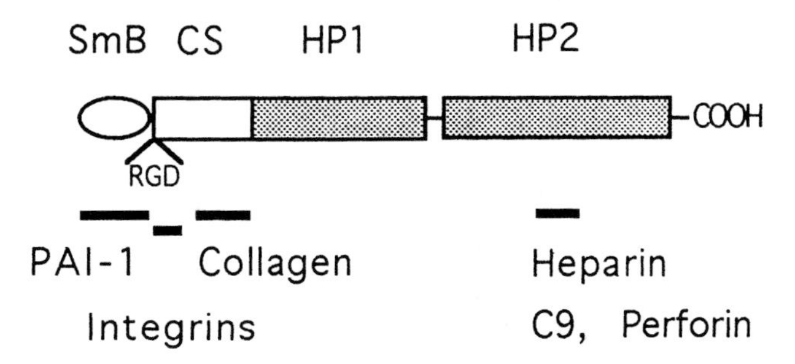

Figure 3. 4 Schematic model of the primary structure of vitronectin. Vitronectin contains two types of structural units homologous to motifs found in other proteins: the somatomedin B domain (SmB) and the hemopexin-like domains (HP1 and HP2). In addition, the common integrin-binding sequence RGD is present in the connecting segment (CS). The bars indicate the location of identified binding sites for the components listed below.

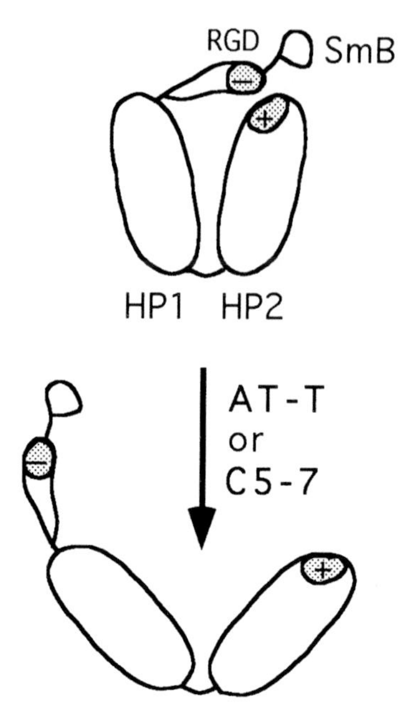

Figure 3.5 Schematic model illustrating the conformational transition of vitronectin from a folded state to an open form. The transition is induced by interactions with ligands such as antithrombin-thrombin complexes (AT-T) and lytic complement complexes (C5-7). The basic region in the hemopexin-like domain 2 is thought to interact with the acidic region in the CS region in the folded state, and to become available for binding to heparin-like polysaccharides after the conformational transition.

A number of posttranslational modification reactions of vitronectin are known. In addition to glycosylation (Yoneda *et al.*, 1993) and tyrosine sulfation (Jenne *et al.*, 1989) these include phosphorylation at several positions (McGuire *et al.*, 1988; Mehringer *et al.*, 1991), proteolytic cleavage (Tomasini and Mosher, 1986; Tollefsen *et al.*, 1990), and multimerization through disulfide bonds (Preissner *et al.*, 1989; Stockmann *et al.*, 1993). The function of these naturally occurring modifications are as yet poorly understood. Formation of covalent multimeres of vitronectin can also be generated by factor XIII and tissue transglutaminase *in vitro* (Sane *et al.*, 1988; Sane *et al.*, 1991), probably involving Gln93 and two other glutamines in the connecting segment (Skorstengaard *et al.*, 1990). It is not known if such crosslinking of vitronectin to itself, or to other matrix components occur *in vivo*.

Vitronectin Mediates Cell Adhesion *in vitro*

For most cells cultured in the presence of serum the initial adhesion to the dish is mainly mediated via vitronectin. Vitronectin is a very efficient cell adhesion protein, probably due to its strong tendency to become adsorbed to plastic and glass surfaces (Hayman *et al.*, 1985; Bale *et al.*, 1989) and to the abundant expression of vitronectin receptors on many types of cells. At least four different cellular receptors of the integrin family ($\alpha_v\beta_1$, $\alpha_v\beta_3$, $\alpha_v\beta_5$, and $\alpha_{IIB}\beta_3$) are known to recognize vitronectin, all of which bind to the RGD site in the protein (Hynes, 1992; Felding and Cheresh, 1993). However, considering the relatively small amount of vitronectin present in extracellular matrices and the other functions ascribed to the protein (discussed below), it is possible that anchoring of cells is not a primary physiological function of vitronectin.

Vitronectin Regulates Proteolysis

Interaction of Vitronectin with PAI-1

Vitronectin has important functions in the regulation of extracellular proteolysis in blood and in tissues. It binds with high affinity ($K_D \approx 0.3$ nM) to PAI-1 (Seiffert and Loskutoff, 1991), a member of the serine protease inhibitor (serpin) family present in plasma and extracellular matrix. A major binding site for PAI-1 in vitronectin has been localised to the somatomedin B-like domain (Seiffert *et al.*, 1994; Sigudardottir and Wiman, 1994). By inhibiting plasminogen activators (tPA and uPA), PAI-1 has a key function in the control of several plasmin dependent events including fibrinolysis, activation of collagenase and other latent proteases, cell migration, tumor metastasis, and implantation of embryos (Flaumenhaft and Rifkin, 1991; Kleiner and Stetler, 1993). PAI-1 is a labile protein which slowly acquires an inactive conformation unless it is stabilized by the interaction with vitronectin (Declerck *et al.*, 1988). Vitronectin is also thought to mediate the deposition of PAI-1 to the matrix (Mimuro and Loskutoff, 1989). The importance of such localized protease protection can be illustrated by the larger size and longer lasting time of cellular focal adhesion sites containing vitronectin, compared to fibronectin-containing focal adhesions, in the presence of PAI-1 (Ciambrone and McKeown, 1990).

A second effect of vitronectin on PAI-1 is a modulation of its specificity for thrombin; the reaction rate of PAI-1 with thrombin is greatly increased in the presence of vitronectin (Ehrlich *et al.*, 1990). This constitutes a direct link between coagulation and fibrinolysis.

Interaction of Vitronectin with Thrombin-Serpin Complexes

Yet another activity of vitronectin is to recognize thrombin in complex with antithrombin or other serpins (Ill and Ruoslahti, 1985; Tomasini *et al.*, 1989), and to mediate endocytosis of the ternary complexes (Panetti and McKeown, 1993). The binding of a thrombin-serpin complex to vitronectin, in the circulation or in the extracellular matrix, induces the heparin-binding conformation of vitronectin (de Boer *et al.*, 1993; Panetti and McKeown, 1993). This allows binding of the complex to (unidentified) heparan sulfate proteoglycans at the cell surface (de Boer *et al.*, 1992). Internalization of the complex for degradation appears to require binding of vitronectin both to the proteoglycan and to integrin $\alpha_v\beta_5$ (Panetti and McKeown, 1993).

Vitronectin Protects against Cytolysis

Interaction of Vitronectin with Complement Factors

Activation of the complement system results in the sequential assembly of components C5b, C6, C7, C8, and several copies of C9 to form a cytolytic pore through membranes (Podack and Tschopp, 1984). In order to assure that only microorganisms and tumor cells, but not the normal host cells, are attacked by this system, protecting factors exist.

Vitronectin is one of the major inhibitors of the lytic complex. It blocks the membrane binding site of the C5b-7 complex and reacts with C9 to prevent its polymerization (Tschopp *et al.*, 1988; Milis *et al.*, 1993). Thus, freely circulating membrane attack complexes will be short-lived, and only those complexes in close vicinity to the complement activating agent will reach its target. Binding of the terminal complement complex to vitronectin induces the heparin binding activity of vitronectin (Hogasen *et al.*, 1992), suggesting that the complex may be removed from the circulation by a similar mechanism as thrombin-serpin complexes.

Interaction of Vitronectin with Perforin

Vitronectin also participates in the control of perforin-mediated killing by cytotoxic T-lymphocytes and NK cells. Like C9, perforin acts by polymerization on membranes of target cells and is inhibited by vitronectin (Tschopp *et al.*, 1988).

PERSPECTIVES

Despite the amount of data that has accumulated on the biology of extracellular matrix glycoproteins, many questions remain regarding the physiological functions even for the most studied molecules. The size, complexity, and flexibility of mosaic proteins like fibronectins, tenascins, laminins etc. make it difficult to learn their detailed structure, to observe subtle conformational changes that may be functionally important, and to sort out artifacts from the numerous interactions and activities ascribed to them. However, new powerful techniques will allow a much better understanding of these proteins in the near future. The need to crystalize proteins to obtain the three-dimensional structures by x-ray diffraction has partly been overcome by the use of high-resolution NMR of proteins in solution. The folding of proteins of up to 150 amino acid length can at present be determined in this way, making it useful for studies of individual modules, or module pairs, of mosaic proteins (Baron *et al.*, 1991). The modules can be produced by peptide synthesis or expression of recombinant DNA in cells. Information from such pieces may then be put together to give a picture of the intact protein. The methodology of gene targeting by homologous recombination has made it possible to destroy an individual gene, or replace it with DNA carrying specific mutations. In combination with the technique of generating transgenic mice from embryonic stem cells, the functions of any protein can be studied *in vivo* in essentially unlimited detail. The impact these techniques will have on our future understanding of extracellular matrix proteins can hardly be overestimated.

References

Adams, J. and Lawler, J. (1993). The thrombospondin family. *Current Biology*, **3**, 188–190.

Alitalo, K., Hovi, T. and Vaheri, A. (1980). Fibronectin is produced by human macrophages. *J Exp Med*, **151**, 602--613.

Aota, S., Nagai, T. and Yamada, K.M. (1991). Characterization of regions of fibronectin besides the arginine-glycine- aspartic acid sequence required for adhesive function of the cell-binding domain using site-directed mutagenesis. *J Biol Chem*, **266**, 15938–43.

Aufderheide, E. and Ekblom, P. (1988). Tenascin during gut development: appearance in the mesenchyme, shift in molecular forms, and dependence on epithelial-mesenchymal interactions. *J Cell Biol*, **107**, 1341–1349.

Aukhil, I., Joshi, P., Yan, Y. and Erickson, H.P. (1993). Cell- and heparin-binding domains of the hexabrachion arm identified by tenascin expression proteins. *J Biol Chem*, **268**, 2542–53.

Bale, M.D., Wohlfahrt, L.A., Mosher, D.F., Tomasini, B. and Sutton, R.C. (1989). Identification of vitronectin as a major plasma protein adsorbed on polymer surfaces of different copolymer composition. *Blood*, **74**, 2698–706.

Banyai, L., Tordai, H. and Patthy, L. (1994). The gelatin-binding site of human 72 kDa type IV collagenase (gelatinase A). *Biochem J*, **298**, 403–407.

Banyai, L., Trexler, M., Koncz, S., Gyenes, M., Sipos, G. and Patthy, L. (1990). The collagen-binding site of type-II units of bovine seminal fluid protein PDC-109 and fibronectin. *Eur J Biochem*, **193**, 801–6.

Barkalow, F.J. and Schwarzbauer, J.E. (1991). Localization of the major heparin-binding site in fibronectin. *J Biol Chem*, **266**, 7812–8.

Barnes, D.W. and Reing, J. (1985). Human spreading factor: synthesis and response by HepG2 hepatoma cells in culture. *J Cell Physiol*, **125**, 207–14.

Baron, M., Norman, D.G. and Campbell, I.D. (1991). Protein modules. *Trends Biochem Sci*, **16**, 13–17.

Beck, K., Spring, J., Chiquet-Ehrismann, R., Engel, J. and Chiquet, M. (1991). Structural motifs of the extracellular matrix proteins laminin and tenascin. In *Patterns in Protein Sequence and Structure.*, W.R. Taylor (ed), 231–256. Berlin, Springer-Verlag.

Bernfield, M., Kokenyesi, R., Kato, M., Hinkes, M.T., Spring, J., Gallo, R.L. and Lose, E.J. (1992). Biology of syndecans: A family of transmembrane heparan sulfate proteoglycans. *Ann Rev Cell Biol*, **8**, 365–393.

Blystone, S.D. and Kaplan, J.E. (1993). The role of fibronectin in macrophage fibrin binding: a potential mechanism for high affinity, high capacity clearance of circulating fibrin. *Blood Coagul Fibrinolysis*, **4**, 769–81.

Bork, P. and Doolittle, R.F. (1992). Proposed acquisition of an animal protein domain by bacteria. *Proc Natl Acad Sci U S A*, **89**, 8990–4.

Borsi, L., Carnemolla, B., Nicolo, G., Spina, B., Tanara, G. and Zardi, L. (1992). Expression of different tenascin isoforms in normal, hyperplastic and neoplastic human breast tissues. *Int J Cancer*, **52**, 688–92.

Bourdon, M.A., Wikstrand, C.J., Furthmayr, H., Matthews, T.J. and Bigner, D.D. (1983). Human glioma-mesenchymal extracellular matrix antigen defined by monoclonal antibody. *Cancer Res*, **43**, 2796–805.

Bristow, J., Tee, M.K., Gitelman, S.E., Mellon, S.H. and Miller, W.L. (1993). Tenascin-X: a novel extracellular matrix protein encoded by the human XB gene overlapping P450c21B. *J Cell Biol*, **122**, 265–78.

Carnemolla, B., Leprini, A., Allemanni, G., Saginati, M. and Zardi, L. (1992). The inclusion of the type III repeat ED-B in the fibronectin molecule generates conformational modifications that unmask a cryptic sequence [published erratum appears in J Biol Chem 1993 Apr 5;268(10):7602]. *J Biol Chem*, **267**, 24689–92.

Carsons, S., Lavietes, B.B., Diamond, H.S. and Berkowitz, E. (1989). Role of fibronectin in rheumatic diseases. In *Fibronectin*, D.F. Mosher (ed), 327–361. New York, Academic Press.

Chernousov, M.A., Fogerty, F.J., Koteliansky, V.E. and Mosher, D.F. (1991). Role of the I-9 and III-1 modules of fibronectin in formation of an extracellular fibronectin matrix. *J Biol Chem*, **266**, 10851–8.

Cherny, R.C., Honan, M.A. and Thiagarajan, P. (1993). Site-directed mutagenesis of the arginine-glycine-aspartic acid in vitronectin abolishes cell adhesion. *J Biol Chem*, **268**, 9725–9.

Chiquet, E.R. (1991). Anti-adhesive molecules of the extracellular matrix. *Curr Opin Cell Biol*, **3**, 800–4.

Chiquet, E.R., Mackie, E.J., Pearson, C.A. and Sakakura, T. (1986). Tenascin: an extracellular matrix protein involved in tissue interactions during fetal development and oncogenesis. *Cell*, **47**, 131–9.

Chiquet, E.R., Matsuoka, Y., Hofer, U., Spring, J., Bernasconi, C. and Chiquet, M. (1991). Tenascin variants: differential binding to fibronectin and distinct distribution in cell cultures and tissues. *Cell Regul*, **2**, 927–38.

Chiquet, M. and Fambrough, D.M. (1984). Chick myotendinous antigen. II. A novel extracellular glycoprotein complex consisting of large disulfide-linked subunits. *J Cell Biol*, **98**, 1937–46.

Ciambrone, G.J. and McKeown, L.P. (1990). Plasminogen activator inhibitor type I stabilizes vitronectin-dependent adhesions in HT-1080 cells. *J Cell Biol*, **111,** 2183–95.

Constantine, K.L., Ramesh, V., Banyai, L., Trexler, M., Patthy, L. and Llinas, M. (1991). Sequence-specific ^{1}H NMR assignments and structural characterization of bovine seminal fluid protein PDC-109 domain b. *Biochemistry*, **30**, 1663–72.

Crossin, K.L., Hoffman, S., Grumet, M., Thiery, J.P. and Edelman, G.M. (1986). Site-restricted expression of cytotactin during development of the chicken embryo. *J Cell Biol*, **102**, 1917–30.

Dahlback, K., Lofberg, H., Alumets, J. and Dahlback, B. (1989). Immunohistochemical demonstration of age-related deposition of vitronectin (S-protein of complement) and terminal complement complex on dermal elastic fibers. *J Invest Dermatol*, **92**, 727–33.

de Boer, H.C., de Groot, P.G., Bouma, B.N. and Preissner, K.T. (1993). Ternary vitronectin-thrombin-antithrombin III complexes in human plasma. Detection and mode of association. *J Biol Chem*, **268**, 1279–83.

de Boer, H.C., Preissner, K.T., Bouma, B.N. and de Groot, P.G. (1992). Binding of vitronectin-thrombin-antithrombin III complex to human endothelial cells is mediated by the heparin binding site of vitronectin. *J Biol Chem*, **267**, 2264–8.

Declerck, P.J., De, M.M., Alessi, M.C., Baudner, S., Paques, E.P., Preissner, K.T., Muller, B.G. and Collen, D. (1988). Purification and characterization of a plasminogen activator inhibitor 1 binding protein from human plasma. Identification as a multimeric form of S protein (vitronectin). *J Biol Chem*, **263**, 15454–61.

Delannet, M., Martin, F., Bossy, B., Cheresh, D.A., Reichardt, L.F. and Duband, J.-L. (1994). Specific roles of the $\alpha V\beta 1$, $\alpha V\beta 3$ and $\alpha V\beta 5$ integrins in avian neural crest cell adhesion and migration on vitronectin. *Development*, **120**, 2687–2702.

Dickinson, C.D., Veerapandian, B., Dai, X.P., Hamlin, R.C., Xuong, N.H., Ruoslahti, E. and Ely, K.R. (1994). Crystal structure of the tenth type III cell adhesion module of human fibronectin. *J Mol Biol*, **236**, 1079–92.

Dzamba, B.J., Bultmann, H., Akiyama, S.K. and Peters, D.M. (1994). Substrate-specific binding of the amino terminus of fibronectin to an integrin complex in focal adhesions. *J Biol Chem*, **269**, 19646–52.

Dzamba, B.J., Wu, H., Jaenisch, R. and Peters, D.M. (1993). Fibronectin binding site in type I collagen regulates fibronectin fibril formation. *J Cell Biol*, **121**, 1165–72.

Ehrlich, H.J., Gebbink, R.K., Keijer, J., Linders, M., Preissner, K.T. and Pannekoek, H. (1990). Alteration of serpin specificity by a protein cofactor. Vitronectin endows plasminogen activator inhibitor 1 with thrombin inhibitory properties. *J Biol Chem*, **265**, 13029–35.

Ekblom, M., Fassler, R., Tomasini, J.B., Nilsson, K. and Ekblom, P. (1993). Downregulation of tenascin expression by glucocorticoids in bone marrow stromal cells and in fibroblasts. *J Cell Biol*, **123**, 1037–45.

Erickson, H.P. (1993). Tenascin-C, tenascin-R and tenascin-X: a family of talented proteins in search of functions. *Curr Opin Cell Biol*, **5**, 869–76.

Erickson, H.P. and Carrell, N.A. (1983). Fibronectin in extended and compact conformations. Electron microscopy and sedimentation analysis. *J Biol Chem*, **258**, 14539–44.

Felding, H.B. and Cheresh, D.A. (1993). Vitronectin and its receptors. *Curr Opin Cell Biol*, **5**, 864–8.

Ffrench, C.C. and Hynes, R.O. (1989). Alternative splicing of fibronectin is temporally and spatially regulated in the chicken embryo. *Development*, **106**, 375–88.

Ffrench, C.C., Van, d.W.L., Dvorak, H.F. and Hynes, R.O. (1989). Reappearance of an embryonic pattern of fibronectin splicing during wound healing in the adult rat. *J Cell Biol*, **109**, 903–14.

Flaumenhaft, R. and Rifkin, D.B. (1991). Extracellular matrix regulation of growth factor and protease activity. *Curr Opin Cell Biol*, **3**, 817–23.

Fleischmajer, R. and Timpl, R. (1984). Ultrastructural localization of fibronectin to different anatomic structures of human skin. *J Histochem Cytochem*, **32**, 315–21.

Fogerty, F.J. and Mosher, D.F. (1990). Mechanisms for organization of fibronectin matrix. *Cell Differ Dev*, **32**, 439–50.

Fuss, B., Wintergerst, E.S., Bartsch, U. and Schachner, M. (1993). Molecular characterization and in situ mRNA localization of the neural recognition molecule J1-160/180: a modular structure similar to tenascin. *J Cell Biol*, **120**, 1237–49.

Gebb, C., Hayman, E.G., Engvall, E. and Ruoslahti, E. (1986). Interaction of vitronectin with collagen. *J Biol Chem*, **261**, 16698–703.

George, E.L., Georges, L.E., Patel, K.R., Rayburn, H. and Hynes, R.O. (1993). Defects in mesoderm, neural tube and vascular development in mouse embryos lacking fibronectin. *Development*, **119**, 1079–91.

Grumet, M., Hoffman, S., Crossin, K.L. and Edelman, G.M. (1985). Cytotactin, an extracellular matrix protein of neural and non-neural tissues that mediates glia-neuron interaction. *Proc Natl Acad Sci U.S.A.*, **82**, 8075–9.

Guettier, C., Hinglais, N., Bruneval, P., Kazatchkine, M., Bariety, J. and Camilleri, J.P. (1989). Immunohistochemical localization of S protein/vitronectin in human atherosclerotic versus arteriosclerotic arteries. *Virchows Arch A Pathol Anat Histopathol*, **414**, 309–13.

Guidry, C., Miller, E.J. and Hook, M. (1990). A second fibronectin-binding region is present in collagen alpha chains. *J Biol Chem*, **265**, 19230–6.

Hauzenberger, D., Martin, N., Johansson, S. and Sundquist, K.-G. (1995). Characterzation of lymphocyte fibronectin. *Exp. Cell Res.* In press

Hayman, E.G., Pierschbacher, M.D., Ohgren, Y. and Ruoslahti, E. (1983). Serum spreading factor (vitronectin) is present at the cell surface and in tissues. *Proc Natl Acad Sci U.S.A.*, **80**, 4003–7.

Hayman, E.G., Pierschbacher, M.D., Suzuki, S. and Ruoslahti, E. (1985). Vitronectin—a major cell attachment-promoting protein in fetal bovine serum. *Exp Cell Res*, **160**, 245–58.

Hedman, K., Johansson, S., Vartio, T., Kjellen, L., Vaheri, A. and Hook, M. (1982). Structure of the pericellular matrix: association of heparan and chondroitin sulfates with fibronectin-procollagen fibers. *Cell*, **28**, 663–71.

Heinegård, D. and Oldberg, Å. (1993). Glycosylated matrix proteins. In *Connective Tissue and its Heritable Disorders*, P.M. Royce and B. Steinmann (eds), 189–209. New York, Wiley-Liss, Inc.

Hetland, G., Pettersen, H.B., Mollnes, T.E. and Johnson, E. (1989). S-protein is synthesized by human monocytes and macrophages in vitro. *Scand J Immunol*, **29**, 15–21.

Hoffman, S., Crossin, K.L. and Edelman, G.M. (1988). Molecular forms, binding functions, and developmental expression patterns of cytotactin and cytotactin-binding proteoglycan, an interactive pair of extracellular matrix molecules. *J Cell Biol*, **106**, 519–32.

Hogasen, K., Mollnes, T.E. and Harboe, M. (1992). Heparin-binding properties of vitronectin are linked to complex formation as illustrated by in vitro polymerization and binding to the terminal complement complex. *J Biol Chem*, **267**, 23076–82.

Homandberg, G.A. (1987). Inaccessibility to ligands of the amino-terminal region of plasma fibronectin. *Thromb Res*, **48**, 321–7.

Hook, M., Switalski, L.M., Wadström, T. and Lindberg, M. (1989). Interactions of pathogenic microorganisms with fibronectin. In *Fibronectin*, D.F. Mosher (ed), 295–308. New York, Academic Press.

Hormann, H. (1982). Fibronectin—mediator between cells and connective tissue. *Klin Wochenschr*, **60**, 1265–77.

Humphries, M.J., Komoriya, A., Akiyama, S.K., Olden, K. and Yamada, K.M. (1987). Identification of two distinct regions of the type III connecting segment of human plasma fibronectin that promote cell type-specific adhesion. *J Biol Chem*, **262**, 6886–92.

Hynes, R.O. (1990). *Fibronectins*, New York, Springer-Verlag.

Hynes, R.O. (1992). Integrins: versatility, modulation, and signaling in cell adhesion. *Cell*, **69**, 11–25.

Ill, C.R. and Ruoslahti, E. (1985). Association of thrombin-antithrombin III complex with vitronectin in serum. *J Biol Chem*, **260**, 15610–5.

Ingham, K.C., Brew, S.A. and Atha, D.H. (1990). Interaction of heparin with fibronectin and isolated fibronectin domains. *Biochem J*, **272**, 605–11.

Ingham, K.C., Brew, S.A. and Migliorini, M.M. (1989). Further localization of the gelatin-binding determinants within fibronectin. Active fragments devoid of type II homologous repeat modules. *J Biol Chem*, **264**, 16977–80.

Izumi, M., Shimo, O.T., Morishita, N., Ii, I. and Hayashi, M. (1988). Identification of the collagen-binding domain of vitronectin using monoclonal antibodies. *Cell Struct Funct*, **13**, 217–25.

Jenne, D., Hille, A., Stanley, K.K. and Huttner, W.B. (1989). Sulfation of two tyrosine-residues in human complement S-protein (vitronectin). *Eur J Biochem*, **185**, 391–5.

Jenne, D. and Stanley, K.K. (1985). Molecular cloning of S-protein, a link between complement, coagulation and cell-substrate adhesion. *Embo J*, **4**, 3153–7.

Johansson, S. (1985). Demonstration of high affinity fibronectin receptors on rat hepatocytes in suspension. *J Biol Chem*, **260**, 1557–61.

Johansson, S., Hedman, K., Kjellen, L., Christner, J., Vaheri, A. and Hook, M. (1985). Structure and interactions of proteoglycans in the extracellular matrix produced by cultured human fibroblasts. *Biochem J*, **232**, 161–8.

Johansson, S. and Smedsrod, B. (1986). Identification of a plasma gelatinase in preparations of fibronectin. *J Biol Chem*, **261**, 4363–6.

Jones, F.S., Hoffman, S., Cunningham, B.A. and Edelman, G.M. (1989). A detailed structural model of cytotactin: protein homologies, alternative RNA splicing, and binding regions. *Proc Natl Acad Sci U S A.*, **86**, 1905–9.

Joshi, P., Chung, C.Y., Aukhil, I. and Erickson, H.P. (1993). Endothelial cells adhere to the RGD domain and the fibrinogen-like terminal knob of tenascin. *J Cell Sci*, **106**, 389–400.

Kar, L., Lai, C.S., Wolff, C.E., Nettesheim, D., Sherman, S. and Johnson, M.E. (1993). ^{1}H NMR-based determination of the three-dimensional structure of the human plasma fibronectin fragment containing inter-chain disulfide bonds. *J Biol Chem*, **268**, 8580–9.

Klein, G., Beck, S. and Muller, C.A. (1993). Tenascin is a cytoadhesive extracellular matrix component of the human hematopoietic microenvironment. *J Cell Biol*, **123**, 1027–35.

Kleiner, D.J. and Stetler, S.W. (1993). Structural biochemistry and activation of matrix metalloproteases. *Curr Opin Cell Biol*, **5**, 891–7.

Kleinman, H.K., Wilkes, C.M. and Martin, G.R. (1981). Interaction of fibronectin with collagen fibrils. *Biochemistry*, **20**, 2325–2330.

Knox, P., Crooks, S. and Rimmer, C.S. (1986). Role of fibronectin in the migration of fibroblasts into plasma clots. *J Cell Biol*, **102**, 2318–23.

Kornblihtt, A.R., Umezawa, K., Vibe, P.K. and Baralle, F.E. (1985). Primary structure of human fibronectin: differential splicing may generate at least 10 polypeptides from a single gene. *Embo J*, **4**, 1755–9.

Kruse, J., Keilhauer, G., Faissner, A., Timpl, R. and Schachner, M. (1985). The J1 glycoprotein—a novel nervous system cell adhesion molecule of the L2/HNK-1 family. *Nature*, **316**, 146–8.

Kuusela, P., Vaheri, A., Palo, J. and Ruoslahti, E. (1978). Demonstration of fibronectin in human cerebrospinal fluid. *J Lab Clin Med*, **92**, 595–601.

La, F.M., Beaulieu, A.D., Kreis, C. and Poubelle, P. (1987). Fibronectin gene expression in polymorphonuclear leukocytes. Accumulation of mRNA in inflammatory cells. *J Biol Chem*, **262**, 2111–5.

Laterra, J. and Culp, L.A. (1982). Differences in hyaluronate binding to plasma and cell surface fibronectins. Requirement for aggregation. *J Biol Chem*, **257**, 719–26.

Leahy, D.J., Hendrickson, W.A., Aukhil, I. and Erickson, H.P. (1992). Structure of a fibronectin type III domain from tenascin phased by MAD analysis of the selenomethionyl protein. *Science*, **258**, 987–91.

LeBaron, R.G., Esko, J.D., Woods, A., Johansson, S. and Hook, M. (1988). Adhesion of glycosaminoglycan-deficient chinese hamster ovary cell mutants to fibronectin substrata. *J Cell Biol*, **106**, 945–52.

Lightner, V.A. and Erickson, H.P. (1990). Binding of hexabrachion (tenascin) to the extracellular matrix and substratum and its effect on cell adhesion. *J Cell Sci*, **95,** 263–77.

Lilja, H., Oldbring, J., Rannevik, G. and Laurell, C.B. (1987). Seminal vesicle-secreted proteins and their reactions during gelation and liquefaction of human semen. *J Clin Invest*, **80**, 281–5.

Linde, A., Johansson, S., Jonsson, R. and Jontell, M. (1982). Localization of fibronectin during dentinogenesis in rat incisor. *Arch Oral Biol*, **27**, 1069–73.

Lynch, G.W., Slayter, H.S., Miller, B.E. and McDonagh, J. (1987). Characterization of thrombospondin as a substrate for factor XIII transglutaminase. *J Biol Chem*, **262**, 1772–8.

Mackie, E.J., Thesleff, I. and Chiquet, E.R. (1987). Tenascin is associated with chondrogenic and osteogenic differentiation in vivo and promotes chondrogenesis in vitro. *J Cell Biol*, **105,** 2569–79.

Mackie, E.J., Tucker, R.P., Halfter, W., Chiquet, E.R. and Epperlein, H.H. (1988). The distribution of tenascin coincides with pathways of neural crest cell migration. *Development*, **102**, 237–50.

Main, A.L., Harvey, T.S., Baron, M., Boyd, J. and Campbell, I.D. (1992). The three-dimensional structure of the tenth type III module of fibronectin: an insight into RGD-mediated interactions. *Cell*, **71**, 671–8.

Matsumoto, K., Saga, Y., Ikemura, T., Sakakura, T. and Chiquet, E.R. (1994). The distribution of tenascin-X is distinct and often reciprocal to that of tenascin-C. *J Cell Biol*, **125**, 483–93.

McDonagh, R.P., McDonagh, J., Petersen, T.E., Thøgersen, H.C., Skorstensgaard, K., Sottrup-Jensen, L. and Magnusson, S. (1981). Amino acid sequence of the factor XIIIa acceptor site in bovine plasma fibronectin. *FEBS Lett*, **127**, 174–178.

McDonald, J.A., Quade, B.J., Broekelmann, T.J., LaChance, R., Forsman, K., Hasegawa, E. and Akiyama, S. (1987). Fibronectin's cell-adhesive domain and an amino-terminal matrix assembly domain participate in its assembly into fibroblast pericellular matrix. *J Biol Chem*, **262**, 2957–67.

McGavin, M.J., Gurusiddappa, S., Lindgren, P.E., Lindberg, M., Raucci, G. and Hook, M. (1993). Fibronectin receptors from Streptococcus dysgalactiae and Staphylococcus aureus. Involvement of conserved residues in ligand binding. *J Biol Chem*, **268**, 23946–53.

McGuire, E.A., Peacock, M.E., Inhorn, R.C., Siegel, N.R. and Tollefsen, D.M. (1988). Phosphorylation of vitronectin by a protein kinase in human plasma. Identification of a unique phosphorylation site in the heparin-binding domain. *J Biol Chem*, **263**, 1942–5.

McKeown, L.P. and Mosher, D.F. (1985). Interaction of the 70,000-mol-wt amino-terminal fragment of fibronectin with the matrix-assembly receptor of fibroblasts. *J Cell Biol*, **100**, 364–74.

Mehringer, J.H., Weigel, C.J. and Tollefsen, D.M. (1991). Cyclic AMP-dependent protein kinase phosphorylates serine378 in vitronectin. *Biochem Biophys Res Commun*, **179**, 655–60.

Milis, L., Morris, C.A., Sheehan, M.C., Charlesworth, J.A. and Pussell, B.A. (1993). Vitronectin-mediated inhibition of complement: evidence for different binding sites for C5b-7 and C9. *Clin Exp Immunol*, **92**, 114–9.

Mimuro, J. and Loskutoff, D.J. (1989). Purification of a protein from bovine plasma that binds to type 1 plasminogen activator inhibitor and prevents its interaction with extracellular matrix. Evidence that the protein is vitronectin. *J Biol Chem*, **264**, 936–9.

Moon, K.Y., Shin, K.S., Song, W.K., Chung, C.H., Ha, D.B. and Kang, M.S. (1994). A candidate molecule for the matrix assembly receptor to the N-terminal 29-kDa fragment of fibronectin in chick myoblasts. *J Biol Chem*, **269**, 7651–7.

Morla, A. and Ruoslahti, E. (1992). A fibronectin self-assembly site involved in fibronectin matrix assembly: reconstruction in a synthetic peptide. *J Cell Biol*, **118**, 421–9.

Morla, A., Zhang, Z. and Ruoslahti, E. (1994). Superfibronectin is a functionally distinct form of fibronectin. *Nature*, **367**, 193–6.

Mosesson, M.W. and Umfleet, R.A. (1970). The cold-insoluble globulin of human plasma. *J Biol Chem*, **245**, 5728–5736.

Mosher, D.F. (1976). Action of fibrin-stabilizing factor on cold-insoluble globulin and α2-macroglobulin in clotting plasma. *J Biol Chem*, **251**, 1639–1645.

Mould, A.P. and Humphries, M.J. (1991). Identification of a novel recognition sequence for the integrin alpha 4 beta 1 in the COOH-terminal heparin-binding domain of fibronectin. *Embo J*, **10**, 4089–95.

Mould, A.P., Komoriya, A., Yamada, K.M. and Humphries, M.J. (1991). The CS5 peptide is a second site in the IIICS region of fibronectin recognized by the integrin alpha 4 beta 1. Inhibition of alpha 4 beta 1 function by RGD peptide homologues. *J Biol Chem*, **266**, 3579–85.

Murphy, U.J., Lightner, V.A., Aukhil, I., Yan, Y.Z., Erickson, H.P. and Hook, M. (1991). Focal adhesion integrity is downregulated by the alternatively spliced domain of human tenascin [published erratum appears in J Cell Biol 1992 Feb;116(3):833]. *J Cell Biol*, **115**, 1127–36.

Nies, D.E., Hemesath, T.J., Kim, J.H., Gulcher, J.R. and Stefansson, K. (1991). The complete cDNA sequence of human hexabrachion (Tenascin). A multidomain protein containing unique epidermal growth factor repeats. *J Biol Chem*, **266**, 2818–23.

Norenberg, U., Wille, H., Wolff, J.M., Frank, R. and Rathjen, F.G. (1992). The chicken neural extracellular matrix molecule restrictin: similarity with EGF-, fibronectin type III-, and fibrinogen-like motifs. *Neuron*, **8**, 849–63.

Ocklind, G., Talts, J., Fassler, R., Mattsson, A. and Ekblom, P. (1993). Expression of tenascin in developing and adult mouse lymphoid organs. *J Histochem Cytochem*, **41**, 1163–9.

Onda, H., Poulin, M.L., Tassava, R.A. and Chiu, I.M. (1991). Characterization of a newt tenascin cDNA and localization of tenascin mRNA during newt limb regeneration by in situ hybridization. *Dev Biol*, **148**, 219–32.

Owens, R.J. and Baralle, F.E. (1986). Mapping the collagen-binding site of human fibronectin by expression in Escherichia coli. *Embo J*, **5**, 2825–30.

Panetti, T.S. and McKeown, L.P. (1993). Receptor-mediated endocytosis of vitronectin is regulated by its conformational state. *J Biol Chem*, **268**, 11988–93.

Panetti, T.S. and McKeown, L.P. (1993). The alpha v beta 5 integrin receptor regulates receptor-mediated endocytosis of vitronectin. *J Biol Chem*, **268**, 11492–5.

Patel, R.S., Odermatt, E., Schwarzbauer, J.E. and Hynes, R.O. (1987). Organization of the fibronectin gene provides evidence for exon shuffling during evolution. *Embo J*, **6**, 2565–72.

Patthy, L. (1988). Detecting distant homologies of mosaic proteins. Analysis of the sequences of thrombomodulin, thrombospondin complement components C9, C8 alpha and C8 beta, vitronectin and plasma cell membrane glycoprotein PC-1. *J Mol Biol*, **202**, 689–96.

Patthy, L. (1991). Modular exchange principles in proteins. *Curr Opin Struct Biol*, **1**, 351–361.

Paul, J. and Hynes, R.O. (1984). Multiple fibronectin subunits and their post-translational modifications. *J Biol Chem*, **259**, 13407–13487.

Pearson, C.A., Pearson, D., Shibahara, S., Hofsteenge, J. and Chiquet, E.R. (1988). Tenascin: cDNA cloning and induction by TGF-beta. *Embo J*, **7**, 2977–82.

Perris, R. and Johansson, S. (1987). Amphibian neural crest cell migration on purified extracellular matrix components: a chondroitin sulfate proteoglycan inhibits locomotion on fibronectin substrates. *J Cell Biol*, **105,** 2511–21.

Perris, R. and Johansson, S. (1990). Inhibition of neural crest cell migration by aggregating chondroitin sulfate proteoglycans is mediated by their hyaluronan-binding region. *Dev Biol*, **137**, 1–12.

Pesheva, P., Gennarini, G., Goridis, C. and Schachner, M. (1993). The F3/11 cell adhesion molecule mediates the repulsion of neurons by the extracellular matrix glycoprotein J1-160/180. *Neuron*, **10**, 69–82.

Pierschbacher, M.D. and Ruoslahti, E. (1984). Cell attachment activity of fibronectin can be duplicated by small synthetic fragments of the molecule. *Nature*, **309**, 30–33.

Podack, E.R. and Tschopp, J. (1984). Membrane attack by complement. *Mol Immunol*, **21**, 589–603.

Preissner, K.T. (1991). Structure and biological role of vitronectin. *Annu Rev Cell Biol*, **7,** 275–310.

Preissner, K.T., Holzhuter, S., Justus, C. and Muller, B.G. (1989). Identification of and partial characterization of platelet vitronectin: evidence for complex formation with platelet-derived plasminogen activator inhibitor-1. *Blood*, **74**, 1989–96.

Preissner, K.T. and Muller, B.G. (1987). Neutralization and binding of heparin by S protein/vitronectin in the inhibition of factor Xa by antithrombin III. Involvement of an inducible heparin-binding domain of S protein/vitronectin. *J Biol Chem*, **262**, 12247–53.

Prieto, A.L., Edelman, G.M. and Crossin, K.L. (1993). Multiple integrins mediate cell attachment to cytotactin/tenascin. *Proc Natl Acad Sci U S A.*, **90**, 10154–8.

Rathjen, F.G., Wolff, J.M. and Chiquet, E.R. (1991). Restrictin: a chick neural extracellular matrix protein involved in cell attachment co-purifies with the cell recognition molecule F11. *Development*, **113**, 151–64.

Rocco, M., Infusini, E., Daga, M.G., Gogioso, L. and Cuniberti, C. (1987). Models of fibronectin. *Embo J*, **6**, 2343–9.

Ruggeri, Z.M. (1993). von Willebrand factor and fibrinogen. *Curr Opin Cell Biol*, **5**, 898–906.

Ruoslahti, E., Hayman, E.G., Engvall, E. and Cothran, W.C.B., W. T. (1981). Alignment of biologically active domains in the fibronectin molecule. *J Biol Chem*, **256**, 7277–7281.

Ruoslahti, E. and Pierschbacher, M.D. (1986). Arg-Gly-Asp: a versatile cell recognition signal. *Cell*, **44**, 517–8.

Ruoslahti, E. and Vaheri, A. (1975). Interaction of soluble fibroblast surface antigen with fibrinogen and fibrin. *J Exp Med*, **141**, 497–501.

Saga, Y., Tsukamoto, T., Jing, N., Kusakabe, M. and Sakakura, T. (1991). Murine tenascin: cDNA cloning, structure and temporal expression of isoforms. *Gene*, **104**, 177–85.

Saga, Y., Yagi, T., Ikawa, Y., Sakakura, T. and Aizawa, S. (1992). Mice develop normally without tenascin. *Genes Dev*, **6**, 1821–31.

Sanchez, A.P., Dominguez, J.C. and Garcia, P.A. (1994). Activation of the alpha 4 beta 1 integrin through the beta 1 subunit induces recognition of the RGDS sequence in fibronectin. *J Cell Biol*, **126**, 271–9.

Sane, D.C., Moser, T.L. and Greenberg, C.S. (1991). Vitronectin in the substratum of endothelial cells is cross-linked and phosphorylated. *Biochem Biophys Res Commun*, **174**, 465–9.

Sane, D.C., Moser, T.L., Pippen, A.M., Parker, C.J., Achyuthan, K.E. and Greenberg, C.S. (1988). Vitronectin is a substrate for transglutaminases. *Biochem Biophys Res Commun*, **157**, 115–20.

Schwarzbauer, J.E. (1991). Identification of the fibronectin sequences required for assembly of a fibrillar matrix. *J Cell Biol*, **113**, 1463–73.
Schwarzbauer, J.E., Patel, R.S., Fonda, D. and Hynes, R.O. (1987). Multiple sites of alternative splicing of the rat fibronectin gene transcript. *Embo J*, **6**, 2573–80.
Schwarzbauer, J.E., Spencer, C.S. and Wilson, C.L. (1989). Selective secretion of alternatively spliced fibronectin variants. *J Cell Biol*, **109**, 3445–3453.
Seiffert, D., Ciambrone, G., Wagner, N.V., Binder, B.R. and Loskutoff, D.J. (1994). The somatomedin B domain of vitronectin. Structural requirements for the binding and stabilization of active type 1 plasminogen activator inhibitor. *J Biol Chem*, **269**, 2659–66.
Seiffert, D., Keeton, M., Eguchi, Y., Sawdey, M. and Loskutoff, D.J. (1991). Detection of vitronectin mRNA in tissues and cells of the mouse. *Proc Natl Acad Sci U S A.*, **88**, 9402–6.
Seiffert, D. and Loskutoff, D.J. (1991). Evidence that type 1 plasminogen activator inhibitor binds to the somatomedin B domain of vitronectin. *J Biol Chem*, **266**, 2824–30.
Sekiguchi, K., Fukuda, M. and Hakomori, S. (1981). Domain structure of hamster plasma fibronectin. Isolation and characterization of four functionally distinct domains and their unequal distribution between subunit polypeptides. *J Biol Chem*, **256**, 6452–6462.
Shaffer, M.C., Foley, T.P. and Barnes, D.W. (1984). Quantitation of spreading factor in human biologic fluids. *J Lab Clin Med*, **103**, 783–91.
Sigudardottir, O. and Wiman, B. (1994). Identification of the PAI-1 binding site in vitronectin. *Biochim Biophys Acta*, **1208**, 104–110.
Skorstengaard, K., Halkier, T., Hojrup, P. and Mosher, D. (1990). Sequence location of a putative transglutaminase cross-linking site in human vitronectin. *Febs Lett*, **262**, 269–74.
Skorstengaard, K., Jensen, M.S., Sahl, P., Petersen, T.E. and Magnusson, S. (1986). Complete primary structure of bovine plasma fibronectin. *Eur J Biochem*, **161**, 441–53.
Skorstengaard, K., Thogersen, H.C. and Petersen, T.E. (1984). Complete primary structure of the collagen-binding domain of bovine fibronectin. *Eur J Biochem*, **140**, 235–43.
Skorstengaard, K., Thogersen, H.C., Vibe, P.K., Petersen, T.E. and Magnusson, S. (1982). Purification of twelve cyanogen bromide fragments from bovine plasma fibronectin and the amino acid sequence of eight of them. Overlap evidence aligning two plasmic fragments, internal homology in gelatin-binding region and phosphorylation site near C terminus. *Eur J Biochem*, **128**, 605–23.
Smilenov, L., Forsberg, E., Zeligman, I., Sparrman, M. and Johansson, S. (1992). Separation of fibronectin from a plasma gelatinase using immobilized metal affinity chromatography. *Febs Lett*, **302**, 227–30.
Smith, A., Tatum, F.M., Muster, P., Burch, M.K. and Morgan, W.T. (1988). Importance of ligand-induced conformational changes in hemopexin for receptor-mediated heme transport. *J Biol Chem*, **263**, 5224–9.
Sottile, J., Schwarzbauer, J., Selegue, J. and Mosher, D.F. (1991). Five type I modules of fibronectin form a functional unit that binds to fibroblasts and Staphylococcus aureus. *J Biol Chem*, **266**, 12840–3.
Sriramarao, P. and Bourdon, M.A. (1993). A novel tenascin type III repeat is part of a complex of tenascin mRNA alternative splices. *Nucleic Acids Res*, **21**, 163–8.
Sriramarao, P., Mendler, M. and Bourdon, M.A. (1993). Endothelial cell attachment and spreading on human tenascin is mediated by alpha 2 beta 1 and alpha v beta 3 integrins. *J Cell Sci*, **105,** 1001–1012.
Stenman, S. and Vaheri, A. (1978). Distribution of a major connective tissue protein, fibronectin, in normal human tissue. *J Exp Med*, **147**, 1054–1064.
Stockmann, A., Hess, S., Declerck, P., Timpl, R. and Preissner, K.T. (1993). Multimeric vitronectin. Identification and characterization of conformation- dependent self-association of the adhesive protein. *J Biol Chem*, **268**, 22874–82.
Suzuki, S., Oldberg, A., Hayman, E.G., Pierschbacher, M.D. and Ruoslahti, E. (1985). Complete amino acid sequence of human vitronectin deduced from cDNA. Similarity of cell attachment sites in vitronectin and fibronectin. *Embo J*, **4**, 2519–24.
Suzuki, S., Pierschbacher, M.D., Hayman, E.G., Nguyen, K., Ohgren, Y. and Ruoslahti, E. (1984). Domain structure of vitronectin. Alignment of active sites. *J Biol Chem*, **259**, 15307–14.
Tamkun, J.W. and Hynes, R.O. (1983). Plasma fibronectin is synthesized and secreted by hepatocytes. *J Biol Chem*, **258**, 4641–7.
Tamkun, J.W., Schwarzbauer, J.E. and Hynes, R.O. (1984). A single rat fibronectin gene generates three different mRNAs by alternative splicing of a complex exon. *Proc Natl Acad Sci U.S.A.*, **81**, 5140–4.
Tan, S.S., Crossin, K.L., Hoffman, S. and Edelman, G.M. (1987). Asymmetric expression in somites of cytotactin and its proteoglycan ligand is correlated with neural crest cell distribution. *Proc Natl Acad Sci U.S.A.*, **84**, 7977–81.
Tollefsen, D.M., Weigel, C.J. and Kabeer, M.H. (1990). The presence of methionine or threonine at position 381 in vitronectin is correlated with proteolytic cleavage at arginine 379. *J Biol Chem*, **265**, 9778–81.
Tomasini, B.R. and Mosher, D.F. (1986). On the identity of vitronectin and S-protein: immunological crossreactivity and functional studies. *Blood*, **68**, 737–42.

Tomasini, B.R. and Mosher, D.F. (1991). Vitronectin. *Prog Hemost Thromb*, B. Coller (ed), **10**, 269–305.
Tomasini, B.R., Owen, M.C., Fenton, J.2. and Mosher, D.F. (1989). Conformational lability of vitronectin: induction of an antigenic change by alpha-thrombin-serpin complexes and by proteolytically modified thrombin. *Biochemistry*, **28**, 7617–23.
Tomasini, J.B., Ruoslahti, E. and Pierschbacher, M.D. (1993). A 30 kD sulfated extracellular matrix protein immunologically crossreactive with vitronectin. *Matrix*, **13**, 203–14.
Tomasini-Johansson, B.R. (1993). Vitronectin in inflammatory conditions: localization in rheumatoid arthritic synovia. In *Biology of Vitronectins and Their Receptors.*, K.T. Preissner, S. Rosenblatt, C. Kost, J. Wegerhoff and D.F. Mosher (eds), 223–228. Amsterdam, Exerpta Medica.
Tooney, N.M., Mosesson, M.W., Amrani, D.L., Hainfeld, J.F. and Wall, J.S. (1983). Solution and surface effects on plasma fibronectin structure. *J Cell Biol*, **97**, 1686–92.
Tschopp, J., Masson, D., Schafer, S., Peitsch, M. and Preissner, K.T. (1988). The heparin binding domain of S-protein/vitronectin binds to complement components C7, C8, and C9 and perforin from cytolytic T-cells and inhibits their lytic activities. *Biochemistry*, **27**, 4103–9.
Tucker, R.P., Hammarback, J.A., Jenrath, D.A., Mackie, E.J. and Xu, Y. (1993). Tenascin expression in the mouse: in situ localization and induction in vitro by bFGF. *J Cell Sci*, **104**, 69–76.
Tucker, R.P. and McKay, S.E. (1991). The expression of tenascin by neural crest cells and glia. *Development*, **112**, 1031–9.
Tucker, R.P., Spring, J., Baumgartner, S., Martin, D., Hagios, C., Poss, P.M. and Chiquet, E.R. (1994). Novel tenascin variants with a distinctive pattern of expression in the avian embryo. *Development*, **120**, 637–47.
Vaheri, A., Keski-Oja, J. and Vartio, T. (1989). Fibronectin and malignant transformation. In *Fibronectin*, D.F. Mosher (ed), 255–271. New York, Academic Press.
Vaheri, A., Kurkinen, M., Lehto, V.P., Linder, E. and Timpl, R. (1978). Codistribution of pericellular matrix proteins in cultured fibroblasts and loss in transformation: Fibronectin and procollagen. *Proc Natl Acad Sci USA.*, **75**, 4944–4948.
Vaheri, A., Stenman, S. and Wartiovarra, J. (1977). Changes in the distribution of a major fibroblast protein, fibronectin, during mitosis and interphase. *J Cell Biol*, **74**, 453–467.
Van, E.P., Geerts, A., De, B.P., Lazou, J.M., Vrijsen, R., Sciot, R., Wisse, E. and Desmet, V.J. (1992). Localization and cellular source of the extracellular matrix protein tenascin in normal and fibrotic rat liver. *Hepatology*, **15**, 909–16.
Vaughan, L., Huber, S., Chiquet, M. and Winterhalter, K.H. (1987). A major, six-armed glycoprotein from embryonic cartilage. *Embo J*, **6**, 349–53.
Vibe, P.K., Magnusson, S. and Baralle, F.E. (1986). Donor and acceptor splice signals within an exon of the human fibronectin gene: a new type of differential splicing. *Febs Lett*, **207**, 287–91.
Volker, W., Hess, S., Vischer, P. and Preissner, K.T. (1993). Binding and processing of multimeric vitronectin by vascular endothelial cells. *J Histochem Cytochem*, **41**, 1823–32.
Weller, A., Beck, S. and Ekblom, P. (1991). Amino acid sequence of mouse tenascin and differential expression of two tenascin isoforms during embryogenesis. *J Cell Biol*, **112**, 355–62.
Wennerberg, K., Lohikangas, L., Gullberg, D., Pfaff, M., Johansson, S. and Fāssler, R. (1995). β_1 integrin-dependent and -independent polymerization of fibronectin. *J. Cell Biol.* In press.
Williams, E.C., Janmey, P.A., Ferry, J.D. and Mosher, D.F. (1982). Conformational states of fibronectin. Effects of pH, ionic strength, and collagen binding. *J Biol Chem*, **257**, 14973–8.
Williams, M.J., Phan, I., Harvey, T.S., Rostagno, A., Gold, L.I. and Campbell, I.D. (1994). Solution structure of a pair of fibronectin type 1 modules with fibrin binding activity. *J Mol Biol*, **235**, 1302–11.
Woods, A. and Couchman, J.R. (1994). Syndecan 4 heparan sulfate proteoglycan is a selectively enriched and widespread focal adhesion component. *Mol Biol Cell*, **5**, 183–92.
Woods, A., Couchman, J.R., Johansson, S. and Hook, M. (1986). Adhesion and cytoskeletal organisation of fibroblasts in response to fibronectin fragments. *Embo J*, **5**, 665–70.
Woods, A., Johansson, S. and Hook, M. (1988). Fibronectin fibril formation involves cell interactions with two fibronectin domains. *Exp Cell Res*, **177**, 272–83.
Wu, C., Fields, A.J., Kapteijn, B.A.E. and McDonald, J.A. (1995). The role of $\alpha_4\beta_1$ integrin in cell mobility and fibronectin matrix assembly. *J Cell Sci*, **108**, 821–829.
Yamada, H., Watanabe, K., Shimonaka, M. and Yamaguchi, Y. (1994). Molecular cloning of brevican, a novel brain proteoglycan of the aggrecan/versican family. *J Biol Chem*, **269**, 10119–26.
Yamada, K.M., Kennedy, D.W., Kimata, K. and Pratt, R.M. (1980). Characterization of fibronectin interactions with glycosaminoglycans and identification of active proteolytic fragments. *J Biol Chem*, **255**, 6055–6063.
Yamada, S., Ichida, T., Matsuda, Y., Miyazaki, Y., Hatano, T., Hata, K., Asakura, H., Hirota, N., Geerts, A. and Wisse, E. (1992). Tenascin expression in human chronic liver disease and in hepatocellular carcinoma. *Liver*, **12**, 10–16.
Yamagata, M., Shinomura, T. and Kimata, K. (1993). Tissue variation of two large chondroitin sulfate proteoglycans (PG- M/versican and PG-H/aggrecan) in chick embryos. *Anat Embryol (Berl)*, **187**, 433–44.

Yamagata, M., Yamada, K.M., Yoneda, M., Suzuki, S. and Kimata, K. (1986). Chondroitin sulfate proteoglycan (PG-M-like proteoglycan) is involved in the binding of hyaluronic acid to cellular fibronectin. *J Biol Chem*, **261**, 13526–35.

Yang, J.T., Rayburn, H. and Hynes, R.O. (1993). Embryonic mesodermal defects in alpha 5 integrin-deficient mice. *Development*, **119**, 1093–105.

Yatohgo, T., Izumi, M., Kashiwagi, H. and Hayashi, M. (1988). Novel purification of vitronectin from human plasma by heparin affinity chromatography. *Cell Struct Funct*, **13**, 281–92.

Yoneda, A., Ogawa, H., Matsumoto, I., Ishizuka, I., Hase, S. and Seno, N. (1993). Structures of the N-linked oligosaccharides on porcine plasma vitronectin. *Eur J Biochem*, **218**, 797–806.

Young, S.L., Chang, L.Y. and Erickson, H.P. (1994). Tenascin-C in rat lung: distribution, ontogeny and role in branching morphogenesis. *Dev Biol*, **161**, 615–25.

Zhang, Z., Morala, A.O., Vouri, K., Bauer, J.S., Juliano, R.L. and Ruoslahti, E. (1993). The $\alpha_v\beta_1$ integrin functions as a fibronectin receptor but does not support fibronectin matrix assembly and cell migration of fibronectin. *J Cell Biol*, **122**, 235–242.

Zhao, Y. and Sane, D.C. (1993). The cell attachment and spreading activity of vitronectin is dependent on the Arg-Gly-Asp sequence. Analysis by construction of RGD and domain deletion mutants. *Biochem Biophys Res Commun*, **192**, 575–82.

Zisch, A.H., D'Alessandri, L., Ranscht, B., Falchetto, R., Winterhalter, K.H. and Vaughan, L. (1992). Neuronal cell adhesion molecule contactin/F11 binds to tenascin via its immunoglobulin-like domains. *J Cell Biol*, **119**, 203–13.

4 Elastic Tissue, Elastin and Elastin Associated Microfibrils

Edward G. Cleary and Mark A. Gibson

Department of Pathology, University of Adelaide, Adelaide, South Australia, Australia.

INTRODUCTION

The terms "elastic tissue" and "elastic fibre" refer to fibrous matrix components, with characteristic histochemical staining properties, that are predominantly composed of the rubber-like protein, elastin. The latter has a characteristic amino acid composition and is the most hydrophobic of all known proteins. Elastin is found in all vertebrates, except the cyclostomes, and is not found in invertebrates (Sage and Gray, 1979).

Elastic fibres are remarkable for the diverse range of tissues in which they are found in mammals. They can be identified in the extracellular matrix of many tissues as solid, branching and unbranching, fine and thick, rod-like fibres (as in nuchal and other elastic ligaments), or they occur as concentric sheets or lamellae (in blood vessels), or they are arranged in three dimensional meshworks of fine fibrils, as in elastic cartilage of the ear and larynx, or they may occur as combinations of these, as is seen in the skin and the lungs. In all of these tissues, the protein elastin can be shown, immunohistologically, to be localized within the elastic fibres. In general, elastic fibres are found in tissues that are subject to repetitive distending/relaxing or passive lengthening/shortening movements and the elastic fibres are considered to endow tissues with flexibility and rubber-like extensibility. However, they are also found in elastic cartilage and periosteum, which are not obviously exposed to repeated stresses and strains.

Electron microscopy has provided additional insights into the composition and structure of elastic fibres. Ultrastructurally, elastic fibres are seen to consist of two components, amorphous material and microfibrils, 10–12 nm in diameter (see Figure 4.1). The predominant, amorphous component co-distributes with antibodies specific for the protein elastin (Figure 4.1b). It has an affinity for anionic stains such as phosphotungstate, as well as for tannic acid, with which it gives a very electron dense reaction. It is resistant to digestion

Extracellular Matrix, Volume 2, Molecular Components and Interactions, edited by Wayne D. Comper.

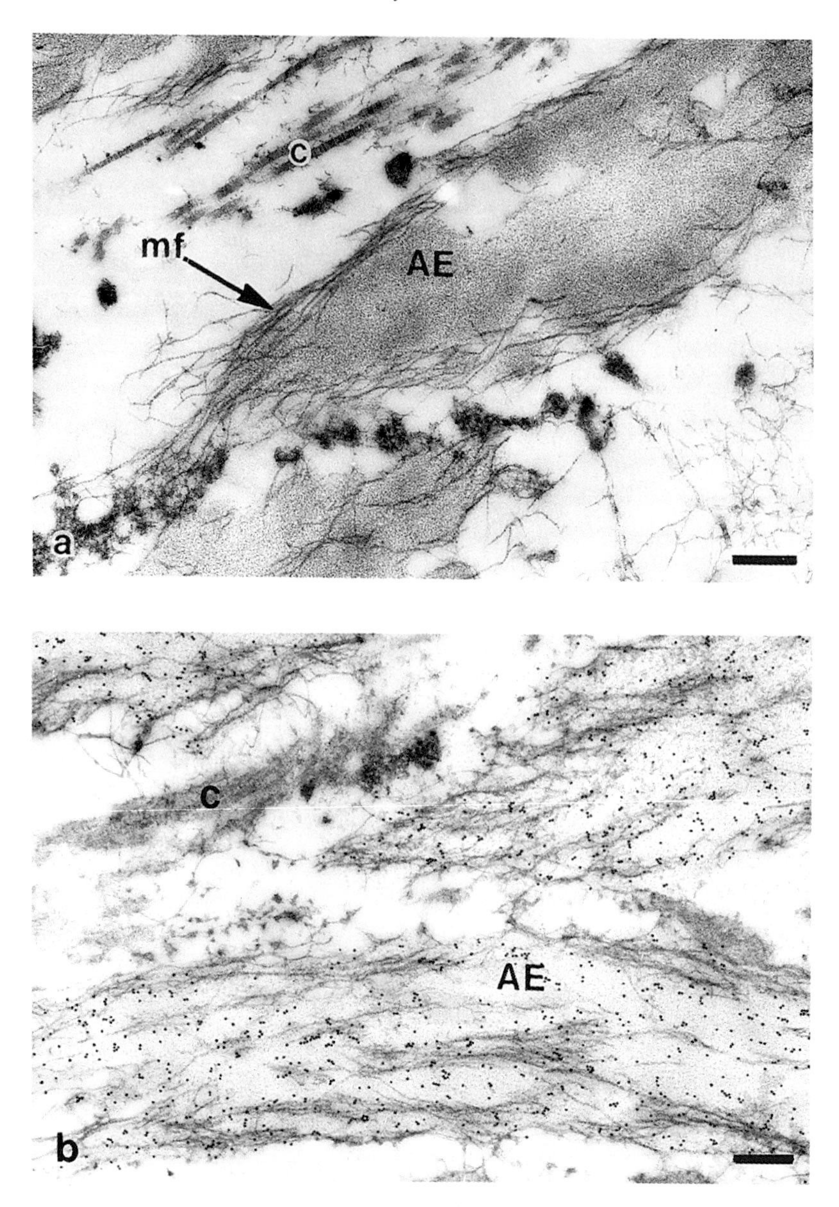

Figure 4.1 Electron micrographs showing elastic fibres in 230-day-old foetal bovine nuchal ligament. In each section, the amorphous elastin (AE) core has a peripheral coating of 10–12 nm diameter microfibrils (mf). Adjacent collagen fibres (c) are also visible. Bar: 0.25 microns. Magnification: × 36,000.

Figure 4.1a Section embedded in Spurrs medium. Portions of a thin cell process can be seen within a groove in the elastic fibre.

Figure 4.1b Section embedded in LR White medium. Immunostained with monoclonal antibody, BA4, specific for an elastin-specific peptide. With this embedding resin, the amorphous elastin (AE) is not so well defined. Collections of electron densely staining microfibrils can be seen as inclusions within the amorphous elastic material. The anti-elastin antibody, labelled with 12 nm gold particles, is bound selectively to the AE, with an additional preference for AE/mf junctions (Courtesy of Dr J. S. Kumaratilake).

with general proteases but is susceptible to elastases, which are mostly serine proteases with relatively broad specificity. The amorphous elastin is surrounded by collections of 10–12 nm diameter, hollow, unbranching microfibrils, which have an affinity for cationic stains, including uranyl, lead and ruthenium red (Greenlee *et al.*, 1966). They also characteristically exhibit staining properties of glycoproteins (Fanning and Cleary, 1988). In addition, some microfibrils are seen, in conventionally stained sections, within small electron dense inclusions in the amorphous elastic material. The elastin-associated microfibrils are susceptible to digestion with trypsin and pepsin, and are resistant to elastase. They have been shown to be composed of a number of microfibrillar proteins, all of which have a very different composition from elastin. The proportion of microfibrils associated with elastic fibres tends to be greatest during the rapid growth phases of late foetal and post-natal development and subsequently it diminishes with ageing (Cleary and Gibson, 1983).

The elastin-associated microfibrils are now known to be highly complex structures, made up predominantly of one or both members of the family of proteins, called fibrillins (Sakai *et al.*, 1991), in association with a smaller glycoprotein, MAGP (microfibril associated glycoprotein) (Gibson *et al.*, 1989). A number of smaller glycoproteins, identified as MP78, MP70, and MP25 (Gibson *et al.*, 1989), AMP (Horrigan *et al.*, 1992) and emilin (which may be specific to chicks) (Bressan *et al.*, 1993) may also be microfibrillar components (see below). Elastin-associated microfibrils have been shown to bind increasing amounts of many other proteins (including Serum Amyloid P, vitronectin, and fibronectin) from the serum and tissues, with increasing age (Dahlback *et al.*, 1989) This latter property has complicated the understanding of the composition of microfibrils and may be functionally significant, if the bound proteins alter the physical properties of the microfibrils. The absence of such proteins from the newly formed microfibrils, during early tissue development, indicates that they are adsorbed materials and not integral to the microfibrillar structure.

The availability of antibodies, specific for the microfibrillar components has allowed recognition of the fact that similar microfibrils (i.e. with apparently identical morphology and immunoreactivity) are widely distributed within the tissues. In some locations they occur in relation to only small amounts of amorphous elastic tissue (as in periodontal ligament) while in others they are seen in the absence of any immunologically recognizable elastin, as in the ocular zonule and in the glomerular mesangium (Kumaratilake *et al.*, 1989; Sakai *et al.*, 1986).

In some tissues there are fibres that are reactive with elastic tissue stains, only after prior oxidation. These are referred to as oxytalan (Fullmer and Lillie, 1958) or elaunin fibres (Gawlik, 1965). In general, the oxytalan staining property relates to microfibrillar collections which lack immunoreactive elastin (as in the ocular zonule), and elaunin staining fibres consist of collections of microfibrils in which there are small amounts of immunoreactive amorphous elastin (as in the periodontal ligament). The extent to which the relative content of amorphous elastin and microfibrils influences histochemical elastic tissue stains in different tissues is poorly understood and poorly documented. For this reason, it is unwise to attempt to interpret apparent changes in relative amounts of amorphous elastin and microfibrils in elastic fibres, by histological methods. Electron microscopy can overcome many of these uncertainties, but obtaining representative sampling is always a major concern with this technique. In addition, small elastic fibres can be difficult to recognize in conventionally stained fields in the electron microscope (Cleary and Gibson, 1983).

Tannic acid staining is often used to facilitate identification of small amounts of amorphous elastin in the electron microscope (Kajikawa *et al.*, 1975), but tannic acid causes such electron dense staining of amorphous elastin as to mask structure within elastic fibres and may also affect interpretation of their microfibrillar content.

As the ocular zonule, which consists of fine, thread-like collections of the 12 nm microfibrils, is responsible for effecting changes in the shape of the ocular lens in response to contraction of the muscle fibres of the ciliary body, it may be inferred that its 12 nm microfibrils are relatively rigid, inextensible structures. Such properties are highly appropriate also for the dermal microfibrillary bundles which connect the dermal elastin network to the dermal-epidermal basement membrane and thence to the epidermis. Thus, one can envisage microfibrils at the periphery of elastic fibres serving as "grappling hooks" that connect extensible elastic tissue networks to nearby structures. However, it will be apparent that the relationship, within elastic fibres, between the amorphous elastin and the microfibrils will have a major influence on the physical properties of those fibres. The factors that determine elastic tissue morphology, and that doubtless also influence its physical properties, have been the subject of little direct research to date and are poorly understood.

Changes in Elastic Fibres During Development

In the earliest stages of elastic fibre formation, as exemplified in the early foetal nuchal ligament, aggregates of aligned 12 nm diameter microfibrils were seen immediately adjacent to elastogenic fibroblasts, often within embayments of their cell membrane (Fahrenbach *et al.*, 1966). Amorphous, immunoreactive elastin was subsequently deposited on these microfibrillar collections, in increasing amounts, to form progressively thicker elastic fibres. The microfibrils appeared to act as a template, on which the amorphous elastin was deposited, and they appeared to determine the orientation of the deposited elastin in the elastic fibres. No specific alterations in this sequence of events have been recognized to date in tissues such as aorta, lung and even elastic ear cartilage to account for the different morphology of the elastic fibres in these tissues. The fate of the microfibrils is not clear. In most tissues, they appear to be displaced progressively to the periphery of the forming elastic fibres and in the process, occasional microfibril collections appear to become entrapped within the amorphous elastin, which accounts for the microfibril-containing electron dense inclusions within many elastic fibres.

ELASTIN

Chemistry

Elastin is extremely inert to protein solvents and is resistant to hydrolysis by even relatively strong alkali, so that it is isolated from tissues as the residue, after everything else had been dissolved, resulting from the action of hot alkali or extraction with denaturing solvents. In general, repeated autoclaving yields a reasonably pure product from adult nuchal ligament, but not from foetal ligament, nor from aorta, lung or skin. With these more complex adult tissues, boiling for one or more 45 minute periods in 0.1 N NaOH, was shown to give quantitatively reproducible yields of an apparently purified elastin, with a characteristic amino acid composition (Lansing, 1951). This product was for many years the "gold standard", against which the purity of other preparations was judged (Jackson and Cleary, 1967). Fortunately, from most tissues, this residue has a composition very similar to that of the soluble precursor form of elastin, which has been isolated and is called tropoelastin.

The insoluble elastins produced by most isolation procedures were found to have undergone some degree of hydrolysis or they have residual peptide contamination, as indicated by an increase in endgroups (Rasmussen *et al.*, 1975). Foetal tissues gave reduced yields as a result of increased amounts of hydrolysis as well as being highly contaminated with polar groups. The complexity of the contaminants within a given tissue, from which the elastin is to be isolated, is thus a determinant of the optimal isolation procedure (Cleary *et al.*, 1967). A combination of treatment with collagenase and extraction with chaotropic or reducing agents, and digestion with cyanogen bromide was found to give acceptably purified insoluble elastin residues in many tissues (Rasmussen *et al.*, 1975).

Prolonged digestion at 45°C in anhydrous (89%) formic acid was also shown to yield a purified elastin residue, but this treatment resulted in dissolution of the insoluble elastin, at a rate of about 2–3% of the dry weight of product per 24 hours of digestion, during the purification process (Hass, 1942). An advantage of this isolation method was that it could be applied to whole tissues, of reasonably large dimension (such as 8–10 cm cylindrical segments of artery or vein), to yield an elastin residue that maintains the morphology and dimensions that it exhibited in the original tissue before digestion (Ayer *et al.*, 1958). Thus, this procedure has been useful for studying the 3-dimensional morphology of elastic tissue in viscera and vessels at macroscopic, light and scanning electron microscopic levels.

Elastin and tropoelastin have a most unusual amino acid composition. Elastin is an extremely non polar protein with a content of alanine that is about 25% greater than that of any other known protein and it has one of the highest valine contents (17%) of any protein. About 1/3 of the amino acids are glycines and about 1/9 are prolines, but unlike collagen it has only a small (~1–2%) content of hydroxyproline (Sandberg *et al.*, 1981). It contains no tryptophan, and this was used as the basis for determining when hot alkaline purification was complete (Lansing, 1951). The degree of hydrophobicity of elastin is a critical determinant of its rubber-like physical properties. Insoluble elastin is also characterized by containing a number of lysine-derived covalent crosslinks. These include fluorescent, pyridinium compounds derived from 4 lysine residues, called desmosine and isodesmosine (Thomas *et al.*, 1963). These products are found only in elastin (with the exception of egg shell membrane) and have been used as the basis of assays for elastin and elastin degradation. Such assays are based on the assumption that the content of desmosines in elastin is constant, which for some tissues and ages is only an approximation (King *et al.*, 1980). Crosslinks derived from 2- and 3-lysine derivatives respectively are also present within the insoluble elastin residue, but in lesser amounts. These crosslinks form, post-translationally and extracellularly, under the influence of an enzyme called lysyl oxidase (Pinnell and Martin, 1968), from lysine residues located within alanine-rich regions of the elastin molecule. These regions are interspersed between hydrophobic peptides within the elastin molecule.

Soluble Forms of Elastin

Historically, the soluble precursor form of elastin was very difficult to isolate. Partridge *et al.*, noting the relatively small content of aspartic and glutamic acid in insoluble elastin, correctly predicted that hydrolysis of the acid-labile peptide bonds involving these residues would solubilize this cross-linked form of elastin and yield relatively large elastin peptides (Partridge *et al.*, 1955). The lyophilized product was a fine pale yellow powder, which they called α-elastin. It was soluble in water and exhibited the property of heat coacervation. This meant that with modest increases in temperature or ionic strength, α-elastin was shown to undergo a process of reversible phase separation to form a rubbery mass which shared many physical properties in common with elastin (Wood, 1958). Subsequently, tropoelastin was also shown to undergo heat coacervation — as do synthetic, hydrophobic, polypentapeptide sequences that have been shown to occur frequently in tropoelastin, so that the process is seen to have relevance to fibre formation *in vivo* (Urry *et al.*, 1974). Coacervation of α-elastin has been shown, ultrastructurally, to involve formation of fine fibres (Cleary and Cliff, 1978; Gotte *et al.*, 1974).

Biosynthesis of Elastin

Copper-deficient young pigs were shown, many years ago, to die suddenly due to aortic rupture. Extracts of the aorta of such pigs were shown to contain soluble material, that could be made to coacervate and which proved to have an amino acid composition similar to insoluble elastin, with the exception that the lysine content was higher and the desmosine crosslinks were absent (Sandberg *et al.*, 1969). This protein, of about 68 kDa, was called tropoelastin, because it was seen to be the soluble precursor form of insoluble elastin. Failure to form covalent crosslinks in the absence of copper was considered to be responsible for its extractability. Tropoelastin is synthesized in the rough endoplasmic reticulum of a range of fibroblastic cells, including aortic smooth muscle cells, myofibroblasts, fibroblasts of ligamentum nuchae, lung and skin, as well as by endothelial cells and chondrocytes of elastic cartilage. It undergoes only relatively minor post-translational modification before it is deposited in the pericellular environment, in relation to previously formed microfibrils, to become amorphous elastic material. A minor proportion (3–8%) of its proline residues are hydroxylated but tropoelastin is not glycosylated. Thus it is not able to use sugar residues as an intracellular trafficking signal. However, there is evidence pointing to the existence of a 67 kDa transporter protein which binds avidly to the tropoelastin within the cell and subsequently acts as a transmembrane cell surface receptor for the tropoelastin, which may in turn serve to facilitate interaction with the highly glycosylated microfibrillar components (Mecham and Heuser, 1991). Interaction between the tropoelastin molecule and the microfibrils seems likely to be served by two cysteine residues, unique to the carboxy-terminal end of the tropoelastin molecule (Cecila *et al.*, 1985). These residues could react with the acidic, cysteine-rich microfibrillar glycoprotein, MAGP (Gibson *et al.*, 1986), which has been shown to bind tropoelastin *in vitro* (Brown-Augsburger *et al.*, 1994).

The rate of tropoelastin synthesis has been shown, in the lung and nuchal ligament of foetal sheep and in chick aorta to be closely correlated with the level of messenger RNA (mRNA) for elastin in those tissues. It seems likely, therefore, that the rate of synthesis of tropoelastin is controlled by the mRNA steady-state level (Burnett *et al.*, 1982; Davidson *et al.*, 1984).

Tryptic peptide analysis of tropoelastin, showed that the molecule contained somewhat variable segments rich in hydrophobic amino acids, and highly conserved alanine-rich segments, which contained one or two lysine residues and which were thought to be involved in cross-link formation. These studies allowed Gray *et al.* to predict that tropoelastin was made up of alternating hydrophobic and cross-linking domains. The alanine-rich regions were postulated to form α-helices, from the same side of which pairs of lysine residues projected, thereby facilitating intrachain condensations between oxidised lysines and subsequent condensations with interchain lysines to form desmosine and isodesmosine (Gray *et al.*, 1973).

These unique covalent crosslinks, derived from lysine, are formed in the extracellular space under the influence of a copper-dependent enzyme, lysyl oxidase, which has been shown by immunoelectron microscopy to be bound on or near the elastin associated microfibrils in the pericellular environment (Kagan *et al.*, 1986). The initial step in the formation of the lysine-derived crosslinks is oxidative deamination of the ϵ-amino group of lysine side chains to form reactive aldehyde groups (α-amino adipic δ-semialdehyde, allysine), which then condense spontaneously with another suitably aligned aldehyde to form allysine aldol, or alternatively the aldol reacts with an ϵ-amino group of an appropriate lysine to form a Schiff base. Subsequently, desmosine and isodesmosine are formed by further condensation of lysine-derived aldehydes from a second elastin chain or alanine-rich segment (Eyre *et al.*, 1984). All except 5–8 of the 38 lysine residues of bovine tropoelastin are modified during elastin maturation, but not all the 26 residues that are oxidised can be accounted for as known crosslinks. The fate of the other oxidised lysine residues is not known, but there remains the possibility that there are other, as yet unidentified, cross links in insoluble elastin. Severe dietary copper deficiency will cause defective cross-linking, as does administration of agents with carbonyl functions, such as β-aminopropionitrile and semicarbazides (Levene *et al.*, 1992).

Cross-linked elastin becomes incorporated into the amorphous material of an elastic fibre, forming on a microfibrillar template. In several elastin-rich tissues in humans, such as aorta and lung, insoluble elastin has been shown to be a metabolically stable structural framework. Although there can be some additional minor elastin formation after the rapid growth phases associated with late uterine development and puberty, the evidence points to there being no appreciable further synthesis during adulthood, indicating that there is little remodelling of the elastin framework, certainly in lung (Shapiro *et al.*, 1991) and aorta (Powell *et al.*, 1992). There is however, clear evidence that other human tissues retain the capability to increase expression of the elastin gene and to deposit considerable amounts of new elastic tissue, given appropriate stimuli, cf. the response of dermal fibroblasts to actinic radiation, leading to solar elastosis of skin (Bernstein *et al.*, 1994)

The Elastin Gene

Molecular Cloning and Primary Structure of Tropoelastin

The complete primary structures of tropoelastin from human (Indik *et al.*, 1987), bovine ((Raju and Anwar, 1987), rat (Pierce *et al.*, 1990) and chick (Bressan *et al.*, 1987) have been determined by cDNA cloning. In each of the above species, the tropoelastin mRNA has been determined to be around 3.5 kb in size and to contain a coding region of between 2.28 kb (bovine) and 2.61 kb (rat). The encoded tropoelastin sequences contain a maximum of 786 amino acids in human, 760 amino acids in bovine, 870 amino acids in rat and 750 amino acids in chick. However, extensive alternative splicing of tropoelastin mRNA occurs in each species, resulting in many slightly smaller tropoelastin isoforms (see below). The calculated molecular sizes of the encoded isoforms are consistent with the observed values for tissue-isolated tropoelastins, which have molecular weights of 65–70 kDa (Indik *et al.*, 1990; Parks and Deak, 1990).

Tropoelastin was confirmed to consist almost entirely of two types of structurally distinct domains, (a) hydrophobic regions, rich in valine, glycine and proline, and (b) cross-linking regions, rich in alanine along with 2 or 3 lysine residues. The hydrophobic and cross-linking domains alternate, except at the ends of the molecule. Individual domains are relatively small (usually less than 30 amino acids) and tropoelastin contains 14–16 of each type. The hydrophobic regions, which often contain short repeat motifs of 3, 5 or 6 amino acids, show considerable species variation in sequence. For instance, the sequence VPGVG is found 11 times near the centre of bovine and chick tropoelastins, but this region is more irregular in the human molecule (Indik *et al.*, 1990). In rat elastin, this repeat sequence is replaced by IPGVG (Pierce *et al.*, 1990). The elastin-receptor binding sequence, VGVAPG, is found 7 times and 5 times in human and bovine tropoelastins respectively, but is absent from the rat and chick molecules. Overall, chick tropoelastin is the most divergent in structure, exemplified by the unique presence of the polytripeptide $(VPG)_{12}$ and the polyheptapeptide $(GGLV/APGV/A)_7$ (Bressan *et al.*, 1987). The extent of the species variation in tropoelastin structure indicates that rigid sequence conservation is not necessary for the function of the molecule. However, some of the repeat motifs have been shown to form unusual β-spiral structures which are considered to be important for the elastic properties of elastin (Urry, 1983). The cross-linking domains are more highly conserved, usually containing two lysine residues which are separated by two or three alanine residues.

Molecular cloning has also identified a highly basic sequence at the carboxyl-terminus of tropoelastin which is highly conserved between the species. This sequence, G G A C L G K A C G R K R K in human tropoelastin, is the only domain of the secreted molecule to contain cysteine residues (Indik *et al.*, 1990; Parks and Deak, 1990). As we have seen, this domain may be important during elastic fibrogenesis for the alignment of the tropoelastin molecule onto a scaffold of the elastin-associated microfibrils, perhaps through the cysteine-rich microfibrillar glycoprotein, MAGP (Gibson *et al.*, 1986).

The Structure of the Elastin Gene

All available evidence indicates that there is only one elastin gene in mammalian and avian genomes (Indik *et al.*, 1990). The human and bovine genes have been extensively characterized and their exon-intron boundaries have been determined (Bashir *et al.*, 1989; Indik *et al.*, 1990; Yeh *et al.*, 1989). In both species, individual hydrophobic and cross-linking domains of tropoelastin are encoded by separate exons (see Figure 4.2). The human gene contains 34 exons spanning around 40 kbp of genomic DNA whereas the bovine gene has 36 exons contained in a 38 kbp genomic sequence. The human gene lacks the exons, numbered 34 and 35 in the bovine gene. These encode, respectively, a cross-linking domain and a hydrophobic domain. Each exon is very small (27–186 bp) with the exception of the final exon containing the 3′ untranslated region, which is over 1 kbp in length. Thus the intron: exon ratio is very large, around 15:1. Each exon is a multiple of three nucleotides and the intron-exon boundary is characterized by the exon beginning and ending with 2/3 and 1/3 split codons, respectively. This facilitates alternative splicing of individual exons from the

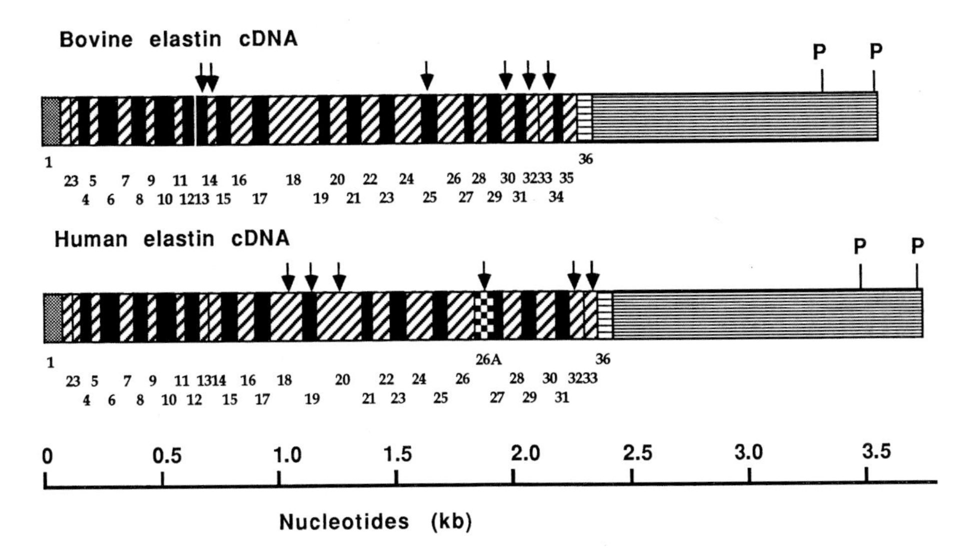

Figure 4.2 The exon structure of cDNAs for bovine and human elastin.

The exons have been categorized according to the nature of their encoded amino acid sequences and are numbered in three distinct tiers for ease of identification. Diagonal stripes: hydrophobic sequences. Black: crosslinking sequences. Grey: signal sequence. Checked: hydrophilic exon 26A. Horizontal stripes: exon 36, encoding the C-terminal sequence (coarse stripes) and containing a large untranslated region (fine stripes). Arrows indicate which exons may be removed from the elastin pre-mRNA by alternative splicing. The positions of polyadenylation signals (P) are also shown. (Adapted from Indik *et al.*, 1990, with permission from Academic Press.)

primary transcript, without disruption of the reading frame. An exon is either included or removed, in a cassette-like manner (see below).

Alternative Exon Usage in Tropoelastin mRNA

Evidence from cDNA sequencing of human, bovine, rat and chick elastin genes has indicated that a range of exons may be differentially spliced from the tropoelastin pre-mRNAs, mostly using the cassette-like mechanism (Bashir *et al.*, 1989; Baule and Foster 1988; Indik *et al.*, 1987; Pierce *et al.*, 1990; Raju and Anwar 1987; Yeh *et al.*, 1989). In human tropoelastin mRNAs, exons encoding 5 hydrophobic domains (exons 22, 24, 26, 32 and 33) and one cross-linking region (exon 23) have been found to be alternatively spliced (Bashir *et al.*, 1989; Indik *et al.*, 1987). A different exon usage pattern occurs in bovines where exons encoding four hydrophobic sequences (exons 14, 30, 32 and 33) and two cross-linking regions (exons 13 and 27) have been found to be alternatively deleted in some transcripts (Raju and Anwar, 1987; Yeh *et al.*, 1989). In the rat, tropoelastin mRNA sequences corresponding to bovine exons 13, 14, 15 and 33 are differentially utilized (Heim *et al.*, 1991). Further analysis of DNA sequences flanking these exons indicated that they contained splicing signals which were distinct from those around constitutively utilized exons (Pierce *et al.*, 1992a). However the mechanism of selection of exons for deletion remains to be elucidated.

The frequency of splicing of individual exons appears to be developmentally regulated, in most cases. Splicing events in bovine tropoelastin mRNA have been quantitated, using an S1 nuclease digestion assay, at several stages in the development of the elastin-rich nuchal ligament, from the 175-day foetus to the adult (Yeh *et al.*, 1989). Tissue-extracted mRNA was hybridized with an appropriate end-labelled tropoelastin cDNA probe and splicing events were detected by enzymatic cleavage of the probe at the sites of exon deletion. The proportion of each splicing event was quantitated by densitometry of the different sized fragments of the cDNA probe, separated by gel electrophoresis. Four exons (13,14, 27 and 33) were found to be alternatively spliced to a significant extent in foetal and adult tissue. Exon 33 was found to be most frequently spliced, at a relatively constant rate (around 50%) at all ages. The other three exons were deleted at a frequency of below 20%. The splicing out of exons 13 and 14 appears to peak at 210-days of foetal development, whereas exon 27 is most often removed at around 270 days. In the adult, significant splicing of five additional exons was observed (Yeh *et al.*, 1989). Developmental differences in alternative splicing of tropoelastin mRNA have also been observed in chick aorta (Pollock *et al.*, 1990). The extensive alternative splicing of the mRNA explains the variation in size and number of tropoelastin isoforms found in tissue extracts and cell cultures from a range of tissues during development (Foster *et al.*, 1980; Parks *et al.*, 1988; Rich and Foster, 1987; Wrenn *et al.*, 1987).

The functional significance of the alternative splicing is not yet known. It is likely that variation in the number and position of cross-linking sequences in tropoelastin affects the tightness of the elastin polymer formed within a tissue and influences the shape of the resulting elastic fibre. Distinct tropoelastin isoforms may also be involved at the different stages of fibrogenesis, during which the elastic fibre grows from a template of glycoprotein microfibrils into a structure consisting predominantly of polymeric elastin (see below). Perhaps regions of tropoelastin necessary for interaction with, or alignment along, the microfibrils are later substituted with sequences required for lateral aggregation of tropoelastin, as the fibre grows in diameter. In turn these sequences may be replaced by domains which interact with macromolecules of the surrounding extracellular matrix to limit the size of the fibre. This idea is supported by findings by Pollock *et al.* (1990) who monitored the expression of a specific tropoelastin isoform, which contains a sequence corresponding to an alternatively-spliced exon, during the development of embryonic chick aorta. Expression of the isoform did not increase during the period of rapid elastic fibrogenesis. But it appeared to be located preferentially at sites of elastin interaction with the microfibrils, suggesting that this isoform plays a specific role in alignment of the nascent elastin onto these microfibrils (Pollock *et al.*, 1990).

The Promoter Region of the Elastin Gene

The 5′ flanking sequence of the elastin gene has been studied in human and bovine to determine which promoter sequences are important for initiation and control of transcription (Bashir *et al.*, 1989; Manohar *et al.*, 1991; Yeh *et al.*, 1989). The promoter region is over 80% homologous between the two species and it contains several features unusual in a highly-regulated gene. The region is very rich in G and C residues, with a high frequency of CpG dinucleotides characteristic of ‘CpG islands’. Such ‘CpG islands’ are found at the 5′ end of many genes, including collagens and fibrillin (Corson *et al.*, 1993; Sandell and Boyd, 1990). They are considered to be regions protected from methylation by bound proteins and are common in constitutively-expressed genes, but less often in genes controlled in a tissue-specific manner (Larsen *et al.*, 1992). There is no obvious canonical TATA consensus sequence required for the initiation of transcription in a wide range of genes, although a TATA-like sequence, ATAAAA, is present about 40 bp upstream of the ATG codon. The lack of a TATA box is characteristic of genes which have multiple sites of initiation of transcription. This has been shown to be the case with the human elastin gene, at least in foetal aorta, where three major clusters of initiation were identified, by S1 nuclease mapping and primer extension experiments (Bashir *et al.*, 1989). The significance of the different initiation sites is unknown, but it is possible that they are linked to distinct patterns of alternative splicing of the elastin message.

A number of putative cis-regulatory elements have been identified in the elastin gene promoter (see Figure 4.3), indicating that control of its transcription is likely to be complex. In the human gene, the 2200 bp sequence immediately upstream of the translation initiation site contains several possible SP-1 and AP2 binding sites, three glucocorticoid-responsive motifs and sequences which resemble cAMP-responsive and phorbol ester-inducible elements. Two CAAT sequences also occur but their positions suggest that they are non-functional (Bashir *et al.*, 1989; Kahari *et al.*, 1990). Delineation of cis-acting regulatory elements has been achieved by transfection of rat aortic smooth muscle cell cultures with chimeric gene constructs (Kahari *et al.*, 1990). These were made by coupling various 5′ flanking sequences of the elastin gene to the chloramphenicol acetyl transferase (CAT) gene which was used to measure the promoter activity. The basic promoter region was found to reside close to the initiation codon (between –128 to –1). Negative, down-regulatory activity was detected in two sequences, between positions –2260 to –1554 and –986 to –476. The region between –495 and –129 exhibited strong up-regulatory activity, which could be explained by the presence of multiple SP-1 and AP2 binding sites, which

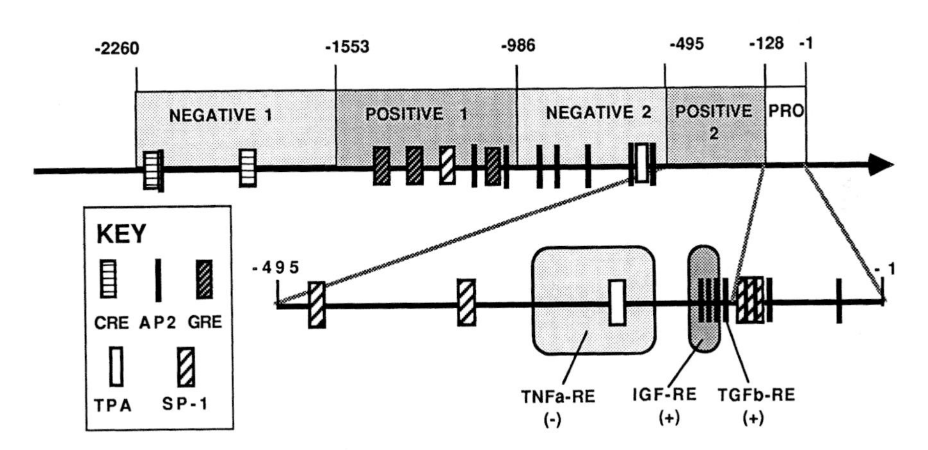

Figure 4.3 The structure of the elastin gene promoter.

The region of the human elastin gene (2260 base pairs) immediately upstream of the ATG translation initiation site is shown. The basic promoter (PRO) is found between positions –128 to –1. Two predominantly upregulatory (positive 1 & positive 2) and two predominantly down regulatory (negative 1 & negative 2) regions are present further upstream (Bashir *et al.*, 1989; Kahari *et al.*, 1990). Present are several possible SP-1 and AP2 binding sites, 3 putative glucocorticoid-responsive motifs (GRE), and sequences which resemble cAMP-responsive (CRE) and phorbol ester-inducible (TPA) elements (see key). The region corresponding to bases –495 to –1 has been expanded to illustrate the positions of a negative regulatory element responsive to tumour necrosis factor-α (TNFα-RE), and positive elements controlled by insulin-like growth factor (IGF-RE) and transforming growth factor-β (TGFβ-RE). (Kahari *et al.*, 1990; Marigo *et al.*, 1994; Wolfe *et al.*, 1993).

can act as cis-acting enhancer elements. Interestingly, the promoter constructs were also shown to be active in cells which do not normally express elastin, including HeLa and NIH-3T3 cells. This suggested that some elements necessary for the tissue-specific control of elastin gene expression are not present in the 2.2 kb 5′-flanking region.

Modulation of Elastin Gene Expression by Soluble Signal Molecules

Elastin gene expression has been shown to be modulated by a variety of growth factors, hormones, vitamins and cytokines. This indicates that control of the gene is likely to be complex and sensitive to soluble signal molecules during processes such as tissue development, inflammation and wound healing. Factors which can stimulate elastin gene expression include: transforming growth factor-β (TGF-β) (Marigo *et al.*, 1993; Marigo *et al.*, 1994), insulin-like growth factor-I (IGF-I) (Rich *et al.*, 1992; Wolfe *et al.*, 1993), interleukin-1β (IL-1β) (Mauviel *et al.*, 1993), glucocorticoids (Mecham *et al.*, 1984c) and retinoic acid (Liu *et al.*, 1993). In each case, the major mechanism appears to be up-regulation of elastin gene transcription.

TGF-β has been shown to stimulate elastin synthesis by a variety of cell types in a tissue specific manner (Kahari *et al.*, 1992b; Liu and Davidson, 1988). The involvement of a transcriptional control mechanism has been shown by analysis of chick embryo aortic cells, transfected with various human elastin promoter/CAT reporter gene constructs (Marigo *et al.*, 1993). Following addition of TGF-β, a maximum increase in expression of 2–4 fold was observed. In chick tendon fibroblasts, which normally make little elastin, no stimulation of the elastin promoter by TGF-β was observed. The TGF-β control element was identified by DNase footprinting with the sequence –138 to –127 of the human elastin promoter (Marigo *et al.*, 1994). The region shared homology with AP2 binding sites, but was shown not to bind AP2. Interestingly, the sequence was very similar to a new TGF-β activation element identified in the rat α1(I) collagen promoter, suggesting that the effect of TGF-β on type I collagen and elastin may be mediated by the same factors. However gel shift assays showed that the TGF-β responsive element of the elastin promoter bound proteins in the aortic cells which were different from those bound in tendon cells. This suggested that a complex mechanism is involved in the tissue-specific control of this element. It may explain why, in another study, TGF-β stimulation of elastin synthesis in human skin fibroblasts was found to be predominantly under a different, post-transcriptional control mechanism (Kahari *et al.*, 1992b).

TGF-β has also been found to stimulate elastin production in porcine vascular smooth muscle cells and skin fibroblasts (Davidson *et al.*, 1993). In this study the mitogenic cytokines, basic fibroblast growth factor (bFGF) and transforming growth factor-a (TGF-a) were shown to be potent antagonists of this TGF-β 1-mediated stimulus.

IGF-I has been shown to result in a 5-fold increase in steady-state tropoelastin mRNA levels when added to cultures of neonatal rat aortic smooth muscle cells (Wolfe *et al.*, 1993). Transient transfections of the cells with human elastin promoter/CAT constructs identified the IGF-I responsive area to sequences between –195 and –136 which contains a negative-regulatory element. Using gel retardation assays and footprint analysis it was demonstrated that IGF-I causes the dissociation of a binding protein from the 28 bp sequence, –165 to –137, which is distinct from other known regulatory elements. Thus IGF-I appears to up-regulate elastin gene expression in aortic smooth muscle cells by a mechanism of derepression. Interestingly, in a related study, lung fibroblasts did not exhibit an elastogenic response to IGF even though the cells were shown to possess an IGF-I receptor. In addition no increase in reporter activity was found when the cells were transfected with the elastin promoter/CAT gene constructs. Thus IGF-I stimulation of the elastin gene appears to occur in a tissue-specific manner (Rich *et al.*, 1993).

Human recombinant IL-1β has been demonstrated to elevate, 3–4 fold, elastin mRNA steady-state levels when added to cultures of human skin fibroblasts (Mauviel *et al.*, 1993). The cytokine caused a similar stimulus when applied to cells incorporating human elastin promoter/CAT gene constructs. Thus IL-1β also appears to increase elastin production by up-regulation of transcription. However, an IL-1β regulatory element in the elastin promoter has yet to be identified.

Soluble signal molecules which repress tropoelastin expression include tumour necrosis factor-α (TNF-α), the phorbol ester tetradecanoylphorbol acetate (TPA), interferon-γ and vitamin D3 (Kahari *et al.*, 1992a; Pierce *et al.*, 1992b). TNF-α and TPA have been demonstrated to regulate the transcription of a number of genes by activating protein kinase C. This leads to a transient increase in synthesis of c-jun and c-fos heterodimers, which bind to AP-1 recognition elements of responsive gene promoters (Curran and Franza, 1988). Recently TNF-α and TPA has each been shown to reduce dramatically the abundance of elastin mRNA, when added to cultured human skin fibroblasts and rat aortic smooth muscle cells. Similar down-regulatory effects of TNF-α and TPA were observed on human elastin promoter/CAT gene constructs, which had been transfected into these cell types. This indicated that the factors exerted their effects at the transcriptional level. Analysis of partially deleted constructs indicated that the TNF-α regulatory element was located in the region between –290 and –198 of the human elastin promoter, which contains an AP-1 binding site. Further experiments confirmed that the mechanism of the TNF-α activity does involve the up-regulation the c-jun and c-fos components of transcription factor AP-1, which in turn complexes with the above AP-1 binding site to suppress elastin gene expression. In the same study interferon-γ was also found to down-regulate elastin gene expression in human skin fibroblasts, but the effect appeared to be due to a post-transcriptional method of regulation (Kahari *et al.*, 1992a).

In another study, TPA addition was shown to cause a 10-fold decrease in tropoelastin mRNA in bovine ear cartilage chondrocytes. However, the compound had no influence on the transcription of a human elastin promoter/CAT construct, following its transfection into the cells. Thus in this instance, TPA appears to be acting via a distinct, post-translational mechanism. TPA has been shown to regulate some genes by causing increased degradation of specific mRNAs, and this may be the mechanism occurring here (Parks *et al.*, 1992). In a related study, vitamin D3 was also shown to down-regulate elastin gene expression in the chondrocytes and in rat lung fibroblasts, by an unidentified post-translational mechanism. The authors postulated that the down-regulation of tropoelastin by vitamin D3 is controlled by decay of the mRNA and that this may involve cis-acting sequences in its 3′ untranslated region, which is highly conserved between species (Pierce *et al.*, 1992b).

Congenital Disease and the Elastin Gene

Until very recently a congenital disorder had not been conclusively linked to a primary defect in the elastin gene, despite extensive study into genetic disorders such as cutis laxa and pseudoxanthoma elasticum (Neldner, 1993; Uitto *et al.*, 1993). However, there is now convincing evidence for linkage of the elastin gene to a relatively uncommon disorder, supravalvular aortic stenosis (SVAS). This syndrome is characterized by haemodynamically significant narrowing of large elastic blood vessels, such as the aorta and the pulmonary, coronary and carotid arteries. The condition is associated with progressive hypertrophy and hyperplasia of smooth muscle cells in the intimal layer and eventual fibrosis. If untreated, SVAS may cause myocardial hypertrophy, heart failure and death. The disease has an incidence of around 1 in 25,000 live births and is inherited in an autosomal dominant manner. SVAS is often associated with Williams syndrome, which has extended features including additional connective tissue defects and mental retardation. Interest centred on the elastin gene as a candidate for the SVAS gene when linkage analysis localized the latter to the long arm of chromosome 7 (Ewart *et al.*, 1993b; Morris *et al.*, 1993; Olson *et al.*, 1993). The locus of the human elastin gene had previously been identified on this chromosome at 7q11.1–21.1 (Fazio *et al.*, 1991). Very recently two separate families of SVAS have been associated with a translocation (Curran *et al.*, 1993) and a deletion (Ewart *et al.*, 1994), respectively, involving the 3′ end of the elastin gene.

A mutation involving an allelic deletion of the entire elastin gene has also been identified with Williams syndrome (Ewart *et al.*, 1993a). While the deletion associated with Williams syndrome appears to include other genes, the SVAS deletion is much smaller (100 kb) and appears to involve only the elastin gene — in a way that is critical for vascular development (Ewart *et al.*, 1993a). The 5′ break point of the SVAS deletion was localized to the intron 27 (between exons 27 and 28) with the remainder of the 3′ end of the elastin gene being deleted. Interestingly, 300 bp "Alu repeat" sequences were identified at breakpoints in intron 27 and at the 3′ end of the mutation. It has been suggested that repetitive elements, such as "Alu repeats", are involved in the mechanism of deletion (Calabretta *et al.*, 1982) and it is likely that they are involved in this instance. "Alu repeats" are relatively abundant in the elastin gene, where they occur at about four times the expected frequency, and it had previously been predicted that this situation might result in genetic instability of the human elastin gene, within the population (Indik *et al.*, 1987; Indik *et al.*, 1990).

It is yet to be established that the mutant SVAS gene is expressed as a truncated tropoelastin, which lacks the carboxyl-terminal region containing two cross-linking domains and a putative microfibril-binding site. It is possible that the deletion results in a non-functional mRNA being transcribed from one allele, due to the concurrent deletion of the polyadenylation signal sequences in the 3′ untranslated region. Polyadenylation is necessary for the normal processing of the mRNA and lack of the signal sequences would presumably lead to a large reduction in the production of tropoelastin. This view is supported by the finding of SVAS in affected members of a Williams syndrome family, in which one allele of the elastin gene is completely deleted (Ewart *et al.*, 1993a).

ELASTIN-ASSOCIATED MICROFIBRILS

Introduction

An extensive and detailed review of the literature relating to early attempts to isolate and characterize the components of elastin associated microfibrils has been undertaken previously (Cleary and Gibson, 1983). Most of these studies were carried out on the nuchal ligaments of third trimester, foetal bovine calves, as this age was shown to correspond with the period of most rapid elastin deposition in this developing structure (Cleary *et al.*, 1967). Tissues were mostly extracted with strong guanidine solutions, in the presence of a reducing agent, along with a cocktail of protease inhibitors. This combination of reagents had been shown by electron microscopic monitoring to be most effective in extracting microfibrils (Ross and Bornstein, 1969). Further advances followed on the demonstration that these extracts, which contain a complex mixture of glycoproteins, could elicit production of antibodies which localized relatively specifically to elastin associated microfibrils (Kewley *et al.*, 1977). Subsequent work in the authors' laboratory showed that specific microfibril-related immunoreactivity resided in both small and large molecular weight components of such reductive guanidine extracts. The small molecule weight component was isolated and characterized as MAGP, a 31 kDa glycoprotein (Gibson *et al.*, 1986). MAGP has since been cloned (Gibson *et al.*, 1991) and the mouse and bovine genes have been characterized (Bashir *et al.*, 1994; Chen *et al.*, 1993). Latest evidence suggests that MAGP acts as a tropoelastin-binding protein on the surface of the microfibrils and may play a role in cross-linking of these structures (Brown-Augsburger *et al.*, 1994) (see below).

Antibodies specific for a 340 kDa microfibrillar protein (MP 340) were also shown, by immunoelectron microscopy, to relate specifically to elastin associated microfibrils, but this protein was best isolated from reductive saline extracts of developing elastic tissues. These reductive saline extracts also contained three other microfibril-related polypeptide species, with apparent molecular weights of 78000, 70000 and 25000, which were thus named MP78, MP70 and MP25, respectively (Gibson *et al.*, 1989). They have been shown to be structurally distinct from other known microfibrillar proteins.

Sakai and coworkers produced monoclonal antibodies against extracts of pepsin-treated human placenta and showed that one of these recognized elastin-associated microfibrils in human skin as well as 12 nm microfibrils in non-elastic tissues. Immunoprecipitation with this antibody, from human skin fibroblast cultures, yielded a 350 kDa glycoprotein, which was named fibrillin (Sakai *et al.*, 1986). Subsequently, fibrillin has been shown to be a major structural component of the 12 nm microfibrils (Maddox *et al.*, 1989; Sakai *et al.*, 1991). Fibrillin has been cloned and its gene has been partly characterized (Corson *et al.*, 1993; Maslen *et al.*, 1991; Pereira *et al.*, 1993). This fibrillin is now known as fibrillin 1, following the recent discovery of a gene for a closely related glycoprotein, which has been named fibrillin 2 (Zhang *et al.*, 1994) (see below). Mutations in the genes for fibrillin 1 and fibrillin 2 have been linked to the congenital disorders, Marfan syndrome and congenital contractural arachnodactyly (CCA), respectively (Lee *et al.*, 1991; Maslen and Glanville, 1993). Amino acid sequences from peptides of bovine MP340 have been shown to be homologous with sequences found in fibrillin 1, indicating that MP340 is the bovine form of fibrillin (Gibson, M.A. and Cleary, E.G., unpublished data).

Recently, cDNA for another structurally-related protein, provisionally named fibrillin-like protein (FLP), has also be described. Antibodies to FLP sequences were shown to be immunoreactive with elastin-associated microfibrils, suggesting that FLP is the third member of the fibrillin family of glycoproteins (Gibson *et al.*, 1993). Several other proteins have been isolated and characterized and antibodies specific for these proteins have been shown to localize to the 12 nm diameter microfibrils of mammalian elastic tissues as well as to ocular zonule in both chicks and mammals. These include the recently cloned 57 kDa protein, named associated-microfibril protein (AMP) (Horrigan *et al.*, 1992) and elastin microfibril interface located protein or emilin (previously known as GP115). Emilin, as its name indicates, was localized at the junction between the microfibrils and the elastin core of elastic fibres in a range of elastic and microfibrillar tissues in chicks (Bressan *et al.*, 1993).

It will be apparent that the composition of the elastin-associated, and related, microfibrils is very complex. It is proposed now to review what is known of the distribution, structure and function of these components, including an examination of the changes in structure and function that have been related to genetic diseases of the individual component proteins. The interrelationships between the individual components within microfibrils in different tissues will be reviewed, along with data relating to the role of microfibrils and its component molecules in elastic fibrogenesis.

Fibrillins

Tissue Distribution

The tissue distribution and ultrastructural location of fibrillin 1 have been determined, using monoclonal antibodies recognising pepsin-resistant fragments of the protein. Fibrillin 1 was identified with microfibrils, 10–14 nm in diameter, in the extracellular matrix of a wide range of tissues, either as components of elastic fibres or as elastin-free bundles (Gibson *et al.*, 1989; Keene *et al.*, 1991; Sakai *et al.*, 1986; Wallace *et al.*, 1991). Immunoreactive tissues included skin, aorta, ear cartilage, bronchioles and alveolar connective tissue of the lung, mesangium of renal glomerulus, muscle perimyseum and endomyseum, tendon, placenta and ocular tissues such as Descemet's membrane of the cornea and the suspensory ligament of lens, the ciliary zonule. Immunoelectron microscopic examination revealed that the monoclonal antibodies bound to the microfibrils with a periodicity of 67 nm, indicating that fibrillin occurs as ordered aggregates within these structures (Sakai *et al.*, 1986).

Antibodies have been developed to recombinant peptides from the glycine-rich, region C (the so-called hinge region) of fibrillin 2 and to peptides from the related, but proline-rich region of fibrillin 1. These have been shown to co-localize to typical elastin associated microfibrils in nuchal ligament and aorta of the developing bovine embryo. However, there was differential staining, by immunohistological assessment, in a number of locations. In foetal bovine aorta, anti-fibrillin 1 localized across all three layers of the wall, while staining with anti-fibrillin 2 antibody was more intense in the media, where elastic fibres are most abundant. In elastic ear cartilage, fibrillin 1 staining was strongest in the connective tissue adjacent to the central elastic cartilaginous core, but the core itself was unstained. By contrast this core region stained strongly with anti-fibrillin 2 antibody. These differences in distribution in the ear cartilage were considered to be consistent with the flattening of the helix and the crumpled appearance of the antihelix of the ear cartilage in patients with congenital contractual arachnodactyly. There were other differences in immunohistostaining with the two antibodies, notably in hyaline cartilage and in the type II collagen-rich region of Descemet's membrane in the cornea. Unfortunately electron microscopic data relating to most of these locations, especially the latter differences, were not presented so that it was not confirmed that the localization in these tissues is microfibrillar. It is thus difficult to evaluate the basis of the differences in staining, at this time. On balance, there appears to be sufficient evidence to support the conclusion reached by the authors, that the two fibrillin molecules have distinct, but related, functions in the formation and maintenance of the extracellular elastin associated microfibrils of elastin rich tissues. Their postulate that microfibrillar aggregates (with or without amorphous elastic tissue) function to influence biomechanical properties to a variety of tissues and organ systems is also consistent with the findings (Zhang *et al.*, 1994).

Shape of the Fibrillin Molecule

The shape of the fibrillin molecule has been studied by rotary shadowing in the electron microscope and compared with that of the 12 nm microfibrils of zonular and elastic tissue. Three structurally distinct fibrillin fragments were purified from pepsin digests of human placenta. Two of these, PF1 and PF2, were visualised as short rods. Rotary shadowed preparations of the third, larger fragment, PF3, revealed a unique structure composed of a dense central globular domain with several protruding arms. The results suggested that fibrillin was, at least in part, a rod-shaped molecule and that these rods formed aggregates connected via bead-like structures (Maddox *et al.*, 1989). Similar structures have since been identified in intact microfibrils (Keene *et al.*, 1991; Wallace *et al.*, 1991). Fibrillin 1 has also been purified, in intact form, from fibroblast cultures. The monomer that was isolated was shown, in rotary shadowed images, to be an extended flexible rod approximately 148 nm long and 2.2 nm wide (Sakai *et al.*, 1991).

Molecular Structure of Fibrillins

The amino acid sequences and domain structures of human fibrillins 1 and 2 have been determined by cDNA cloning (Corson *et al.*, 1993; Maslen *et al.*, 1991). Messenger RNA for fibrillin 1 (9.7 kbs) encodes a polypeptide of 2,871 amino acids with a predicted molecular mass of 347 kDa (Pereira *et al.*, 1993). These figures compare well with the previously estimated sizes of the fibrillin mRNA (10 kb) and of fibrillin protein (350 kDa) (Maslen *et al.*, 1991; Sakai *et al.*, 1986).

The domain structure of fibrillin 1 (see Figure 4.4) was shown to consist almost entirely of 56 cysteine-rich repeat structures, which explains the very high levels of cysteine found in hydrolysates of the purified protein (Gibson *et al.*, 1989; Sakai *et al.*, 1991). Forty-seven of the repeat motifs, termed EGF-like repeats, contain 6 cysteine residues and have homology with similar motifs found in epidermal growth factor (EGF) precursor and a range of other proteins including extracellular matrix components, blood coagulation factors and developmentally important agents (Corson *et al.*, 1993; Pereira *et al.*, 1993). Interspersed among this series of EGF-like repeats is a single sequence of 9 cysteine residues (which is unique to fibrillin) and eight examples of 8-cysteine motifs, previously found only in TGF-β1-binding protein (Kanzaki *et al.*, 1990). The 8-cysteine motifs are of two types. One is structurally similar to the 9-cysteine motif and the remainder, termed TGF-BP repeats, contain the unique CCC tripeptide sequence. Series of EGF-like repeats have been shown to correspond to extended arm-like regions in other proteins. Thus the predicted domain structure of fibrillin 1 indicates that the molecule is rod-like and this correlates well with its shape observed by rotary shadowing (Sakai *et al.*, 1991).

Interestingly, 43 of the EGF-like motifs also contain a sequence which has consensus for both hydroxylation of a specific asparagine or aspartate residue and for calcium binding (Corson *et al.*, 1993; Pereira *et al.*, 1993). The role of the bound calcium in EGF-like

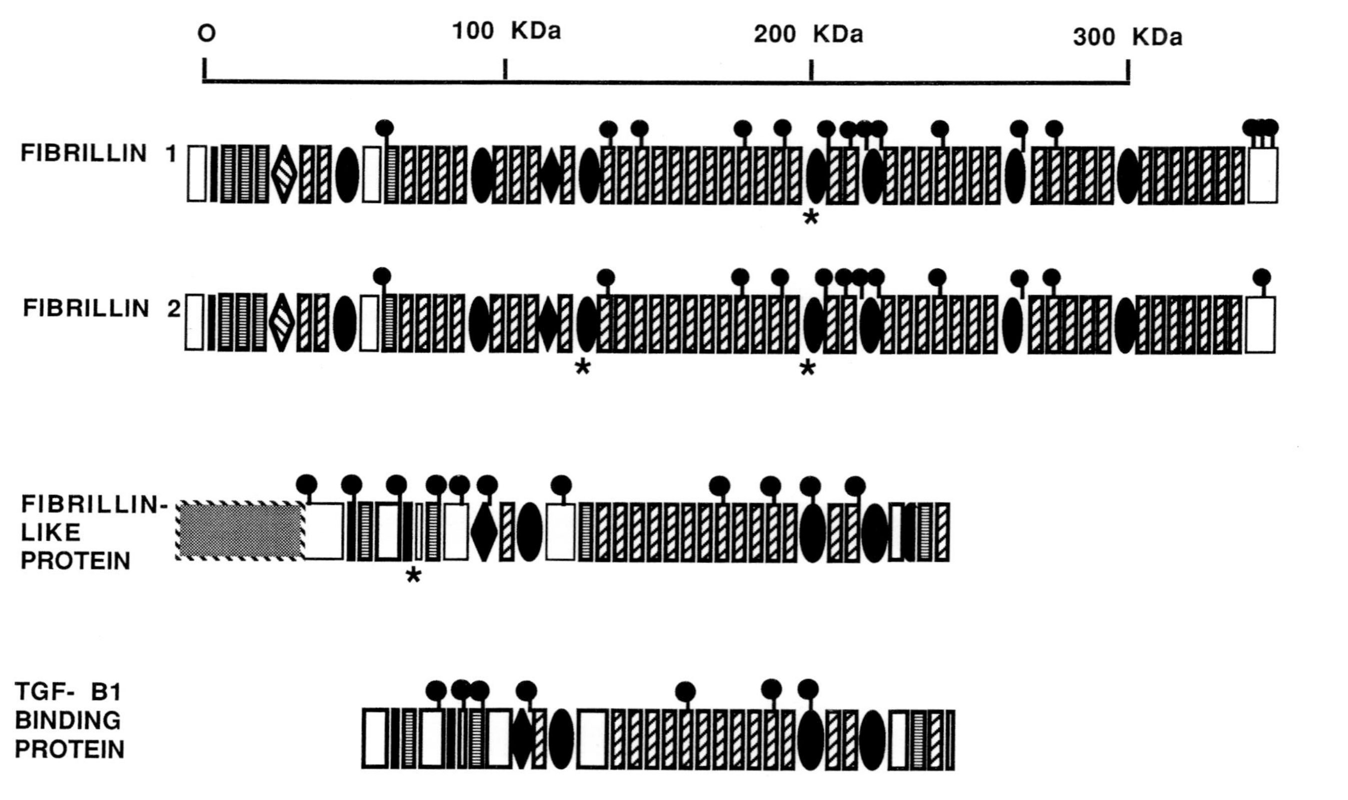

Figure 4.4 The domain structure of fibrillins and related proteins.

Fibrillins are large, rod-shaped proteins consisting predominantly of multiple cysteine-rich repeat sequences. Human fibrillins 1 and 2 are compared to the structurally similar proteins, bovine fibrillin-like protein (FLP) and rat TGF-β- binding protein as elucidated from cDNA clones (Gibson *et al.*, 1993; Pereira *et al.*, 1993; Tsuji *et al.*, 1990; Zhang *et al.*, 1994).

Horizontally-striped rectangles: EGF-like motifs lacking calcium-binding consensus. Diagonally-striped rectangles: EGF-like motifs with calcium-binding consensus. Black ovoids: 8-cysteine repeats. Black diamonds: Hybrid 8-cysteine motifs intermediate in structure between the EGF-like repeats and the 8-cysteine repeats. Diagonally-striped diamond: 9-cysteine motif. Black rectangles: 4-cysteine motifs. Half-ovoid: unique FLP 4-cysteine sequence. White boxes: cysteine-free regions. Shaded box: region of FLP with unknown structure. Black circles: potential N-glycosylation sites. Stars: RGD (potential cell-binding) sequences.

repeats is not fully understood but, in proteins such as notch and factor IX, the cation is necessary for specific protein-protein interactions (Handford *et al.*, 1991). It is unclear whether the calcium is involved in these interactions directly, or indirectly, by stabilizing protein confirmation. Recently, in separate studies, fibrillin has been demonstrated to bind calcium *in vitro* (Corson *et al.*, 1993) and the morphology of fibrillin-containing microfibrils has been shown to be grossly disrupted following incubation with calcium-chelating reagents, EDTA and EGTA. It was suggested the effect may be due to breaking of calcium bridges between adjacent EGF-like repeats involving lateral packing and alignment of fibrillin molecules (Kielty and Shuttleworth, 1993). It is also possible that binding of calcium to fibrillin provides the extra conformational rigidity needed to maintain its rod-like structure and that its removal results in a much more flexible molecule, leading to swelling of the microfibril.

It is evident that intermolecular disulphide cross-links are important for the stabilization of fibrillin 1 molecules within the microfibrils, since the protein can only be solubilized from microfibrils by disruption with reducing agents (Gibson *et al.*, 1989). In addition, the protein readily forms disulphide bonded aggregates *in vitro* (Sakai 1990). The identification of the cysteine residues involved in cross-link formation has been hampered by the large amount of this amino acid in fibrillin. However, major candidates for this role include 4 cysteines in the amino-terminal region, 2 cysteines in the carboxyl-terminal region and the 9-cysteine motif. The 8-cysteine motifs may also be involved, as the pairing pattern of their cysteines has yet to be elucidated. Fibrillin 1 contains 15 potential n-glycosylation sites, indicating that the protein is heavily glycosylated (Pereira *et al.*, 1993). Also present in the middle of fibrillin 1 is a single RGD sequence which appears to be the binding site for fibrillin 1 to nuchal ligament cells via the integrin, $\alpha_v\beta_3$ receptor complex (Sakamoto, H., Broekelmann, T., Cheresh, D.A., Ramirez, F., Rosenbloom, J. and Mecham, R.P. unpublished observations).

Analysis of the primary structure of human fibrillin 2 shows that the protein has almost the same size (2,889 amino acids) as fibrillin 1, and contains the same basic domain structure with an identical pattern of cysteine-repeat motifs (Zhang *et al.*, 1994) (Figure 4.4). Overall, there is about 80% homology of sequence between the two proteins. The major differences in composition between the proteins occur in the three regions lacking repeat structures. They include the extreme amino- and carboxyl- terminal domains and, of particular note, a short region in the amino-terminal quarter of the molecule which interrupts the series of cysteine-repeat motifs. This region has been postulated to act as a hinge, which allows the molecule to bend during its alignment, during microfibril assembly (Pereira *et al.*, 1993). The region is proline-rich in fibrillin 1 but glycine-rich in fibrillin 2 suggesting that the hinge may be more flexible in fibrillin 2, which in turn may produce more compliant microfibrils. Fibrillin 2 also contains an additional RGD sequence, but it is not yet known if it functions as an integrin-binding site (Zhang *et al.*, 1994).

Structure of the Fibrillin 1 Gene

The human fibrillin 1 gene (FBN 1) has been shown to be about 110 kbp in size and to contain at least 65 exons, most of which begin and end with 2/3 and 1/3 split codons, respectively. Forty nine of the 56 repeat motifs are encoded by a single exon, as is the 4 cysteine motif at the amino terminus. Five of the 8-cysteine motifs are each encoded by

2 exons and a single exon encodes two EGF-like repeats. Four mosaic exons code for regions of structural transition. Analysis, by primer extension and nuclease protection assays, has determined that the 5′ end of fibrillin 1 mRNA is 134 bases upstream of the ATG translation initiation site in exon 1. Gene sequences immediately upstream were found to lack TATA and CCAAT motifs.

Recent evidence, from another study, suggests that at least three additional upstream exons may be present in the fibrillin 1 gene (Corson *et al.*, 1993). In a small proportion of transcripts, one of the additional exons appears to be alternatively spliced onto exon 1. The full significance of this finding is still unclear. However, it suggests that several fibrillin 1 isoforms may occur, with different amino acid sequences at their amino termini. In addition, it indicates that promotional control of fibrillin gene expression is likely to be exceptionally complex and this may have implications for the pathogenesis of forms of the congenital disorder, Marfan syndrome (see below).

Congenital Disease and Fibrillin Genes

Considerable medical interest was generated by the simultaneous announcement, by two independent groups, of the linkage of the fibrillin 1 gene to the gene responsible for the Marfan syndrome (Dietz *et al.*, 1991; Lee *et al.*, 1991), which had already been located on chromosome 15 (Kainulainen *et al.*, 1990). In addition, one of the groups presented evidence for the linkage of a gene for a second fibrillin (now known as fibrillin 2) to a related condition, congenital contractural arachnodactyly (Lee *et al.*, 1991). The fibrillin 1 gene (FBN 1) has been precisely located to chromosome 15 at 15q21.1 (Magenis *et al.*, 1991), whereas the locus for the fibrillin 2 gene (FBN 2) is found on chromosome 5 between q23 and q31 (Lee *et al.*, 1991). In a separate study, genetic linkage was established between the FBN 1 gene and ectopia lentis, a disorder clinically related to Marfan syndrome. However another congenital disease of connective tissue, annuloaortic ectasia, was found not to be related to either of the fibrillin genes (Tsipouras *et al.*, 1992). There is also a report of a familial Marfan-like disorder which did not link to the fibrillin genes (Boileau *et al.*, 1993). It is possible that these conditions result from defects in other microfibrillar proteins, such as MAGP. The molecular basis of Marfan syndrome has been the subject of two recent reviews (Maslen and Glanville, 1993; Ramirez *et al.*, 1993).

Marfan syndrome is a potentially fatal connective tissue disorder that is inherited as an autosomal dominant trait. The condition is surprisingly common, with a prevalence of around 1 in 10,000 live births. Marfan syndrome is characterized by defects in the cardiovascular, skeletal and ocular systems. Cardiovascular abnormalities include mitral valve prolapse and dilatation of the aortic root, which if untreated can develop into congestive heart failure and aortic aneurysms. However, the skeletal defects which lead to a characteristic marfanoid habitus are the most obvious signs of the syndrome. These include overgrowth of the long bones of the limbs, curvature of the spine, chest deformities, long thin fingers and toes (arachnodactyly), flat feet and loose joints. Ocular features can include lens dislocation, myopia and retinal detachment (Pyeritz and Francke, 1993). Although Marfan syndrome shows high penetrance, the severity of the condition can be very variable, even within the same family (Dietz *et al.*, 1992).

Congenital contractural arachnodactyly (CCA) shares many of the skeletal features, but generally lacks the cardiovascular and ocular abnormalities, associated with Marfan syndrome. Defects characteristic of CCA, but not usually found in Marfan syndrome, are congenital contractures of the joints, most obviously those of the fingers and toes, and abnormal ears (Ramirez *et al.*, 1993). The latter may be explained, as noted above, by the preponderance of fibrillin 2 over fibrillin 1 within the matrix of the developing ear cartilage (Zhang *et al.*, 1994).

Evidence from genetic linkage analysis now indicates that the overwhelming majority of Marfan syndrome cases can be linked to the FBN 1 gene (Tsipouras *et al.*, 1992). A large, and increasing, number of mutations in the FBN 1 gene has been identified, although only about 20% are thought to be being detected by current methods (Aoyama *et al.*, 1994; Dietz *et al.*, 1993a). In contrast, no specific defects of the FBN 2 gene in CCA patients have, as yet, been reported. Most of the defects identified in the FBN 1 gene are missense mutations that involve a single base change in one allele, which results in the substitution of an amino acid residue in the fibrillin 1 molecule. However nonsense mutations, including a 4 base pair insertion and several examples of deletions, have also been reported (Kainulainen *et al.*, 1992). In one instance, the deletion of an entire exon from the mRNA was shown to result from a point mutation at a splice junction (Dietz *et al.*, 1993b). Amino acid substitutions have been identified in a range of calcium-binding, EGF-like domains and in some of the other cysteine-repeat structures. Cysteine is the amino acid that is most commonly substituted. This is not surprising since the absence of a cysteine would result in the disruption of the highly ordered pattern of disulphide bonds within the repeat motif and also would have possible consequences for the folding of adjacent domains. Substitutions of other amino acids which are considered important for calcium binding have also been found, which is consistent with the view (cited above) that this binding is necessary for the structural integrity of fibrillin molecules within the microfibrils.

In order to correlate the relationship between the genotype and phenotype in Marfan syndrome, several groups have investigated fibrillin 1 biosynthesis in skin fibroblast cultures from affected individuals (Aoyama *et al.*, 1993; Kielty *et al.*, 1994; Kielty and Shuttleworth, 1994; Milewicz *et al.*, 1992). From these studies it is clear that most Marfan syndrome cases have some detectable abnormality in the production and/or processing of fibrillin 1, which is consistent with earlier reports of decreased fibrillin 1 immunofluorescence staining in cultures of this type (Hollister *et al.*, 1990). For instance, in one study of 26 probands, 7 produced fibrillin at levels about 50% below normal; in 7, production levels were normal but protein secretion was impaired; and in 8, synthesis and secretion of fibrillin was normal but incorporation into the matrix was reduced (Milewicz *et al.*, 1992). A similar study suggested there was a link between missense mutations, involving substitution of a cysteine residue, and lower rates of secretion of fibrillin from cells due to improper folding of the molecule (Aoyama *et al.*, 1993).

Another approach has been to undertake ultrastructural analysis of the fibrillin-containing microfibrils produced by the fibroblasts in culture, using rotary shadowing (Kielty *et al.*, 1994; Kielty and Shuttleworth, 1994) Interestingly, although most of the examined Marfan cell lines was shown to make abundant microfibrils, a wide range of morphological abnormalities were observed, with each cell line exhibiting a distinct microfibril morphology. This suggested that different mutations in the fibrillin gene may result in microfibrils of characteristically abnormal shape.

Future investigations, which co-ordinate the genetic, biosynthetic and ultrastructural approaches, should help in the prediction of the phenotypic severity in Marfan syndrome, and will lead to greater understanding of the molecular mechanisms involved in the function of fibrillin 1 and its assembly into microfibrils.

Microfibril-associated Glycoprotein (MAGP)

Tissue Distribution and Ultrastructural Localization

MAGP has been shown, using immunofluorescence and immunoelectron microscopy, to be consistently and specifically associated with the elastin-associated microfibrils in both developing and mature elastic tissues from bovines and humans (Gibson and Cleary, 1987; Gibson *et al.*, 1986; Kumaratilake *et al.*, 1989). In addition, anti-MAGP antibodies have been demonstrated to localize to morphologically identical 10–12 nm microfibrils, found without elastin, in a range of tissues similar to that in which fibrillin had been demonstrated. MAGP was not found in association with other constituents of the extracellular matrix such as collagen fibres, type VI collagen microfibrils (4 nm) or basement membranes (Kumaratilake *et al.*, 1989). These results were originally obtained with affinity-purified polyclonal antibodies and more recently they have been confirmed with 3 monoclonal

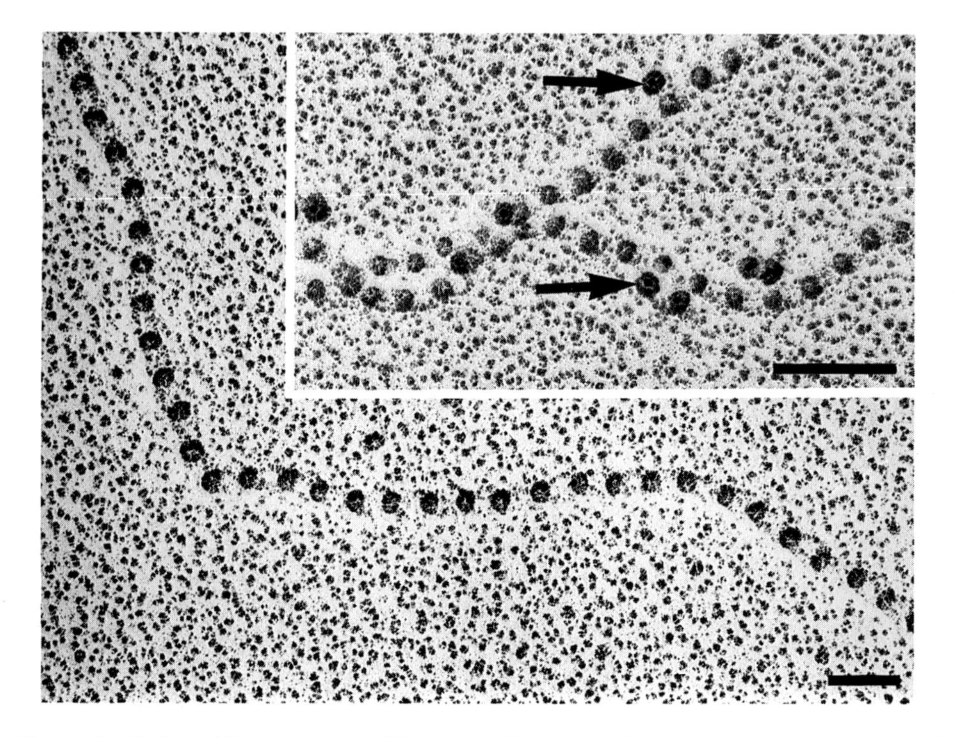

Figure 4.5 The beaded filament structure of 12 nm microfibrils revealed by the rotary shadowing technique and immunogold localization of MAGP to the bead regions. The bead-to-bead period is about 50 nm. The images were obtained from homogenates of bovine ocular zonule of which the microfibrils are the only morphologically identifiable components. Note the resemblance to "beads on a string".

The inset shows two microfibrils which have been incubated with affinity-purified anti-MAGP antibodies. The antibody specifically localized to the bead regions, as visualized by binding of 10 nm particles of protein A-gold conjugate (arrows). Bar = 0.1 microns. Magnifications: × 97,000. Inset: × 168,000 (Courtesy of Mr. R. Polewski).

antibodies which have no cross-reactivity with fibrillin. The anti-MAGP monoclonal antibodies were shown, using pre-embedding labelling techniques, to bind to the microfibrils at periodic intervals, yielding a pattern of staining similar to that reported for anti-fibrillin monoclonal antibodies (see above). However using the rotary shadowing technique, anti-MAGP monoclonal antibodies were shown to label only the bead regions of the 'beads on a string' structure of the microfibril revealed by this technique (see Figure 4.5). These observations are consistent with MAGP being a structural component of the bead regions of the microfibrils (Gibson, 1993).

Molecular Structure of MAGP

Molecular cloning has established that MAGP is synthesized as a 21 kDa polypeptide of 183 amino acids from a 1.1 kb mRNA, although the M_r for MAGP had previously been measured as 31,000 by polyacrylamide gel electrophoresis. The difference appears to be due mainly to anomalous migration on gels during electrophoresis rather than to extensive glycosylation. Sequence analysis of MAGP cDNAs showed that the encoded protein consisted of two structurally dissimilar regions, an amino-terminal segment containing high levels of glutamine, proline, and acidic amino acids and a carboxyl-terminal segment containing all 13 of the cysteine residues and most of the basic amino acids (Gibson *et al.*, 1991). More recent computer analysis has suggested that MAGP may consist of three domains, with a small globular amino-terminal domain connected to a larger disulphide-bonded globular domain by an extended proline-rich region (Bashir *et al.*, 1994). Overall, the amino acid profile predicted from the cDNA is virtually identical to that obtained from hydrolysates of MAGP (Gibson *et al.*, 1986). MAGP was shown to have little homology with any other known protein sequence. No potential sites for N-glycosylation are present in MAGP, which indicates all carbohydrate side chains are O-linked (Gibson *et al.*, 1991). This finding is consistent with the carbohydrate analysis previously obtained from hydrolysates of the glycoprotein. Interestingly, consensus sequences for tyrosine sulphation and a polyglutamine region, with potential for transglutaminase-catalysed cross-linking, were identified in MAGP (Gibson *et al.*, 1991; Tomasini-Johansson *et al.*, 1993). The presence of sulphated tyrosine in MAGP was later confirmed by biosynthetic labelling of the recombinant protein with radiolabelled sulphate, followed by identification of the modified amino acid after hydrolysis of the protein (Brown-Augsburger *et al.*, 1994). The purpose of this post-translational modification of MAGP is currently unknown. In the same study, MAGP was shown to act as a substrate for liver transglutaminase, resulting in large cross-linked aggregates. This finding suggests that MAGP contains glutamine and lysine residues which may be involved in covalent cross-linkage of the MAGP, either to other MAGP molecules or to fibrillin, within the microfibril. Recently, murine MAGP has been cloned and shown to have over 85% homology of amino acid sequence with the bovine protein (Chen *et al.*, 1993).

MAGP Gene Structure

The preliminary characterization of the genes for murine and bovine MAGP has shown them to have similar structures (Bashir *et al.*, 1994; Chen *et al.*, 1993). The murine gene was shown to be entirely contained in a 10 kb genomic fragment and to have 9 exons, with the ATG translation initiation codon residing in exon 2. The intron/exon boundaries were characterized by split codons in the 1/2 configuration, with one exception — where the split was 2/1. Preliminary characterization of the putative promoter region revealed it to be GC-rich, but lacking TATA and CCAAT sequences, and thus it has similarities with the putative promoter region of the FBN 1 gene. This type of promoter is found in genes with broad patterns of expression. The murine MAGP gene was mapped to chromosome 4, D3-E1, which has a conserved synteny with the p36–p35 region of human chromosome 1, to which the human gene locus has been assigned (Faraco *et al.*, 1993). To date, no congenital cardiovascular or skeletal diseases have been mapped to these regions.

Other Proteins Associated with the Microfibrils

A number of additional proteins have been identified in association with the microfibrillar component of elastic fibres, but the nature and function of the interactions remains to be elucidated.

Microfibrillar Proteins MP25, MP78 and MP70

Three other glycoprotein species, provisionally named MP25 (25 kDa), MP78 (78 kDa) and MP70 (70 kDa), have also been shown to be specifically extracted from developing nuchal ligament, together with fibrillin 1 and MAGP (see above). In addition, antibodies raised to purified MP25 and to a mixture of MP78/MP70 both localized specifically to the elastin-associated microfibrils, suggesting that the glycoproteins are also closely associated with these structures (Gibson *et al.*, 1989). MP25 has recently been cloned, using expression screening of a nuchal ligament cDNA library. Database searches revealed that MP25 had little homology with other known proteins with the exception of MAGP. The two proteins were found to be over 50% homologous in a cysteine-rich sequence of 60 amino acids, suggesting that MP25 may have functional similarities to MAGP and that the two proteins represent a new family of extracellular matrix proteins. In the same study, peptide mapping and sequencing have shown that MP78 and MP70 are structurally similar to each other, but distinct from other microfibrillar proteins (Gibson M.A., Sandberg, L.B. and Cleary, E.G. unpublished observations). The peptide sequences have extensive homology to a recently cloned, but poorly characterised 70 kDa protein, beta-IG-H3, which is expressed in a adenocarcinoma cell line, following treatment with TGF-β (Skonier *et al.*, 1992). However, the precise relationship between these proteins remains to be elucidated.

Fibrillin-like Protein

A 290–310 kDa fibrillin-like protein (FLP) from developing elastic tissues has recently been identified and cloned using a monoclonal antibody to fibrillin 1 (Gibson *et al.*, 1993). FLP contains multiple 6- and 8-cysteine repeat sequences characteristic of fibrillins, although the protein more closely resembles TGF-β-binding protein in structure (Kanzaki *et al.*, 1990). Two antibodies, to amino acid sequences unique to FLP, have been shown to localize to the microfibrils. In addition, the protein has been found to be extracted, together with fibrillin 1 and MAGP, from developing nuchal ligament tissue using reductive saline treatment. This procedure is relatively specific for solubilization of microfibrils, indicating that FLP is closely associated with these structures. Measurement of FLP mRNA steady-state levels during elastic tissue development showed significant increases in FLP expression during elastic fibrogenesis, suggesting that the protein had a function in this process. The chromosome locus of the human FLP gene has been determined as 14q 24.3 (Gibson, M.A., Baker, L., Sutherland, G. and Mecham, R.P. unpublished observations). Interestingly, this is the same locus as that for the TGF-β3 gene (Barton *et al.*, 1988). It remains to be established if the major role of FLP is that of a microfibril-associated protein or a binding/chaperone protein for a member of the TGF-β family.

Associated Microfibril Protein

Associated microfibril protein (AMP) was identified following expression screening of a whole chick embryo cDNA library with polyclonal antibodies to bovine ocular zonule (Horrigan *et al.*, 1992). Antibodies to a synthetic AMP peptide were shown to localize specifically to microfibrils in the elastic fibres of aorta and nuchal ligament, and to those in ocular zonule. AMP was shown to be synthesized as a soluble 54 kDa polypeptide which appears to be processed to a 32 kDa protein. The 32 kDa protein was identified in a range of bovine and chick elastic tissues, from which it could be extracted with a solution of 6M urea plus a reducing agent. It appeared to be the same 32 kDa protein previously identified with fibrillin in extracts of bovine ciliary zonule (Streeten and Gibson, 1988). Since the 32 kDa protein lacks cysteine residues, necessary for disulphide bond formation, it is doubtful that reducing agent is necessary for its solubilization. However it is possible that AMP is covalently bound to the microfibrils, by an as yet unidentified reducible cross-linking system, or other unknown mechanism.

Emilin

Emilin, previously known as GP115, is a 115 kDa glycoprotein identified in developing chick elastic tissues (Bressan *et al.*, 1993; Colombatti *et al.*, 1988; Colombatti *et al.*, 1987). Monoclonal and polyclonal antibodies to emilin have both been shown to localize specifically to typical elastic fibres in tissues such as aorta. In addition, labelling was also evident on microfibril-containing structures lacking amorphous elastin, but immuno-staining for tropoelastin. Such structures included the 'oxytalan' fibres of the corneal stroma and Descemet's membrane. In elastic fibres, emilin was found mainly in regions where the microfibrils and the elastin core are in close contract. Addition of anti-emilin antibodies to aortic cells in culture was shown to alter severely the process of elastic fibrogenesis (Bressan *et al.*, 1993). The findings suggested that emilin is important for the normal deposition and alignment of tropoelastin molecules onto the microfibrils during elastic fibre assembly.

Microfibril Structure

The high resolution ultrastructure of elastin associated microfibrils was originally determined from negatively stained preparations of non-denatured microfibrils, extracted from foetal bovine nuchal ligament. The microfibrils were shown to be filaments with cylindrical segments 13 nm in diameter and 13 nm long, separated by electron opaque regions, 10 nm long and 15 nm wide (Fahrenbach *et al.*, 1966). These findings have been extended by applying both negative staining and rotary shadowing techniques to extracts of a variety of tissues (Keene *et al.*, 1991; Sakai *et al.*, 1991). In each tissue, individual microfibrils were visualised as globular structures connected by fine filaments resembling a series of beads on a string. In a careful study of guanidine extracts of amnion, of unhomogenized vitreous humour and of tissue homogenates of nuchal ligament, aorta, skin and ciliary zonule, it was confirmed that similar beaded structures could be demonstrated in each, and that these were all immunoreactive with anti-fibrillin antibody. The diameter of the beaded segments was estimated to be 22 nm by rotary shadowing and 15 nm by negative staining. The periodicity of the globular portions of the microfibrils varied. Average periodicities ranged from 33 nm to 94 nm, but individual periodicities up to 165 nm were measured in stretched preparations. In negatively stained preparations of microfibrils, extracted from nuchal ligament under non-denaturing conditions, fibrillin antibody was shown to bind consistently to one side of the beaded portion of the structure, indicating that the globular domain is asymmetric. For this asymmetric binding to occur, it was reasoned, the beads must contain either the amino terminal end of one molecule and the carboxy terminal end of an adjacent molecule (head to tail alignment), or the central region of the fibrillin molecule so that the amino terminal region of the molecule is on one side and the carboxy terminal region on the other side of the bead (Keene *et al.*, 1991). This study further established that the antibody-labelled beaded strings could be converted to antibody-labelled microfibrils by processing the beaded strings through a standard TEM fixation, dehydration, embedding and staining. The product had a 52 nm period, similar to that seen in microfibrils in standard tissue biopsy sections. It was concluded that this was likely to represent the period present in relaxed tissue following standard processing for TEM. The periodicity in tissues probably varies with the organization of the surrounding tissues. A typical shadowed preparation of bovine zonular microfibrils is shown in Figure 4.5. The beads on a string arrangement can be readily appreciated.

As we have seen above, in shadowed preparations, the monomeric form of fibrillin 1 exists as an extended flexible rod approximately 148 nm long and 2.2 nm wide (Sakai *et al.*, 1991). The fibrillin monomer is thus over twice as long as the 50–67 nm periodicity observed by labelling of the microfibrils with monoclonal antibodies to fibrillin (Sakai *et al.*, 1986). This evidence suggested that extensive overlapping and/or folding of the individual fibrillin molecules occurred within the microfibrillar substructure. When two monoclonal antibodies, to different regions of fibrillin 1 (PF1 and PF3), were applied individually or together to tissue sections, the same labelling period was maintained, indicating that fibrillin 1 molecules are arranged within the microfibrils as bundles in a parallel, head to tail arrangement. These bundles appear to correspond to the inter-bead filaments, since antibodies to both PF1 and PF3 bind to this region of the microfibrils at different, but adjacent points. It was further concluded, from the immunoreactivity of the PF3 fragment, that the terminal domains of fibrillin are associated with the bead regions (Maddox *et al.*, 1989; Sakai *et al.*, 1986). It remains to be established if there is overlapping of the ends of fibrillin molecules in the beads.

It is not yet known if distinct fibrillin 1- and fibrillin 2-containing microfibrils occur as separate structures, or if the two proteins can be incorporated into the same microfibrils. The latter is possible since both proteins have been localized by immunoelectron microscopy to the elastin associated microfibrils of bovine foetal aorta and nuchal ligament.

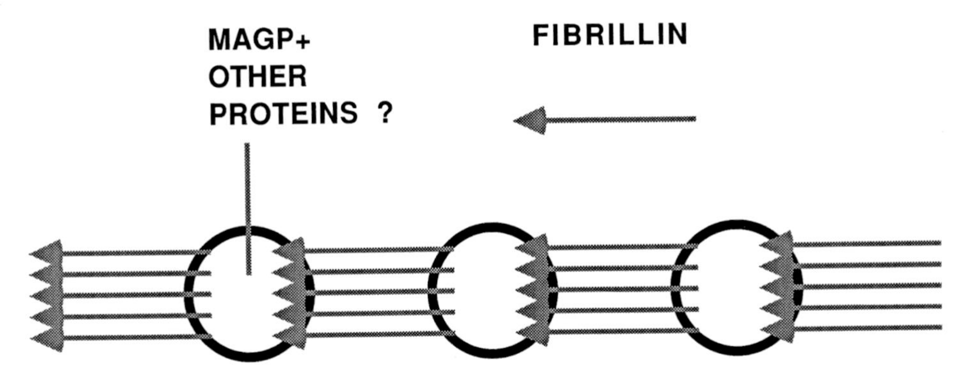

Figure 4.6 Diagrammatical representation of the structure of a microfibril. The technique of rotary shadowing has revealed the morphology of the microfibril to be a beaded filament reminiscent of "beads on a string ". Current evidence suggests that each interbead region is made up of several fibrillin molecules, arranged in parallel (arrows). These lateral aggregates are linked in a head-to-tail manner via the bead regions, which appear to contain MAGP and possibly other proteins. It is yet to be established if the fibrillin molecules overlap within the bead regions.

However, as we have seen above, they appear to be differentially distributed in other tissues (Zhang *et al.*, 1994).

There is more recent evidence that at least one other protein, MAGP, is a component of the beads of isolated zonular microfibrils, as can be seen in the inset to Figure 4.5, from the immunodecoration of the bead section of bovine ocular zonule microfibrils with anti-MAGP antibody. Anti-MAGP antibody localizes to all of the 12 nm microfibrils in which fibrillin has been identified and it too gives a repeat periodicity in tissue sections, indicating that it has a regular distribution within the microfibrils. The available evidence concerning the disposition of these components within the 12 nm microfibrils is summarised diagrammatically in Figure 4.6.

The Role of Microfibrillar Proteins in Elastic Fibre Formation

Currently, little is known about the functions of individual macromolecular constituents in the alignment and deposition of tropoelastin molecules onto the microfibrils during the important process of elastic fibre formation. We summarize the present state of knowledge in this field in cartoon form in Figure 4.7.

Progress in this field has been hampered by difficulties in obtaining sufficient quantities of purified, biologically active microfibrillar proteins from tissues. This is mainly due to the requirement for disruption of disulphide bridges, for solubilization of microfibrils, and the need to use denaturants to prevent aggregation (Gibson *et al.*, 1989). The recent molecular cloning of fibrillins 1 and 2 and of MAGP should facilitate the production of recombinant proteins in sufficient quantities for the study of their molecular interactions. Already two recombinant versions of bovine MAGP have been shown to bind to tropoelastin in *in vitro* enzyme-linked immunosorbent (ELISA) assays (Bashir *et al.*, 1994; Brown-Augsburger *et al.*, 1994). This confirms preliminary observations of an interaction between tissue-extracted MAGP and tropoelastin (Gibson *et al.*, 1989) and is consistent with the contention that MAGP is involved in the binding and alignment of tropoelastin onto the surface of the elastin-associated microfibrils. The interaction appears to involve the cysteine-rich

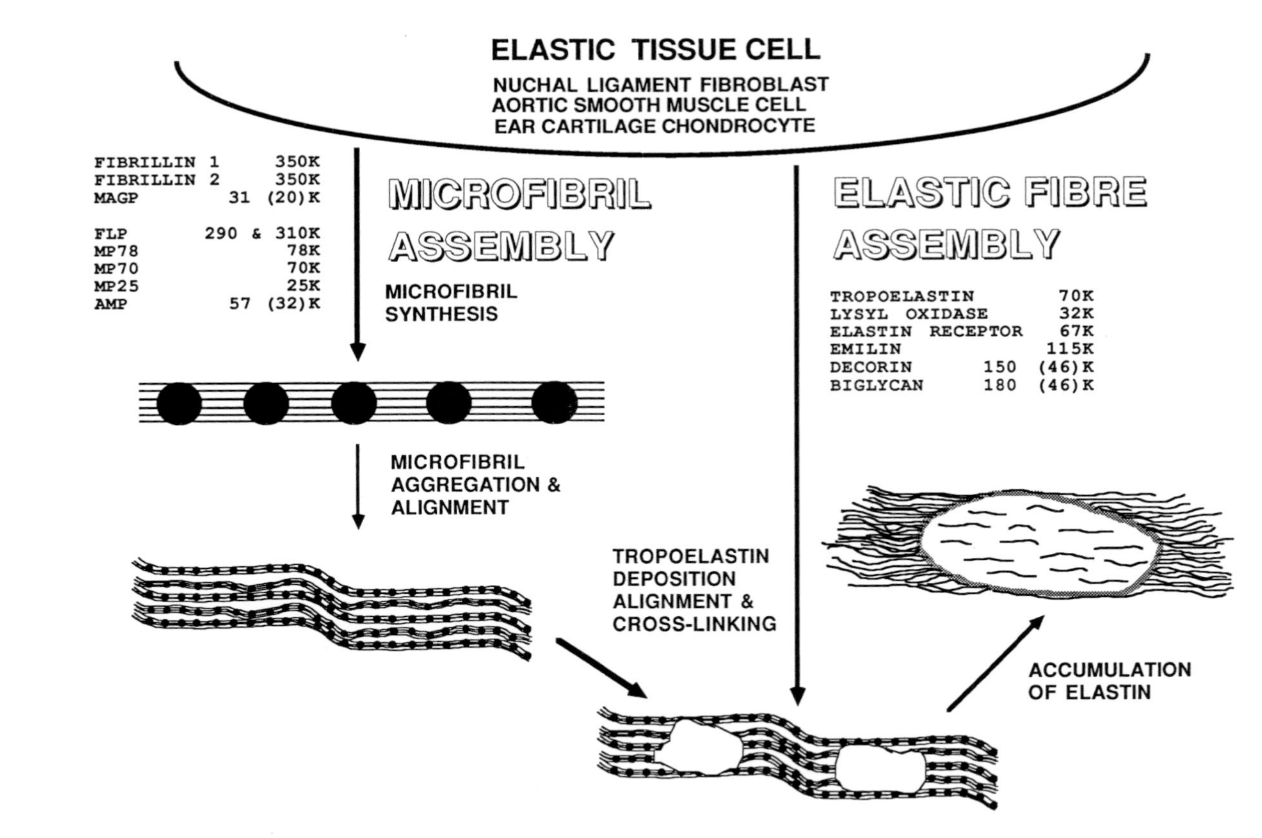

Figure 4.7 Schematic representation of the stages of elastic fibre formation. Morphological evidence suggests that elastic fibre formation occurs via an ordered series of events, the molecular bases of which are just beginning to be elucidated. Elastin-synthesizing cell types appear initially to assemble fibrillin and MAGP into microfibrils, which are laid down as precisely aligned, parallel aggregates close to the cell surface. These aggregates are then considered to acts as scaffolding for the binding of tropoelastin molecules, which are then cross-linked by lysyl-oxidase to form polymeric elastin. Elastin is first visualized as small coalescences of amorphous material which gradually increase in size to become the major component of the elastic fibre. The microfibrillar component, although diminished, persists as a mantle surrounding the fibre and as electron-dense inclusions within the amorphous elastin core. Molecules which are candidates for involvement in microfibril formation are listed on the left and additional molecules with known, or possible, roles in elastin deposition are listed on the right.

domain of MAGP and the carboxyl-terminal region of tropoelastin, which is highly conserved and which contains the only two cysteines in the molecule (Brown-Augsburger *et al.*, 1994).

Sakai has reported a study of the biosynthesis and assembly of fibrillin in embryonic chick organ cultures. In this she showed that fibrillin was synthesized within 30 minutes and the monomers quickly polymerized, within hours of initial secretion, into large multimers stabilized by intermolecular disulphide bonds. These multimers were estimated to contain as many as 23 fibrillin molecules (Sakai, 1990).

Another approach used to study the function of microfibrillar proteins in elastinogenesis has been to correlate the expression of fibrillins and MAGP with that of elastin by measuring their mRNA steady-state levels at intervals during bovine elastic tissue development (Mariencheck *et al.*, 1994). In both nuchal ligament and aorta, the period of highest elastin synthesis was shown to coincide with increases in the expression of fibrillin 1 and fibrillin 2. Interestingly, the increase in fibrillin 1 expression was found to be greater in ligament, in which the elastic fibres are linear, and the rise in fibrillin 2 expression was demonstrated to be more marked in aorta, in which the elastic fibres form sheets. These findings are consistent with each fibrillin playing distinct roles in elastinogenesis and having a function in the determination of fibre shape. In contrast, MAGP was reported to be strongly expressed at relatively constant levels throughout foetal development in both tissues. This observation suggests that MAGP may have other functions, in addition to its role in elastic fibre assembly. This concept is supported by the recent observation that a putative form of MAGP is secreted by cancer cells of lymphoid origins (Tomasini-Johansson *et al.*, 1993).

There is evidence to suggest that two small dermatan sulphate proteoglycans (PGs), with genetically distinct core protein chains of similar size, are involved in elastic fibrogenesis. One, called decorin, has been shown previously to have a role in regulating collagen fibrillogenesis in tendon (Scott, 1990). It contains a single glycosaminoglycan (GAG) chain while the other, biglycan, has two such GAG chains of similar composition. Both have been shown, by immunoelectron microscopy, to localize to elastic fibres. Anti-decorin antibody localized predominantly, but at relatively low density, in relation to the peripheral and included microfibrils, while anti-biglycan was interpreted to localize, at even lower density, in relation to the amorphous elastin component of the fibre (Baccarani-Contri *et al.*, 1990). Unfortunately, positive control sections were not shown and as very little collagenous matrix was concluded in the electron micrographs published, one cannot be sure what is the reason for the low antibody staining of these sections. It has been postulated that, within the fibre, the GAG chains bind readily to newly formed, positively charged, free ϵ-amino groups in tropoelastin and, by virtue of their capability for hydration, serve to prevent the spontaneous aggregation of hydrophobic tropoelastin (Fornieri *et al.*, 1987). As a complication to this story, it has been shown that galactosamine-containing GAGs, such as these, have a capability to interfere with the binding of elastin to an elastin binding protein on the surface of elastin secreting cells, and thereby to disturb elastic fibrogenesis (Hinek *et al.*, 1991).

As we have seen above, antibody to lysyl oxidase, the enzyme responsible for the oxidative deamination of the ϵ-amino groups of lysine in elastin (and collagen) and thus for covalent cross link formation (including desmosine and isodesmosine), has also been shown to localize in relation to microfibrils in tissues in which elastic fibre deposition is actively proceeding (Kagan *et al.*, 1986). This localization is consonant with the known extracellular reactivity of this enzyme with tropoelastin, and compatible with a potential role of crosslink formation in elastic fibrogenesis.

Elastin Receptor, Cell Matrix Interactions and Fibrogenesis

As we have seen above, there is evidence for the presence on the surface of elastogenic fibroblasts of an elastin-binding protein that appears to be involved in elastic fibrogenesis. The first evidence for such an elastin receptor came with the demonstration that monocytes and fibroblasts could respond chemotactically to tropoelastin and to specific peptide fragments of elastin (Senior *et al.*, 1982). Subsequently, it was shown that nuchal ligament fibroblasts acquired chemotactic behaviour during foetal development, and that acquisition of this property coincided with the onset of elastin production (Mecham *et al.*, 1984a). A major part of the chemotactic activity of elastin was shown to reside in one of its repeating peptides (VGPAPG) and a monoclonal antibody (BA-4) to this peptide was shown to inhibit most of the chemotactic response to elastin peptides (Wrenn *et al.*, 1986). The elastin receptor was isolated and identified as a complex of three proteins with M_rs of 67 kDa, 61 kDa and 55 kDa. The 55 kDa and 61 kDa subunits are transmembrane proteins, whereas the 67 kDa protein is extracellular and was shown to bind laminin as well as elastin. It was also shown to contain a protein binding site that recognizes a hydrophobic sequence (VGVAPG) in elastin and a separate carbohydrate binding site that interacts with galactoside sugars (Hinek *et al.*, 1988). In the absence of sugar, elastin is bound avidly to this binding protein, but in the presence of sugar, affinity for elastin is greatly reduced (Mecham *et al.*, 1991).

Immunolocalization studies with elastin producing cells established that tropoelastin and the 67 kDa protein co-localize intracellularly, which was interpreted to suggest that tropoelastin was bound to the 67 kDa protein, within the secretory pathway. It was thus postulated that it was functioning to provide transport signals as well as acting to maintain tropoelastin in solution until reaching the extracellular space. At the plasma membrane, tropoelastin was postulated to remain bound to the 67 kDa protein until an interaction with a microfibril-associated galactoside sugar induced the transfer to an acceptor site on the microfibril. In this way it could also serve to facilitate fibre assembly (Mecham and Heuser, 1991). Other cell surface binding proteins have also been described. These include elastonectin, a 120 kDa protein, whose presence on the surface of smooth muscle cells is inducible by elastin peptides (Hornebeck *et al.*, 1986), and a 59 kDa protein found on the surface of tumour cells, that is coupled to protein kinase C (Blood and Zetter, 1989). A role for these receptor proteins in elastic fibrogenesis is unclear.

Of particular interest in relation to elastin receptors, are the observations by Mecham *et al.* that early gestation nuchal ligament fibroblasts, that have not yet begun to express tropoelastin, begin to do so when cultured on an extracellular matrix substratum prepared from a late gestation nuchal ligament. Such differentiation was not seen when these cells were grown on a nuchal ligament substratum derived from a young foetus, in which elastin synthesis was not yet established. This matrix-induced differentiation occurred in the entire population of cells in culture and was shown to be stable and heritable. Further, when elastin producing foetal nuchal ligament fibroblasts were grown in cell culture they lost their secretory phenotype after repeated doublings, and elastin production fell. The secretory phenotype could be restored if the cells were allowed to come into physical contact with older nuchal ligament matrix, but not with matrix of undifferentiated younger nuchal ligament. It was concluded that there was compelling evidence that contact between nuchal ligament fibroblasts and the extracellular matrix, directs the acquisition of the differentiated state of the cells and specifies or permits elastin synthesis in cells committed to elastin production (Mecham *et al.*, 1984b). More recently it has been reported that chondrocytes, from the hyaline cartilage of the knee joint of bovine late gestation foetuses, when grown in monolayer culture on plastic, acquired an elastin secreting phenotype. This conversion in phenotype was associated with induction or modulation of all known components of elastic fibres, but at least some of the cells retained their chondrogenic phenotype in that they continued to produce type II collagen. *In vivo*, these cells do not produce any elastin. It would appear likely, in this situation, that cell-matrix interactions are serving to repress tropoelastin expression (Lee *et al.*, 1994). Understanding of the mechanism(s) involved in controlling these processes may lead to the capability to repress or stimulate elastin synthesis, as appropriate, in disease states in humans.

Physical Properties of Elastin and Elastic Fibres

Elastin Structure

Elastic tissues are characterized by exhibiting rubber-like properties, including long range extensibility and the capacity to return to their original dimensions when the stress is removed. Several structural models have been proposed for elastin, based on its primary and higher order molecular structure, to explain its physical and thermodynamic properties. The classical kinetic theory of rubber elasticity provided the first explanation of the long-range elastic properties of elastin (Flory, 1969). This model envisages a network of random chains, arranged so as to allow free movement of molecules within the structure. At rest, this network is arranged in maximum disorder, so that entropy is at a maximum. When the network is stretched, order is increased and so entropy is decreased. Thus when the stretching or distending force is removed, the network returns to the maximum entropy state spontaneously. A modification of this model, to take account of interchain crosslinking, has been shown to predict many of the physical properties of elastin as well as being compatible with the thermodynamic data (Fleming *et al.*, 1980). However, classic rubber theory incompletely describes elastin structure, as it does not take into account the effects of hydrophobic interaction forces that can be demonstrated in stretched elastin networks (Gosline, 1976). Furthermore, the sequencing of elastin has brought the knowledge that parts of the elastin molecule can and do exhibit regions of relatively long-range order (Gray *et al.*, 1973) and several workers have demonstrated a fibrillar substructure for elastin, electron microscopically, in sonicated preparations and in elastin prepared by grinding (see review by Gosline (1976)).

At least 60% of the amino acid side chains in elastin are non-polar so that the molecule is extremely hydrophobic. Thus it is likely to take on a globular conformation in aqueous solution. This property was the basis of the corpuscular model proposed by Partridge (Partridge, 1967) and of a more ordered, "oiled coil" model proposed by Gray *et al.* (1973). In this latter model, the hydrophobic regions are arranged in a broad coil consisting of numerous β-turns, in succession. These structures alternate with polyalanine regions, which form α-helices and contain lysine residues that are envisaged as forming covalent cross links between chains (Gray *et al.*, 1973). In these two models, the non-polar regions of the molecule seek one another out to exist in a disordered resting state from which water is excluded. When the structure is stretched, the hydrophobic groups come into contact with water and the repulsive forces thus generated provide the force for recoil, which involves reaggregration of the hydrophobic regions and expulsion of the water after the extending force is removed. However, nuclear magnetic resonance data (Torschia and Piez, 1973) are held to demonstrate that the polymer chains in the elastin network are kinetically free and thus they exclude the possibility that any fixed structures, such as the "oiled coil", exist in the elastin network (Gosline, 1976). In his review, Gosline concluded that "elastin contains a nonhomogeneous network of kinetically free protein chains in which hydrophobic regions (from which water is *partially* excluded) are interspersed between regions of fully hydrated protein chains".

As we have seen above, solubilized elastin preparations exhibit the property of coacervation, or reversible precipitation into a fine fibrillar form, within which are some regions of α-helix (Gotte *et al.*, 1974). These α-helices were postulated to be the alanine-rich regions, which are also the location of lysine-derived crosslinks (Gray *et al.*, 1973). Structural analyses have shown that elastin contains significant amounts of secondary structure and that the hydrophobic regions of the elastin molecule contain a number of repeat peptides with sequences that favour the formation of β-turns which, in series, form a helix called a β-spiral (Urry *et al.*, 1974).

These observations led Urry to propose for elastin a three-component, two or three chain, fibrillar model with a sequential arrangement of β-spirals and α-helices (Urry, 1983). In this structure, polyhexapeptides that are adjacent to the α-helical regions are arranged such that their hydrophobic side chains are on the outer side of the spiral. These can thus associate with other molecules by hydrophobic interactions and serve to align the chains while cross-linking takes place between α-helices. The major portion of the structure is composed of hydrophobic regions arranged in β-turns, in series, to form the β-spiral, a water containing helix. In modeling this structure, the β-turns function as spacers between the turns of this large helix and the contacts between turns are hydrophobic in nature, which allows for substantial flexibility. The presence of water inside the β-spirals is important for permitting the mobility required for elasticity. This model explains many of the physical and thermodynamic properties of elastin, but only about 15–25% of the molecule is able to form this ordered structure. The advantages of the model are that it readily explains the coacervation phenomenon, is predominantly fibrillar in nature, and it provides an explanation for the calcification that occurs when elastin is exposed to normal serum. Calcium is thus seen to be able to bind to a non-charged, binding site within a repeating polypentatide within tropoelastin (Urry *et al.*, 1976). It seems likely that elastin in tissues includes elements of both the fibrillar model and the random chain network model.

The presence within the elastin molecule of β-turn structure is relevant to ageing and disease of elastin. The β-turn favours interaction with lipids, which in turn has been shown to decrease markedly the tensile strength of fibres of elastin and greatly to enhance the susceptibility of the protein to elastase (Tamburro, 1981). Deposition of cholesterol, cholesterol esters, and other lipids on elastic fibres in the arterial wall in the early stages of atherosclerosis could therefore serve to weaken this region of the vessel wall. Gosline has noted that these lipid deposits might alter the hydration of the elastin network, resulting in increased hysteresis, that could contribute to the mechanical breakdown of the elastic lamellae in the long term (Gosline, 1976).

Microfibrils and the Physical Properties of Elastic Fibres

Point mutations in the fibrillin 1 gene, as has been noted above, produce the clinical features of Marfan syndrome. Examination of this condition provides some interesting insights into the relationship between the physical properties of affected tissues and their elastic fibre substructure and composition. Dislocation of the ocular lens is a common feature of cases of Marfan syndrome, as is lengthening of long bones. The ocular zonular fibres consist solely of collections of typical microfibrils, which transmit the stresses that regulate the focal length of the ocular lens, so that their microfibrils must be inherently inextensible. As these microfibrils are morphologically and immunochemically indistinguishable from elastin associated microfibrils, it is pertinent to examine the periosteal elastic fibres for morphological differences from other elastic tissues, particularly in relation to the number, morphology and disposition of their microfibrillar components.

Ultrastructurally, each adult periosteal elastic fibre consists of a typical amorphous core surrounded by a thick layer of electron densely staining material, which has been shown to exhibit the specific staining characteristics of glycoproteins. The nature of this material is not known, but it is deposited in the region where microfibrils would normally be visible in elastic fibres in other tissues (Kumaratilake *et al.*, 1991). Examination of periosteal elastic fibres during foetal and early post-natal development shows that the periphery of the periosteal elastic fibres contains large numbers of typical elastin associated microfibrils immunoreactive with anti-fibrillin and anti-MAGP antibodies. These findings are illustrated in Figure 4.8. With further development, the relatively thick layer of microfibrils gradually develops a coating of the glycoprotein material, which masks the microfibril content of adult fibres. Closer examination of the amorphous elastic core of the young periosteal elastic fibres reveals a substructure of parallel collections of immunoreactive microfibrils throughout the amorphous elastin (Figure 4.8b). Microfibrils appear to persist within the amorphous elastic material after foetal development, so that they can be demonstrated within quite mature elastic fibres (Cleary E.G. and Kumaratilake J., unpublished observations). Such findings, taken together with the observed excessive lengthening of the long bones in patients with point mutations in fibrillin 1 gene, suggest that periosteal elastic tissue is not normally extensible to the same degree as is elastic tissue in other locations. The periosteal elastic fibres appear to be responsible, under normal conditions, for restricting the lengthening of long bones during rapid growth phases. Defects in fibrillin structure appear to result in weakening of the young periosteal elastic fibres so that they are unable to restrict bone lengthening during growth, thus producing the typical habitus of Marfan syndrome. There is experimental support for this proposal from surgical studies in which stripping the periosteum from the shaft of long bones in young animals has been shown to promote bone elongation (McLain and Vig, 1983).

These findings raise more general questions as to what factors control the synthesis and organization of the two major morphological components of elastic fibres. Does the relative contents of fibrillin 1 and fibrillin 2, and of MAGP and the other microfibrillar components influence the physical properties of the microfibrils or the interaction of microfibrils with elastin? Do these factors affect the formation and physical properties of the elastic fibres? A survey of the distribution of the many components within apparently similar 10–12 nm diameter microfibrils, in different microfibril-rich tissues, might reveal interesting differences in composition associated with different locations and functional requirements.

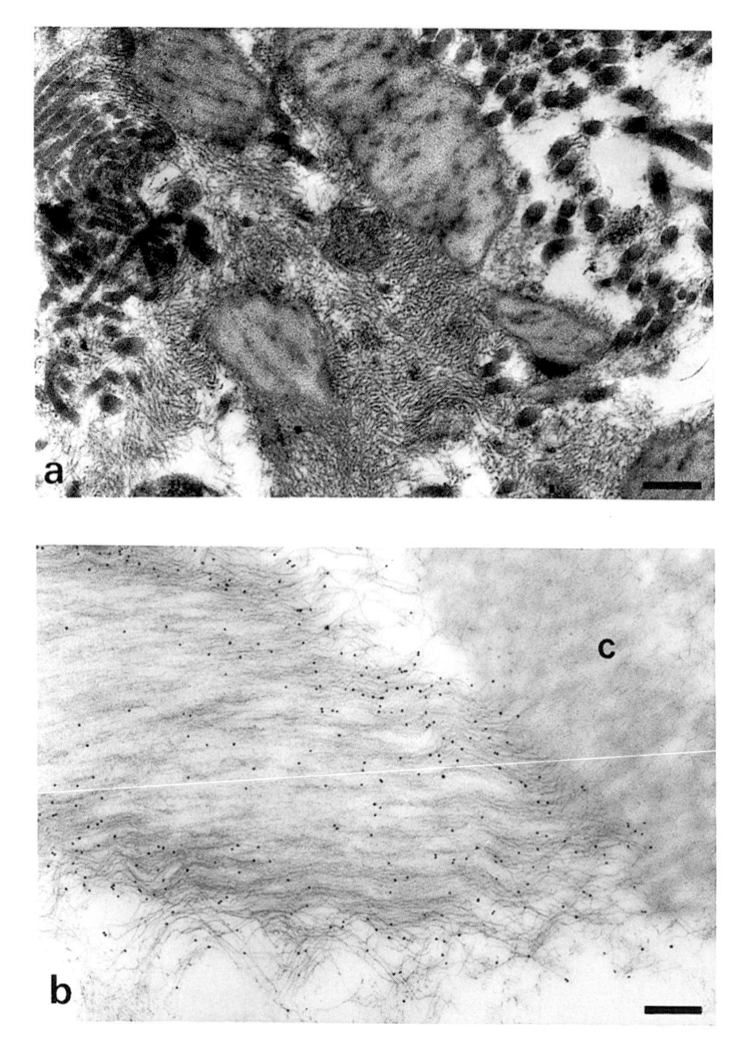

Figure 4.8 Electron micrograph of periosteal elastic fibres from long bone of 230-day-old bovine foetus. Bar = 0.31 microns. Magnification: × 32,000

Figure 4.8a Section embedded in Spurrs medium. The elastic fibres are sectioned transverse to their long axis. The amorphous component of the fibres is more densely staining than that in nuchal ligament elastic fibres (Figure 4.1) and there is already a rim of more electron dense staining, Masses of 12 nm diameter microfibrils are seen in proximity to the elastic fibres, but are not evenly distributed around the perimeter of the fibres. Relatively, periosteal elastic fibres have a higher proportion of microfibrils than those in skin or aorta. Loose collections of collagen fibres, in transverse and longitudinal section, are seen in the surrounding tissues.

Figure 4.8b Section embedded in LR White medium. Immunostained with affinity purified polyclonal anti-MAGP antibody, labelled with 12 nm diameter gold. The antibody localizes specifically to the 12 nm microfibrils. The elastic fibre is sectioned along the longitudinal axis of the fibre. The amorphous component is surrounded by microfibrils, which are aligned with the axis of the fibre and appear to persist throughout the amorphous material, where there is intermittent antibody binding, especially in relation to small relatively electron densely staining inclusions. A bundle of collagen fibres (c), sectioned obliquely is stained relatively weakly, as is usual in sections from LR White blocks (Courtesy of Dr. J.S. Kumaratilake).

DISEASES AFFECTING ELASTIC TISSUES

As has been noted above, elastic tissue is present in mammalian tissues in a variety of forms, many of which appear to be related to the functions subserved by the tissue. The last decade has seen major advances in our understanding of the molecule structure and biosynthetic control mechanisms for the protein elastin and for the component proteins that make up elastin associated microfibrils. However, along with this new knowledge has come a realization that the elastic fibre system is immensely more complicated than was perceived 10 years ago. For many years, elastic tissue researchers have searched for genetic diseases affecting elastin and elastic tissue components, in the hope that such a disease would shed light on how structural changes in the molecules affect function and morphology. This deficiency is now remedied, first with the recognition that Marfan syndrome and CCA are related to mutations in fibrillin genes, and more recently with the identification of defects in the elastin gene in SVAS and the related Williams Syndrome, as noted above.

It may be helpful to readers new to elastic tissue research to review briefly a selection of acquired diseases that affect elastic tissues which, in the opinion of these reviewers, have the potential to provide additional insights into structure/function relationships deriving from disturbances of normal elastic tissue biochemistry. No attempt has been made to cover all the conditions associated with changes in elastic tissues.

Conditions Associated with Overproduction of Elastic Tissue

Although, as seen above, there is evidence from stable radioisotopic studies to indicate that there is only minimal turnover of elastin in human lung and aorta, there is histochemical, biochemical and molecular biological evidence for new elastin and microfibrillar synthesis and deposition in a variety of tissues, including the aorta and blood vessels, in response to altered physiological stimuli and in disease.

Elastin is synthesized and deposited in arteries in response to increased blood pressure. This is true of aorta of weanling rabbits (Cleary and Moont, 1976), and of rats (Keeley and Alatawi, 1991), of the pulmonary arteries of new born calves (Prosser *et al.*, 1989) and of mesenteric resistance vessels in rats (King *et al.*, 1992). There is also deposition of new elastic tissue (as assessed by histochemical staining) in the venous side of arterio-venous fistulae in humans (Holman, 1937) and new elastin synthesis has been demonstrated, immunohistochemically and by in situ hybridization, in the wall of the venous segment of large iliac vessel fistulae in young pigs (Cleary E.G. and Faris, I.B., unpublished observations) although chronic carotid-jugular fistulae in sheep (whose arteries tend to have a higher elastin content than other species) resulted in a decrease in elastin content of the wall of the venous segment (Davis *et al.*, 1989). A series of interesting experiments has been reported recently detailing increases in elastic tissue, consequent upon induction of hypertension in the foetus by ligation of the ductus arteriosus. From these studies, it was concluded that elastin biosynthesis in large arteries in response to raised pressure is more active in foetal and neonatal tissues than in more mature tissues (Belik *et al.*, 1994).

New elastic tissue formation and deposition is a feature of the dermal response to actinic ultraviolet (UV) radiation. In this situation, it is clear that new elastin is accumulating in mature adult skin, in spite of a long-standing, widely held belief that new elastin synthesis is unusual after puberty. The newly forming elastic fibres contain many typical microfibrils and amorphous elastic material. However, with prolonged exposure, the amorphous elastin undergoes degradation and appears granular, and the number of microfibrils decreases (Lavker, 1979). Although it has now been shown to retain its immunoreactivity with anti-elastin antibodies, the abnormal staining properties and ultrastructural appearances of this dermal elastic tissue led to it being called "elastotic" and the process "solar elastosis". Increased levels of mRNA for both elastin and fibrillin 1 have been demonstrated in sun damaged human skin (Bernstein *et al.*, 1994). Ultraviolet B radiation has been held to be mostly responsible, although there is evidence to suggest that UVA may also have a role to play in this elastogenic response (Zheng and Kligman, 1993). In the hairless mouse, in response to predominantly UVB exposure, the relative rates of increased deposition of elastin in the superficial papillary and deeper dermis were similar but, surprisingly, microfibrils were deposited at a much greater rate in the deeper dermis compared to both elastin in this region and to microfibrils in the superficial dermis (Kumaratilake and Cleary, unpublished observations).

Excess elastin deposition is also a feature of a desmoplastic reaction in some cancers, especially of breast, but also of stomach and colon. New collagen deposition also occurs during this response. In situ hybridization studies indicate that most of the new elastin is secreted by fibroblasts within the cancerous mass, although the stimulus for elastogenesis has not been identified. The cancer cells themselves have also been shown to be responsible for both elastogenesis and collagenesis in some scirrhous breast cancers (Krishnan and Cleary, 1990). No histological abnormality could be demonstrated in these elastic fibres, which contain amorphous elastic material and microfibrils in normal relationships.

Masses of abnormal fibres, with staining characteristics of elastic tissue, are typical of a lesion called an elastofibroma. For many years these were held to be tumours of the connective tissue of the subscapular space. However, recent evidence indicates that they occur in response to chronic periosteal irritation, and that the cells responsible for the secretion of the abnormal elastic fibres are periosteal fibroblasts. The lesion consists of masses of highly disordered elastic fibres, with varying proportions of amorphous elastin (immunoreactive with anti-elastin antibodies) and microfibrils (immunoreactive with anti-fibrillin and anti-MAGP antibodies). Some fibres consist of masses of amorphous elastin alone, while others contain only microfibrils (Kumaratilake *et al.*, 1991). The reason for the disordered elastic fibrogenesis remains obscure. There is an exceptionally high incidence of this condition among the inhabitants of Okinawa, suggesting linkage to a specific gene (Nagamine *et al.*, 1982).

Disordered Cross-linking of Elastin

It has been noted above that cross-linking of elastin requires lysyl oxidase activity. This is a copper dependent enzyme, and copper deficiency has been reported to be associated with reduced lysyl oxidase activity and reduced cross-linking of elastin. The reduction in cross-links appears to be responsible for vascular aneurysm formation and rupture of the aorta in copper deficient pigs (Waisman *et al.*, 1969). It was from the aorta of these animals that tropoelastin was originally characterized (Sandberg *et al.*, 1969). Arterial aneurysms and rupture occur in immature, rapidly growing rats (Ponseti and Baird, 1952) and in birds (Simpson *et al.*, 1980) exposed to toxic doses of lathyrogens, which act as competitive substrates for lysyl oxidase and thus inhibit cross linkage formation in both elastin and collagen (Levene *et al.*, 1992). In both copper deficiency and in lathyrism in animals, there is defective morphology of the vascular elastic lamellae of the aorta and large arteries (Simpson *et al.*, 1980). These changes affect the ultrastructural appearances of both the amorphous elastic material and the microfibrils in lathyrism (Pasquali-Ronchetti *et al.*, 1981), and in copper deficiency (Fanning and Cleary, 1974).

Menkes Syndrome in humans is an X-linked disorder associated with decreased functional activity of lysyl oxidase and thus with reduced cross-linking of elastin and to a lesser extent of collagen. The condition is thought to be secondary to disturbances in intracellular copper translocation and consequently impaired incorporation of copper into lysyl oxidase. Typically patients are diagnosed at about 3–6 months of age with severe neurological symptoms and deteriorate progressively thereafter. At autopsy, the elastic lamellae of the aorta and large arteries are fragmented and disrupted. Ultrastructurally, the elastic fibres are disordered and there is proliferation of microfibrils. These data have been reviewed recently by Danks (1993). An apparently identical condition is observed in mice with X-linked mutations affecting coat colour. Danks argues that the defect in these mice (of which the brindled mutant has been most studied) is homologous with that in humans (Danks, 1993). In a recent study of the expression of matrix proteins and lysyl oxidase in Menkes fibroblasts and fibroblasts from mottled mice, it was shown that levels of lysyl oxidase mRNA transcripts were less than 15% of levels for corresponding controls. The level of elastin mRNA transcripts were also markedly lower in both cells lines in comparison to controls, although the levels of procollagen type I mRNA were equal to those in the controls. It was concluded that the connective tissue defects associated with Menckes syndrome and those occuring in mottled mouse mutants involve more than abnormal copper utilization in the formation of lysyl oxidase holoenzyme. The production of lysyl oxidase and elastin also appeared to be altered at the level of transcription or mRNA turnover (Gacheru *et al.*, 1993).

Calcification of Elastic Tissues

There are a number of diseases typically, and apparently specifically, associated with calcification of elastic fibres. Some are congenital and some acquired. In evaluating this evidence it is important to recall that aortic elastic lamellae have been shown to bind radioactive calcium with extreme avidity (Martin *et al.*, 1963) and Urry has confirmed that specific sequences in elastin have a marked affinity for calcium (Urry *et al.*, 1976). Microfibrils also have the capacity to bind calcium, as fibrillins have been shown to contain calcium binding consensus sequences within most of the many EGF-like repeats that are typical of this family of proteins. Gross disruption of the microfibrils, in the interbead domains, results if they are incubated with calcium chelating agents, such as EDTA or EGTA (Kielty and Shuttleworth, 1993). However, there are, as yet, no diseases in which microfibrillar binding of calcium has been demonstrated to be aetiologically important.

Ageing of human aorta has long been known to be associated with progressive increase in calcification. Some of this is visible to the naked eye as calcified plaques occur in atherosclerotic lesions. Some forms of peripheral vascular disease of the lower limbs are also associated with increased calcification and here the plaques lie in the muscular medial layer. In addition to these changes, there is a process of medial calcification within the aorta, which Lansing (on the basis of microincineration data) noted to be unrelated to atherosclerosis and to increase with age (Lansing, 1959). It is more prominent in the thoracic aorta than in the abdominal aorta, unlike calcification associated with atherosclerosis. By modern standards, this process (which Lansing called elastocalcinosis) and its precise localization have not been adequately documented. There has been only one recent biochemical study of human aortic calcification, the data from which are consonant with the earlier literature (Elliott and McGrath, 1994).

Calcification of elastic fibres of the skin, eyes and cardiovascular system are characteristic and diagnostic features of a congenital disease called pseudoxanthoma elasticum, or PXE (Neldner, 1993). High resolution electron microscopic examination has revealed no specific abnormalities of the amorphous or microfibrillar components of the elastic fibres, either in lesional or non-lesional tissues, although individual cases are reported with some abnormalities in one or the other (Lebwohl *et al.*, 1993). Elastin, purified from lesional skin in PXE patients, has been shown to be more resistant to elastase action than that from non-lesional skin, and pretreatment of the lesional elastin with testicular hyaluronidase made it more susceptible to elastase (Schwartz *et al.*, 1991). Recently an elastin gene mutation was excluded as a cause of this disorder in an affected family (Raybould *et al.*, 1994), and increased expression of lysyl oxidase has been excluded as the underlying abnormality in a group of PXE patients (Yeowell *et al.*, 1994). The cause of this disorder remains unknown.

Pulmonary Emphysema — a Disease Associated with Elastolysis

Elastin plays a key role in the functioning of the lung. Destruction of elastin in the alveolar walls is a characteristic feature of pulmonary emphysema, a common and important human disease. Neutrophil and macrophage elastases have been considered for many years to be the principal agents inducing this pulmonary elastolysis, but the human 72 kDa and 92 kDa metalloproteinases (type IV collagenases) have also been shown to be capable of degrading elastin (Senior *et al.*, 1991). Naturally occurring antiproteases, such as α-1-antitrypsin, counter the action of elastase and the extent of elastolysis in a tissue is determined by the balance achieved between protease and antiprotease activity (see review by Janoff, 1985). Emphysema has been shown to occur with increased frequency in patients with a congenital deficiency of α1-antitrypsin since the original observation of this condition (Laurel and Eriksson, 1963). Increased levels of elastin degradation peptides are demonstrable in the serum in a proportion of emphysema patients, especially high serum levels being found in smokers (Dillon *et al.*, 1992), who in general have an increased susceptibility to emphysema. The recent review by Snider provides access to the relevant literature (Snider, 1992).

CONCLUSION

Elastin and microfibrils have been difficult to work with as proteins, because of their inertness and their insolubility, respectively. It will be apparent that research in the field of elastic tissue is in a highly active and productive phase at present. The impetus has been provided principally by advances in molecular biology and the recognition of several heriditary diseases affecting elastin and microfibrillar proteins. But of course, most of these advances have relied on the prior painstaking, and hard-won advances in protein biochemistry and ultrastructure. Much still remains to be done, including: determination of the factors that regulate the elastin gene; determination of the extent of alternative splicing of the elastin gene in different tissues and during development and the effects this has on tissue morphology and function; identification of additional mutations in the elastin and microfibrillar protein genes and their consequences; determination of the factors that regulate the process of elastic fibrogenesis; determination of the role of individual microfibrillar components in elastic fibrogenesis and in other processes; the extent and consequences of interactions, within tissues, of elastic tissues with other components of the extracellular matrix; the mechanisms that regulate phenotypic differentiation and expression of elastogenetic cells within the matrix; and the role of elastic tissue proteins in disease aetiology and in tissue repair.

Research into microfibrillar components has also benefitted from the general increase in activity in this field. Some interesting responses to disease have been reviewed and these highlight the issue of tied regulation of microfibrillar and elastin biosynthesis. How is this regulated? It is pertinent to ask, do microfibrils in different tissues have different physical properties, and if so what is the basis of the differences? How do changes in composition of the microfibrils influence physical properties? How will changes in persistence of microfibrils in elastic fibres be regulated, and what effect will this have on the physical properties?

The answers to many of these questions are likely to come from advances in recombinant technology, now that most of the protein components of elastic fibres have been cloned and their genes characterized.

Acknowledgements

Unpublished studies cited in this review have been supported over the years by grants in aid from the National Health and Medical Research Council of Australia, the Australian Research Council, the National Heart Foundation of Australia, Sandoz Foundation, the Skin and Cancer Institute, the Anti-Cancer Foundation of South Australia and the Clive and Vera Ramaciotti Foundation. The electron micrographs were kindly provided by Dr. J.S. Kumaratilake, to whom we are also grateful for helpful and constructive criticisms and discussion. The photographs were prepared by Mr. D. Caville.

References

Aoyama, T., Francke, U., Dietz, H.C. and Furthmayr, H. (1994) Quantitative differences in biosynthesis and extracellular deposition of fibrillin in cultured fibroblasts distinguish five groups of Marfan syndrome patients and suggest distinct pathogenetic mechanisms. *J Clin Invest*, **94**, 130–137.

Aoyama, T., Tynan, K., Dietz, H.C., Francke, U. and Furthmayr, H. (1993) Missense mutations impair intracellular processing of fibrillin and microfibril assembly in Marfan syndrome. *Hum Mol Genet*, **2**, 2135–2140.

Ayer, J.P., Hass, G.M. and Philpott, D.E. (1958) Aortic elastic tissue. Isolation with use of formic acid and discussion of some of its properties. *Arch Pathol*, **65**, 519–544.

Baccarani-Contri, M., Vincenzi, D., Cicchetti, F., Mori, G. and Pasquali-Ronchetti, I. (1990) Immunocytochemical localization of proteoglycans within normal elastin fibers. *Eur J Cell Biol*, **53**, 305–312.

Barton, D.E., Foellmer, B.E., Du, J., Tamm, J., Derynck, R. and Francke, U. (1988) Chromosomal mapping of genes for transforming growth factors beta 2 and beta 3 in man and mouse: dispersion of TGF-beta gene family. *Oncogene Res*, **3**, 323–331.

Bashir, M.M., Indik, Z., Yeh, H., Ornstein, G., N, Rosenbloom, J.C., Abrams, W., Fazio, M., Uitto, J. and Rosenbloom, J. (1989) Characterization of the complete human elastin gene. Delineation of unusual features in the 5′-flanking region. *J Biol Chem*, **264**, 8887–8891.

Bashir, M.M., Abrams, W.R., Rosenbloom, J., Kucich, U., Bacarra, M., Han, M.D., Brown-Augsburger, P.,

Mecham, R. and Rosenbloom, J. (1994) Microfibril-associated glycoprotein: Characterization of the bovine gene and of the recombinantly expressed protein. *Biochemistry*, **33**, 593–600.

Baule, V.J. and Foster, J.A. (1988) Multiple chick tropoelastin mRNAs. *Biochem Biophys Res Commun*, **154**, 1054–1060.

Belik, J., Keeley, F.W, Baldwin, F., Rabinowitch, M. (1994) Pulmonary hypertension and vascular remodeling in fetal sheep. *Am J Physiol*, **266**, H2303–H2309.

Bernstein, E., Chen, Y., Tamai, K., Shepley, K., Resnik, K., Zhang, H., Tuan, R., Mauviel, A. and Uitto, J. (1994) Enhanced elastin and fibrillin gene expression in chronically photodamaged skin. *J Invest Dermatol*, **103**, 182–186.

Blood, C.H. and Zetter, B.R. (1989) Membrane-bound protein kinase C modulates receptor affinity and chemotactic responsiveness of Lewis lung carcinoma sublines to an elastin-derived peptide. *J Biol Chem*, **264**, 10614–10620.

Boileau, C., Jondeau, G., Babron, M.-C., Coulon, M., Alexandre, J.-A., Sakai, L., Melki, J., Delorme, G., Dubourg, O., Bonaiti-Pellie, C., Bourdarais, J.-P. and Junien, C. (1993) Autosomal dominant Marfan-like connective-tissue disorder with aortic dilation and skeletal anomalies not linked to the fibrillin genes. *Am J Hum Genet*, **53**, 46–54.

Bressan, G.M., Daga, G., D, Colombatti, A., Castellani, I., Marigo, V. and Volpin, D. (1993) Emilin, a component of elastic fibers preferentially located at the elastin-microfibrils interface. *J Cell Biol*, **121**, 201–212.

Bressan, G.M., Argos, P. and Stanley, K.K. (1987) Repeating structure of chick tropoelastin revealed by complementary DNA cloning. *Biochemistry*, **26**, 1497–1503.

Brown-Augsburger, P., Broekelmann, T., Mecham, L., Mercer, R., Gibson, M.A., Cleary, E.G., Abrams, W.R., Rosenbloom, J. and Mecham, R.P. (1994) Microfibril associated glycoprotein (MAGP) binds to the carboxy-terminal domain of tropoelastin and is a substrate for transglutaminase. *J Biol Chem*, **269**, 28443–28449.

Burnett, W., Finnigan-Bunick, A., Yoon, K. and Rosenbloom, J. (1982) Analysis of elastin gene expression in the developing chick aorta using cloned elastin cDNA. *J Biol Chem*, **257**, 1569–1572.

Calabretta, B., Robberson, D.L., Berrera-Saldana, H.A., Lambrou, T.P. and Saunders, G.F. (1982) Genome instability in a region of human DNA enriched in Alu repeat sequences. *Nature*, **296**, 219–225.

Chen, Y., Faraco, J., Yin, W., Germiller, J., Francke, U. and Bonadio, J. (1993) Structure, chromosomal localization, and expression pattern of the murine MAGP gene. *J Biol Chem*, **268**, 27381–27389.

Cicila, G., May, M., Ornstein-Goldstein, N., Indik, Z., Morrow, S., Yeh, H.S., Rosenbloom, J., Boyd, C., Rosenbloom, J. and Yoon, K. (1985) Structure of the 3′ portion of the bovine elastin gene. *Biochemistry*, **24**, 3075–3080.

Cleary, E.G. and Cliff, W.J. (1978) The substructure of elastin. *Exp Mol Pathol*, **28**, 227–246.

Cleary, E.G. and Gibson, M.A. (1983) Elastin-associated microfibrils and microfibrillar proteins. In *International Review of Connective Tissue*, D.A. Hall and D.S. Jackson (ed.), pp. 97–209. New York: Academic Press.

Cleary, E.G. and Moont, M. (1976) Hypertension in weanling rabbits. *Adv Exp Med Biol*, **79**, 477–490.

Cleary, E.G., Sandberg, L.B. and Jackson, D.S. (1967) The changes in the chemical composition during development of the bovine nuchal ligament. *J Cell Biol*, **33**, 469–479.

Colombatti, A., Bonaldo, P., Volpin, D. and Bressan, G.M. (1988) The elastin associated glycoprotein GP115. Synthesis and secretion by chick cells in culture. *J Biol Chem*, **263**, 17534–17540.

Colombatti, A., Poletti, A., Bressan, G.M., Carbone, A. and Volpin, D. (1987) Widespread codistribution of glycoprotein GP115 and elastin in chick eye and other tissues. *Coll Rel Res*, **7**, 259–275.

Corson, G.M., Chalberg, S.C., Dietz, H.C., Charbonneau, N.L. and Sakai, L.Y. (1993) Fibrillin binds calcium and is coded by cDNAs that reveal a multidomain structure and alternatively spliced exons at the 5′ end. *Genomics*, **17**, 476–484.

Curran, M.E., Atkinson, D.L., Ewart, A.K., Morris, C.A., Leppert, M.F. and Keating, M.T. (1993) The elastin gene is disrupted by a translocation associated with supravalvular aortic stenosis. *Cell*, **73**, 159–168.

Curran, T. and Franza, B.J. (1988) Fos and Jun: the AP-1 connection. *Cell*, **55**, 395–397.

Dahlback, K., Lofberg, H., Alumets, J. and Dahlback, B. (1989) Immunohistochemical demonstration of age-related deposition of vitronectin (S-protein of complement) and terminal complement component complex on dermal elastic fibres. *J Invest Dermatol*, **92**, 727–733.

Danks, D.M. (1993) Disorders of copper transport: Menkes disease and the occipital horn syndrome. In *Connective Tissue and its Heritable Disorders.*, P.M. Royce and B. Steinman (ed.), pp. 487–505. New York: Wiley-Liss.

Davidson, J.M., Zoia, O. and Liu, J.M. (1993) Modulation of transforming growth factor-beta 1 stimulated elastin and collagen production and proliferation in porcine vascular smooth muscle cells and skin fibroblasts by basic fibroblast growth factor, transforming growth factor-alpha, and insulin-like growth factor-I. *J Cell Physiol*, **155**, 149–156.

Davidson, J.M., Shibahara, S., Boyd, C., Mason, M.L., Tolstoshev, P. and Crystal, R.G. (1984) Elastin mRNA levels during foetal development of sheep nuchal ligament and lung. *Biochem J*, **220**, 653–663.

Davis, P., Ryan, P.A., Osipowicz, J., Anderson, M.J., Sweeney, A. and Stehbens, W.E. (1989) The biochemical

composition of hemodynamically stressed vascular tissue — The insoluble elastin of experimental arteriovenous fistulae. *Exp Mol Pathol*, **51**, 103–110.
Dietz, H.C., Cutting, G.R., Pyeritz, R.E., Maslen, C.L., Sakai, L.Y., Corson, G.M., Puffenberger, E.G., Hamosh, A., Nanthakumar, E.J., Curristin, S.M., *et al.* (1991) Marfan syndrome caused by a recurrent de novo missense mutation in the fibrillin gene. *Nature*, **352**, 337–339.
Dietz, H.C., McIntosh, I., Sakai, L.Y., Corson, G.M., Chalberg, S.C., Pyeritz, R.E. and Francomano, C.A. (1993a) Four novel FBN1 mutations: significance for mutant transcript level and EGF-like domain calcium binding in the pathogenesis of Marfan syndrome. *Genomics*, **17**, 468–475.
Dietz, H.C., Pyeritz, R.E., Puffenberger, E.G., Kendzior, R.J., Corson, G.M., Maslen, C.L., Sakai, L.Y., Francomano, C.A. and Cutting, G.R. (1992) Marfan phenotype variability in a family segregating a missense mutation in the epidermal growth factor-like motif of the fibrillin gene. *J Clin Invest*, **89**, 1674–1680.
Dietz, H.C., Valle, D., Francomano, C.A., Kendzior, R.J., Pyeritz, R.E. and Cutting, G.R. (1993b) The skipping of constitutive exons in vivo induced by nonsense mutations. *Science*, **259**, 680–683.
Dillon, T.J., Walsh, R.L., Scicchitano, R., Eckert, B., Cleary, E.G. and McLennan, G. (1992) Plasma elastin-derived peptide levels in normal adults, children, and emphysematous subjects. Physiologic and computed tomographic scan correlates. *Am Rev Resp Dis*, **146**, 1143–1148.
Elliott, R.J. and McGrath, L.T. (1994) Calcification of the human thoracic aorta during aging. *Calcif Tissue Int*, **54**, 268–273.
Ewart, A.K., Jin, W., Atkinson, D., Morris, C.A. and Keating, M.T. (1994) Supravalvular aortic stenosis associated with a deletion disrupting the elastin gene. *J Clin Invest*, **93**, 1071–1077.
Ewart, A.K., Morris, C.A., Atkinson, D., Jin, W., Sternes, K., Spallone, P., Stock, A.D., Leppert, M. and Keating, M.T. (1993a) Hemizygosity at the elastin locus in a developmental disorder, Williams syndrome. *Nat Genet*, **5**, 11–16.
Ewart, A.K., Morris, C.A., Ensing, G.J., Loker, J., Moore, C., Leppert, M. and Keating, M. (1993b) A human vascular disorder, supravalvular aortic stenosis, maps to chromosome 7. *Proc Natl Acad Sci USA.*, **90**, 3226–3230.
Eyre, D.R., Paz, M. A. and Gallop, P.M. (1984) Cross-linking in collagen and elastin. *Annu Rev Biochem*, **53**, 717–748.
Fahrenbach, W.H., Sandberg, L.B. and Cleary, E.G. (1966) Ultrastructural studies on early elastogenesis. *Anat Rec*, **155**, 563–575.
Fanning, J.C. and Cleary, E.G. (1974) Connective tissues in copper-deficient sheep. *Eighth International Congress on Electron Microscopy*, **II**, 696–697.
Fanning, J.C. and Cleary, E.G. (1988) Identification of glycoproteins associated with elastin-associated microfibrils. *J Histochem Cytochem*, **33**, 287–294.
Faraco, J., Milatovitch, A., Mecham, R., Bashir, M.M., Rosenbloom, J. and Francke, U. (1993) Assignment of the microfibril-associated glycoprotein (MAGP) gene to chromosome 1p36. *Am J Med Genet*, **47**, 148.
Fazio, M., Mattei, M., Passage, E., Chu, M., Black, D., Solomon, E., Davidson, J. and Uitto, J. (1991) Human elastin gene: new evidence for localization to the long arm of chromosome 7. *Am J Hum Genet*, **48**, 696–703.
Fleming, W.W., Sullivan, C.E. and Torchia, D.A. (1980) Characterization of molecular motions in ^{13}C-labelled elastin ^{12}C-^{1}H magnetic double resonance. *Biopolymers*, **94**, 191–196.
Flory, P.J. (1969) *Statistical Mechanics of Chain Molecules*, New York: Wiley-Interscience.
Fornieri, C., Baccarini-Contri, M., Quaglino, D. and Pasquali-Ronchetti, I. (1987) Lysyl oxidase activity and elastin/glycosaminoglycan interactions in growing chick and rat aortas. *J Cell Biol*, **105**, 1463–1469.
Foster, J.A., Rich, C.B., Fletcher, S., Karr, S.R. and Przybyla, A. (1980) Translation of chick aortic elastin ribonucleic acid. Comparison to elastin synthesis in chick organ culture. *Biochemistry*, **19**, 857–864.
Fullmer, H.M. and Lillie, R.D. (1958) The oxytalan fiber. A previously undescribed connective tissue fiber. *J Histochem Cytochem*, **6**, 425–430.
Gacheru, S., McGee, C., Uriu-Hare, J.Y., Kosonen, T., Packman, S., Tinker, D., Krawetz, S.A., Reiser, K., Keen, C.L. and Rucker, R.B. (1993) Expression and accumulation of lysyl oxidase, elastin, and type I procollagen in human Menkes and mottled mouse fibroblasts. *Arch Biochem Biophys*, **301**, 325–329.
Gawlik, H. (1965) Morphological and morphochemical properties of the elastic system in the motor organ of man. *Folia Histochem Cytochem*, **3**, 233–251.
Gibson, M.A., Kumaratilake, J.S. and Cleary, E.G. (1989) The protein components of the 12-nanometer microfibrils of elastic and nonelastic tissues. *J Biol Chem*, **264**, 4590–4598.
Gibson, M.A., Sandberg, L.B., Grosso, L.E. and Cleary, E.G. (1991) Complementary DNA cloning establishes microfibril-associated glycoprotein (MAGP) to be a discrete component of the elastin-associated microfibrils. *J Biol Chem*, **266**, 7596–7601.
Gibson, M.A. (1993) Elastin associated microfibrils: Fibrillin aggregates or multi-component glycoprotein complexes? *Am J Med Genet*, **47**, 148.
Gibson, M.A. and Cleary, E.G. (1987) The immunohistochemical localisation of microfibril-associated glycoprotein (MAGP) in elastic and non-elastic tissues. *Immunol Cell Biol*, **65**, 345–356.

Gibson, M.A., Davis, E., Filiaggi, M. and Mecham, R.P. (1993) Identification and partial characterisation of a new fibrillin-like protein (FLP). *Am J Med Genet*, **47**, 148.

Gibson, M.A., Hughes, J.L., Fanning, J.C. and Cleary, E.G. (1986) The major antigen of elastin-associated microfibrils is a 31-kDa glycoprotein. *J Biol Chem*, **261**, 11429–11436.

Gosline, J.M. (1976) The physical properties of elastic tissue. In *International Review of Connective Tissue Research*, D.A. Hall and D.S. Jackson (ed.), pp. 211–249. New York: Academic Press Inc.

Gotte, L., Giro, M.G., Volpin, D. and Horne, R.W. (1974) The ultrastructural organisation of elastin. *J Ultrastruct Res*, **46**, 23–33.

Gray, W.R., Sandberg, L.B. and Foster, J.A. (1973) Molecular model for elastin structure and function. *Nature*, **246**, 461–466.

Greenlee, T.K., Ross, R. and Hartman, J.L. (1966) The fine structure of elastic fibres. *J Cell Biol*, **30**, 59–71.

Handford, P.A., Mayhew, M., Baron, M., Winship, P.R., Campbell, I.D. and Brownlee, G.G. (1991) Key residues involved in calcium-binding motifs in EGF-like domains. *Nature*, **351**, 164–167.

Hass, G.M. (1942) Elastic tissue. -1. Description of a method for the isolation of elastic tissue. *Arch Pathol*, **34**, 807–819.

Heim, R.A., Pierce, R.A., Deak, S.B., Riley, D.J., Boyd, C.D. and Stolle, C.A. (1991) Alternative splicing of rat tropoelastin mRNA is tissue-specific and developmentally regulated. *Matrix*, **11**, 359–366.

Hinek, A., Mecham, R.P., Keeley, F. and Rabinovitch, M. (1991) Impaired elastin fiber assembly related to reduced 67-kD elastin-binding protein in fetal lamb ductus arteriosus and in cultured aortic smooth muscle cells treated with chondroitin sulfate. *J Clin Invest*, **88**, 2083–2094.

Hinek, A., Wrenn, D.S., Mecham, R.P. and Barondes, S.H. (1988) The elastin receptor is a galactoside binding protein. *Science*, **239**, 1539–1541.

Hollister, D.W., Godfrey, M., Sakai, L.Y. and Pyeritz, R.E. (1990) Immunohistologic abnormalities of the microfibrillar-fiber system in the Marfan syndrome. *N Engl J Med*, **323**, 152–159.

Holman, E. (1937) *Arteriovenous Aneurysm*, London: MacMillan.

Hornebeck, W., Tixier, J.M. and Robert, L. (1986) Inducible adhesion of mesenchymal cells to elastic fibers: Elastonectin. *Proc Natl Acad Sci USA.*, **83**, 5517–5520.

Horrigan, S.K., Rich, C.B., Streeten, B.W., Li, Z.Y. and Foster, J.A. (1992) Characterization of an associated microfibril protein through recombinant DNA techniques. *J Biol Chem*, **267**, 10087–10095.

Indik, Z., Yeh, H., Ornstein-Goldstein, N., Sheppard, P., Anderson, N., Rosenbloom, J.C., Peltonen, L. and Rosenbloom, J. (1987) Alternative splicing of human elastin mRNA indicated by sequence analysis of cloned genomic and complementary DNA. *Proc Natl Acad Sci USA.*, **84**, 5680–5684.

Indik, Z., Yeh, H., Ornstein-Goldstein, N. and Rosenbloom, J.C. (1990) Structure of the elastin gene and alternative splicing of elastin mRNA. In *Genes for Extracellular Matrix Proteins*, L. Sandell and C. Boyd (ed.), pp. 221–250. New York: Academic Press.

Jackson, D.S. and Cleary, E.G. (1967) The determination of collagen and elastin. In *Methods of Biochemical Analysis*, D. Glick (ed.), pp. 25–76. New York: Interscience.

Janoff, A. (1985) Elastases and emphysema. Current assessment of the protease-antiprotease hypothesis. *Am Rev Respir Dis*, **132**, 417–433.

Kagan, H.M., Vaccaro, C.A., Bronson, E., Tang, S.-S. and Brody, J.S. (1986) Ultrastructural immunolocalization of lysyl oxidase in vascular connective tissue. *J Cell Biol*, **103**, 1121–1128.

Kahari, V.M., Chen, Y.Q., Bashir, M.M., Rosenbloom, J. and Uitto, J. (1992a) Tumor necrosis factor-alpha down-regulates human elastin gene expression. Evidence for the role of AP-1 in the suppression of promoter activity. *J Biol Chem*, **267**, 26134–26141.

Kahari, V.M., Fazio, M.J., Chen, Y.Q., Bashir, M.M., Rosenbloom, J. and Uitto, J. (1990) Deletion analyses of 5′-flanking region of the human elastin gene. Delineation of functional promoter and regulatory cis-elements. *J Biol Chem*, **265**, 9485–9490.

Kahari, V.M., Olsen, D.R., Rhudy, R.W., Carrillo, P., Chen, Y.Q. and Uitto, J. (1992b) Transforming growth factor-beta up-regulates elastin gene expression in human skin fibroblasts. Evidence for post-transcriptional modulation. *Lab Invest*, **66**, 580–588.

Kainulainen, K., Pulkkinen, L., Savolainen, A., Kaitila, I. and Peltonen, L. (1990) Location on chromosome 15 of the gene defect causing Marfan syndrome. *N Engl J Med*, **323**, 935–939.

Kainulainen, K., Sakai, L.Y., Child, A., Pope, F.M., Puhakka, L., Ryhanen, L., Palotie, A., Kaitila, I. and Peltonen, L. (1992) Two mutations in Marfan syndrome resulting in truncated fibrillin polypeptides. *Proc Natl Acad Sci USA.*, **89**, 5917–5921.

Kajikawa, K., Yamaguchi, T., Katsuda, S. and Miwa, A. (1975) An improved electron stain for elastic fibers using tannic acid. *J Electron Micros*, **24**, 287–289.

Kanzaki, T., Olofsson, A., Moren, A., Wernstedt, C., Hellman, U., Miyazono, K., Claesson, W., L and Heldin, C.H. (1990) TGF-beta 1 binding protein: a component of the large latent complex of TGF-beta 1 with multiple repeat sequences. *Cell*, **61**, 1051–1061.

Keeley, F.W. and Alatawi, A. (1991) Response of aortic elastin synthesis and accumulation to developing hypertension and the inhibitory effect of colchicine on this response. *Lab Invest*, **64**, 499–507.

Keene, D.R., Maddox, B.K., Kuo, H.J., Sakai, L.Y. and Glanville, R.W. (1991) Extraction of extendable beaded structures and their identification as fibrillin-containing extracellular matrix microfibrils. *J Histochem Cytochem*, **39**, 441–449.

Kewley, M.A., Steven, F.S. and Williams, G. (1977) Preparation of a specific antiserum towards the microfibrillar protein of elastic tissues. *Immunology*, **32**, 483–489.

Kielty, C.M., Phillips, J.E., Child, A.H., Pope, F.M. and Shuttleworth, C.A. (1994) Fibrillin secretion and microfibril assembly by Marfan dermal fibroblasts. *Matrix Biol*, **14**, 191–199.

Kielty, C.M. and Shuttleworth, C.A. (1993) The role of calcium in the organization of fibrillin microfibrils. *FEBS Lett*, **336**, 323–326.

Kielty, C.M. and Shuttleworth, C.A. (1994) Abnormal fibrillin assembly by dermal fibroblasts from two patients with Marfan syndrome. *J Cell Biol*, **124**, 997–1004.

King, G.S., Mohan, V.S. and Starcher, B.S. (1980) Radioimmunoassay for desmosine. *Connect Tissue Res*, **7**, 263–267.

King, R.A., Smith, R.M., Krishnan, R. and Cleary, E.G. (1992) Effects of enalapril and hydralazine treatment and withdrawal upon cardiovascular hypertrophy in stroke-prone spontaneously hypertensive rats. *J Hypertens*, **10**, 919–928.

Krishnan, R. and Cleary, E.G. (1990) Elastin gene expression in elastotic human breast cancers and epithelial cell lines. *Cancer Res*, **50**, 2164–2171.

Kumaratilake, J.S., Gibson, M.A., Fanning, J.C. and Cleary, E.G. (1989) The tissue distribution of microfibrils reacting with a monospecific antibody to MAGP, the major glycoprotein antigen of elastin-associated microfibrils. *Eur J Cell Biol*, **50**, 117–127.

Kumaratilake, J.S., Krishnan, R., Lomax-Smith, J. and Cleary, E.G. (1991) Elastofibroma: Disturbed elastic fibrillogenesis by periosteal-derived cells. *Hum Pathol*, **22**, 1017–1029.

Lansing, A.I. (1951) Chemical morphology of elastic fibers. In *Transactions Second Conference on Connective Tissue.*, C. Ragan (ed.), pp. New York: Josiah Macy, Jr. Foundation.

Lansing, A.I. (1959) Elastic Tissue. In *The Arterial Wall*, A.I. Lansing (ed.), pp. 136–160. Baltimore: Williams and Wilkins Co.

Larsen, F., Gundersen, G., Lopez, R. and Prydz, H. (1992) CpG islands as gene markers in the human genome. *Genomics*, **13**, 1095–1107.

Laurel, C.B. and Eriksson, S. (1963) The electrophoretic alpha 1-globulin pattern of serum in alpha 1-antitryspin deficiency. *Scand J Clin Invest*, **15**, 132–140.

Lavker, R.M. (1979) Structural alterations in exposed and unexposed aged skin. *J Invest Dermatol*, **73**, 59–66.

Lebwohl, M., Schwartz, E., Lemlich, G., Lovelace, O., Shaikh-Bahai, F. and Fleischmajer, R. (1993) Abnormalities of connective tissue components in lesional and non-lesional tissue of patients with pseudoxanthoma elasticum. *Arch Dermatol Res*, **285**, 121–126.

Lee, B., Godfrey, M., Vitale, E., Hori, H., Mattei, M.G., Sarfarazi, M., Tsipouras, P., Ramirez, F. and Hollister, D.W. (1991) Linkage of Marfan syndrome and a phenotypically related disorder to two different fibrillin genes. *Nature*, **352**, 330–334.

Lee, K.A., Pierce, R.A., Davis, E.C., Mecham, R.P. and Parks, W.C. (1994) Conversion to an elastogenic phenotype by fetal hyaline chondrocytes is accompanied by altered expression of elastin-related macromolecules. *Dev Biol*, **163**, 241–252.

Levene, C.I., Sharman, D.F. and Callingham, B.A. (1992) Inhibition of chick embryo lysyl oxidase by various lathyrogens and the antagonistic effect of pyridoxal. *Int J Exp Pathol*, **73**, 613–624.

Liu, B., Harvey, C. and McGowan, S. (1993) Retinoic acid increases elastin in neonatal rat lung fibroblast cultures. *Am J Physiol*, **265**, L430–L437.

Liu, J.M. and Davidson, J.M. (1988) The elastogenic effect of recombinant transforming growth factor-beta on porcine aortic smooth muscle cells. *Biochem Biophys Res Commun*, **154**, 895–901.

Maddox, B.K., Sakai, L.Y., Keene, D.R. and Glanville, R.W. (1989) Connective tissue microfibrils. Isolation and characterization of three large pepsin-resistant domains of fibrillin. *J Biol Chem*, **264**, 21381–21385.

Magenis, R.E., Maslen, C.L., Smith, L., Allen, L. and Sakai, L.Y. (1991) Localization of the fibrillin (FBN) gene to chromosome 15, band q21.1. *Genomics*, **11**, 346–351.

Manohar, A., Shi, W. and Anwar, R. (1991) Partial characterization of bovine elastin gene; comparison with the gene for human elastin. *Biochem Cell Biol*, **69**, 185–192.

Mariencheck, M.C., Davis, E.C., Zhang, H., Ramirez, F., Rosenbloom, J., Gibson, M.A., Parks, W.C. and Mecham, R.P. (1994) Fibrillin-1 and fibrillin-2 show temporal and tissue-specific regulation of expression in developing elastic tissues. *Connect Tissue Res*, **31**, 1–11.

Marigo, V., Volpin, D. and Bressan, G.M. (1993) Regulation of the human elastin promoter in chick embryo cells. Tissue-specific effect of TGF-beta. *Biochim Biophys Acta*, **1172**, 31–36.

Marigo, V., Volpin, D., Vitale, G. and Bressan, G.M. (1994) Identification of a TGF-beta responsive element in the human elastin promoter. *Biochem Biophys Res Commun*, **199**, 1049–1056.

Martin, G.R., Schiffman, E., Bladen, H.A. and Mylen, M. (1963) Chemical and Morphological studies on the *in vitro* calcification of aorta. *J Cell Biol*, **16**, 243–252.

Maslen, C.L., Corson, G.M., Maddox, B.K., Glanville, R.W. and Sakai, L.Y. (1991) Partial sequence of a candidate gene for the Marfan syndrome. *Nature*, **352**, 334–347.

Maslen, C.L. and Glanville, R.W. (1993) The molecular basis of Marfan syndrome. *DNA Cell Biol*, **12**, 561–572.

Mauviel, A., Chen, Y.Q., Kahari, V.M., Ledo, I., Wu, M., Rudnicka, L. and Uitto, J. (1993) Human recombinant interleukin-1 beta up-regulates elastin gene expression in dermal fibroblasts. Evidence for transcriptional regulation in vitro and in vivo. *J Biol Chem*, **268**, 6520–6524.

McLain, J.B. and Vig, P.S. (1983) Transverse periosteal sectioning and femur growth in the rat. *Anat Rec*, **207**, 339–348.

Mecham, R.P., Griffin, G.L., Madaras, J.G. and Senior, R.M. (1984a) Appearance of chemotactic responsiveness to elastin peptides by developing fetal bovine ligament fibroblasts parallels the onset of elastin production. *J Cell Biol*, **98**, 1813–1816.

Mecham, R.P. and Heuser, J.E. (1991) The elastic fiber. In *Cell Biology of Extracellular Matrix*, E.D. Hay (ed.), pp. 79–109. New York: Plenum Press.

Mecham, R.P., Madaras, J.G. and Senior, R.M. (1984b) Extracellular matrix-specific induction of elastogenic differentiation and maintenance of phenotypic stability in bovine nuchal ligament fibroblasts. *J Cell Biol*, **98**, 1804–1812.

Mecham, R.P., Morris, S.L., Levy, B.D. and Wrenn, D.S. (1984c) Glucocorticoids stimulate elastin production in differentiated bovine ligament fibroblasts but do not induce elastin synthesis in undifferentiated cells. *J Biol Chem*, **259**, 12414–12418.

Mecham, R.P., Whitehouse, L., Hay, M., Hinek, A. and Sheetz, M. (1991) Ligand affinity of the 67 kDa elastin/laminin binding protein is modulated by the protein's lectin domain: visualization of elastin/laminin receptor complexes with gold-tagged ligands. *J Cell Biol*, **113**, 187–194.

Milewicz, D.W., Pyeritz, R.E., Crawford, E.S. and Byers, P.H. (1992) Marfan syndrome: defective synthesis, secretion, and extracellular matrix formation of fibrillin by cultured dermal fibroblasts. *J Clin Invest*, **89**, 79–86.

Morris, C.A., Loker, J., Ensing, G. and Stock, A.D. (1993) Supravalvular aortic stenosis cosegregates with a familial 6; 7 translocation which disrupts the elastin gene. *Am J Med Genet*, **46**, 737–744.

Nagamine, N., Nohara, Y. and Ito, E. (1982) Elastofibroma in Okinawa. A clinicopathologic study of 170 cases. *Cancer*, **50**, 1794–1805.

Neldner, K.H. (1993) Pseudoxanthoma Elasticum. In *Connective Tissue and its Heritable Disorders.*, P.M. Royce and B. Steinman (ed.), pp. 425–436. New York: Wiley-Liss Inc.

Olson, T.M., Michels, V.V., Lindor, N.M., Pastores, G.M., Weber, J.L., Schaid, D.J., Driscoll, D.J., Feldt, R.H. and Thibodeau, S.N. (1993) Autosomal dominant supravalvular aortic stenosis: localization to chromosome 7. *Hum Mol Genet*, **2**, 869–873.

Parks, W.C. and Deak, S.B. (1990) Tropoelastin heterogeneity: implications for protein function and disease. Am J Respir *Cell Mol Biol*, **2**, 399–406.

Parks, W.C., Kolodziej, M.E. and Pierce, R.A. (1992) Phorbol ester-mediated downregulation of tropoelastin expression is controlled by a posttranscriptional mechanism. *Biochemistry*, **31**, 6639–6645.

Parks, W.C., Secrist, H., Wu, L.C. and Mecham, R.P. (1988) Developmental regulation of tropoelastin isoforms. *J Biol Chem*, **263**, 4416–4423.

Partridge, S.M. (1967) Elastin Structures and biosynthesis. In *Symposium on Fibrous Proteins*, W.G. Crewther (ed.), pp. 246–264. Canberra, Australia: Butterworth's.

Partridge, S.M., Davis, H.F. and Adair, G.S. (1955) Soluble proteins derived from partial hydrolysis of elastin. *Biochem J*, **61**, 11–21.

Pasquali-Ronchetti, I., Fornieri, C., Castellani, I., Bressan, G.M. and Volpin, D. (1981) Alterations of the connective tissue components induced by ß-aminopropionitrile. *Exp Mol Pathol*, **35**, 42–56.

Pereira, L., D'Alessio, M., Ramirez, F., Lynch, J.R., Sykes, B., Pangilinan, T. and Bonadio, J. (1993) Genomic organization of the sequence coding for fibrillin, the defective gene product in Marfan syndrome. *Hum Mol Genet*, **2**, 961–968.

Pierce, R.A., Alatawi, A., Deak, S.B. and Boyd, C.D. (1992a) Elements of the rat tropoelastin gene associated with alternative splicing. *Genomics*, **12**, 651–658.

Pierce, R.A., Deak, S.B., Stolle, C.A. and Boyd, C.D. (1990) Heterogeneity of rat tropoelastin mRNA revealed by cDNA cloning. *Biochemistry*, **29**, 9677–9683.

Pierce, R.A., Kolodziej, M.E. and Parks, W.C. (1992b) 1,25-Dihydroxyvitamin D3 represses tropoelastin expression by a posttranscriptional mechanism. *J Biol Chem*, **267**, 11593–11599.

Pinnell, S.R. and Martin, G.R. (1968) The cross-linking of collagen and elastin: Enzymatic conversion of lysine in peptide linkage to a α-aminoadipic-∂-semialdehyde (allysine) by an extract from bone. *Proc Natl Acad Sci USA.*, **61**, 708–716.

Pollock, J., Baule, V.J., Rich, C.B., Ginsburg, C.D., Curtiss, S.W. and Foster, J.A. (1990) Chick tropoelastin isoforms. From the gene to the extracellular matrix. *J Biol Chem*, **265**, 3697–3702.

Ponseti, I.V.B. and Baird, W.A. (1952) Scoliosis and dissecting aneurysm of the aorta in rats fed with lathyrus odoratus seeds. *Am J Pathol*, **28**, 1059–1077.

Powell, J.T., Vine, N. and Crossman, M. (1992) On the accumulation of D-aspartate in elastin and other proteins of the ageing aorta. *Atherosclerosis*, **97**, 201–208.
Prosser, I.W., Stenmark, K.R., Suthar, M., Crouch, E.C., Mecham, R.P. and Parks, W.C. (1989) Regional heterogeneity of elastin and collagen gene expression in intralobar arteries in response to hypoxic pulmonary hypertension as demonstrated by in situ hybridization. *Am J Pathol*, **135**, 1073–1088.
Pyeritz, R.E. and Francke, U. (1993) The Second International Symposium on the Marfan Syndrome. *Am J Med Genet*, **47**, 127–135.
Raju, K. and Anwar, R.A. (1987) Primary structure of bovine elastin a, b and c deduced from the sequences of cDNA clones. *J Biol Chem*, **262**, 5755–5762.
Ramirez, F., Pereira, L., Zhang, H. and Lee, B. (1993) The fibrillin-Marfan syndrome connection. *Bioessays*, **15**, 589–594.
Rasmussen, B.L., Bruenger, E. and Sandberg, L.B. (1975) A new method for purification of mature elastin. *Anal Biochem*, **64**, 255–259.
Raybould, M.C., Birley, A.J., Moss, C., Hulten, M. and McKeown, C.M. (1994) Exclusion of an elastin gene (ELN) mutation as the cause of pseudoxanthoma elasticum (PXE) in one family. *Clin Genet*, **45**, 48–51.
Rich, C.B., Ewton, D.Z., Martin, B.M., Florini, J.R., Bashir, M.M., Rosenbloom, J. and Foster, J.A. (1992) IGF-I regulation of elastogenesis: comparison of aortic and lung cells. *Am J Physiol*, **263**, L276–L282.
Rich, C.B., Goud, H.D., Bashir, M.M., Rosenbloom, J. and Foster, J.A. (1993) Developmental regulation of aortic elastin gene expression involves disruption of an IGF-I sensitive repressor complex. *Biochem Biophys Res Commun*, **196**, 1316–1322.
Rich, C.B. and Foster, J.A. (1987) Evidence for the existence of three chick lung tropoelastins. *Biochem Biophys Res Commun*, **146**, 1291–1285.
Ross, R. and Bornstein, P. (1969) The elastic fiber: The separation and partial characterization of its macromolecular components. *J Cell Biol*, **40**, 366–381.
Sage, H. and Gray, W.R. (1979) Studies on the evolution of elastin - 1. Phylogenetic distribution. *Comp Biochem Physiol*, **64B**, 313–327.
Sakai, L.Y., Keene, D.R., Glanville, R.W. and Bachinger, H.P. (1991) Purification and partial characterization of fibrillin, a cysteine-rich structural component of connective tissue microfibrils. *J Biol Chem*, **266**, 14763–14770.
Sakai, L.Y., Keene, D.R. and Engvall, E. (1986) Fibrillin, a new 350-kD glycoprotein, is a component of extracellular microfibrils. *J Cell Biol*, **103**, 2499–2509.
Sakai, L.Y. (1990) Disulfide bonds crosslink molecules of fibrillin in the connective tissue space. In *Elastin: Chemical and Biological Aspects*, A.M. Tamburro and J.M. Davidson (ed.), pp. 214–227. Galatina, Italy: Congedo Editore.
Sandberg, L.B., Soskel, N.T. and Leslie, J.B. (1981) Elastin structure, biosynthesis and relation to disease states. *New Engl J Med*, **304**, 566–579.
Sandberg, L.B., Weissman, N. and Smith, D.W. (1969) The purification and partial characterization of a soluble elastin–like protein from copper-deficient porcine aorta. *Biochemistry*, **8**, 2940–2945.
Sandell, L.J. and Boyd, C.D. (1990) Conserved and divergent sequence and functional elements within collagen genes. In *Extracellular Matrix Genes*, L.J. Sandell and C.D. Boyd (ed.), pp. 1–56. New Brunswick: Academic Press, Inc.
Schwartz, E., Thieberg, M., Cruickshank, F.A. and Lebwohl, M. (1991) Elastase digestion of normal and pseudoxanthoma elasticum lesional skin elastins. *Exp Mol Pathol*, **55**, 190–195.
Scott, J.E. (1990) Proteoglycan:collagen interactions and subfibrillar structure in collagen fibrils. Implications in the development and ageing of connective tissues. *J Anat*, **169**, 23–35.
Senior, R.M., Griffin, G.L., Fliszar, C.J., Shapiro, S.D., Goldberg, G.I. and Welgus, H.G. (1991) Human 92-kilodalton and 72-kilodalton type IV collagenases are elastases. *J Biol Chem*, **266**, 7870–7875.
Senior, R.M., Griffin, G.L. and Mecham, R.P. (1982) Chemotactic response of fibroblasts to tropoelastin and elastin-derived peptides. *J Clin Invest*, **70**, 614–618.
Shapiro, S.D., Endicott, S.K., Province, M.A., Pierce, J.A. and Campbell, E.J. (1991) Marked longevity of human lung parenchymal elastic fibers deduced from prevalence of D-aspartate and nuclear weapons-related radiocarbon. *J Clin Invest*, **87**, 1828–1834.
Simpson, C.F., Boucek, R.J. and Noble, N.L. (1980) Similarity of aortic pathology in Marfan's syndrome, copper deficiency in chicks and β-aminopropionitrile toxicity in turkeys. *Exp Mol Pathol*, **32**, 81–90.
Skonier, J., Neubauer, M., Madisen, L., Bennett, K., Plowman, G.D. and Purchio, A.F. (1992) cDNA cloning and sequence analysis of beta IG-H3, a novel gene induced in a human adenocarcinoma cell line after treatment with transforming growth factor-beta. *DNA Cell Biol*, **11**, 511–522.
Snider, G.L. (1992) Emphysema — the first two centuries — and beyond — a historical overview, with suggestions for future research .2. *Am Rev Respir Dis*, **146**, 1615–1622.
Streeten, B.W. and Gibson, S.A. (1988) Identification of extractable proteins from the bovine ocular zonule: major zonular antigens of 32 kD and 250 kD. *Curr Eye Res*, **7**, 139–146.

Tamburro, A.M. (1981) Elastin: Molecular and supramolecular structure. In *Connective Tissue Research: Chemistry, Biology, and Physiology.*, Z. Dehl and M. Adam (ed.), pp. 45–62. New York: Alan R. Liss, Inc.

Thomas, J.T., Elsden, D.F. and Partridge, S.M. (1963) Degradation products from elastin. Partial structure of two major degradation products from the cross-linkages in elastin. *Nature*, **200**, 651–652.

Tomasini-Johansson, B.R., Ruoslahti, E. and Pierschbacher, M.D. (1993) A 30 kD sulfated extracellular matrix protein immunologically crossreactive with vitronectin. *Matrix*, **13**, 203–214.

Torschia, D.A. and Piez, K.A. (1973) Mobility of elastin chains as determined by ^{13}C nuclear magnetic resonance. *J Mol Biol*, **76**, 419–424.

Tsipouras, P., Del-Mastro, R., Sarfarazi, M., Lee, B., Vitale, E., Child, A.H., Godrey, M., Devereux, R.B., Hewett, D. and Steinmann, B. (1992) Genetic linkage of the Marfan syndrome, ectopia lentis, and congenital contractural arachnodactyly to the fibrillin genes on chromosomes 15 and 5. *New Engl J Med*, **326**, 905–909.

Tsuji, T., Okada, F., Yamaguchi, K. and Nakamura, T. (1990) Molecular cloning of the large subunit of transforming growth factor type masking protein and expression of the mRNA in various rat tissues. *Proc Natl Acad Sci USA.*, **87**, 8835–8839.

Uitto, J., Fazio, M.J. and Christiano, A.M. (1993) Cutis Laxa and premature aging syndromes. In *Connective Tissue and its Heritable Disorders.*, P.M. Royce and B. Steinman (ed.), pp. 409–423. New York: Wiley-Liss.

Urry, D.W. (1983) What is elastin; what is not. *Ultrastruct Pathol*, **4**, 227–251.

Urry, D.W., Long, M.M., Cox, B.A., Oshnishi, T., Mitchell, L.W. and Jacobs, M. (1974) The synthetic polypentapeptide of elastin coacervates and forms filamentous aggregrates. *Biochim Biophys Acta*, **371**, 597–602.

Urry, D.W., Long, M.M., Hendrix, C.F. and Okamoto, K. (1976) Cross-linked polypentapeptide of tropoelastin: An insoluble, serum calcifiable matrix. *Biochemistry*, **15**, 4089–4094.

Waisman, J., Cancilla, P.A. and Coulson, W.F. (1969) Cardiovascular studies on copper-deficient swine. XIII. The effect of chronic copper deficiency on the cardiovascular system of miniature pigs. *Lab Invest*, **21**, 548–554.

Wallace, R.N., Streeten, B.W. and Hanna, R.B. (1991) Rotary shadowing of elastic system microfibrils in the ocular zonule, vitreous, and ligamentum nuchae. *Curr Eye Res*, **10**, 99–109.

Wolfe, B.L., Rich, C.B., Goud, H.D., Terpstra, A.J., Bashir, M.M., Rosenbloom, J., Sonenshein, G.E. and Foster, J.A. (1993) Insulin-like growth factor-I regulates transcription of the elastin gene. *J Biol Chem*, **268**, 12418–12426.

Wood, G.C. (1958) The reconstitution of elastin from a soluble protein derived from ligamentum nuchae. *Biochem J*, **69**, 539–544.

Wrenn, D.S., Griffin, G.L., Senior, R.M. and Mecham, R.P. (1986) Characterization of biologically active domains on elastin: identification of a monoclonal antibody to a cell recognition site. *Biochemistry*, **25**, 5172–5176.

Wrenn, D.S., Parks, W.C., Whitehouse, L.A., Couch, E.C., Kuchih, U., Rosenbloom, J. and Mecham, R.P. (1987) Identification of multiple tropoelastins secreted by bovine cells. *J Biol Chem*, **262**, 2244–2249.

Yeh, H., Anderson, N., Ornstein, G., N, Bashir, M.M., Rosenbloom, J.C., Abrams, W., Indik, Z., Yoon, K., Parks, W., Mecham, R. *et al.* (1989) Structure of the bovine elastin gene and S1 nuclease analysis of alternative splicing of elastin mRNA in the bovine nuchal ligament. *Biochemistry*, **28**, 2365–2370.

Yeowell, H.N., Marshall, M.K., Walker, L.C., Ha, V. and Pinnell, S.R. (1994) Regulation of lysyl oxidase mRNA in dermal fibroblasts from normal donors and patients with inherited connective tissue disorders. *Arch Biochem Biophys*, **308**, 299–305.

Zhang, H., Apfelroth, S.D., Hu, W., Davis, E.C., Sanguineti, C., Bonadio, J., Mecham, R.P. and Ramirez, F. (1994) Structure and expression of fibrillin-2, a novel microfibrillar component preferentially located in elastic matrices. *J Cell Biol*, **124**, 855–863.

Zheng, P. and Kligman, L.H. (1993) UVA-induced ultrastructural changes in hairless mouse skin: A comparison to UVB-induced damage. *J Invest Dermatol*, **100**, 194–199.

5 Hyaluronan

J. Robert E. Fraser[1] and Torvard C. Laurent[2]

[1]*Department of Biochemistry and Molecular Biology, Monash University, Clayton, Victoria, Australia.*

[2]*Medical and Physiological Chemistry Department, University of Uppsala, Uppsala, Sweden.*

HISTORICAL AND BIOLOGICAL PERSPECTIVES

Hyaluronan was first isolated from the vitreous body of the eye by Meyer & Palmer (1934), and named hyaluronic acid according to its origin (from the Greek *hyalos*, meaning glassy), and its uronic acid content. Shortly afterwards, it was identified in umbilical cord, the aqueous humour of the eye, the synovial fluid of joints, and skin (for review, see Meyer, 1947).

Hyaluronan proved to be the archetype for a new class of heteropolysaccharide found primarily in the extracellular matrix. Meyer and others subsequently determined its structure as a polymer of a glucuronic acid *N*-acetylglucosamine disaccharide (hyalobiuronic acid). When the chondroitin sulphates and keratan sulphate were later established as its congeners, all were grouped together with the heparins as acid mucopolysaccharides. This term is now superseded by their more explicit classification as glycosaminoglycans (Jeanloz, 1960), which includes keratan sulphate where galactose replaces uronic acid, and incidentally excludes polyuronic acids of plant origin that are of little interest in mammalian physiology. Hyaluronan has more recently been adopted as a generic term for hyaluronic acid or hyaluronate to avoid any need to specify the salt form or counter ions *in vivo* or in untreated extracts, and to conform with the nomenclature of other polysaccharides (Balazs *et al.*, 1986).

Before the discovery of hyaluronan, a brilliant series of studies in experimental pathology initiated by Duran-Reynals in 1928 (see Duran-Reynals, 1942)[1] had defined a variety of spreading factors that increased tissue permeability. These were found initially in

[1] Much of the earlier work on the permeability of ground substance cited by Duran-Reynals in 1942 could not be expressed at that time in terms of specific matrix components, but remains a source of valuable data if interpreted in the light of present knowledge.

Extracellular Matrix, Volume 2, Molecular Components and Interactions, edited by Wayne D. Comper.

aqueous testicular extracts, and subsequently in venoms and bacterial culture media. They promoted the spread of viruses and bacteria, and the diffusion of dyes, colloids and fluid injected into skin and other tissues. Some were later identified as various classes of hyaluronidase. The testicular enzyme was found to degrade both chondroitin sulphate and hyaluronan, whereas the bacterial enzymes attacked only hyaluronan and sulphate-depleted chondroitin. Spreading activity was detected in extracts from most normal and malignant tissues from most animal species but it was not invariably identifiable with hyaluronidase. Similar effects were also induced by ascorbic acid and other reducing agents known to lower the viscosity of hyaluronan. Collagenolytic and other enzymes that might degrade matrix were considered but could not be readily studied at that time.

This work provided, in a very clear and simple manner, an early illustration of one of the outstanding functions of extracellular matrix, and in particular, hyaluronan; namely, the capacity to retard diffusion and by inference, hydrodynamic transport, which was later confirmed by direct measurement. The implications for the dissemination of infective agents and tumour cells were immediately apparent and examined intensively, though in the second case, inconclusively.

It is self-evident that each chemically distinct endogenous component of the body is likely to have some biologically unique function. Nevertheless, hyaluronan stands apart from other macromolecular elements of the extracellular matrix in the conservation of its structural simplicity, its wide distribution in nature, its molecular size, and its prime functions. Hyaluronan appears first in phylogenesis, for example, as the main constituent in certain bacterial capsules; chiefly those of *Streptococcus pyogenes*. In contrast, other bacterial capsules, such as those of *S. pneumoniae*, contain heteropolysaccharides largely unrelated to those of the mammalia (although both species and other bacterial genera secrete hyaluronidases). Other glycosaminoglycans have not been found in orders lower than Porifera, and their appearance in phylogenesis has been related to the emergence of cellular aggregation (Cássaro and Dietrich, 1977). Indeed the restricted occurrence of hyaluronan in a few mammalian bacterial pathogens suggests the capture of the essential part of its synthase genetic code from their hosts.

Hyaluronan dominates the first stage in the generation of extracellular matrix by most kinds of mammalian cells and in most tissues or organs: during early embryonic development, regeneration, and metamorphosis (Toole, 1976, 1981), and in repair of injury such as fractures of bone (Antonopoulos *et al.*, 1965). As the growth of the cell mass slows and structural and functional differentiation ensues, the proteoglycans and collagens develop to reach an equilibrium characteristic of the tissue or scar; in most cases with a concurrent fall in the proportion of hyaluronan. The same sequence — namely, hyaluronan before proteoglycans (other than cell-associated heparan sulphates) or collagens, with regression to histiotype as growth slows — is also reproduced in serially passaged cell cultures from various sources, not only mesodermal but also epithelial (Lamberg *et al.*, 1986). It appears to be a fundamental feature of developmental biology.

After tissue injury, hyaluronan appears in greater abundance during the inflammatory response and the clearance of cell and matrix débris, before reconstitution of the matrix begins. It is similarly prominent in other forms of inflammation and immunological reactions where there is little tissue damage, and in certain kinds of neoplasia. These circumstances are strong presumptive evidence that it is not merely a favourable physical environment for cellular movement and proliferation but specifically influences the behaviour of infiltrating cells. At the same time it can have equally profound effects on the function of the resident cells of the tissue, both mesenchymal and parenchymal; not simply through its modification of their environment but by the agency of specific receptors.

Developments in cell and tissue culture, organ perfusion and whole-body studies combined with physical, biochemical and histochemical methodology have generated a rapidly growing body of experimental data on the behaviour of hyaluronan *in situ*, reinforced by information from experimental and natural pathology. The recognition of a variety of cellular receptors and specific binding sites in other elements of matrix has added a complexity to its physiological role that is not immediately obvious in its structural simplicity.

Even so, much of our present understanding of the subject has arisen from the study of hyaluronan *ex vivo*, provoked in the first place by striking natural phenomena such as the viscosity of synovial fluid in joints and the gelatinous character of vitreous humour in the eye. This work continues to provide fresh information essential to the interpretation of observations *in vivo*. Although conceptual advances continue to appear from both directions, it is appropriate first to consider the intrinsic properties of hyaluronan, particularly for those seeking an introduction to the subject from unrelated backgrounds, since these still remain essential to understanding its more complex biological activities.

This review is primarily concerned with the place of hyaluronan in the structure and function of extracellular matrix. Although it is not intended to encompass metabolism or turnover, some consideration of these subjects and of methodology will be necessary for interpretation of certain aspects.

PHYSICOCHEMICAL NATURE OF HYALURONAN

Primary Structure

Saccharide constitution. Hyaluronan (HA) is a linear polymer of hyalobiuronic acid, the disaccharide referred to above, with GlcNAc(β1–4)GlcUA and GlcUA(β1–3)GlcNAc linkages. Its carboxyl groups are fully ionised in the prevailing pH of extracellular fluids, whether normal or pathological. In contrast with other unmodified glycosaminoglycans (GAG), it has no sulphate substitution or covalently linked peptide, and its chain length and molecular weight at its origins are usually very much greater.

Earlier analyses indicated homogeneity in the polymer but some uncertainty has been expressed or inferred from time to time, and the difficulty in recovering HA completely free from protein still attracts comment. In the light of the variability discovered in proteoglycans, such doubts are understandable and will be reviewed at this point since they are pertinent to understanding the behaviour of hyaluronan.

For example, reduction of HA polymers to an apparent limiting value in the order of 65 kD has been observed on exposure to reductive conditions (Pigman *et al.*, 1961; Swann, 1967). This has been taken to imply some distinct form of linkage between polymer units of a particular size (Swann, 1969). Moreover, the results of saccharide analysis (see, for example, Preston *et al.*, 1965) have not always been consistent with the primary structure cited above, and recent work has shown unsuspected variations in the saccharide sequence in the main chains of other GAG, remote from the established variation in the region of their peptide attachment (see Volume 2, Chapter 6).

A variant saccharide order in HA oligosaccharides can appear after digestion with hyaluronidase (Aronson and Davidson, 1967; Cashman *et al.*, 1969), or with acetylhexosaminidase and glucuronidase (Longas & Meyer, 1981). This, however, results from enzymatic transglycosylation, which is due at least in part to ionic conditions of digestion (Gorham *et al.*, 1975). In earlier work, the concerted action of testicular hyaluronidase with these two enzymes *in vitro* yielded the expected monosaccharides with a residue of hyalobiuronic acid (Linker *et al.*, 1955), which was confirmed by later work in Meyer's laboratory (Reeves and Meyer, 1981). In our own experience, isotopically labelled HA purified by ultracentrifugation in dissociative conditions repeatedly gives a virtually complete yield of the expected hexa- and tetrasaccharide residues when exhaustively digested by *Streptomyces* hyaluronidase, which is active only against hyaluronan.

The question of covalently linked peptide. Although early studies indicated otherwise (Mörner, 1889), glycosaminoglycans were generally thought to be protein-free after their primary structures were finally established. Using gentle forms of extraction from tissue, Shatton and Schubert (1954) then showed that sulphated GAG remained firmly bound to protein. The covalent nature and precise structure of the carbohydrate-peptide linkages were shortly established in a series of studies by Muir (1958), Anderson *et al.*, (1963), Gregory *et al.* (1964) and others (see Volume 2, Chapter 6).

It was therefore logical to consider a similar structure in hyaluronan. The possibility of a peptide component had been raised earlier when HA extracted from synovial fluid by ultrafiltration (Ogston & Stanier, 1950, 1951, 1953) was consistently found to contain some 20 to 25% protein, which resisted phenolic extraction (Denborough & Ogston, 1965). The protein content of these residues was, however, identified immunochemically with certain plasma proteins, and could be changed in character by varying the ionic conditions (Curtain, 1955), or preventing the generation of an HA gel on the filter (Fraser *et al.*, 1977). Several plasma proteins have been associated with HA extracted from synovial fluid of normal and inflamed joints by other gentle means such as gel permeation chromatography. Similar complexes can be readily created from purified material *in vitro* (see, for example, Hutadilok *et al.*, 1988). Two recent studies have nevertheless indicated that a secondary covalent linkage with proteins is possible (Yoneda *et al.*, 1990; Huang *et al.*, 1993). Any selective association of protein, with or without covalent linkage, will clearly influence the functions of hyaluronan, and is discussed again later.

The existence of a smaller intrinsic peptide content was proposed when radioactivity was detected in HA synthesised by the aneuploid 3T3 strain of mouse fibroblasts cultured with labelled leucine (Mikuni-Takagaki and Toole, 1981). Incubation of cultured rat chondrosarcoma cells with ^{3}H-labelled serine and leucine for varied periods and in pulse-chase sequence failed to show any incorporation in newly synthesised hyaluronan, either extracellular or cell-associated (Mason *et al.*, 1982). Subsequently, isotope was specifically identified in the acetyl groups of hyaluronan when synovial fibroblasts in the diploid growth phase were incubated with leucine either uniformly labelled with ^{14}C, or ^{3}H-substituted at C4 and C5. It appeared that catabolism of leucine by fibroblasts generated both acetoacetylCoA and acetylCoA as sources of acetate, in common with fat, muscle and brain cells (Fraser & Baxter, 1984). No other evidence of direct amino acid incorporation has appeared since.

The most telling argument against a peptide component in hyaluronan now lies in the elucidation of its synthesis. Whereas the proteoglycans are generated by initial protein synthesis and partial glycosylation in endoplasmic reticulum, with subsequent saccharide chain elongation in the Golgi bodies, HA is formed entirely by a synthase resident in the plasma membrane (Phillipson and Schwartz, 1984; Prehm, 1984). The polymer is developed at the reducing end by addition of its alternating monosaccharides from their UDP conjugates (Prehm, 1983) and is immediately extruded, which precludes the participation of a peptide primer, core or linking units, and allows no scope for late structural modification such as occurs in the heparins.

Summary. Chemical studies have uncovered no evidence of potential branch points, either saccharide or peptide, which indicates that HA is an unbranched linear polysaccharide chain. Fessler and Fessler (1966)* have confirmed this by electron microscopy with the Kleinschmidt technique and shown that the length of the visualised chains corresponds to the length calculated from the degree of polymerisation measured independently, and from the length of the disaccharide (about 1.0 nm). Although similar homogeneity of saccharide content and uniformity of sequence have not been formally established in HA from all its varied sources, it has so far shown remarkable conservation in phylogeny, and the primary structure outlined above can certainly be assumed in mammalian physiology. There are no grounds for the occasional references to HA as a proteoglycan which still appear from time to time.

MOLECULAR DIMENSIONS AND SECONDARY STRUCTURE

Molecular Weight *(Relative molecular mass; M_r)*

Sources of error in sample preparation. Estimates of its molecular weight in the native state must be accepted with reservation unless the mode of prior purification is specified. Ease of extraction obviously varies greatly between, say, joint fluid and cartilage, but even in a single tissue such as skin it shows enough variation to indicate distinct hyaluronan compartments (Myer and Stern, 1994), consistent with evidence from its rates of turnover (Laurent and Reed, 1991). The potential causes of degradation will be reviewed here, since virtually all of them can also operate naturally, at least in disease.

Endogenous hyaluronidases appear to offer little danger if pH is maintained ≥7.5 during extraction; except for the testicular enzyme those in extracellular fluids are almost entirely of lysosomal origin and active only in a low pH range (de Salegui and Pigman, 1967; Vaes, 1965; Stephens *et al.*, 1975). Nevertheless the risk cannot be entirely dismissed in all circumstances. A hyaluronidase active at neutral pH has been suspected in embryonic tissues.

HA polymers can be disrupted by intense mechanical shearing, and more readily by high frequency sonication (see for illustration, Fraser *et al.*, 1989); by electromagnetic radiation from gamma (Ragan *et al.*, 1947) through to ultraviolet (Balazs and T C Laurent, 1951); to β radiation (Balazs *et al.*, 1959); heat (Jeanloz and Forchielli, 1950; Wik *et al.*,

* Studies with rotary shadowing electron microscopy show a more complex orientation (Scott *et al.*, 1991), but this reflects the retention of secondary and tertiary structure by the different method of preparation.

1979); and chemical hydrolysis. Free radical activity, especially OH•, is a contributory or immediate cause in several of these circumstances (Balazs *et al.*, 1967a) and can be generated inadvertently in preparation; for example, with catalysis by Fe^{2+} (Halliwell, 1978). Similar effects arise from high redox activity generated by use of cysteine in conjunction with papain digestion, or from high levels of ascorbate. Conversely, radiolysis of HA labelled with low-energy β-emitting isotopes is largely suppressed by $OH^{\cdot}$ scavengers, such as dimethyl sulphoxide. HA can be separated from other macromolecules in body liquids and extracts by numerous methods, singly or in combination: filtration, proteolytic digestion, precipitations with alcohol or detergents, ion exchange chromatography, preparative electrophoresis and perhaps with least risk, isopycnic ultracentrifugation (Silpananta *et al.*, 1967) in dissociative conditions. Incomplete recovery, with possible selective loss, can arise from absorption to certain gel matrices. This can be countered by neutral (personal observation) or zwitterionic detergents (Ng *et al.*, 1992).

Some of these hazards still occur in handling naturally fluid samples or culture media that do not require fractionation before M_r analysis. All reported estimates of its molecular weight in the native state therefore carry some element of uncertainty.

Methods of measurement. Absolute methods include light-scattering, sedimentation and diffusion, sedimentation equilibrium and in the low molecular weight range, osmometry. For calculation they utilise characteristics such as refractive increment and partial specific volume that are independently determined.

Relative methods include viscometry and gel permeation chromatography (Wik *et al.*, 1979), which require calibration with fractions of independently established molecular weight. Chromatography is the only method that can be frequently used without prior separation, since it can be combined with a specific HA assay (Laurent and Granath, 1983). It depends critically on a stable gel of high porosity, rigorous control of elution and accurately defined reference standards. Results are remarkably similar to those of direct methods (Table 5.1).

The molecular weight of HA shows considerable physiological variation. This can be portrayed succinctly and usefully in the expression of mean M_r as the weight-average (Mw) or number-average (Mn), calculated as follows.

$$M\mathrm{w} = \frac{\Sigma_i C_i \cdot M_i}{\Sigma_i C_i} \quad \text{or} \quad \frac{\Sigma_i N_i \cdot M^2_i}{\Sigma_i C_i \cdot M_i}$$

$$M\mathrm{n} = \frac{\Sigma_i N_i \cdot M_i}{\Sigma_i N_i} \quad \text{or} \quad \frac{\Sigma_i C_i}{\Sigma_i (C_i / M_i)}$$

where C is concentration as mass/volume, and N as moles/volume.

Identity of Mw and Mn thus indicates uniform polymer size, and their divergence is an index of polydispersity. The polydispersity of native HA was first demonstrated, however, by separation into fractions shown to have different molecular weights (see Laurent *et al.*, 1960; Cleland *et al.*, 1968; Wik, 1979).

Normal tissues. Extensive information is available from synovial fluid, which is essentially a liquid form of extracellular matrix and permits simplified study. This is quoted in detail in

Table 5.1 Molecular weight of hyaluronan from various mammalian sources

Source	*Mw (weight-average molecular weight* $\times 10^{-6}$*)*	*Technique*	*Reference*
Human umbilical cord	2.8–4.3	Light scattering	Laurent and Gergely, 1955
Bovine synovial fluid	8 –9	Sedimentation and viscometry	Ogston and Stanier, 1951
Human synovial fluid			
Normal	6.0	Light-scattering	Balazs *et al.*, 1967b
	6.3–7.6	Gel chromatography	Dahl *et al.*, 1989a
Arthritic	2.7–4.5	Light-scattering	Balazs *et al.*, 1967b
	3.2–6.8	Gel chromatography	Dahl *et al.*, 1989a
Adult bovine	0.3–0.5	Light-scattering	Laurent, 1955
vitreous humor	0.5–0.8	Gel chromatography	Laurent and Granath, 1983
Monkey vitreous humor	2.9	Viscosity	Balazs and Denlinger, 1984
Human thoracic lymph*	1.4	Gel chromatography	Tengblad *et al.*, 1986
Human blood serum	0.2	Gel chromatography	Tengblad *et al.*, 1986
Human urine	0.004–0.012	Gel chromatography	Laurent *et al.*, 1987

*For data on pre- and post nodal lymph, see Volume 1, Chapter 5.

Volume 1, Chapter 11. We shall merely note here for comparison that the mean *M*w of HA in normal human fluid is ~7×10^6 and that in fluid from rheumatoid joints ~5×10^6 (Dahl *et al.*, 1989a); estimates that are reproduced in HA synthesised by synovial cells in culture.

The molecular weight of HA has also been closely studied in the eye; initially, in the vitreous body of cattle because of its ready availability. In this species its mean M_r ranges between 3 and 5×10^5 (Laurent, 1955) and the material is remarkably polydisperse (Laurent *et al.*, 1960). Values of 3 to 4.5×10^6 have been observed as viscosity-average M_r in humans and monkeys (Balazs and Denlinger, 1984). The difference may be due to an intrinsic tendency to non-enzymatic degradation in the bovine vitreous (Sundblad and Balazs, 1966). Laurent and Granath (1983) studied HA in rabbit eyes with newly developed methods of chromatography and microanalysis that eliminate the need for removal of protein, and with inclusion of isotopically labelled HA to detect any concurrent degradation. In the aqueous humor, *M*w and *M*n were respectively 5.1×10^6 and 1.0×10^6; the corresponding values in the vitreous body were 2 to 3×10^6 and 0.2 to 0.3×10^6, thus proving their separate origins.

In skin extracts, with labelled HA likewise used as an internal control, figures of 0.35 to 1.1×10^6 have been obtained (Reed *et al.*, 1988). These are matched by the synthetic products of epidermis and dermis in culture (Tammi *et al.*, 1991).

In articular cartilage from human knee joints, an identical rise in HA/uronate ratio was found in whole-cartilage extracts and in the separated proteoglycan aggregates with increasing age from the first to the ninth decade; in the extracts, M_r fell concurrently from 2×10^6 to 3×10^5. In contrast, the peak value in newly synthesised cartilage HA remained extremely high without apparent increase in polydispersity in tissue from donors in their eighties, indicating that the fall in polymer size with age was a postsecretory modification (Holmes *et al.*, 1988).

Estimates of a similar order can be made by electron microscopic measurements of the length of hyaluronan in proteoglycan aggregates from bovine tissues (Rosenberg and Buckwalter, 1986).
Assuming that 1 nm = 400 D, these yield the following results.

Articular cartilage	
Mature	250 kD
Calf	450 kD
Nasal cartilage	
Mature	650 kD
Calf	730 kD
Fetal epiphyseal cartilage	~ 1,000 kD

Pulse-chase studies in tissue culture of skin (Tammi *et al.*, 1991) and bovine articular cartilage have both revealed a rapid progressive postsynthetic reduction in M_r. In cartilage (Ng *et al.*, 1992), this continued through the 12 days of the study while the total HA content and its M_r profile remained constant, the latter at a lower range than the freshly synthesised material. The degradation exhibited random-hit kinetics at a rate consistent with free-radical activity rather than enzymatic, and was cell-dependent. In contrast with skin cultures very little was recovered from the medium, indicating turnover by catabolism. The size of HA in proteoglycan aggregates seems therefore inversely related to age. Species differences should be viewed with caution, since the appearance of some animal breeds, especially beef cattle, suggests dyschondroplasia.

The polymer range of HA associated with cell surfaces is broadly similar to that remote from the surface. A wider range might be expected, since some of the chains should be incomplete and still attached to the membrane synthase. A small fraction of ~60,000 Dalton has been found in the pericellular aggregate released by trypsin treatment of cultured human synovial cells (Baxter, Clarris and Fraser, unpublished). Most of it shows the same high molecular weight as HA released into the culture medium.

Circulating body fluids. M_r of HA is distinctly lower after passage through lymph nodes and lower again after entry to the blood stream, which reflects selective absorption of larger polymers by the metabolic receptor in lymph nodes and liver, possible exposure to free radicals in lymph nodes, and other influences during its circulation (see Volume 1, Chapter 5).

It is appropriate to note here that the range found in peripheral (prenodal) lymph of the popliteal and prefemoral drainage regions in the sheep (mean Mw 2.33 and 2.11 × 10^6, and mean Mn 0.15 and 0.18 × 10^6, respectively) is close to that in skin. Although this lymph includes drainage from other tissues including joints, where the mean M_r is much higher, skin is probably the main source of its HA content. We do not yet know whether the lymphatic content consistently reflects the polymer profile in its tissue of origin, but in current studies with RNP Cahill and WG Kimpton, we have injected isotopically labelled HA into joints and found its Mw in the draining lymph to be identical with that remaining in the joint. A rigorous comparison of polydispersity in tissue and regional prenodal lymph will be necessary to determine this issue, however, since the same might not be true of more densely aggregated HA or of pools that are restrained by proteoglycans and collagen bundles or by specific binding peptides alone.

In amniotic fluid of sheep, HA is polydisperse, with a high *M*w (1×10^6) early in pregnancy, which falls in mid-term and tends to rise later (Dahl *et al.*, 1986). In human amniotic fluid, *M*w is 0.33×10^6 at 16 weeks but at 40 weeks shows a bimodal or unimodal range with a change in *M*w from 1.1 to 3.6×10^6 in samples with higher concentrations (Dahl *et al.*, 1986). These variations may reflect changes in the sources and turnover of HA at different stages of pregnancy (Dahl *et al.*, 1989b). In human umbilical cord, *M*w ranges between 2.8 and 4.3×10^6 (Laurent and Gergely, 1955).

In human cerebrospinal fluid, the mean *M*w of HA is 3×10^6 (UBG Laurent, personal communication). No M_r data are available for HA in the fluids of the serous cavities.

Disease. The effects of non-infective inflammation on the polymer profile of HA have been most thoroughly studied by comparative analysis of rheumatoid and normal synovial fluid, referred to earlier. The lower M_r observed in fluid from inflamed joints might arise from synthesis of smaller polymers, from partial degradation after synthesis, or both. HA produced by cultured rheumatoid synovial fibroblasts has a lower viscosity-determined M_r and is synthesised at a higher rate than that of corresponding normal cells (Castor & Dorstewitz, 1966). These changes were reproduced by normal cells treated with a tissue-derived activator, and thus appeared due in the rheumatoid cells to prior activation *in vivo* (Castor *et al.*, 1971, 1972). Vuorio *et al.* (1977) reported similar changes in HA synthesised by cultured rheumatoid synovial fibroblasts, and also found that the molecular size of exogenous HA was reduced after addition to the cultures. The only forms of hyaluronidase likely to occur in these cultures are serum- or cell-derived, neither of which is active in extracellular pH. Free radical activity generated at the cell surface (Prehm, 1990) provides an alternative explanation despite its short effective range.

In the body, free radical degradation of HA is probably common in inflammatory exudates of the joints and other tissues, since free radicals are copiously generated by activation of neutrophils and during phagocytic activity by macrophages. Both kinds of cell are numerous in joint fluid in most kinds of arthritis. Macrophages are prominent in delayed hypersensitivity types of reaction in skin and lung, where the matrix or exudate has an increased HA content, and in many other forms of acute and chronic inflammation. To our knowledge, its polymer range has not been examined in this kind of reaction.

In our experience with cultured normal synovial fibroblasts, reduced M_r is not intrinsically linked with an increased rate of HA synthesis. There is neither reduction in the modal M_r (7×10^6 from these cells) nor increased polydispersity with as much as a five-fold increase in output induced by cholera enterotoxin or dibutyryl cyclic adenosine monophosphate. The macrophage-related Type A cell, however, is the more numerous cell type in both the normal and inflamed synovial lining *in situ* (Barland *et al.*, 1962) and in primary culture (Fraser *et al.*, 1985a); if activated they might cause a secondary reduction in M_r by free radicals.

In a case of mesothelioma, where tumour cells are likely to be the direct source, the molecular parameters of HA extracted from ascitic fluid by Scott and Lowther were similar to those in bovine synovial HA (Preston *et al.*, 1965).

Summary. Hyaluronan is secreted in most parts of the body as large polymers, especially in synovial joints, though it invariably shows some polydispersity. Postsecretory reduction in molecular weight has been observed in cartilage, and during circulation in lymph and blood. This can be partly explained by preferential absorption of larger polymers by metabolic receptors, and less certainly by mechanical shearing or free radical activity. The effects of inflammation have not been intensively studied except in joint disease, where there is evidence that M_r is reduced both during synthesis and after shedding from the cell.

Molecular Domain

Hyaluronan is exceptional amongst extracellular macromolecules for the size of its molecular domain relative to molecular mass (Figure 5.1). As noted earlier, in electron microscopy with the Kleinschmidt technique the polymer appears as an extended single unbranched chain. Its behaviour in dilute solution, however, reflects the adoption of a stiffened random-coil configuration.

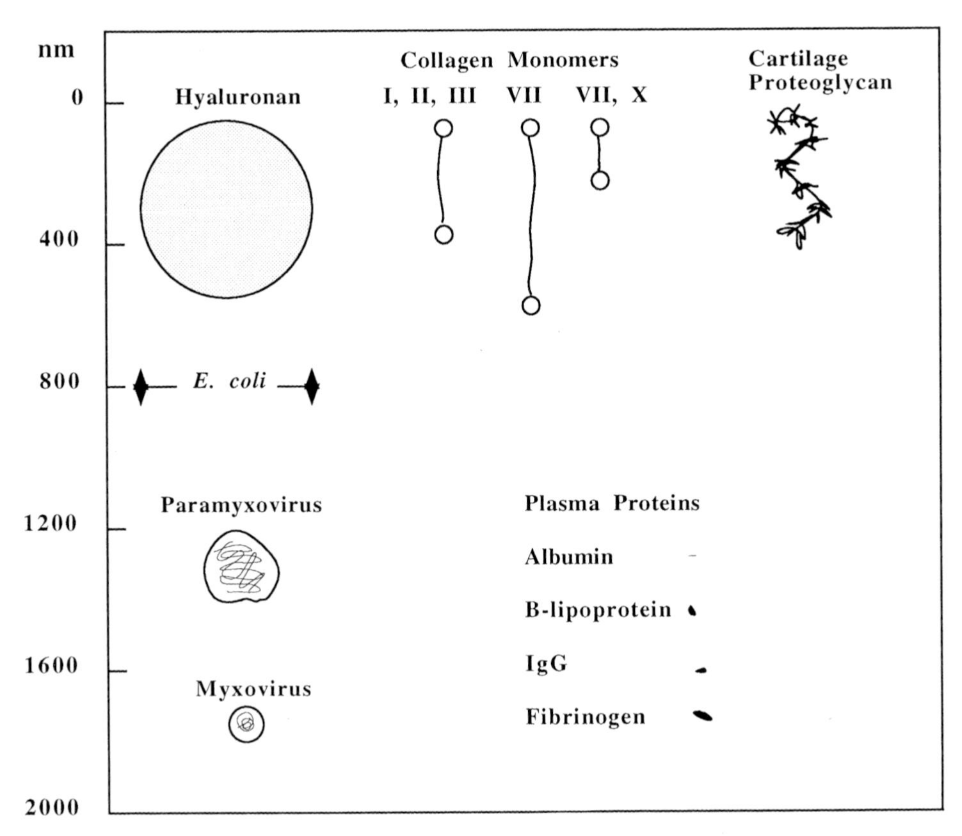

Figure 5.1 Schematic comparison of HA molecular domain (M_r 4 × 10^6) with those of other matrix components and of micro organisms. The width of Escherichia coli is indicated. It should be noted that HA will only occupy such a volume in low concentrations before entanglement occurs (see Text).

Evidence for this is as follows.

(i) The angular variation of the light-scattering intensity corresponds to such a configuration (Laurent *et al.*, 1960; Preston *et al.*, 1965).

(ii) A plot of intrinsic viscosity versus molecular weight follows the relationship expected for a random coil (Laurent *et al.*, 1960; Cleland and Wang, 1970; Wik, 1979).

(iii) Ogston and Stanier (1951) have summarised the hydrodynamic results and concluded that HA must be a voluminous nearly spherical molecule, compatible with an extended random coil configuration. Preston *et al.* (1965) later considered the HA-protein complex as a "cage" with apparent branching or cross-linking. After removal of protein the radius of gyration increased from 2500 Å to 3800 Å, which presumably eliminated an interaction with the protein (see later).

(iv) Using small-angle X-ray scattering to determine the chain structure, Cleland (1977) found that it behaved as a worm-like molecule.

(v) The molecular dimensions change with ionic strength and pH as expected in a flexible polyelectrolyte (Cleland, 1968).

Table 5.2 Relationship between molecular weight and molecular domain

Molecular weight $\times 10^{-6}$	*Radius of gyration (nm)*	*Reference*
13–14	380	Preston *et al.*, 1965
3–4	240	Laurent and Gergely, 1955
1.6	143	Laurent *et al.*, 1960
1.3	157	Laurent *et al.*, 1960
0.4	63	Laurent *et al.*, 1960
0.3	61	Cleland, 1968

The size of the random coil increases as expected with molecular weight (Table 5.2). In the higher range, it is comparable to the dimensions of some bacteria, and the molecular domains begin to overlap at concentrations as low as 0.5 to 1 g/L, levels often exceeded in tissues and normal joint fluids. At this point some 50% of the solvent water will be entrained within them. In contrast, the overlap concentration for large proteoglycans is some 10 times higher (Comper and Laurent, 1978). Although the internal domain is available to solutes of small M_r, other macromolecules are partly excluded, mainly in proportion to their size (Laurent, 1964, 1970).

Higher-Ordered Structure

In the investigation of unexpected differences in the susceptibility of various GAG to periodate oxidation (Scott, 1968), Scott postulated interference by an intramolecular hydrogen bond (Scott & Tigwell, 1978) and marshalled evidence from nuclear magnetic resonance and other studies for several kinds, including one with a bridging water molecule (Heatley and Scott, 1988). This secondary structural modification clearly provides another basis for the stiffness in the random-coil configuration of HA which is additional to internal electrostatic repulsion alone (Wik *et al.*, 1979; Scott, 1992). The twist imparted to the saccharide chain creates series of eight contiguous CH groups exposed on alternate sides (Figure 5.2), thus conferring an amphiphilic character not apparent in its primary structure.

The phenomenon is also consistent with the reversible simulation of a branched structure with partial cross-linking (Preston *et al.*, 1965). Other potential consequences are numerous (Scott, 1989) and may well influence much of the behaviour of HA in the aqueous state; not only in terms of specific binding sites and receptors, but in a wider range of nonspecific interactions with lipid membranes, other GAG and "free" proteins.

With the support of computer simulation, the concept has been applied to self-association in solution, where rotary shadowing electron microscopy shows aggregation in a continuous network at much lower concentrations than the calculated level of overlap; several forms of duplex were shown to be feasible (Scott *et al.*, 1991). Possible tertiary structures based on H-bonds, hydrophobic stacking or combinations thereof have been derived (Mikelsaar & Scott, 1994) to explain the network formation with molecular modelling from the secondary structures. The intensity of such bonding will obviously increase with concentration.

INTRINSIC FUNCTIONAL PROPERTIES

The primary functional roles of HA in its free colloidal state have virtually all been related to its physicochemical structure. The same characteristics remain relevant in many respects to its interactions with other components of the extracellular matrix and with cells, whether nonspecific or modified by specific binding.

Viscoelasticity

The most distinctive of its properties in the hydrated state is its *viscoelastic character*. These have been discussed with reference to synovial fluid in Volume 1 Chapter 11, but are equally pertinent to its space-filling role and other functions in the softer connective tissues where the applied stresses are commonly in a different range and form, and to its influence on their gross physical qualities. Both properties are positively related to concentration and molecular weight.

The importance of these attributes becomes apparent when the distribution of HA in various tissues is closely examined. In its common molecular weight of millions, hyaluronan is distinct from all other GAG, whose chain length seems invariably <50,000; and its usual molecular domain (Figure 5.1) exceeds that of the largest intact proteoglycans, which are in any case structurally immobilised.

The viscosity of HA in synovial fluid (Ogston and Stanier, 1953) and in the pure state (Gibbs *et al.*, 1968) is of the pseudoplastic non-Newtonian kind; that is, its viscosity at a given concentration is not a fixed component of resistance to flow but varies inversely with the rate of shear, or velocity gradient.

Shear-dependence develops at the overlap concentration of 0.5–1.0 g/L (Morris *et al.*, 1981). It will not therefore influence any slight viscous effect on the flow of lymph or blood even at the highest abnormal levels (Table 5.3). The interplay of concentration, molecular weight and shear rate in determining viscosity at higher concentrations has

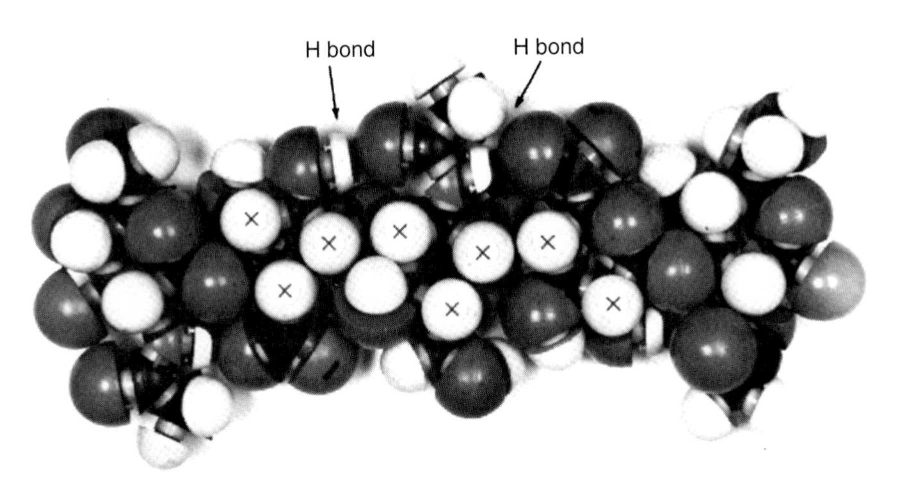

Figure 5.2 Model of a hyaluronan saccharide based upon secondary structure porposed by Heatley and Scott (1988) using Courtald space-filling units. Intramolecular hydrogen bonds orient the molecule so that sequences of contiguous CH groups are created to provide hydrophobic segments that alternate with hydrophilic segments.
Reproduced from Scott (1989), The Biology of Hyaluronan; editors David Evered and Julie Whelan; published by John Wiley and Sons, Chichester from the Ciba Foundation. Illustration kindly provided by Prof. J.E. Scott.

been systematically studied by Bothner and Wik (1991; Table 5.4).

At first sight the reduction of viscosity with rapid movement seems an ideal lubricant property, but it applies only to the frictional dissipation of the applied force within the fluid itself. In the lubrication of joints it is secondary to other factors in reducing the friction between hard load-bearing cartilaginous surfaces, although a thin film of hyaluronan can sustain heavy loads (Ogston and Stanier, 1953; see Volume 1, Chapter 11). It appears to play a more direct part in the lubrication of the soft tissue lining (Radin *et al.*, 1971), which probably applies also to tendon sheaths, bursae and the less structured planes of movement between individual muscles and adjacent skin or bone; possibly again with the participation of the phospholipid micelles discussed in Vol. 1. Hyaluronan occurs notably within skeletal and cardiac muscle, which move rapidly, but not in the mature smooth muscle of the gut (see later).

Synovial fluid and solutions of pure HA also show shear-dependent elasticity (Ogston *et al.*, 1950; Ogston and Stanier, 1953; Gibbs *et al.*, 1968; Balazs and Gibbs, 1970), expressed in rheological terms as normal force, or storage modulus. In contrast with viscosity, this increases with rate of shear or oscillatory frequency (Figure 5.3). The net result of these two rheological properties is that viscosity is maximum at rest, and elasticity greatest in rapid movement. The quantitative distribution of an applied force in these two forms is discussed in terms of synovial fluid and joints in Vol. 1. Their application in other hyaluronan-rich tissues can be reasonably deduced and in some cases simply demonstrated. Where the applied stresses are normally low and steady, (for example, in the normal flux of extracellular fluid at rest) the maximum viscosity at low shear will oppose displacement of the HA and consequently of the water entrained within it. On the other hand, elastic recoil will ensure minimum displacement and quick restoration of the tissue geometry after the rapid brief jolts to which most parts of the body are exposed.

Table 5.3 Concentration of hyalyronan in different tissues and body fluids

Species	*Tissue*	*Conc. mg/l or mg/kg*	*References*
Human	Umbilical cord	4,100	Meyer *et al.*, 1977
	Synovial fluid	1,420–3,600	Sundblad, 1965
	Synovial fluid	1,400–3,100	Balazs *et al.*, 1967b
	Vitreous body	140–338	Balazs, 1965
	Dermis	200	Pearce & Grimmer, 1972
	Thoracic lymph	8.5–18	Tengblad *et al.*, 1986
	Amniotic fluid:		
	– 16 weeks	21.4±8.8	Dahl, L *et al.*, 1983, 1986
	– at term	1.1±0.5	Dahl, L *et al.*, 1983, 1986
	Aqueous humor	0.3–2.2	Laurent, 1983
	Skin blister	0.8–5.6	Juhlin *et al.*, 1986
	Urine	0.1–0.3	Laurent *et al.*, 1987
	Serum (Normal)	0.01–0.10	Engström-Laurent *et al.*, 1985b
	(Abnormal)	up to 1.0 or higher	Personal observation
Sheep	Vitreous body	260	Balazs, 1965
	Lung	98–243	Lebel *et al*, 1988
	Pleural fluid	7±3	Broaddus *et al.*, 1988
	Amniotic Fluid	0.6–22.9	Dahl, LB *et al.*, 1989b
	Lymph*		
	– thoracic	1–34	Lebel *et al.*, 1989
	– lung	1–23	Lebel *et al.*, 1988, 1989
	– lumbar	2.6	Laurent & Laurent, 1981
	– popliteal	11	Tengblad *et al.*, 1986
	– intestinal	40–53	
	– prescapular	1±0.8	Tengblad *et al.*, 1986
	Plasma	0.124–0.2455	Engström-Laurent *et al.*, 1985b
Pig	Vitreous humor	124	Balazs, 1965
Rabbit	Synovial fluid	3,890	Sundblad, 1965
	Renal papillae	250	Farber & Van Praag, 1970
	Renal cortex	4	Van Praag *et al.*, 1972
	Pericardial fluid	82	Honda *et al.*, 1986
	Brain	54–76	Laurent & Tengblad, 1980
	Vitreous body	29	Laurent & Granath, 1983
	Muscle	26–28	Laurent & Tengblad, 1980
	Aqueous humor	0.84	Laurent & Granath, 1983
	Liver	1.4–1.6	Laurent & Tengblad, 1980
	Serum	0.019–0.086	Engström-Laurent *et al.*, 1985b

* For comparison in pre- and post-nodal lymph, see Volume 1, Chapter 5.

The interplay of elastic recoil and displacement by flow can be illustrated in skin swollen with fluid (oedema). If it is briefly indented to a certain depth with a finger, it recoils quickly. If the pressure is sustained, a hollow develops which disappears slowly on release of pressure (pitting). It is an empirical observation that pitting occurs less readily in myxoedema or in oedema caused by lymphatic obstruction, where HA content is likely to be higher, and especially in the latter when increased fibrous tissue develops. Its high viscosity at low shear and low permeability to water, coupled with the flow resistance of the fibrous elements in skin, will also retard downward movement under gravitational stress.

By these viscoelastic attributes, the abundance of HA in soft tissues can minimise disturbances of the cellular environment and contribute pliability and resilience to preservation

Table 5.4 Shear-dependent viscosity of hyalyronan

Effects of concentration and molecular weight on limiting value at low shear (η_0) and at high shear rate (1000 sec^{-1}, η_{1000}). From Bothner & Wik 1991. With kind permission of the authors.

Concentration (mg/ml)	*Molecular weight*	η_0	η_{1000}	η_0/η_{1000}
5	4,300,000	17,000	39	440
5	1,000,000	180	33	5.5
5	500,000	45	25	1.8
5	100,000	5	5	1.0
10	4,300,000	410,000	110	3,700
10	1,000,000	2,500	100	25
10	500,000	420	77	5.5
10	100,000	15	15	1.0
15	4,300,000	4,600,000	310	15,000
15	1,000,000	40,000	310	130
15	500,000	5,800	280	21
15	100,000	56	56	1.0

of form. Even in fibrous tissues, it may well buffer the cellular microenvironment against mechanical deformation applied lateral to the orientation of their collagen bundles.

The viscoelastic contribution of HA to the gross physical properties of tissues is perhaps the most distinctive expression of its intrinsic character, since it depends on its exceptional molecular weight. It can only be described in a general way, however, since its effects will also depend on the nature, the relative proportions and the orientation of firmly anchored and less compliant matrix components, particularly collagen fibres and the larger proteoglycans.

Osmotic Properties

In common with other GAG, HA exhibits non-ideal osmotic behaviour. At higher concentrations HA develops osmotic pressures comparable to that of albumin at the same concentration and many times greater in molar terms; the osmotic activity is further magnified when the two components are mixed (Laurent and Ogston, 1963). The physiological significance of this property lies in the total effect of all macromolecular species within a particular tissue and each kind of tissue must therefore be considered individually regardless of its GAG profile (see for example, Comper, 1991). One organ in which HA presumably functions as an osmotic buffer is the kidney, where it is most abundant in the medulla and papillae, the sites of considerable flux of water and solutes as the glomerular filtrate is concentrated.

Permeability

The bulk of HA in the body, even in fluid matrix such as that of synovial joints, is largely immobilised by a variety of restraints; membranous, fibrous, specific binding to other matrix components and self-association. This is illustrated by comparison of the levels

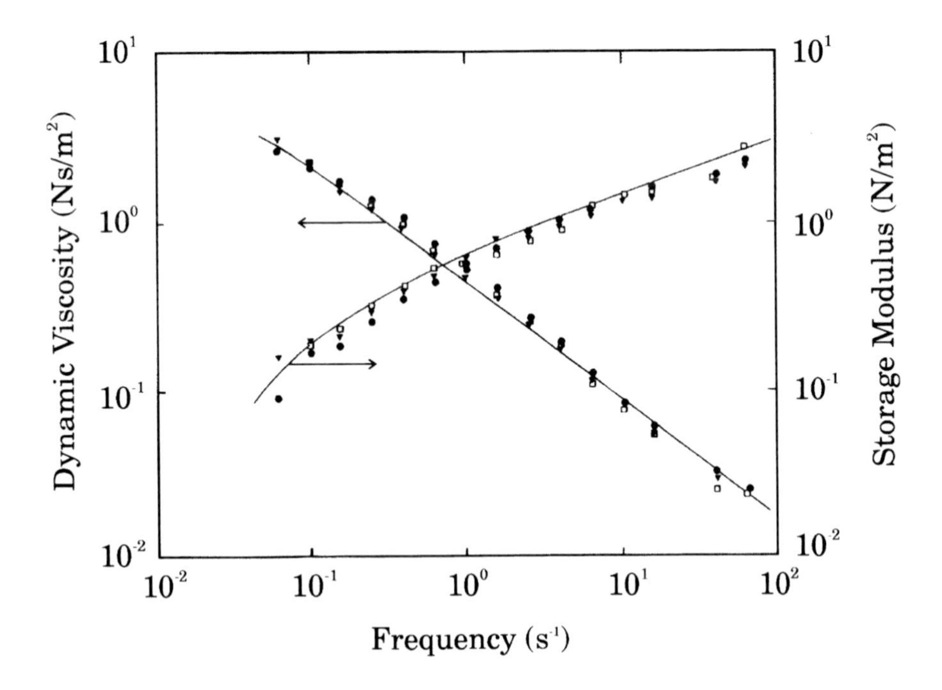

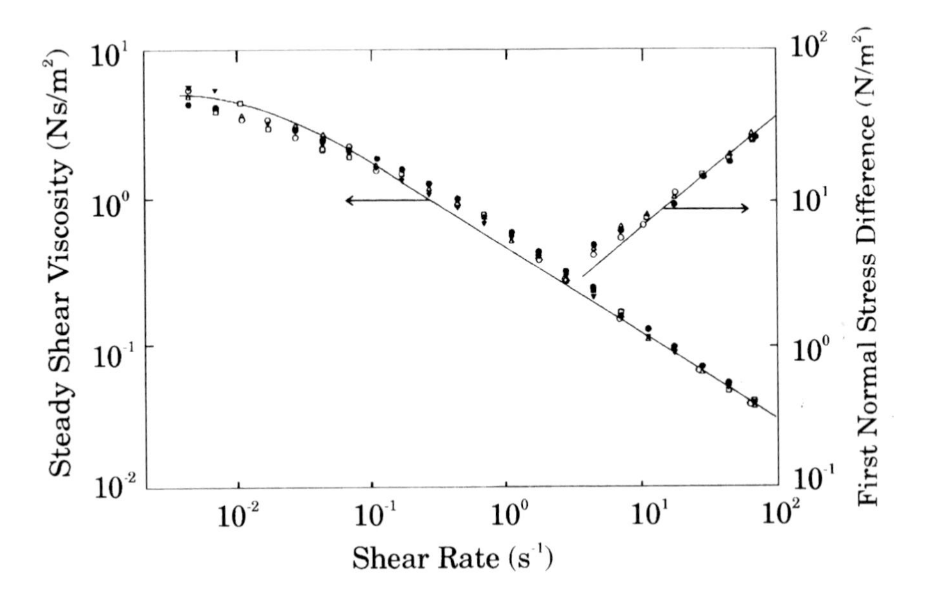

Figure 5.3 Viscoelastic behaviour of normal bovine synovial fluid with variations in steady shear rate (upper panel) and oscillatory movement (Lower Panel). Viscosity is higher at low rates of shear or oscillatory frequency, whereas elasticity increases with rapid movement. See Tirtaatmadja *et al.*, 1984.

found in peripheral lymph with those in the tissues it drains. In the main, extracellular water flows through it, not with it. Under raised pressure, a greater fraction is eluted into lymph (Volume 1, Chapter 5).

The considerable capacity of HA for retardation of water flow has been directly measured (Ogston and Sherman, 1961). To exclude error from its tendency to concentrate on restraining semipermeable membranes, it has also been estimated from its permeability coefficient, determined by ultracentrifugal sedimentation (Preston *et al.*, 1965). The increased tissue permeability that follows injection of limited amounts of hyaluronidase is due to the breakdown of the matrix structure and convection rather than a change in friction between water and polysaccharide (Fessler, 1960a, 1960b). However, other structural and hydrodynamic determinants of flow (Levick, 1994; Volume 1, Chapter 11) in the tissues may be more important than the local concentration of HA. The magnitude of the controlling effect of HA on water flow *in vivo* has nevertheless been obvious from the work on spreading factors cited in the Introduction, from the gross effect of hyaluronidases on tissue permeabilility (Day, 1950) and in their past clinical use to facilitate subcutaneous infusion of fluid or local anaesthetics.

Transport Through Hyaluronan-Rich Compartments

The very low diffusion constant of hydrated HA itself is a further restraint on its displacement from its origins, especially above its overlap concentration. The effects of HA on retention and movements of water and on small solutes are considered in depth elsewhere. Translation of particles ≤ 1 μm diameter (Clarris and Fraser, 1968) and viruses (Clarris *et al.*, 1970) through the HA-rich pericellular zone depends on Brownian movement. Despite their being much larger than excluded proteins (Figure 5.1), their movement is merely retarded but not prevented by the high viscosity of the matrix. The enhanced spread of non-motile bacteria by the action of their hyaluronidases has already been mentioned.

Steric Exclusion

The first evidence of this phenomenon was provided by equilibration across a membrane that retained HA and allowed free passage of albumin. Albumin was partly excluded from the HA compartment and enriched in the buffer compartment (Ogston and Phelps, 1961). The concept of steric exclusion implied by this phenomenon was developed and applied to the distribution of the normal free macromolecular content of interstitial matrix by Laurent (1970). It is primarily based on molecular size and contributes to the relatively lower levels of immunoglobulins in extravascular fluids. However, steric exclusion counters the reduction of antibody concentrations (Hellsing, 1969) by raising their activities in the non-excluded aqueous phase external to that in HA. Specific neutralising IgM antibodies are often generated within the matrix at the site of persistent virus infection, and enhancement of their activity by steric exclusion may explain the common difficulty in recovering viable virus by culture from diseased tissue. The influence of steric exclusion in other aspects of pathology will be discussed later.

Ionic Activity

As noted earlier, HA should be fully ionised in physiological conditions except within lysosomal vacuoles. Effects on Gibbs-Donnan equilibria and water/salt distribution are discussed elsewhere in this Volume. Salt linkage with calcium is less likely than with sulphated GAG since the separation of the carboxyl groups is greater than the van der Waals radius of this cation. Ionic interactions may become significant at the cell membrane, where high concentrations of HA (see note on permeability) and of some sulphated GAG are most likely, and where ion flux is critical in cellular activity, including membrane signalling dependent on ion channels. This should be borne in mind in interpreting the effects of removing HA from the pericellular matrix.

Free Radical Scavenging

The reactivity of HA with free radicals may have a physiological role as a protection for cells, particularly on body surfaces exposed to solar radiation. This proposition is consistent with the localisation of HA in the skin where it is concentrated between the cells of the epidermis (Tammi *et al.*, 1988) and in the most superficial part of the dermis (Wang *et al.*, 1992). In the skin of rats, ultra-violet irradiation causes an increase in sulphated GAG and HA, but in animals treated with tocopherol, the increase is confined to HA (Longas *et al*, 1993). At least in the short term, the cellular response to this form of irradiation is protective, and possibly determined by inflammatory mediators.

Presti and Scott (1994) have directly demonstrated the effectiveness of HA in protecting cultured tendon fibroblasts against damage by OH•. They found that this activity was not solely due to the primary and secondary structure of HA, as expected, but also related to molecular weight. They propose that the HA meshwork excludes potentially damaging enzymes from the immediate cellular environment.

DISTRIBUTION OF HYALURONAN IN MAMMALIAN TISSUES

Before considering the more complex roles of HA, it is appropriate to review its distribution within the body. Hyaluronan has been found in virtually all parts of the body where it has been sought with appropriate methods, being conspicuous in some and scarce in others. It is present during development and maturity in tissues of neuroectodermal origin as well as in the connective tissues of mesodermal origin. It is likely therefore that the capacity to synthesise HA is manifest by most cells at some point in their natural cycles of activity.

In mature soft tissues, in less florid and persistent inflammation and often in tumours, HA is the dominant GAG in amorphous areas of ground substance between the fibrous elements of the intercellular matrix. It is preserved by organic and usually by aqueous aldehyde fixatives, presumably by entrapment with fixed proteins, but may be partly lost in the latter leaving clear-gap artefacts. Typically it stains lightly with routinely used stains such as hematoxylin and eosin. Its metachromatic reactions are weak; with toluidine blue, for example, it gives a purple tinge in contrast with the stronger reddish hues of the sulphated GAG, and is readily obscured by the latter. Selective staining of HA and other GAG by cationic dyes can be achieved with rigorous control of ionic conditions (Scott and Hughes, 1983), though discrimination of HA from glycoproteins is still often difficult. Until recently histological identification of HA otherwise relied on stains with little specificity, and the use of testicular hyaluronidase which also digests other GAG, though cautious interpretation of several methods could give a reasonable degree of identification. The introduction of improved fixation (Hellström *et al.*, 1991) and of hyaluronan-binding peptides (HABP) either chromogen-coupled (Ripellino *et al.*, 1985; Knudson and Toole, 1985a) or combined with immunochemical staining (Delpech and Delpech, 1984; Delpech *et al.*, 1989), now provide an absolutely specific histochemical tool.

The variations in HA distribution and in its association with cells and other components of each tissue not only illustrate its obvious structural and functional roles but also reflect more subtle influences on both cells and matrix.

Gross Distribution

Earlier measurements of hyaluronan relied mainly on standard colorimetric reactions for *N*-acetylglucosamine and glucuronic acid, which required a rigorous separation of the extracted hyaluronan from other glycosaminoglycans and proteins in tissue extracts, and especially from monosaccharides, glycoproteins and plasma proteins in body fluids, which interfere with these reactions. Information gained by these means is limited by the sensitivity of the final assays and losses in recovery. The range of data has been greatly expanded by methods that utilise hyaluronan-binding peptides, which combine specificity with a high sensitivity (~10–1000 g/L) applicable to material of low content.

Concentrations in various tissues and body fluids are shown in Table 5.3, with estimates from whole-body analysis of rats in Table 5.5.

There are inevitably differences, reflecting in part the species (organ/body weight ratios, etc.), possibly age or sampling site within a tissue, and methods of measurement. For example, the apparent fall in the HA content of skin with age may be due to firmer binding and resistance to extraction (Myer and Stern, 1994). Nevertheless the data from the rat should

Table 5.5 Distribution of hyaluronan in various organs and tissues

Source	*Weight (g)*	*Total HA (mg)*	*Concentration (μg/g)*
1. Whole body, rat	201	61	300
Skin	40	34	840
Muscles	36	5	130
Skeleton, joints, etc.	58	16	280
Stomach and gut	16	0.5	32
Other viscera	43	5	120
2. Whole body, rat	265	58	220
Brain	1.5	0.11	74
Lungs	1.2	0.04	34
Kidney	2.1	0.06	30
Gut	24	1.03	44
Liver	12	0.05	4

Data for the rat (two individual rats) are adapted from those of Reed *et al.*, 1988.

be broadly applicable; more than half the body content resides in skin, about a quarter in skeletal structures with attached joints, about a tenth in muscles. With the exception of cartilage, most of it resides in soft tissues, usually with lymphatic drainage. The concentrations in renal medulla and brain are also noteworthy; as pointed out by Delpech *et al.* (1989), in areas of high solute and ionic exchange in both.

Localisation of Hyaluronan Within Various Tissues

Local concentrations in extracellular matrix will inevitably be higher than in the whole tissue where cells, proteoglycans or collagen fibres constitute a large proportion of the tissue mass. The new histochemical methods have given a much clearer picture of this (Figure 5.4). Though not quantitative, the results agree where direct measurement has been done. Some examples follow.

Connective tissues. With the exception of the fluids in synovial joints and the lining cavities of chest, heart and abdomen, hyaluronan usually lies in the interstices of fibrous collagen networks. These may be sparse and delicate as in the vitreous of the eye, and in the interstitium of the synovial lining. In the dermis, collagen is organised into dense broad bands which effectively restrict hyaluronan to the space between them (Figure 5.5). In organs and tissues where collagen and elastin form enveloping capsules and fascia, or supporting bands or trabeculae that penetrate the substance of the tissue, hyaluronan also tends to be more abundant in the planes of movement, and in the intercellular ground substance in the main substance of the tissue.

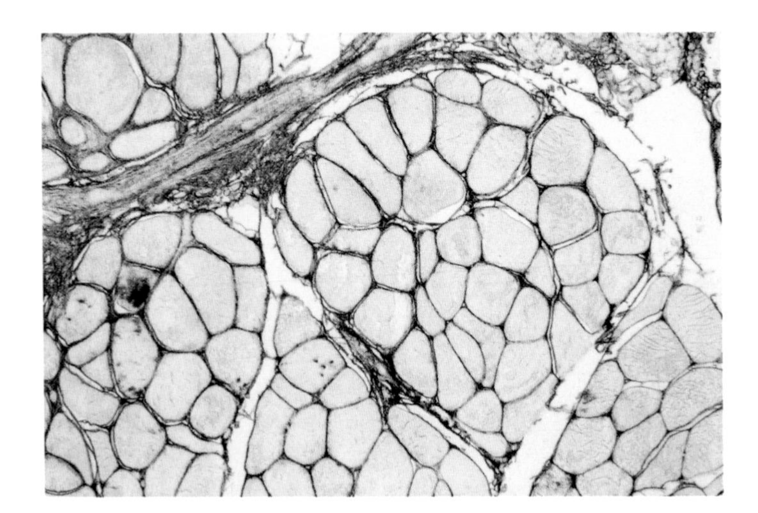

Figure 5.4 Localisation of HA in various tissues by application of biotinylated HA-binding peptide. Reaction with avidin-peroxidase leaves a brown stain.

A. **Human skeletal (quadriceps) muscle.** HA is abundant round individual muscle fibres, and in loose connective tissue between small and larger groups of muscle fibres.

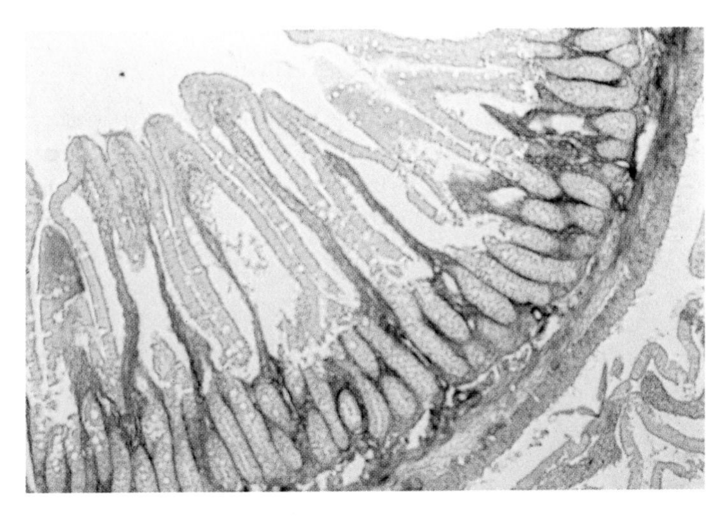

Figure 5.4 B. **Wall of rat jejunum (small bowel).** HA is prominent in the lacteals (see Text) in the centres of the villi projecting into the lumen of the bowel, and in the interstitium at their bases.

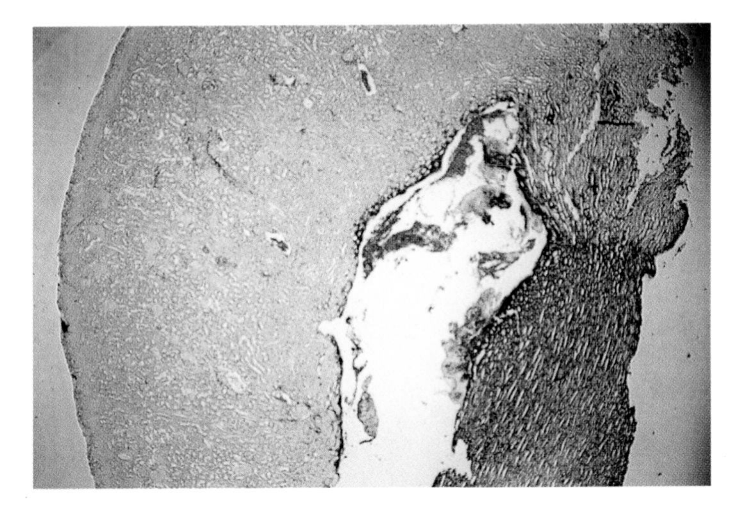

Figure 5.4 C. **Rat kidney.** The intensely stained region is the medulla. Very little stain in the broader band of cortical tissue shown in this section.

Muscles. In large skeletal muscles, HA is found round individual fibres, in septa between bundles and in the enclosing fibrous sheath (Laurent *et al*., 1991). In small muscles of eye and ear, it is mainly around single fibres. It is also found around the muscle fibres of the atria and ventricles of the heart, but not in smooth muscle of gut, aorta or bronchus. The functional significance of this distribution is obvious (see discussion of viscoelasticity). The absence of HA in mature smooth muscle contrasts with its synthesis by proliferating smooth muscle cells in culture, which is a focus of intense interest in arterial disease. Wells *et al*. (1990a) noted HA in sclerotic arteries in biopsies of rejected kidney grafts.

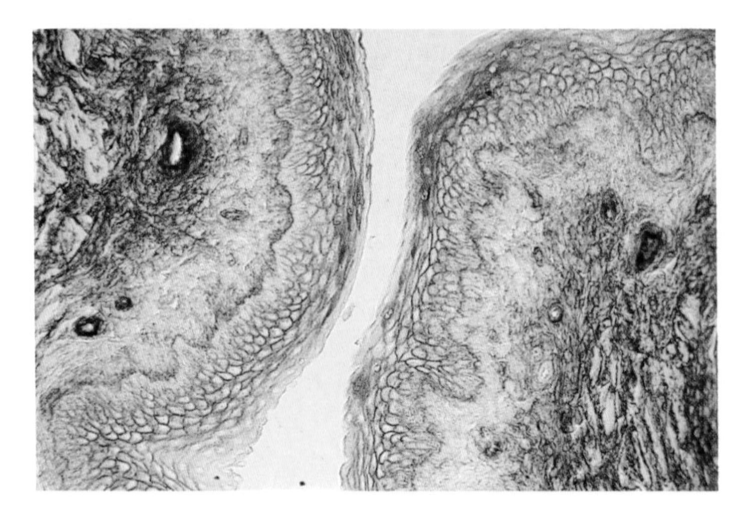

Figure 5.4 D. **Section of vagina from non-pregnant rat.** The densest staining is in the walls of blood vessels. Within the squamous epithelium, HA stains most intensely round the cells in the central layer and in the adjacent cells that show a flattened orientation parallel to the surface. The basal epidermal cells, which lie with a vertical orientation just above the wavy basal lamina, show very little staining.

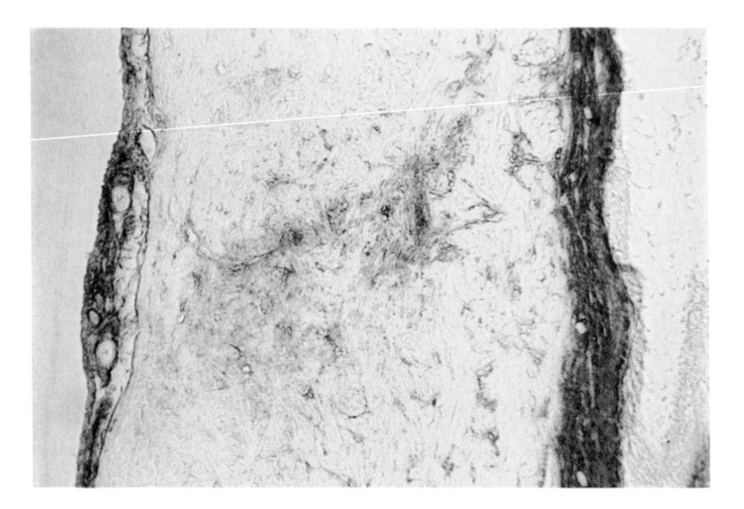

Figure 5.4 E. **Wall of ductus deferens in the rat.** There is very little HA in the smooth muscle which makes up the central of the tissue.

Gut. Laurent *et al.* (1991) observed HA mainly in the lamina propria beneath the villous epithelium, in submucosal connective tissue and in the villous lacteals (peripheral lymphatic vessels; refer Vol. 1, Chapter 5).

Blood vessels. Laurent *et al.* (1991) found HA only in the tunica adventitia (outer coat) of the aorta and of smaller arteries in the gut. Chemical measurements have shown the HA content of the major pulmonary arteries to be higher than that of the aorta but it has not yet been localised by histochemistry (Murata and Yokoyama, 1988). The association of HA

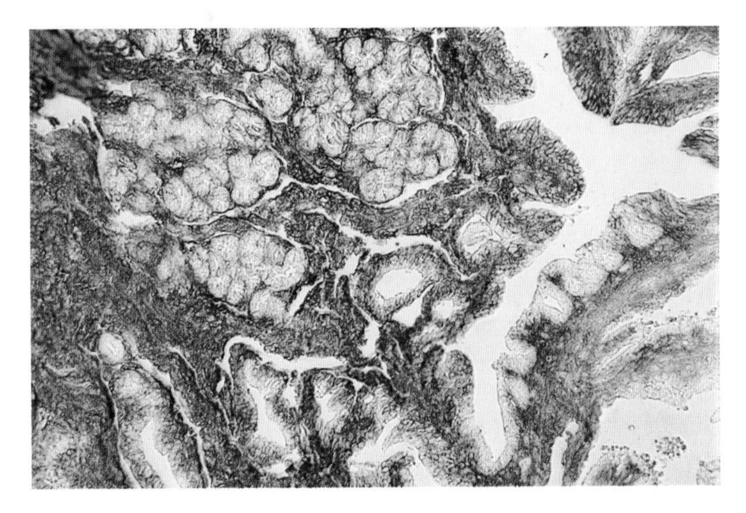

Figure 5.4 F. **Rat prostate.** HA is densely stained in the connective tissue. In contrast with that in blood vessel walls, gut and ductus deferens, smooth muscle is interspersed with HA in this tissue.

Illustration kindly provided by Drs Anna Engström-Laurent and Claude Laurent with permission of Astra Läkemedel AB.

with the outer coat of blood vessels is repeatedly observed in vessels of all sizes, in normal and abnormal tissues, and in newly formed vessels: for example, in normal tissues of the female reproductive tract (Edelstam *et al.*, 1991); in skin affected by impaired thyroid function (Gabrilove and Ludwig, 1957); in normal kidney and synovial tissue (Wells *et al.*, 1990a, 1992); and around capillaries in brain tumours (Delpech *et al.*, 1993).

The constancy of this association suggests that there is a fibroblastic phenotype committed to high levels of HA synthesis, which is specifically associated with blood vessels.

Nervous system. Extraction and measurement show significant regional variation. HA levels are higher in white than grey matter, and in spinal cord, lower brain stem and corpus callosum than in the rest of the brain (Delpech *et al.*, 1989). They are also much higher in fetal brain than adult (Margolis *et al.*, 1975, Delpech *et al.*, loc. cit.). Delpech *et al.* localised HA histochemically round certain groups of neurones and also round the nodes of Ranvier. (See also Bignami *et al.*, 1993.)

Kidney. Histochemical studies (Hällgren *et al.*, 1990b; Wells *et al.*, 1990a) have confirmed the results of chemical extraction (Table 5.4); namely, that the medulla and papillae contain a much higher concentration of HA than the cortex.

FUNCTIONAL EFFECTS OF INTERACTIONS WITH PROTEINS AND CELLS

Rheological Interactions

Earlier physicochemical and viscometric studies failed to detect any change though an increase or a decrease could ensue depending on the nature of the protein (see Fraser *et al.*,

1972). Removal of the small proportion of protein associated with HA in ultrafilter residues greatly reduces the anomalous viscosity, however, without degradation of the HA (Silpananta *et al.*, 1968, 1969). Inclusion of serum with synovial fluid increases the resistance of collagen gels to centrifugal compression above that of synovial fluid alone; the effect is not eliminated by high salt concentrations in either case (Baxter *et al.*, 1971). Direct studies by addition of sucrose and serum plasma proteins to HA in a physiological range of viscosity also showed significant viscous effects (Fraser *et al.*, 1972), which were confirmed by more rigorous shear-controlled observations (Fraser, Boger and Bialkower, unpublished) and by the separation and recombination of HA and protein in bovine synovial fluid (Tirtaatmadja *et al.*, 1984). Compared with synovial fluid, its purified HA at the same concentration showed a reduction of all measured rheological properties, with reduced zero-shear viscosity and less shear thinning. Reduction of steady-shear viscosity was 20% at 100 sec^{-1} and 50% at 0.01 sec^{-1}. Flow properties were restored by recombina-

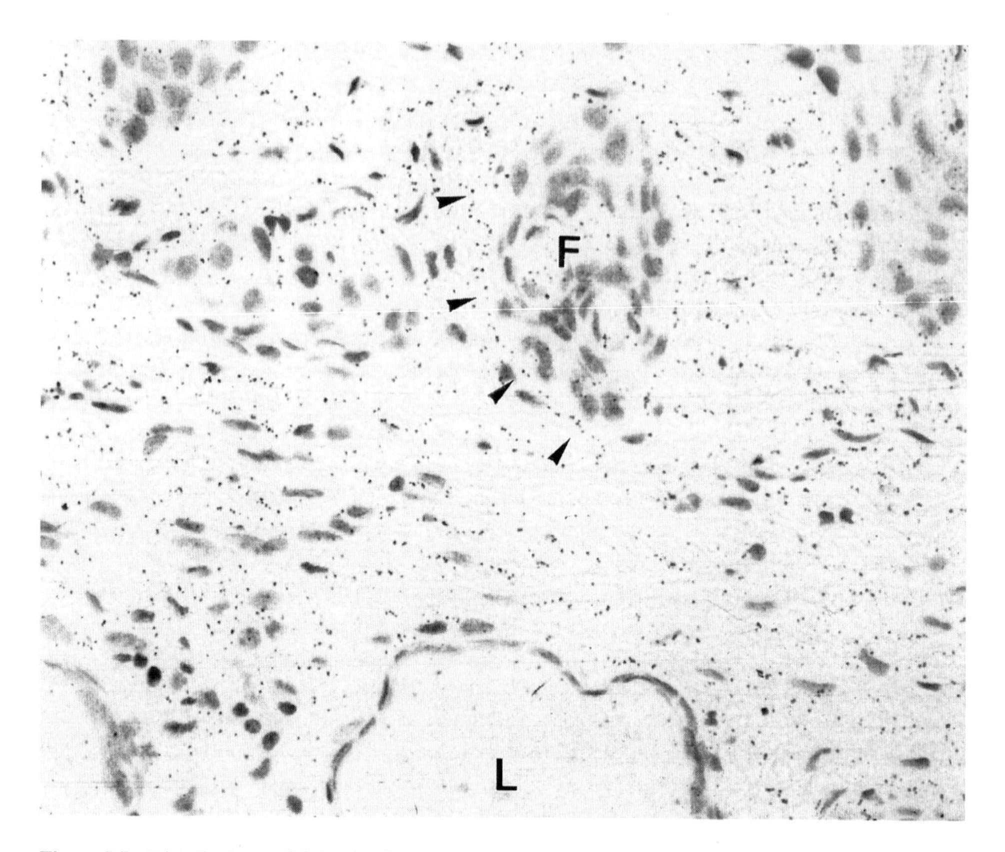

Figure 5.5 Distribution of HA in the dermis. Autoradiography of mouse skin removed and fixed 12h after application of ^{3}H-labelled HA (M_r 400 kD) to the surface of the skin. The localisation of absorbed HA in the dermis is identified by black silver gains. The counterstain is deliberately light and does not reproduce the bundles of collagen fibres clearly. The silver deposits, however, lie between them, arrayed to a large degree in wavy lines as indicated by the arrows. No grains were developed in skin pre-treated with *Streptomyces* hyaluronidase, or in skin treated with unlabelled HA. L: Lymphatic vessel. F: Hair follicle. Magnification approx. 400. T.J. Brown and J.R.E. Fraser; unpublished.

tion, but incompletely, possibly due to changes in conformation of the proteins or HA during separation.

These findings may be significant in inflammation, not only in joints but in soft tissues where similar changes in HA and plasma protein content occur, and may contribute in the latter to the associated increase in tissue turgor.

Secondary Association With Free Proteins

With one exception, we shall consider here proteins without known sequence homology with specific binding peptides, and also include a brief reference to lipid interactions.

Salt-linkage. The interaction that leads to separation of a mucin clot from synovial fluid requires changes in pH, ionic strength or both beyond the range that occurs naturally. Precipitation can occur at physiological pH with polypeptides of high pI such as protamine, but these are not found normally in extracellular fluids.

Looser kinds of association. This is similar to Ogston's ultrafiltration residues from normal synovial fluid have been repeatedly described in rheumatoid synovial fluid. These can usually be broken by density gradient ultracentrifugation or gel chromatography in guanidinium chloride but otherwise resist usual conditions of gel chromatography or filtration.

The first to be described was an association with inter-α trypsin inhibitor after separation of HA by ultrafiltration and passage through hydroxyapatite with IRC50 resin. This association was resistant to 6M urea and high salt concentrations (Sandson and Hamerman, 1964, Sandson *et al.*, 1965). Electrophoretic mobility, ultracentrifugal sedimentation and gelation at pH 4.5 were grossly different from those of a similar preparation of HA from normal synovial fluid.

A strikingly similar result was later obtained by chromatography in Sepharose CL-2B alone (Hutadilok *et al.*, 1988). Protein content of the recovered HA was 1% in all fluid samples from 15 normal joints, in 24/29 from osteoarthritic, 9/10 from gouty, and 59/149 rheumatoid joints. In 67 of the 149 rheumatoid samples, the protein content fell within the highest range (from 4% to >7%). The proteins were identified in the pooled HA fraction as inter-α trypsin inhibitor (33%), haptoglobin (29%) and α-1 protease inhibitor (18%). Similar complexes were recreated by incubation of normal HA with an excess of each protein individually. Protease inhibition was preserved by the associated form of the inter-α trypsin inhibitor but not by that of the α-1 protease inhibitor. Of special interest was the finding that the complexes from rheumatoid fluid and those formed with inter-α trypsin inhibitor or with haptoglobin had an increased resistance to degradation by oxygen-derived free radicals, which are considered important in inflamed joints. Since the association presumably depends on the increased levels of these proteins in the synovial fluid of inflamed joints, it remains to be seen whether they are more protective bound than free.

These complexes have now gained fresh significance with the report that inter-α trypsin inhibitor binds rapidly to TSG-6 (Wisniewski *et al.*, 1994). TSG-6 is a recently discovered glycoprotein secreted by human fibroblasts and peripheral blood mononuclear cells only after treatment with TNF or IL-1, and not at all by endothelial cells or a variety of tumour or transformed cell lines (Lee *et al.*, 1992). It has sequence homology with the binding region of link protein, proteoglycan core and CD44, binds HA specifically and has a domain common to developmentally regulated proteins. It might be found in the 20% of the complexed protein unaccounted for in the HA isolated from inflamed joints.

TNF and IL-1 are both commonly present in rheumatoid joint fluid, which also has a high and variable mononuclear cell content. The striking variability of complexed protein in these fluids might therefore provide some very informative indirect links with other aspects of the inflammatory process.

Although the protein content of the complex is not as high as in the Ogston type of preparation, a comparison of its effect on rheological behaviour could also add to the understanding of rheological changes in rheumatoid fluids.

An intriguing interaction with phospholipid has recently been demonstrated *in vitro* (Ghosh *et al.*, 1994). Physical measurements showed that it increased the chain flexibility and molecular domain (root mean square radius) of pure HA. Any effect on viscoelasticity remains to be determined, but the interaction may be related to the phospholipid micelles.

Selective binding to other proteins in plasma, such as IgM, has been occasionally reported. None takes any part in plasma clearance of hyaluronan, which is mediated by a specific metabolic receptor.

Covalent linkage. A serum protein adsorbed from culture medium to HA synthesised by dermal fibroblasts is identical with the heavy chain of inter-a trypsin inhibitor (Yoneda *et al.*, 1990; Huang *et al.*, 1993). The association resists isopycnic ultracentrifugation and gel chromatography in dissociative conditions and ion-exchange separation.

Whatever the kind of bonding in secondary protein associations with HA, it is clear that they may reflect important influences on the course of rheumatic disease and are worthy of continued study.

Specific Binding and Cell Receptors

Hyaladherins. The structural role of HA was extended figuratively and literally by the recognition of its specific binding to other macromolecular elements of the extracellular matrix, first demonstrated with cartilage proteoglycan (Hardingham and Muir, 1972). The potential for an entirely new range of physiological functions appeared with the later demonstration of specific cellular binding, firstly on transformed fibroblasts during the elucidation of changes in pericellular HA accretion (Underhill and Toole, 1979, 1982), and then in receptor-facilitated endocytosis as the initiating step in its clearance from plasma and subsequent catabolism by hepatic sinusoidal endothelial cells (Laurent *et al.*, 1986).

In both groups, the binding site is primarily a peptide sequence with clusters of basic amino acids, substituted in a diverse range of proteoglycans and cell membrane glycoproteins; some, such as versican, with other binding affinities (LeBaron *et al.*, 1992) and some without. Where HA affinity has been determined, binding has been by and large limited to a minimum HA chain length of six to eight sugars in the receptors and ten to twelve in the matrix binding proteins. Binding also shows varied sensitivity to salt concentrations and pH, which has not been systematically related to the peptide structure or its glycosylation. Finally, binding may be sensitive to orientation of the binding site at the cell membrane, or rate of cycling of the receptor between surface and interior of the cell, which might underlie the rapid changes in binding sometimes observed with induced cell activity.

Some HA-binding peptides may eventually prove by their structure and function to belong to both matrix- and cell-associated classes. Their generic description as hyaladherins (Toole, 1991) avoids any present need for such discrimination.

Matrix binding. The link protein and aggrecan of articular cartilage macroaggregates provided the prototypes for all HA- binding peptides, and now have analogues in numerous other tissues. The resolution of their amino acid structure greatly accelerated the detection and identification of other members of the matrix binding group and of cell receptors. These aspects of the subject are considered in Volume 2, Chapter 6. It is appropriate to note here that some matrix HA-binding proteins may be closely associated with cell surfaces, especially in the brain (Delpech *et al.*, 1989; Bignami *et al.*, 1993).

Cell receptors. The kinds of cell that carry HA receptors are not confined to connective tissues, and indeed some with the most clearly established receptor functions are not mesodermal in origin. Evidence of HA binding to cell surfaces has existed for a long time. Notably, its aggregation of certain lymphoma cell lines (Pessac and Defendi, 1972; Wasteson *et al.*, 1973) but not normal lymphocytes (Wasteson *et al.*, 1973) anticipated the recent demonstration of induced or upregulated receptor expression.

Candidates for HA receptor status or variants of known receptors are appearing in increasing numbers and in some instances it is still too soon to describe their functions with certainty. We shall therefore adopt the operational definition of a receptor, which simply calls for proof of appropriate binding characteristics associated with a cell membrane, rather than the strict functional classification, which requires an active cell response to binding of the ligand. Binding to the cell surface without response may in any case initiate the development of the pericellular matrix of some cell types, or in motile cells form a glycocalyx comparable to those of unicellular organisms, which promotes cell aggregation or adhesion to other binding sites.

Cell receptors identified to date fall into three main groups: CD44, RHAMM, and the most recent addition, ICAM-1. Some have yet to be classified. The first and third of these were already known as cell adhesion molecules with other recognised ligands before their HA binding was discovered.

1. CD44: A Multipurpose Receptor. CD44 is also known as Hermes antigen, H-CAM, Pgp-1, ECMIII and other designations in different fields of biological research. It binds to collagens, hence ECMIII (Carter and Wayner, 1988), and has been cast in the role of a lymphocyte "homing" receptor (Jalkanen *et al.*, 1986). Its recognition as an HA receptor came from two directions; through detection of sequence homology with the cartilage HA-binding peptides (Stamenkovic *et al.*, 1989; Goldstein *et al.*, 1989; Wolffe *et al.*, 1990), and by identification with a monoclonal antibody raised against the cell-surface HA-binding protein of SV-3T3 mouse fibroblasts (Culty *et al.*, 1990). A large amount of data was already available for the latter, including binding characteristics and linkage to actin.

Formal demonstration of HA binding to CD44 soon followed (Aruffo *et al.*, 1990; Lesley *et al.*, 1990; Miyake *et al.*, 1990). CD44 is very widely distributed in the body, however, and is found in cells and tissues where previous tracer studies with labelled HA had shown little evidence of accumulation. This reservation was directly confirmed by the demonstration that while some CD44-positive T and B lymphocyte cell lines showed completely specific HA binding activity others showed none, nor did normal bone marrow myeloid cells and splenic lymphocytes with a high expression of CD44 (Lesley *et al.*, 1994).

These reservations were rapidly resolved by the recognition of distinct isoforms with varied HA binding, and of upregulation by activation, possibly with post-translational modification by glycosylation or other conformational change. HA binding can be rapidly induced in T cells by a monoclonal antibody (Lesley and Hyman, 1992; Lesley *et al.*, 1992), suggesting conformational change, and more slowly in B cells by other means (Murakami *et al.*, 1990, 1991). Further observations on expression of activity have been reviewed by Lesley *et al.* (1993). Binding can be induced in splenic T cytotoxic cells by immunisation with allogeneic cells but HA appears to play no part in their cytotoxic activity (Lesley *et al.*, 1994). A major distinction exists between the common haemopoietic form of ~85 kD (CD44H), found also in mesodermal cells and associated with HA binding, and a larger epithelial form (CD44E) (Stamenkovic *et al.*, 1991). Other variations, not necessarily concerned with HA binding, have been associated with alternative splicing and the expression of at least ten variant exons, and also related to the particular epitopes and to different kinds of malignancy (see for example, Salmi *et al.*, 1993; Wirth *et al.*, 1993; Jackson *et al.*, 1994).

Two sequences in the extracellular domain of CD44H, one close to the free N terminus and the other more central, express the basic amino acid clusters associated with HA binding, which is abolished by mutation of one arginine conserved in all binding proteins (Peach *et al.*, 1993). Deletion of the central segment impairs adhesion to HA-coated surfaces but not uptake (see following).

The outcomes of CD44-HA binding include cell aggregation, adhesion to HA-coated surfaces and high endothelial cells *in vitro*, endocytosis of HA leading to its catabolism in lung macrophages, certain transformed fibroblast lines (Culty *et al.*, 1992) and chondrocytes (Hua *et al.*, 1993), and assembly of pericellular matrices from HA and proteoglycan (Knudson *et al.*, 1993). A study of eleven human breast cancer lines has shown a correlation between CD44 expression, HA uptake and catabolism, and apparent invasiveness (Culty *et al.*, 1994). There is already a large literature on its possible significance in neoplasia (glioma, melanoma and carcinoma, as well as further work on the original observation in lymphoma); in tissue distribution; and its expression in embryogenesis.

Although mature haemopoietic cells do not express an active binding form of CD44 without stimulation, bone marrow is one of the sites where circulating HA is taken up and concentrated (Fraser *et al.*, 1983). The mode and cellular site of its uptake in this tissue have not been studied.

2. RHAMM: A Motility Receptor. A different kind of HA- binding protein has been identified which is conspicuously associated with cell locomotion. It was first isolated from the culture supernatant of chick embryo heart fibroblasts. When the washed cells were enriched with this protein, they exhibited marginal ruffling within a few minutes of adding HA, followed by increasing random movement and release from contact inhibition (Turley 1989). The protein is concentrated in the surface blebs and ruffles of actively moving cells but disperses over the cell and regresses with confluence and quiescence. Ultrastructural and immunochemical studies localised it on the cell surface, confirmed the distribution patterns, showed close association with actin fibrils in cell processes disrupted by cytochalasin D and also an intensified association with the cytoskeleton (Turley *et al.*, 1990).

The amino acid sequence of the protein was determined by cloning from a cell line containing an inducible mutant *ras* gene that controls its expression. It showed the basic amino acid clusters of an HA-binding domain but no further homology with other hyaladherins (Hardwick *et al.*, 1992). It was therefore designated RHAMM (receptor for HA-mediated motility). It lacked a well-defined membrane insertion sequence, which may reflect the rapid variations in its concentration and distribution on the cell surface. Further analysis identified a sequence of seven amino acids flanked by arginine or lysine, found also in link protein and CD44 (Yang *et al.*, 1994). Two isoforms have now been identified, one with an insertion sequence (Entwhistle *et al.*, 1995).

The *ras*-transformed cell line expresses RHAMM and CD44. Although CD44 binds much more HA to this cell than RHAMM does, its HA-induced motility is mediated only by RHAMM (Turley *et al.*, 1993). The rapid turnover of focal adhesion points associated with cell movement is accompanied by transient tyrosine phosphorylation when induced by HA or by low levels of an anti-RHAMM antibody that otherwise blocks it at higher levels (Hall *et al.*, 1994). TGF-β, which is known to stimulate both HA synthesis and cell motility, was shown to induce concurrently transcription, synthesis and membrane expression of RHAMM (Samuel *et al.*, 1993).

RHAMM has now been recognised in diverse kinds of normal and neoplastic cells and in pathological processes. It has been identified (1) in spermatozoa, in which motility is reduced by specific antibody (Kornovski *et al.*, 1994); (2) in thymocytes and T lymphocytes where it is up- or down-regulated by physiological signals, cyclosporin A and during development (Pilarski *et al.*, 1993); (3) variably in normal B lymphocytes according to their location; (4) and more persistently, but again variably, in the B cell malignancies, multiple myeloma and hairy cell leukaemia (Turley *et al.*, 1993). The role of HA and RHAMM in tumorigenesis has been reviewed recently (Hall and Turley, 1995).

3. ICAM-1: A Metabolic Receptor? The receptor that initiates metabolism of plasma HA in liver endothelial cells (LEC) was the second kind to be formally defined by specific binding characteristics (T C Laurent *et al.*, 1986). The cellular response linked to it is well defined and remarkably rapid. The ligand-receptor complex is internalised to form endocytotic vesicles (Eriksson *et al.*, 1983) so quickly that it is difficult to capture at the cell surface in conditions of maximum uptake (Fraser *et al.*, 1985b). Degradation products of labelled HA from cultured LEC are detectable in about 30 min (Smedsrød *et al.*, 1984) but *in vivo* GlcNAc and GlcUA from the labelled polymer have been identified in rabbit liver and in spleen, another site of high uptake, in < 15 min (personal observations). In the sheep fetus, free acetate has been detected ~9 min after intravenous injection (Fraser *et al.*, 1989). Ligand binding of this receptor thus triggers a highly co-ordinated cascade of events that includes the formation of an endocytotic vesicle, its fusion with primary lysosomes, enzymatic digestion to monosaccharides, active transmembrane transport of these sugars to cell sap, phosphorylation of GlcNAc (Fraser *et al.*, 1988, 1989) and enzymatic deacetylation (Rodén *et al.*, 1989; Campbell *et al.*, 1990); all within about 10 min. These observations have been reinforced by demonstration of the rapid intracellular recycling of the LEC receptor (McGary *et al.*, 1989).

Although the rate cannot be measured, similar degradation products of HA have been identified in the lymph node (Fraser *et al.*, 1988). With other evidence, the high extraction rate from low concentrations in lymph also indicates receptor-mediated metabolism (Volume 1, Chapter 5).

In contrast with CD44, the endothelial receptor shows an even higher affinity for chondroitin sulphate of the same molecular weight (T C Laurent *et al.*, 1986), which has been supported by competitive studies *in vivo* (Tzaicos *et al.*, 1989) and complemented by the detection of similar small metabolic endproducts in the lymph node *in vivo* (work in progress). Given that CD44 can mediate degradation by endocytosis, it remains to be seen whether this is as rapid and complete *in vivo* as that effected by the endothelial receptor. In view of its cellular distribution, CD44 might well be responsible for the slower turnover of HA whose primary function is to remain at the surface for cell association or as a glycocalyx.

Yannariello-Brown and Weigel (1992) have ingeniously distinguished another kind of receptor on cultured LEC. It differs from the metabolic receptor in that it is not recycled but holds the ligand at the surface, and the binding is Ca^{2+}-dependent. This is further evidence that particular cells may concurrently express more than one HA receptor, as exemplified by CD44 and RHAMM.

Taken with its linkage to an exquisitely co-ordinated response, the economy in cellular organisation that results from the broader ligand specificity of the metabolic receptor on endothelial cells suggests a distinct evolutionary pathway committed to metabolic turnover. In contrast, absolute specificity would seem obligatory in the binding proteins that form precisely structured aggregates in matrix, and in cell receptors that control specific responses. The basis for specificity, however, is not fully resolved. There may be variations of, or adjacent to, the common binding motif which explain differences in oligosaccharide competition and the effects of ionic conditions. Conformation in the cell membrane might also influence binding; the CD44 receptor of SV 3T3 cells reacts with chondroitin sulphate in detergent extracts and when inserted in liposomes, but not in the cell membrane (Chi-Rosso and Toole, 1987).

The differentiation of the LEC receptor from others has now been confirmed by its further characterisation. After recovery of a homogeneous protein of 90 kD from rat LEC by HA affinity chromatography, a monospecific polyclonal antiserum was raised (Forsberg and Gustafson, 1991). With immunoblotting, proteins of similar size were recovered from spleen, kidney, lymph nodes and thymus. Positive immunohistological staining was detected in these organs, concentrated in the blood vascular sinusoids of liver and spleen, some renal capillaries, lymphatic sinuses of lymph nodes, capillaries of small gut, thymic reticular cells and corneal endothelium (Forsberg *et al.*, 1994). Apart from the last three, these are all sites of high HA uptake and proven or probable catabolism *in vivo*. The distribution of reactive sites was remarkably similar to the patterns found in autoradiographic studies of HA uptake *in vivo* (Fraser *et al.*, 1985b; Fraser and Laurent, 1989), particularly in the lymphatic sinuses and the complex circulation in the spleen. (HA uptake in the rat spleen *in vivo* is an exception for unrelated physiological reasons; Fraser *et al.*, 1985b.) These sites are not well stained with the anti-SV 3T3 monoclonal cross-reactive with the CD44 group.

Using the same monospecific antiserum to monitor its purification through five distinct stages of affinity chromatography, McCourt *et al.* (1994) have extracted the putative receptor in rat liver. The amino acid sequence was determined in four peptides recovered after trypsin cleavage, which contained individually between 7 and 14 amino acids to a total of 44. A precise match for the sequence of each was found in the glycoprotein, ICAM-1, except for one unidentified amino acid that corresponded with a glycosylation point in ICAM-1. These authors point out that ICAM-1 is found at the same sites as the metabolic receptor. Though further direct evidence of identity will no doubt be sought, there is sufficient now to anticipate that ICAM-1, or some variant thereof, will prove to be a third class of HA receptor. In contrast with CD44, existing data for ICAM-1 show little genetic or phenotypic variation within species. The relationship with the LEC HA receptor might provide the first evidence of this kind. It is at least certain that the main metabolic receptor is distinct from the CD44 group.

Whether this receptor influences other cell functions through HA binding as in the case of CD44 remains to be determined. It has already been found on corneal endothelial cells and on mouse mastocytomas (a malignant tumour of mast-cell origin). Gustafson *et al.* (1995) found that this tumour avidly absorbed HA from the blood stream. Staining with receptor antibody was limited to poorly defined vascular structures and collagenous areas but was greatly enhanced by prior treatment with hyaluronidase. This indicates that the receptor also holds HA in high concentration on the cell membrane in contrast with its rapid internalisation in the liver as a metabolic receptor. It may therefore serve two functions in this tumour, either in distinct isoforms or different kinds of membrane anchorage.

ICAM-1 is a cell adhesion molecule classified as a member of the immunoglobulin supergene family in common with other HA binding proteins (e.g., link protein and aggrecan). It is widely distributed through the tissues on endothelial cells, macrophages and other cells, though commonly with low levels of mRNA and cell surface expression that are quickly responsive to endotoxin, TNF and IL-1. Its recognised counter- receptors are the leucocyte integrins, LFA-1 (Makgoba *et al.*, 1988) and Mac-1 (Diamond *et al.*, 1990), which together constitute an important cell adhesive system in inflammation and immune reactions (Kishimoto *et al.*, 1989). Its capacity for quick upregulation, like that of CD44H, clearly marks out a potential role for it in the many pathological processes where HA accumulates.

Existing data for ICAM-1 show little genetic and phenotypic variation comparable to that of CD44, though further exploration of its homology with the LEC hyaluronan receptor might prove to be a move in that direction.

It is now clear that there are two distinct receptors linked with HA catabolism. The ICAM-1-related receptor is likely to be mainly responsible for the clearance of HA from lymph and blood plasma, which accounts for perhaps most of its whole-body turnover (for review, see Laurent and Reed, 1993). Nevertheless, a macrophage-like cell has been observed to internalise HA in lymphatic and splenic sinuses (Fraser and Laurent, 1989; unpublished observations). Normally, a fraction of HA is degraded in its tissue of origin; in synovial joints and skin some 20-30%, and in cartilage probably most of it (Ng *et al.*, 1992). Chondrocytes ingest and metabolise HA through CD44 uptake (Hua *et al.*, 1993). Local turnover almost certainly dominates in other solid tissues, or those without lymphatic drainage. During lung development, macrophages, with upregulated CD44 and with a capacity for HA uptake and degradation, appear at the onset of differentiation when the HA content of the tissue falls sharply. Distinctive cells of monocyte-macrophage lineage are normally found in many tissues; for example, as histiocytes in skin, Kupffer cells in liver and dendritic cells in the immune system. Although Kupffer cells in liver sinusoids normally take little part in HA turnover beside endothelial cells, macrophage-derived cells are probably responsible for turnover in tissue matrix. It remains to be seen how far this function is distributed between the different receptors in normal or abnormal conditions.

It is unlikely that all newly recognised receptors (see for example, Banerjee and Toole, 1992) will be related to these three kinds. Even so, there is now a considerable amount of work being done to delineate the distinctive place in physiology and pathology of those defined at present.

The foregoing account of HA receptors is merely an outline of current developments. It does not, for example, extend to recent observations on their role in embryogenesis, fetal development or wound biology, or indicate fully their involvement in oncology. These matters are covered by reviews in literature dedicated to the respective areas. It has been our prime purpose here to outline the enormous expansion in our perception of the biological role of HA that has followed the recognition of its cell receptors.

Other Relationships With Cellular Activity

Of the numerous cellular interactions with HA recognised before the foregoing developments (see Laurent and Fraser, 1986), many of the positive reactions have since been attributed to receptors. Most of the others are also consistent with the expression or upregulation of receptors. For example, in macrophages, HA may enhance phagocytosis (Ahlgren and Jahlstrand., 1984) or inhibit it, depending on its molecular weight (McNeil *et al.*, 1989). In common with other glycosaminoglycans it stimulates pinocytosis (Cohn and Parks, 1967) which suggests multiple receptors or a receptor with broad specificity. Hyaluronan at a level of 25 to 400 ng/ml stimulates mitosis in corneal epithelial cells (Inoue and Katakami, 1993).

Negative effects, which may be due to physical impedance in some cases, include inhibition of chondroitin sulphate synthesis by chondrocytes, also induced by HA oligosaccharides (Wiebkin and Muir, 1973), and inhibition of response to mitogens, migration of leucocytes and contact-dependent reactions (Darzynkiewicz and Balazs, 1971; Forrester and Balazs, 1980; Fraser and Clarris, 1970). The last two of these depend partly on the force that the cells can exert in their movements. In some circumstances, HA interferes with cell-substrate attachment.

TISSUE MODULATION, CELLULAR PROLIFERATION AND INFILTRATION

Intuitively, the physical environment provided by HA-rich matrix seems ideal for migration and multiplication, whether of normally resident or exogenous inflammatory and neoplastic cells, and especially in embryonic development.

Elucidation of the means by which HA can influence the behaviour of cells and matrix is now clarifying some of the striking associations of HA with cellular activity that have been recognised for a long time. The first to receive close scrutiny was its role in embryonic development, which has proved a most informative model for similar associations that appear later in various circumstances.

In the other examples given in the Introduction, and in inflammation, neoplastic infiltration, skin diseases, and thyroid hormone deficiency, the HA content of the normal matrix increases and the tissue is swollen. Where this is circumscribed in acquired diseases it is customarily described by the descriptive term, mucinosis. This can be misleading for those unfamiliar with pathology, since the mucins of epithelial mucous, from which the term derives, have largely been identified as glycoproteins and there is normally little or no HA in mucous. Mucinous change in extracellular matrix consists primarily of GAG, and is mainly HA. "Mucoid" and "mucinous" remain apt descriptions of gross appearance and tactile qualities, but we suggest that hyaluronosis is a more informative description in pathological histology now that it can be identified with confidence.

SOURCES OF VARIATION IN TISSUE HYALURONAN CONTENT

Local HA content is determined by a balance in the kinetics of synthesis and turnover which can be disturbed in three ways.

Synthesis. Hyaluronan synthesis commonly varies with the rate of cellular multiplication in natural and experimental conditions. Usually, both rise and fall together. An intrinsic coupling of HA synthesis with mitosis has been repeatedly demonstrated by cell culture (see for example, Moscatelli and Rubin, 1975; Cohn *et al.*, 1976), and it has been suggested that it is responsible for the cell detachment necessary for mitosis (Brecht *et al.*, 1986).

Synthesis of HA rises sharply when cells such as synovial cells, dermal fibroblasts and chondrocytes grow rapidly in culture, although the proportions and kinds of other GAG produced continue to reflect their phenotype at a lower level. The balance swings back when the cells become confluent and growth slows. Collagen fibres usually appear at this time. The mechanism of this coupling is not clear though it has been linked with HA synthase activity in different phases of the cell growth cycle (Mian, 1986). It is probably most active in developmental growth, regeneration and the reparative phase of wound recovery.

Increase in HA synthesis can be partly (Fraser *et al.*, 1969) or totally independent of increased cell multiplication. HA synthesis continues in tissues such as synovial joints where the cell population responsible for it is in equilibrium and cell turnover is not conspicuously high.

The number of cytokines, growth factors and other inflammatory mediators that can stimulate synthesis (Laurent and Fraser, 1986) has continued to grow, notably with the addition of TNF (tumour necrosis factor). These agents can be autocrine or paracrine, and are mainly relevant to matrix changes in inflammation and neoplasia; though bFGF (basic fibroblast growth factor), which may be considered autocrine and also stimulates cell growth, has been implicated in embryonic development. It is most abundant when proliferation is most active (Munaim *et al.*, 1988).

Many of the stimulating peptides are known to be released by resident or infiltrating cells in the conditions considered here. Some stimulate release of collagenase and other neutral proteases as well as HA synthesis (for example, IL-1; Dayer *et al.*, 1986; Hamerman and Wood, 1984) which in effect reinforces the softening of the matrix caused by an increased HA content.

Westergren-Thorsson *et al.* (1990) found that TGF-β (transforming growth factor) stimulates HA synthesis in confluent cultured lung fibroblasts but not in skin fibroblasts; whereas Heldin *et al.* (1989) observed increased HA synthesis by skin fibroblasts in both sparse and dense cultures in response to TGF-β, as well as the PDGF-BB isoform of platelet-derived growth factor, bFGF, and EGF (epidermal growth factor). The different response to TGF-β in these two studies might reflect cell source (one embryonic, one foreskin) or cultural conditions. All factors stimulated HA synthesis, but the variations in cellular response show that the effect should not be translated too precisely to tissue behaviour.

Cooperative effects on HA by various combinations of growth factors (Honda *et al.*, 1991; Heldin *et al.*, 1989) and cytokines (Meyer *et al.*, 1990) have been demonstrated. The potential difficulties in resolving cellular responses to HA *in vivo* are illustrated by work on bone marrow-derived macrophages. Through activation of CD44, exposure to HA induced a complex autocrine cascade involving expression of TNF, IL-1 and IGF-1 (insulin-like growth factor) mRNAs; a similar response was provoked by asbestos (Noble *et al.*, 1993). Thus the activities of the macrophage class of cell may promote both the accumulation and the degradation of HA.

The stimulation of GAG synthesis by direct contact of fibroblasts with melanoma, lung carcinoma or pancreatic carcinoma cell lines (Knudson *et al.*, 1984) is a particularly interesting phenomenon. It is effected by a cell membrane protein which remains active in separated cell membranes, extracts thereof and when reconstituted in lipid vesicles (Knudson and Toole, 1988). The increased GAG is mainly HA. An oncofetal protein secreted by Wilms tumour also stimulates HA synthesis (Longaker *et al.*, 1990).

Degradation. Changes in local degradation of HA, apart from the obvious effects of bacterial, leech and venom hyaluronidases, are largely a matter of surmise. As noted earlier, degradation by resident or infiltrating macrophages or by free radicals is more likely to be increased in inflammation. Macrophages have already been incriminated in HA turnover in development of the lung (Allen *et al.*, 1991). Nothing is known about inhibition of cellular degradation.

Outflow. Turnover by lymphatic outflow might be reduced by blockage of lymph vessels, since lymph can coagulate. It is, however, not easily proven unless persistent. Drainage and HA outflow are accelerated by increased tissue fluid flux (see Vol. 1, Chapter 5). This may partly deplete matrix HA at least temporarily in acute inflammation, but in less florid forms such as delayed hypersensitivity immune reactions, the increase in HA synthesis seems to outweigh any increased turnover.

THE DEVELOPMENTAL MODEL

The abundance of HA in fetal tissues and its gradual decline in postnatal development have been firmly established in numerous tissues and organs; for example, in the brain of rats and humans (Margolis *et al.*, 1975; Delpech *et al.*, 1989). It has been shown to dominate the most rapid phase of growth in early embryonic development when cells multiply and migrate to define the ultimate form of the particular tissue or organ. Synthesis of HA is a prime function of undifferentiated cells and as noted repeatedly in this chapter, most cells from mature tissues, no matter how differentiated in form and function, resume or increase the synthesis of HA in natural and experimental conditions if they are released from restraints on their proliferation. This capability is not restricted in mature tissues to cells of mesodermal lineage (see for example, Tammi *et al.*, 1988, 1991).

In the earliest phase of limb bud growth, the inherent link between mitotic activity and HA synthesis (Brecht *et al.*, 1986) is reinforced by high levels of bFGF (basic fibroblast growth factor). bFGF also enhances the development of HA-rich pericellular investments of cells taken from the limb bud after tissue condensation, when this capacity is usually no longer expressed (Toole *et al.*, 1989). Oocytes similarly produce a humoral factor that stimulates HA production in their surrounding cumulus cells after priming by follicle-stimulating hormone before ovulation (Salustri *et al.*, 1990a,b). The HA-rich matrix is common to all tissues at this stage, including those of the neural crest (Toole, 1976).

The subsequent tissue condensation is associated with the differentiation of synthetic functions characteristic of the tissue, and a sharp fall in the relative bulk of the matrix and its HA content during condensation of the cell mass. Thus in the cornea, when mesenchymal cells cease to migrate into the acellular matrix created by epithelial cells, hyaluronidase activity appears, HA content is reduced, the structure shrinks, and the matrix is replaced by the sulphated GAG and collagen that characterise its mature structure (Toole and Trelstad, 1971). This phenomenon was confirmed in the spinal column and limb buds (Toole, 1972), with interesting distinctions between skin and cartilage. It seems common to other tissues; for example, kidney (Belsky and Toole, 1983). Quite localised differences in HA content and cell density have been identified in palatal development (Brinkley and Morris-Wyman, 1987). Other humoral factors may moderate the regression of HA in the matrix during condensation. While the production of HA recedes in deeper tissues, it is sustained in the future subcutaneous layer by the release of transforming growth factor-β from the overlying ectoderm (Toole *et al.*, 1989)

At this stage of embryonic development, the lymphatics have not developed and HA must be removed entirely by local metabolism. As exemplified by smooth muscle, synthesis of HA is reduced or ceases according to the mature cell phenotype. In the fusion of skeletal myoblasts, HA is redistributed at the myotube surface to form a much thinner coating (Orkin *et al.*, 1985).

Close parallels with the developmental sequence have been established in limb regeneration in newts and in tadpole metamorphosis induced by thyroxine (Polansky and Toole, 1976; for a comprehensive review, see Toole, 1976). The fundamental features of the sequence have recently been coupled with the expression of CD44 in development of the hair follicle. In the initial mesenchymal condensation of cells in the hair bulb, CD44 is prominent but HA has disappeared, in contrast with the adjacent interstitium; around the mature follicle, HA has reappeared and the cells no longer express CD44 (Underhill, 1993).

Although no longer subject to the sharply defined temporal changes of embryonic development, the prominence of HA in growth and development continues in the fetal phase and after birth, as demonstrated by the brain (Margolis *et al.*, 1975). This later involvement is probably reflected in the variations of HA in amniotic fluid, which is continuous with the liquids of the the bronchial tree and gut, and it is certainly evident in the exceptionally high turnover of plasma HA in the fetus (Dahl *et al.*, 1989b; Fraser *et al.*, 1989). Considerable changes also occur in the pulmonary and pleural distribution of HA before and after birth (Allen *et al.*, 1991), a question of critical import in neonatal acute respiratory distress.

These observations have provided a model for the healing of fractures and of wounds and much data that is relevant to inflammation and neoplasia. These will be considered in a later section.

HYALURONAN IN THE PERICELLULAR ENVIRONMENT

Much of the work on the control of cellular activity arising from these studies, particularly in the field of receptors and cell-matrix interactions, has been discussed elsewhere in this chapter. One aspect worthy of separate consideration concerns the immediate cellular environment, since it throws light on the secretion of HA as well as the first stage in matrix development.

Many kinds of cell in culture quickly develop a three-dimensional gel around themselves individually or in groups which is invisible in phase-contrast microscopy and undetectable by routine stains. It was first recognised in cultured human synovial fibroblasts by its exclusion of erythrocytes and leucocytes from a clear area that extends as much as 30 μm beyond the cell membrane (Clarris and Fraser, 1967, 1968). Its integrity was found to depend primarily on its content of HA. It was dispersed by hyaluronidase but not by neuraminidase or nuclease. Low concentrations of trypsin caused it to swell but it disappeared after subculture in higher concentrations of trypsin, to appear again within a few hours of reattachment. Its GAG content consisted mainly of HA with a small proportion of chondroitin sulphates but the protein content was not examined. It was shown to impede passage of small non-motile particles, delay virus infection (Clarris *et al.*, 1974), and prevent reactions with allogeneic lymphocytes (Fraser and Clarris, 1970).

Similar pericellular investments that blocked cytolysis by lymphocytes were detected in other normal and malignant cell types (McBride and Bard, 1979), and in the 3T3 strain of mouse fibroblasts (Underhill and Toole, 1982). They also develop in undifferentiated limb bud mesenchymal cells but fail to develop in cells cultured from the chondrogenic and myogenic regions in the stage of condensation. After differentiation into chondrocytes the pericellular coats again develop in culture but with a higher proteoglycan content (Knudson and Toole, 1985b). The cohesion of the investments still depends on HA at that stage (Goldberg and Toole, 1984). The size of the investment is greatly reduced in virus-transformed (SV) 3T3 cells but in contrast with the parent 3T3 cell line they express a high-affinity receptor which can bind exogenous HA and holds endogenous HA at the cell surface (Goldberg *et al.*, 1984).

It has since been shown that HA receptors are responsible for the pericellular matrix of cultured chondrocytes. This matrix can be dispersed by HA hexasaccharides and reconstituted by addition of HA with aggrecan (Knudson, 1993), and can form on other cells providing they bear receptors. Cells transfected with CD44-H or CD44-E also develop a pericellular matrix with HA and aggrecan added together but not with either alone; again HA hexasaccharide disperses it (Knudson *et al.*, 1993). An antibody directed against an as-yet unclassified HA receptor and HA hexasaccharide have each been shown to disperse or block the formation of pericellular matrix in six other cell types, including 3T3 (Yu *et al.*, 1992).

Development of extensive hyaluronan-containing coats that form around cultured normal human mesothelial cells is inhibited by HA dodecasaccharides but not by hexasaccharides, and without altering HA synthesis (Heldin and Pertoft, 1993). These results suggest that the integrity of the matrix in these cells is determined by cohesion within the matrix, either by an HABP of the matrix type or by HA self-adhesion, rather than by binding to the cell surface. The proportions of cultured human skin fibroblasts and mesothelial cells that generate pericellular investments are much reduced (more so in the latter cell type) by the extraction of inter-α trypsin inhibitor from the serum proteins used in the culture medium. They reappear on addition of inter-α trypsin inhibitor in purified form (Blom *et al.*, 1995). It must be noted, however, that some 20% of the dermal fibroblasts used in this study were still found to form pericellular investments in the absence of this protein. Since inter-α trypsin inhibitor can form a complex with HA, directly or indirectly (see *Secondary Association with Free Proteins*), its effect in promoting pericellular accumulation may lie in its HA binding rather than inhibition of proteases.

This interpretation is supported by the behaviour of normal human synovial cells. A striking feature of their pericellular gels is the incongruity between their perimeters and the cell surface; the gel can project as long spikes vertical to the cell membrane in contrast with the more congruous outlines in illustrations of chondrocytes and other cells. Close examination at high magnification shows localised movements of the cell surface and short random movements of the whole cell that are not reflected in the margin of the matrix. This was confirmed by time-lapse microcinematography, which also showed that the cells could suddenly move for relatively long distances, leaving the gel intact to disperse gradually in 1 to 4 h depending on its size (Fraser *et al.*, 1970b). In cultures of these cells and in cultures of oligodendroglioma cells (H Pertoft, personal communication), it is common to see at any time clear areas without cellular content that nevertheless impede the Brownian movement of the indicator particles. These areas are undoubtedly patches of pericellular matrix left behind by cellular migration, as demonstrated by time-lapse ciné recording; again showing that cohesion in the matrix and its adhesion to the culture surface are both stronger than adhesion of the matrix to the cell. It is therefore understandable that the gels are not carried with the cells during subculture with trypsin. Unless they retain a protease, their rate of dissipation indicates that they are held together by the much weaker bonds of hyaluronan networks rather than a high-affinity binding peptide.

It is clear that in these kinds of cell, the matrix is no longer bound to the cell once its HA content is shed from the synthase, and it does not depend on cell binding for its structure. Unlike chondrocytes, these cell types maintain a high level of HA secretion, consistent with their origins. Again in contrast with chondrocytes, the capacity of synovial cells to generate a pericellular matrix of the same character remains unimpaired after repeated subculture.

The development of the pericellular gel in cultured cells therefore remains a useful means to study the initial assembly of different kinds of extracellular matrix in different tissues.

HYALURONAN IN DISEASE

Hyaluronan has commanded close attention in pathology from the time of its discovery. It was considered in the first evidence for the synthetic function of the Golgi bodies (King, 1935); mistakenly in view of the much later location of HA synthase in the cell membrane, though happily the association was not essential in King's conclusions. The mass of information gathered before the developments reviewed here is available in standard texts. Much of the present review has already referred to pathological conditions, and with tumours in particular. The following is a brief summary of selected areas.

Clinical Pathology

The readiest source of information is provided by measurement of HA in serum or urine. Although urinary excretion accounts for only 1 to 2% of plasma HA turnover, it still reflects plasma levels. Twenty-four hour collections may therefore be useful when the plasma changes are intermittent, though not much used as yet.

The liver is the main site of HA elimination from blood (Fraser *et al.*, 1981). Plasma HA can be raised (1) by impairment of liver function or circulation, or (2) by increased input from the tissues, which occurs almost entirely through lymph (Laurent and Laurent, 1981; Tengblad *et al.*, 1986). Although a high proportion is removed by lymph nodes (Fraser *et al.*, 1988), the tissue production is often raised sufficiently to influence plasma despite normal liver function. This has been documented in conditions as diverse as psoriasis, scleroderma, arthritis and cancer (for review, see Engström-Laurent, 1989). In some conditions such as acute elevation of pulmonary venous pressure (Lebel *et al.*, 1988) or septicaemia (Berg *et al.*, 1992), both causes operate. Correlation of plasma HA with the kind and severity of liver disease has been examined intensively (Engström-Laurent *et al.*, 1985; Frébourg *et al.*, 1986; Nyberg *et al.*, 1988; Gibson *et al.*, 1992). Raised plasma HA is particularly likely to be associated with cirrhosis.

Normal hepatic uptake and metabolism of HA are mainly attributable to the fenestrated sinusoidal endothelial cells (Smedsrød *et al.*, 1984; Fraser *et al.*, 1985b), which are little affected by most liver diseases. (For example, their uptake of HA is impaired only by high doses of endotoxin; Alston-Smith *et al.*, 1992). In liver transplant rejection, however, where sinusoidal endothelium is directly involved, there is a highly significant correlation with elevated plasma HA, especially in the chronic but also in the acute phase (Adams *et al.*, 1989). The elevation precedes other biochemical evidence of liver dysfunction by 1–2 days.

The elevation of plasma HA in cirrhosis has been associated with loss of fenestration and appearance of Factor VIII-related antigen and Weibel-Pallade bodies in the sinusoidal endothelial cells, accompanied by development of a basement membrane (Ueno *et al.*, 1993). The impairment of HA uptake is clearly related in this instance to a change in the endothelial cell phenotype, which in turn suggests a loss of receptor expression.

With careful interpretation, plasma measurements of HA provide a distinctive kind of information in several areas of disease research and may prove useful in the critical evaluation of therapeutic measures (see for example, Gibson *et al.*, 1994).

Inflammatory Disease, Injury and Immunological Reactions

Swelling is a classical sign of inflammation and is ultimately caused by exudation from blood vessels. Any accompanying accumulation of HA must derive from the tissues, since the plasma content is so much lower.

The structure of the lungs offers unusual access to interstitial fluid. Most of the pulmonary interstitium and its matrix is separated from the terminal air sacs (alveoli) by the extremely thin alveolar walls. Any increase in interstitial fluid extends into the alveoli, where it can be sampled by bronchopulmonary lavage, a common diagnostic procedure. Although diluted, the content of the fluid can be compared with the washings from normal lung.

By this means, increased HA has been demonstrated in farmer's lung (Bjermer *et al.*, 1987) a recurrent allergic condition; in idiopathic interstitial fibrosis (Bjermer *et al.*, 1989), in which oedema is accompanied by progressive fibrous change; in adult acute respiratory distress syndrome (Hällgren *et al.*, 1989); and in sarcoidosis (Hällgren *et al.*, 1985), a chronic inflammatory disorder of unknown cause notable for profuse infiltration of macrophages.

In acute pulmonary inflammation induced experimentally in rats by bleomycin, the HA and the water content of the lungs rose and fell together, reaching a peak after 3 days and then subsiding toward normal in 30 days (Nettelbladt *et al.*, 1989). This work provides not only evidence of increased HA in acute inflammation, but demonstrates its part in the local retention of fluid mainly responsible for the tissue swelling.

The sacs that surround the lungs (pleura), heart (pericardium) and line the abdominal cavity (peritoneum) are lined by mesothelial cells which secrete HA. Their fluids are normally just a thin sticky film and no estimates of their HA concentrations exist although the total content has been measured in the rabbit pleura (Allen *et al.*, 1992). The concentration of HA in pleural effusions caused by infective and other benign causes, and by several kinds of malignant tumour is always very much higher than in plasma but not distinguished by the cause (Hillerdal *et al.*, 1991). Aseptic peritoneal inflammation induces a 200-fold increase in HA concentration (Edelstam *et al.*, 1992)

Association of HA with rheumatic diseases soon followed its isolation from normal synovial fluid. Effusions into the synovial cavity invariably contain more HA than the normal joint even though it is almost always more dilute. The swelling associated with inflammation of joints commonly involves the surrounding soft tissues and can extend far beyond the joint in gout and virus infection. The HA content in the joint lining and the tissues beneath it has now been studied in several kinds of arthritis with an HABP histochemical probe (Wells *et al.*, 1992). Comparison with normal tissue showed increased HA in tissue deep to the synovial lining, which was more marked in newly vascularised areas and around foci of inflammatory cells (which are predominantly macrophages and lymphocytes in rheumatoid tissue, in contrast with the effusions). A monoclonal antibody to a nuclear antigen expressed in proliferating cells reacted strongly with fibroblasts, and reactive cells were concentrated round groups of lymphocytes and related to the distribution of HA.

Myocardial infarction represents a common form of non-mechanical injury; namely, severe cell damage and death, caused in this instance by sudden cellular hypoxia. In an experimental study (Waldenström *et al.*, 1991), HA content of the injured heart muscle increased within 24 h to reach nearly three times normal after 3 days, and was accompanied by interstitial oedema; changes which together may further impair heart function through altered compliance and raised interstitial pressure. By the third day infiltration by macrophages had begun.

The same association of increased HA content with oedema has been observed in experimental heart and renal transplant rejection (Hällgren *et al.*, 1990a; Hällgren *et al.*, 1990b), and in rejection of human renal transplants (Wells *et al.*, 1990a). In the kidney, the extensive accumulation of HA in the cortex contrasts with its virtual absence in that part of the normal kidney, and the interstitium between the renal tubules is grossly expanded. Precisely the same picture of HA accumulation and interstitial swelling has been found in cases of idiosyncratic drug reactions causing sudden reduction of urinary output (Fraser *et al.*, unpublished observations).

Tuberculin reactions after BCG immunisation typify inflammation in the delayed hypersensitivity type of immune reaction (DTH). In DTH induced by instillation of tuberculin into the bronchial tree, HA has been shown to cause the strong agglutination of macrophages noted in the fluid exudate (Love *et al.*, 1979; Shannon *et al.*, 1981). It is responsible for the sequestration of macrophages by agglutination in peritoneal DTH (Shannon *et al.*, 1980; Shannon and Love, 1980). An increased HA content has been demonstrated in cutaneous DTH (Campbell *et al.*, 1982). These reactions are accompanied by an exceedingly complex interplay of macrophages with lymphocytes and other cells, in which numerous humoral mediators of HA synthesis and regulators of HA binding probably take part. Several cytokines released by activated immunocytes or mononuclear inflammatory cells are potent stimulants of fibroblast HA synthesis (Whiteside and Buckingham, 1989; see also *Hyperthyroidism*). The effects of HA on the participating immunocytes may be equally complex and do not necessarily promote cell contact. For example, HA inhibits rosette formation (binding of activated T lymphocytes to erythrocytes) (Chevrier *et al.*, 1982), and may thus moderate their activity.

Injury of various causes is also accompanied by the early development of a HA-enriched matrix comparable to that associated with cell migration and proliferation in development and growth. This is not surprising in injuries of soft tissues, but HA enrichment also occurs in the initial stages in repair of bone (Antonopoulos *et al.*, 1965) and tendon (Reid and Flint, 1974) even though HA is inconspicuous in the corresponding mature tissues. The infiltrating inflammatory cells are likely through the agency of cytokines to promote these changes by stimulation of HA synthesis in the resident connective tissue cells; thus creating a setting that promotes their own activity and sets the stage for any subsequent regeneration and repair.

Restoration of Damaged Tissue

Where there is little or no tissue damage, the swelling with its high HA content simply recedes and normal tissue structure is restored. In more destructive injury, whatever the cause, higher species do not have the capacity for regeneration possessed, for example, by the newt. In some tissues such as heart and skeletal muscle, the parenchymal cells (those responsible for the function of the tissue as distinct from those of the matrix) have little or no capacity to proliferate once fully differentiated. In these circumstances the recession of HA is followed by development of a scar in which collagenous tissue replaces the parenchymal cells and their normal matrix, then condenses and contracts.

The density of this fibrous replacement varies in other tissues such as skin, and often bridges gliding surfaces by adhesions with consequent loss of mobility. Following his earlier work on its cellular interactions Balazs (Balazs and Denlinger, 1989) developed the concept that an HA-enriched environment can inhibit certain aspects of cellular activity, and in particular, that of matrix cells responsible for fibrous scars. This was demonstrated, for example, in severed and crushed tendons; these healed normally after application of an HA gel, and with significantly better mobility than without it (St Onge *et al.*, 1980; see also Rydell and Balazs, 1971). HA also permits healing of injury with an impressive inhibition of scarring in the delicate tympanic membrane of the ear (C Laurent *et al.*, 1986).

In the fetus, the HA content of the wound remains high longer than in the adult and healing occurs with virtually no fibrous scarring (Longaker *et al.*, 1991). Although there may be other influences operating in the fetus, it appears from the foregoing that HA also moderates development of collagen in replacement of matrix in the adult. Its persistence at this phase might be determined by the kinds of growth factor and of inflammatory cell invoked by the prior injury; the cytokines they produce, and their capacity for catabolism of HA. The subject of fetal wound healing has recently been reviewed in depth by Gailit and Clark (1994).

In tissue damage and inflammation, the increase in HA content appears with the activity of the exogenous inflammatory cells before restoration by the resident cell population begins.

Vascular Disease

The part played by intercellular matrix has been a matter of intense interest in the causation of atherosclerosis, the commonest cause of arterial disease. Whether the initial lesion is due to local injury or accumulation of lipids, it is notable for an early appearance of smooth muscle cells in the intima (inner arterial layer) and an accompanying accumulation of GAG. Smooth muscle cells are not normally found in this layer but are restricted to the tunica media where in the normal mature state there is little HA. The cellular proliferation necessary for intimal invasion presumably leads to release of the capacity to synthesise HA, as in cell culture. The earlier investigation of this phenomenon identified platelet products as potential agents linked with local injury to the inner arterial lining surface (Ross, 1975). Further developments have been reviewed by Ross (1986).

Oncology

The stromal reaction to malignant neoplasia has always been considered critical to understanding their behaviour. At one extreme, carcinomas of several sources are notable for an extensive and dense fibrotic reaction surrounding small islands of tumour. At the other, the tumour cells seem to spread through the tissues with little obvious reaction of any kind. Some attract mononuclear cells, which may be favourable to the outcome as in chronic infections, and might also influence the stromal reaction; others do not.

Early interest in the HA content of tumours was rekindled by the analogy offered in the association of HA with cellular proliferation and migration in embryogenesis. This implication was reinforced by correlation of invasiveness with enrichment of HA in and around a carcinoma (designated V2) in rabbits (Toole *et al.*, 1979). Furthermore, it had been shown that an HA-rich pericellular investment provided protection against cytolysis by allogeneic lymphocytes (Fraser and Clarris, 1970), a phenomenon which could similarly interfere with any cellular immune reaction against tumour. There were thus several reasons to think that HA might promote the spread of malignant tumours.

Malignant tumours of mesodermal origin, such as osteogenic sarcomas, chondrosarcomas, mesotheliomas, and the rare synovioma continue to synthesise in varied degree the matrix components typical of their tissues of origin, often selectively and changed in pattern. Some tumours of epithelial origin secrete large amounts of HA in contrast with their cellular progenitors. Loss or change in differentiated secretory function is a common feature of neoplastic transformation which is no doubt an expression of chromosomal and karyotypic changes.

Newer developments, some of which have already been referred to, have focused on changes in tumour matrix, which might not be apparent in routine histological examination but nevertheless directly influence their invasiveness. These changes are effected through the cells of the matrix and the formation of new blood vessels, which are often defective in structure and function. They have been observed in all kinds of tumour, including lymphoid and erythromyeloid. In addition to the induction of HA synthesis by direct contact between tumour and matrix cells (see *Synthesis*), neoplastic cells can induce profound changes in the motility and morphology of matrix cells through the agency of humoral mediators (Grey *et al.*, 1989). The specific role of hyaluronan outlined here has been expounded in depth by Knudson *et al.*, (1989).

The amounts of collagen and other GAG in more complex forms of tumour matrix may also prove to be influenced at least in part by the multiple binding specificities of HA-binding cell receptors and matrix peptides, and the variable expression of these specificities in their different isoforms or post-translational modifications (see section on *Specific Binding and Cell Receptors*).

A strong association of HA with tumours should not necessarily be viewed as an indication of aggressiveness, since its effects can be influenced by the presence of HA-binding peptide in the matrix. Hyaluronectin (HN) was one of the first of this group of hyaladherins to be found outside cartilage; initially in nervous tissue, and later in numerous extraneural tissues (Delpech *et al.*, 1989). Delpech and his colleagues have now made extensive analysis of the part played by HA and HN in the behaviour of diverse kinds of tumour. An increased content of HA has been identified in numerous types of carcinoma (Bertrand *et al.*, 1992). HN is found in the stroma round nests of cells in basal cell carcinoma of skin (Delpech *et al.*, 1982) and HA has been demonstrated in the same position (Wells *et al.*, 1990b). This tumour is locally invasive but does not establish remote secondary growths by lymphatic or bloodstream spread.

Delpech *et al.* (1993) have made a most informative study of nervous system tumours which illuminates the role of HA and its binding agents. In the benign meningioma tumours, levels of both HA and HN were low. In addition to a more general intercellular distribution a perivascular staining of both was noted in all kinds of tumour, moreso in the walls of small vessels and capillaries. Staining was also seen in endothelial cytoplasm, raising the question of endocytosis.

The malignancy of primary brain tumours is expressed in local invasiveness. Delpech and colleagues have shown in this study that tumours with the highest ratio of HN to HA fall within the less aggressive category, and they relate this finding to reduced cell migration in non-neoplastic conditions where HA becomes associated with an HA-binding peptide. More complex correlations of tumour cell behaviour with other forms of matrix organisation can be expected to appear. It would be imprudent to assume an immediate cause-and-effect relationship in all such associations, however, since they might merely reflect the general tendency for differentiated functions to be retained in less aggressive forms of malignancy.

Endocrine Dysfunction

Diabetes

Apart from a greater risk of the common form of atherosclerosis in large arteries, the complications of diabetes include a distinctive kind of microvascular disease, and renal lesions which might prove to be foreshadowed by subtle early changes in glomerular function and blood flow.

Changes in large arteries. Atherosclerosis in diabetes seems to differ in its earlier onset and severity rather than in any essential feature. The GAG content of the aortic tunica media, the layer that contains smooth muscle, has been analysed in regions with and without sclerotic plaques from non-diabetic subjects and from insulin-dependent and non-insulin-dependent subjects (Heickendorff *et al.*, 1994). A relative and absolute increase was found in the HA content of media overlying normal tissue from insulin-dependent diabetics which was related to duration of diabetes but not to age, and unaccompanied by change in other GAG. There were no differences in any GAG from normal or abnormal regions of aorta in any of the other groups. It is not possible to draw firm conclusions at this point, but in the light of other observations on HA in arterial walls, its relationship to smooth muscle proliferation, and the following observations, further pursuit of this work will be informative.

Early glomerular changes. Early diabetes, both natural and experimental, is remarkable for abnormally raised renal blood flow and glomerular filtration. An increased glomerular production of prostaglandin occurs at the same time which is directly related to glucose concentration, and would explain increased blood flow. In pursuit of their work on the mediation of these changes, Larkins and colleagues considered that prostaglandin, a known stimulant of HA synthesis, might alter the GAG content of the glomerulus, and they have accordingly examined the synthesis of GAG by isolated glomerular cores from diabetic and non-diabetic rats (Mahadevan *et al.*, 1995). These cores consist largely of mesangial cells, which are of mesodermal lineage. They found (1) greater HA synthesis by the diabetic cores in normal and high glucose concentrations, but in the non-diabetic only at high glucose levels; (2) increased HA synthesis by both diabetic and non-diabetic cores in response to prostaglandin E_2 and its suppression by indomethacin; (3) reduced release of sulphated GAG from diabetic cores, from control cores in high glucose levels, and from both kinds of core on exposure to prostaglandin E_2 or exogenous HA. The reduction in sulphated GAG in diabetic glomeruli had previously been observed and considered responsible for loss of selective glomerular permeability by reducing fixed anionic charge.

These findings suggest that HA might be implicated in the initiation also of other diabetic complications, since the metabolic pathways leading to increased prostaglandin activity and HA synthesis are likely to be stimulated elsewhere by raised glucose levels; for example, in smooth muscle, in other connective tissue cells, and in glial and other supporting cells of the nervous system which synthesise HA (for further discussion of the last-mentioned see Delpech *et al.*, 1989; Bignami *et al.*, 1993).

Thyroid dysfunction

Thyroid dysfunction is associated with the most striking excesses of HA seen in diseased tissue.

Hypothyroidism. The more severe effects of reduced thyroid hormone activity, or hypothyroidism, result in a clinical condition which was termed myxoedema in 1878 after gelatinous, or mucinous, swelling was discovered in tissues and organs throughout the body. Its development is related to the degree and duration of impaired thyroid function. It occurs in primary hypothyroidism arising from disease in the thyroid gland itself, and in secondary hypothyroidism caused by a failure of pituitary gland secretion of thyrotrophin (thyroid-stimulating hormone; TSH). It is seen less commonly in the latter, since failure of other endocrine functions probably brings the patient to earlier attention.

On *prima facie* grounds, thyroid hormone deficiency therefore appears the likely proximate cause of HA accumulation in generalised myxoedema, since TSH activity is raised in the primary disease and reduced in the secondary.

Early histochemical studies used metachromasia and less specific stains that merely indicate GAG, but they provide data on precise localisation that can be interpreted as HA in view of later chemical analyses. In contrast with normal skin, serial skin biopsies from myxoedematous patients showed extracellular accumulation in the outer layer of the dermis which distorted the orderly array of collagen bundles, and around vessels and hair follicles (Gabrilove and Ludwig, 1957). These changes were eliminated with thyroid hormone treatment and recurred after its cessation. No difference was found between primary and secondary myxoedema; treatment with TSH restored a normal appearance in secondary cases but had no effect in the primary group.

The first direct analysis of GAG in the localised pretibial form identified HA and sulphated GAG (Watson and Pearce, 1947; see following section). Analyses of skin in primary myxoedema have shown a raised HA content, no change in chondroitin or dermatan sulphates, and reduction of heparan sulphate; treatment with *l*-thyroxine reduced the HA content to normal without significant change in other GAG (Lund *et al.*, 1986). Since treatment is so effective, modern data for other tissues are rare. Tissues from a single case have been compared with those of two control subjects with congestive cardiac failure, which also causes oedema (Parving *et al.*, 1982). Interstitial oedema was found in the tongue, heart and skeletal muscle, and skin of the myxoedematous patient but not in the controls. (The sampling site of skeletal muscle and skin was not stated, but it was presumably not in the dependent parts where oedema occurs in heart failure.) Analysis showed high levels of HA in all myxoedematous tissues except stomach. They were most marked in gut and skin. No consistent changes were found in the sulphated GAG. It is noteworthy that in myxoedema viscous fluid effusions occur in the pleura, pericardium and peritoneum, which normally contain HA. An early study noted the exceptionally high mucin content in the fluid of synovial joints as well as in skin (Ropes *et al.*, 1947).

These findings have been supported experimentally by a study of the skin in rats (Schiller *et al.*, 1962). In the equivalent of primary hypothyroidism, the concentration of HA was elevated and those of sulphated GAG reduced, moreso in the dermatan sulphate fraction. (It is worthy of note that dermatan sulphate has since been closely associated with collagen fibres, which may have some bearing on their appearance in myxoedema.) The half-life of HA in the skin was almost twice that in the controls, which provided an explanation for its accumulation. The changes were reversed by thyroxine treatment, and alternatively induced by pituitary gland removal. TSH did not induce significant changes in normal rats.

The direct effect of thyroid hormone on HA synthesis has been confirmed in cultured skin fibroblasts (Smith *et al.*, 1982). Their synthesis of GAG in thyroid hormone-depleted growth medium (80% as HA) was reduced some 28-60% by tri-iodothyronine, the peripherally active form of thyroid hormone. Seventy-three per cent. of the maximum achieved response occurred at physiological hormone concentrations. No evidence of HA catabolism was found. It can be concluded that thyroid hormone exerts a direct negative control over HA synthesis. The induction of hyaluronidase by thyroid hormone in tadpoles (Polansky and Toole, 1976) might reflect species difference or the reaction of a different kind of cell.

One important factor that has been overlooked but will influence all aspects of HA turnover in hypothyroidism is tissue temperature. This is normally 5 to 6° lower than body core temperature in skin, peripheral joints and resting limb muscles in a comfortable ambient temperature. Core temperature, and especially skin temperature and circulation, are lowered by thyroid deficiency. This will reduce HA synthesis (Castor and Yaron, 1976), local cellular uptake and degradation and flux of tissue fluid, and at the same time increase the viscosity of HA (Bothner-Wik *et al.*, 1991).

Hyperthyroidism. In spontaneous hyperthyroidism HA is also a major constituent of localised myxoedema (usually pretibial myxoedema; that is, in front of the shins) and probably in exophthalmos, a related condition characterised by retro-orbital swelling which causes protrusion of the eyes (exophthalmos) and other consequences. These conditions seem almost entirely restricted to Graves' disease, a particular form of hyperthyroidism with accompanying features of autoimmunity, but are not related to the level of excess thyroid hormone. Pre-tibial myxoedema and exophthalmos frequently occur together, often develop long after the onset of hyperthyroidism and persist when thyroid function returns to normal or below, and sometimes develop without clear evidence of prior hyperthyroidism. Taken with their restricted distribution and the evidence just given, these facts indicate a causation quite different from that of generalised myxoedema.

Histological studies of both conditions show accumulation of GAG and interstitial swelling. Mononuclear cell infiltration, mainly of lymphocytes, is often prominent. In the orbital cavity these changes occur in the ocular muscles with serious secondary functional effects.

Analysis of the skin has shown 6 to 16 times the normal GAG content, which is almost entirely HA (Sisson, 1968). The orbital contents are not so accessible but the close relationship with the cutaneous syndrome, taken with indirect evidence from cell culture, make it reasonable to assume that HA is the dominant GAG, providing yet another instance of the fundamental effect of HA on the distribution of tissue water.

These conditions present two questions of further interest in matrix biology; firstly, the reasons for their peculiar localisation, and secondly, their causation. The first is still not completely answered, though it has led to further recognition of phenotypic distinctions in fibroblasts from different parts of the body (Cheung *et al.*, 1978). The second also remains unsolved. Despite the evidence cited, experimental studies in guinea pigs (Sisson and Miles, 1967) and dogs (Bollet *et al.*, 1961) have shown that TSH causes accumulation of HA or unspecified GAG in orbital tissues. Protease digests of TSH have yielded residues with reduced thyroid-stimulating activity which induce exophthalmos (Kohn and Winand, 1971, 1975). It is difficult to envisage how such a variation in pituitary secretion could arise spontaneously, and these effects may reflect species variation unrelated to the distinctive character of Graves' disease.

Other work suggests instead that these peculiar conditions are a direct outcome of the autoimmune manifestations of Graves' disease. Such an explanation eliminates the obvious difficulty in reconciling the tissue changes quantitatively with the disturbance of thyroid function. The autoimmune components of the disease include a group of antibodies that react with and stimulate the TSH receptor of thyroid cells. An auto-immune pathogenesis also provides a ready explanation for the lymphocytic infiltration in the affected extrathyroidal tissues, even though the specific immunological commitment of these lymphocytes is not yet known. Smith and colleagues (1991) have shown that interferon-gamma increases the output of GAG some 50% in cultured human fibroblasts of retro-orbital origin but not in those of dermal origin. It has now been reported that leukoregulin selectively stimulates HA synthesis in retro-orbital fibroblasts. This cytokine caused a nearly four-fold increase in HA synthesis by dermal fibroblasts, and an eight-fold increase in the HA output of retro-orbital fibroblasts without a change in synthesis of sulphated GAG (Smith *et al.*, 1995). Though these studies were necessarily done in culture, the activated T lymphocytes identified in the retro-orbital infiltrates in thyroid eye disease (De Carli *et al.*, 1993) are an obvious source of these cytokines and possibly others. It is also quite possible that some of the various autoantibodies found in this disease might activate receptors with a direct influence on HA synthesis.

Serum hyaluronan in thyroid dysfunction. Faber *et al.* (1990) have measured the changes in serum HA following correction of hyper- and hypothyroidism. The mean levels were not grossly abnormal before treatment but rose and fell respectively with restoration of normal function. The wide age range does not permit close comparison of the results with normal subjects though some levels in hypothyroidism were slightly but distinctly above the normal range. These observations are consistent with the disturbances of HA metabolism outlined and support an increased total HA turnover in hypothyroidism, since normal liver function moderates any rise from increased tissue input. In the absence of data for hepatic circulation in hypothyroidism, this conclusion must finally be confirmed by direct measurement of turnover.

CONCLUDING COMMENTS

Despite its widespread distribution in nature, HA has preserved its simple primary structure without acquiring the diversity offered by the varied amino acid sequence of conjugated peptides, and indeed without the variations in saccharide content displayed by the sulphated glycosaminoglycans.

This remarkable degree of conservation is unmatched in any other component of the interstitial matrix, or indeed, in the many other kinds of macromolecular material in the secretory products of cells. The structure of the protein cores with which other glycosaminoglycans are conjugated is so varied as to compromise the distinction between proteoglycans and glycoproteins, and it is tempting to conclude that their diversification is linked with that of protein glycosylation.

Potential explanations for the evolutionary survival of hyaluronan without change continue to be uncovered in both its intrinsic physiochemical properties and its biological functions.

Though differing in degree from those of other glycosaminoglycans, its osmotic properties, effects on permeability and diffusion, ionic interactions, and its capacity for steric exclusion and free radical scavenging are not distinctive. The most striking feature of mammalian hyaluronan is that while some of it is firmly bound to cells and other matrix components, most of it remains relatively free in the interstitium. The exceptional length and the stiffened extended orientation of its polysaccharide chain allow it to form extensive molecular meshworks held together by attractive forces of low energy. These meshworks are inherently labile, and thus provide cohesion while at the same time permitting the movements of cells with minimum expenditure of energy, and facilitating their necessary disengagement from the matrix during mitosis. Moreover, the effect of HA on rheological behaviour develops at a much lower concentration than with the largest proteoglycans, which is again a reflection of the much greater polysaccharide chain length achieved by hyaluronan. It is therefore understandable that hyaluronan not only dominates the intercellular matrix in the earliest phase of embryonic development, but also reappears in abundance whenever tissues are reconstituted or massive cellular infiltration occurs from without, and free cellular movement and proliferation are once again required. These characteristics seem to offer the most persuasive reason for its evolutionary persistence.

The same intrinsic attributes also account for its consistent dominance in fluid forms of extracellular matrix such as synovial and serous fluids, and for its prominence in soft tissues where in conjunction with sparser forms of fibrous structure it provides the required degree of plasticity and resilience. (That is, viscosity and elasticity.)

The high-energy kind of association revealed more recently in its specific interactions with other structural components of the matrix and with so many kinds of differentiated cell is almost certainly a later evolutionary development. Although this kind of binding is necessarily reflected by certain common features in the interacting amino acid sequence, the variety of receptors and matrix binding peptides already suggests that they are not confined to the immunoglobulin supergene family of proteins. The related functions of hyaluronan stem from genetic divergence in proteins during cellular differentiation. They are in effect a cellular adaptation that uses the ubiquitous distribution of hyaluronan as a signal for a specific response. Further studies in more primitive organisms should reveal the level in phylogeny at which this form of interaction first appears. This would afford an interesting comparison with the known evolution of immunoglobulins.

The more complex roles of hyaluronan in mammalian physiology largely rest upon specific binding. The expression of specific binding in cells and extracellular matrix has understandably attracted most attention and will continue to do so, not only for its scientific significance but also for possible applications in therapeutic intervention.

References

Adams D.H., Wang L., Hubscher S.G., Neuberger J.M. (1989) Hepatic endothelial cells. Targets in rejection? *Transplantation* **47**: 479–482.

Ahlgren T., Jarlstrand C.. (1984) Hyaluronic acid enhances phagocytosis of human monocytes in vitro. *J Clin Immunol* **4**: 246–249.

Alston-Smith J., Pertoft H., Fraser J.R.E., Laurent T.C. (1992) Effects of endotoxin in hepatic endocytosis of hyaluronan. **In** *Hepatic Endocytosis of Lipids and Proteins*, pp. 119–126. Eds. E. Windler, H. Greten. Publ. W. Zuckschwerdt Verlag.

Allen S.J., Sedin E G, Jonzon A., Wells A.F, Laurent T.C. (1991) Lung hyaluronan during development: a quantitative and morphological study. *Am J Physiol* **260** (Heart Circ. Physiol.): H1449–H1454.

Allen S.J., Fraser J.R.E., Laurent U.B.G., Reed R.K., Laurent T.C. (1992) Turnover of hyaluronan in the rabbit pleural space. *J Appl Physiol* **73**: 1457–1460.

Anderson B., Hoffman P., Meyer K. (1965) The *O*-serine linkage in peptides of chondroitin 4- or 6-sulphate. *J Biol Chem* **240**: 156–167.

Antonopoulos C.A., Engfeldt B., Gardell S., Hjertquist S-O, Solheim K. (1965) Isolation and identification of the glycosaminoglycans from fracture callus. *Biochim Biophys Acta* **101**: 150–156.

Aronson N.N.Jr., Davidson E.A. (1967) Lysosomal hyaluronidase from liver. II. Properties. *J Biol Chem* **242**: 441–444.

Aruffo A., Stamenkovic I., Melnick M., Underhill C.B, Seed B. (1990) CD44 is the principal cell surface receptor for hyaluronate. *Cell* **61**: 1303–1313.

Balazs E.A. (1965) Amino sugar-containing macromolecules in the tissues of the eye and the ear. **In** *The Amino Sugars*, Vol IIA, pp. 401–460. Eds. E A Balazs, J.W. Jeanloz. Academic Press, New York and London.

Balazs E.A., Denlinger J.L. (1984) The Vitreous. **In** *The Eye*. Ed. H. Davson. Ch 4, Vol 1A, pp. 533–589. Academic Press, London.

Balazs E.A., Denlinger J.L. (1989) Clinical uses of hyaluronan. **In** *The Biology of Hyaluronan*, eds. D. Evered, J. Whelan. Ciba Foundation Symposium 143, pp. 265–280. Publ. J Wiley and Sons, Chichester.

Balazs E.A., Gibbs D.A. (1970) The rheological properties and biological function of hyaluronic acid. **In** *Chemistry and Molecular Biology of the Intercellular Matrix*, pp. 1241–1254. Ed. E.A. Balazs. Academic Press, New York.

Balazs E.A., Laurent T.C. (1951) Viscosity function of hyaluronic acid as a polyelectrolyte. *J Polymer Sci* **6**: 665–668.

Balazs E.A., Laurent T.C., Howe A.F., Varga L. (1959) Irradiation of mucopolysaccharides with ultraviolet light and electrons. *Radiation Res* **11**: 149–164.

Balazs E.A., Davies J.V., Phillips G.O., Young M.D. (1967a) Transient intermediates in the radiolysis of hyaluronic acid. *Radiation Res* **31**: 243–2.

Balazs E.A., Watson D., Duff I.F., Roseman S. (1967b) Hyaluronic acid in synovial fluid. I. Molecular parameters of hyaluronic acid in normal and arthritic human fluids. *Arthr Rheum* **10**: 357–376.

Balazs E.A., Laurent T.C., Jeanloz R.W. (1986) Nomenclature of hyaluronic acid. *Biochem J* **235**: 903.

Banerjee S.D., Toole B.P. (1992) Hyaluronan-binding protein in endothelial cell morphogenesis. *J Cell Biol* **119**: 643–652.

Barland P., Novikoff A.B., Hamerman D. (1962) Electron microscopy of the human synovial membrane. *J Cell Biol* **14**: 207–220.

Baxter E., Fraser J.R.E., Harris G.S. (1971) Interaction between hyaluronic acid and serum dispersed in collagen gels. *Ann Rheum Dis* **30**: 419–422.

Belsky E., Toole B.P. (1983) Hyaluronate and hyaluronidase in the developing chick embryo kidney. *Cell Differentiation* **12**: 61–66.

Berg S., Jansson I., Hesselvik F.J., Laurent T.C., Lennquist S., Walther S.(1992) Hyaluronan: relationship to hemodynamics and survival in porcine injury and sepsis. *Crit Care Med* **20**: 1315–1321.

Bertrand P., Girard N., Delpech B., Duval C., d'Anjou J., Daucé J.P. (1992) Hyaluronan (hyaluronic acid) and hyaluronectin in the extracellular matrix of human breast carcinomas. Comparison between invasive and non- invasive areas. *Int J Cancer* **52**: 1–6.

Bignami A., Hosley M., Dahl D. (1993) Hyaluronic acid and hyaluronic acid-binding proteins in brain extracellular matrix. *Anat Embryol* **188**: 419–433.
Bjermer A., Engström-Laurent A., Lundgren R., Rosenhall L., Hällgren R. (1987) Hyaluronic acid and procollagen III peptide in bronchoalveolar lavage fluid as indicators of lung disease activity in farmer's lung. *Brit Med J* **295**: 801–806.
Bjermer A., Lundgren R., Hällgren R. (1989) Hyaluronic acid and procollagen III peptide concentrations in bronchoalveolar lavage fluid in idiopathic pulmonary fibrosis. *Thorax* **44**: 126–131.
Blom A., Pertoft H., Fries E. (1995) Inter-α-inhibitor is required for the formation of the hyaluronan-containing coat of fibroblasts and mesothelial cells. *Biochem J.* (In press.)
Bollet A.J., Beierwaltes W.H., Knopf R.F., Matovinovic J., Clure H.R. (1961) Extraocular muscle, skeletal muscle, and thyroid gland mucopolysaccharide response to thyroid-stimulating hormone. *J Lab Clin Med* **58**: 884–891.
Bothner H., Wik O. (1991) Rheology of intraocular solutions. In *Rheological Studies of Sodium Hyaluronate in Pharmaceutical Preparations*, H. Bothner, Acta Univ Upsal. Abstr. Uppsala Diss. Fac. Pharm. **79**: V3–22.
Bothner-Wik H., Sundelöf L-O, Wik O. (1991) The effect of molecular weight, concentration and temperature on the viscoelastic properties of sodium hyaluronate. In *Rheological Studies of Sodium Hyaluronate in Pharmaceutical Preparations*, H. Bothner, Acta Univ Upsal. Abstr. Uppsala Diss. Fac. Pharm. **79**: III3–26.
Brecht M., Mayer M., Schlosser E., Prehm P. (1986) Increased hyaluronan synthesis is required for fibroblast detachment and mitosis. *Biochem J* **239**: 445–450.
Brinkley L.L., Morris-Wyman J. (1987) Computer-assisted analysis of hyaluronate distribution during morphogenesis of the mouse secondary palate. *Development* **100**: 629–635.
Broaddus V.C., Wiener-Kronish J.P., Laurent T.C., Staub N.C. (1988) Clearance of hyaluronan into the pleural space during high pressure pulmonary edema in sheep. *FASEB J* **2**: A1703.
Campbell P., Thompson J.N., Fraser J.R.E., Laurent T.C., Pertoft H., Rodén L. (1990) N-acetylglucosamine 6-phosphate deacetylase in hepatocytes, Kupffer cells and sinusoidal endothelial cells from rat liver. *Hepatology* **11**: 199–204.
Campbell R.D., Love S.H., Whiteheart S.W., Young B., Myrvik Q.N. (1982) Increased hyaluronic acid is associated with dermal delayed-type sensitivity. *Inflammation* **6**: 235–244.
Carter W.G., Wayner E.A. (1988) Characterization of the class III collagen receptor, a phosphorylated transmembrane glycoprotein expressed in nucleated human cells. *J Biol Chem* **263**: 4193–4201.
Cashman D.C., Laryea J.U., Weissmann B. (1969) The hyaluronidase of rat skin. *Arch Biochem Biophys* **135**: 387–395.
Cássaro C.M.F., Dietrich C.P. (1977) Distribution of sulfated mucopolysaccharides in invertebrates. *J Biol Chem* **252**: 2254–2261.
Castor C.W., Dorstewitz E.L. (1966) Abnormalities of connective tissue cells cultured from patients with rheumatoid arthritis. I. Relative unresponsiveness of rheumatoid synovial cells to hydrocortisone. *J Lab Clin Med* **68**: 300–313.
Castor C.W., Yaron M. (1976) Connective tissue activation. VIII. The effects of temperature studied in vitro. *Arch Phys Med Rehab* **57**: 5–9.
Castor C.W., Dorstewitz E.L., Rowe K., Ritchie J.C. (1971) Abnormalities of connective tissue cells cultured from patients with rheumatoid arthritis. II. Defective regulation of hyaluronate and collagen formation. *J Lab Clin Med* **77**: 65–75.
Castor C.W., Dorstewitz E.L., Ritchie J.C., Smith S.F. (1972) Connective tissue activation. III. Observations on the mechanism of action of connective tissue activating peptide. *J Lab Clin Med* **79**: 285–301.
Cheung H.S., Nicoloff J.T., Kamiel M.B., Spolter L., Nimni M.E. (1978) Stimulation of fibroblast biosynthetic activity by serum of patients with pretibial myxedema. *J Invest Dermatol* **71**: 12–17.
Chevrier A., Girard N., Delpech B., Gilbert D. (1982) Inhibition of active E rosette forming T lymphocytes by hyaluronic acid. Evidence of a receptor for hyaluronic acid on a lymphocyte subpopulation. *Biomedicine* **36**: 1100–103.
Chi-Rosso P., Toole B.P. (1987) Hyaluronate-binding protein of Simian virus 40-transformed 3T3 cells: membrane distribution and reconstitution into lipid vesicles. *J Cell Biol* **33**: 173–184.
Clarris B.J, Fraser J.R.E. (1968) On the pericellular zone of some mammalian cells in vitro. *Exp Cell Res* **49**: 181–193.
Clarris B.J., Fraser J.R.E., Rodda S.J. (1970) Effect of cell-bound hyaluronic acid on infectivity of Newcastle disease virus for human synovial cells in vitro. *Ann Rheum Dis* **33**: 240-242.
Cleland R.L. (1968) Ionic polysaccharides. II. Comparison of polyelectrolyte behaviour of hyaluronate with that of carboxymethyl cellulose. *Biopolymers* **6**: 1519–1529.
Cleland R.L. (1977) The persistence length of hyaluronic acid: an estimate from small-angle X-ray scattering and intrinsic viscosity. *Arch Biochem Biophys* **180**: 57–68.
Cleland R.L., Wang J.L. (1970) Ionic polysaccharides. III. Dilute solution properties of hyaluronic acid fractions. *Biopolymers* **9**: 799–810.

Cleland R.L., Cleland M.C., Lipsky J.J, Lyn V.E. (1968) Ionic polysaccharides. I. Adsorption and fractionation of polyelectrolytes on (diethylamino) ethyl cellulose. *J Am Chem Soc* **90**: 3141–3146.

Cohn R.H., Cassiman J-J., Bernfield M.R. (1976) Relationship of transformation, cell density, and growth control to the cellular distribution of newly synthesized glycosaminoglycan. *J Cell Biol* **71**: 280–294.

Cohn Z.A., Parks E., (1967) The regulation of pinocytosis in mouse macrophages. II. Factors inducing vesicle formation. *J Exp Med* **125**: 213–232.

Comper W.D. (1991) Physicochemical aspects of cartilage extracellular matrix. **In** *Cartilage: Molecular Aspects*, pp. 59–96. Eds. B Hall, S Newman. Publ. CRC Press Inc, Boca Raton, FA, USA.

Comper W.D., Laurent T.C. (1978) Physiological function of connective tissue polysaccharides. *Physiol Rev* **58**: 255–315.

Culty M., Miyake K., Kincade P.W., Sikorski E., Butcher E.C., Underhill C.B. (1990) The hyaluronate receptor is a member of the CD44 (H-CAM) family of cell surface glycoproteins. *J Cell Biol* **111**: 2765–2774.

Culty M., Nguyen A., Underhill C.B. (1992) The hyaluronan receptor (CD44) participates in the uptake and degradation of hyaluronan. *J Cell Biol* **116**: 1055–1062.

Culty M., Shizari M., Thompson E.W., Underhill C.B. (1994) Binding and degradation of hyaluronan by human breast cancer cell lines expressing different forms of CD44: correlation with invasive potential. *J Cell Physiol* **160**: 275–286.

Curtain C.C. (1955) The nature of the protein in the hyaluronic complex of bovine synovial fluid. *Biochem J* **61**: 688–697.

Dahl L.B., Hopwood J.J., Laurent U.B.G., Lilja K., Tengblad A. (1983) The concentration of hyaluronate in amniotic fluid. *Biochem Med* **30**: 280–283.

Dahl L.B., Dahl I.M.S., Børresen A-L. (1986) The molecular weight of sodium hyaluronate in amniotic fluid. *Biochem Med Metab Biol* **35**: 219–226.

Dahl L.B., Dahl I.M.S., Engström-Laurent A., Granath K. (1989a) Concentration and molecular weight of sodium hyaluronate in synovial fluid from patients with rheumatoid arthritis and other arthropathies. *Ann Rheum Dis* **44**: 817–822.

Dahl L.B., Kimpton W.G., Cahill R.N.P., Brown T.J., Fraser J.R.E. (1989b) The origin and fate of hyaluronan in amniotic fluid. *J Devel Physiol* **12**: 209–218.

Darzynkiewicz Z., Balazs E.A. (1971) Effect of connective tissue intercellular matrix on lymphocyte stimulation. I. Suppression of lymphocyte stimulation by hyaluronic acid. *Exp Cell Res* **66**: 113–123.

Day T.D. (1950) Connective tissue permeability and the mode of action of hyaluronidase. *Nature* **166**: 785–786.

Dayer J-M., de Rochemonteix B., Burrus B., Demczuk S., Dinarello C.A. (1986) Human recombinant interleukin-1 stimulates collagenase and prostaglandin E_2. *J Clin Invest* **77**: 645–648.

Delpech A., Delpech B. (1984) Expression of hyaluronic acid-binding glycoprotein, hyaluronectin, in the developing rat embryo. *Devel Biol* **101**: 391–400.

Delpech A., Delpech B., Girard N., Boullié M.C., Lauret P. (1982) Hyaluronectin in human skin and in basal cell carcinoma. *Br J Dermatol* **106**: 561–568.

Delpech B., Delpech A., Brückner G., Girard N., Maingonnat C. (1989) Hyaluronan and hyaluronectin in the nervous system. **In** *The Biology of Hyaluronan*, eds. D. Evered, J. Whelan. Ciba Foundation Symposium 143, pp. 208–232. Publ. J Wiley and Sons, Chichester.

Delpech B., Maingonnat C., Girard N., Chauzy C., Maunoury R., Olivier O., Tayot J., Creissard P. (1993) Hyaluronan and hyaluronectin in the extracellular matrix of human brain tumor stroma. *Europ J Cancer* **29A**: 1012–1017.

Denborough M.A., Ogston A.G. (1965) Failure of phenol to remove residual protein from hyaluronic acid. *Nature* **207**: 4.

De Salegui M., Pigman W. (1967) The existence of an acid-active hyaluronidase in serum. *Arch Biochem Biophys* **120**: 60–67.

Diamond M.S., Staunton D.E., deFougerolles A.R., Stacker S.A., Garcia-Aguilar J., Hibbs M.L., Springer T.A. (1990) ICAM-1 (CD54): a counter-receptor for Mac-1 (CD11b/CD18). *J Cell Biol* **111**: 3129–3139.

Duran-Reynals F. (1942) Tissue permeability and the spreading factors in infection. A contribution to the host: parasite problem. *Bacteriol Rev* **6**: 197–252.

Edelstam G.A., Lundkvist O.E., Wells A.F., Laurent T.C. (1991) Localization of hyaluronan in regions of the human female reproductive tract. *J Histochem Cytochem* **39**: 1131–1135.

Edelstam G.A.B., Laurent U.B.G., Lundqvist Ö.E., Fraser J.R.E., Laurent T.C. (1992) Concentration and turnover of intraperitoneal hyaluronan during inflammation. *Inflammation* **16**: 459–469.

Engström-Laurent A., (1989) Changes in hyaluronan concentration in tissues and body fluids in disease states. **In** *The Biology of Hyaluronan*, eds. D. Evered, J. Whelan. Ciba Foundation Symposium 143, pp. 233–247. Publ. J Wiley and Sons, Chichester.

Engström-Laurent A., Lööf L, Nyberg A., Schröder T. (1985a) Increased serum levels of hyaluronate in liver disease. *Hepatol* **5**: 638–642.

Engström-Laurent A., Laurent U.B.G., Lilja K., Laurent T.C. (1985b) Concentration of sodium hyaluronate in serum. *Scand J Clin Lab Invest* **45**: 497–504.

Entwhistle J., Zhang S., Yang B., Wong C., Hall C.L., Curpen G., Mowat M., Greenberg A.H., Turley E.A. (1995) The molecular characterization of the gene encoding the hyaluronan receptor RHAMM. *Gene*. (In press.)

Eriksson S., Fraser J.R.E., Laurent T.C., Pertoft H., Smedsrød B. (1983) Endothelial cells are a site of uptake and degradation of hyaluronic acid in the liver. *Exp Cell Res* **144**: 223–228.

Faber J., Hørslev-Petersen K., Perrild H., Lorenzen I. (1990) Different effects of thyroid disease on serum levels of procollagen III N-peptide and hyaluronic acid. *J Clin Endocrinol Metab* **71**: 1016–1021.

Farber S.J., van Praag D. (1970) Composition of glycosaminoglycans (mucopolysaccharides) in rabbit renal papillae. *Biochim Biophys Acta* **205**: 219–226.

Fessler J.H. (1960a) A structural function of mucopolysaccharide in connective tissue. *Biochem J* **76**: 126–132.

Fessler J.H. (1960b) Mode of action of testicular hyaluronidase. *Biochem J* **76**: 132–135.

Fessler J.H., Fessler L.I. (1966) Electron microscopic visualization of the polysaccharide hyaluronic acid. *Proc Natl Acad Sci USA* **56**: 141–147.

Forrester J.V., Balazs E.A. (1980) Inhibition of phagocytosis by high molecular weight hyaluronate. *Immunology* **40**: 435–446.

Forsberg N., Gustafson S. (1991) Characterization and purification of the hyaluronan-receptor on liver endothelial cells. *Biochim Biophys Acta* **1078**: 12–18.

Forsberg N., von Malmborg A., Madsen K., Rolfsen W., Gustafson S., (1994) Receptors for hyaluronan on corneal endothelial cells. *Exp Eye Res* **59**: 689–696.

Fraser J.R.E., Baxter E. (1984) Leucine metabolism as a source of acetate in the synthesis of hyaluronic acid. *Conn Tiss Res* **12**: 287–296.

Fraser J.R.E., Clarris B.J. (1970) On the reactions of human synovial cells exposed to homologous leucocytes *in vitro*. *Clin Exp Immunol* **6**: 211–225.

Fraser J.R.E., Laurent T.C. (1989) Turnover and metabolism of hyaluronan. **In** *The Biology of Hyaluronan*, eds. D. Evered, J. Whelan. Ciba Foundation Symposium 143, pp. 41–59. Publ. J Wiley and Sons, Chichester.

Fraser J.R.E., Harris G.S., Clarris B.J. (1969) Influence of serum on secretion of hyaluronic acid by synovial cells. Its possible relevance in arthritis. *Ann Rheum Dis* **28**: 419–423.

Fraser J.R.E., Clarris B.J., Kont L.A. (1970) The morphology and motility of human synovial and their pericellular gels. *Aust J Biol Sci* **23**: 1297–1303.

Fraser J.R.E., Foo W.K., Maritz J.S. (1972) Viscous interactions of hyaluronic acid with some proteins and neutral saccharides. *Ann Rheum Dis* **31**: 513–520.

Fraser J.R.E., Murdoch W.S., Curtain C.C., Watt B.J. (1977) Proteins associated with hyaluronic acid during ultrafiltration of synovial fluid. *Conn Tiss Res* **5**: 61–65.

Fraser J.R.E., Laurent T.C., Pertoft H., Baxter E. (1981) Plasma clearance, tissue distribution and metabolism of hyaluronic acid. *Biochem J* **200**: 415–424.

Fraser J.R.E., Appelgren L-E., Laurent T.C. (1983) Tissue uptake of circulating hyaluronic acid. A whole body autoradiographic study. *Cell Tiss Res* **233**: 285–293.

Fraser J.R.E, Robinson A.D., Clarris B.J. (1985a) Synovial cell interactions in vitro. **In** *Rheumatology-85*, pp. 65–71. Eds. P.M. Brooks and J.R. York. Elsevier Science Publ., Amsterdam.

Fraser J.R.E., Alcorn D., Laurent T.C., Robinson A.D., Ryan G.B. (1985b) Uptake of circulating hyaluronic acid by the rat liver. Cellular localization *in situ*. *Cell Tissue Res* **242**: 505–510.

Fraser J.R.E., Kimpton W.G., Laurent T.C., Cahill R.N.P., Vakakis N. (1988) Uptake and degradation of hyaluronan in lymphatic tissue. *Biochem J* **356**: 153–158.

Fraser J.R.E., Dahl L.B., Kimpton W.G., Cahill R.N.P., Brown T.J., Vakakis N. (1989) Elimination and subsequent metabolism of circulating hyaluronic acid in the fetus. *J Devel Physiol* **11**: 235–242.

Frébourg T., Delpech B., Bercoff E., Senant J., Bertrand P., Deugnier Y., Boureille J. (1986) Serum hyaluronate in liver diseases: study by enzymoimmunological assay. *Hepatol* **6**: 392–395.

Gabrilove J.L., Ludwig A.W. (1957) The histogenesis of myxedema. *J Clin Endocrinol Metab* **17**: 925–932.

Gailit J., Clark R.A.F. (1994) Wound repair in the context of extracellular matrix. *Curr Opin Cell Biol* **6**: 717–725.

Ghosh P., Hutadilok N., Adam N., Lentini A. (1994) Interactions of hyaluronan (hyaluronic acid) with phospholipids as determined by gel permeation chromatography, multi-angle laser-light scattering photometry and ^{1}H-NMR spectroscopy. *Int J Biol Macromol* **16**: 237–244.

Gibbs D.A., Merrill E.W., Smith K.A., Balazs E.A. (1968) The rheology of hyaluronic acid. *Biopolymers* **6**: 777–791.

Gibson P.R., Fraser J.R.E., Brown T.J., Finch C.F., Jones P.A., Colman J.C., Dudley F.J. (1992) Haemodynamic and liver function predictors of serum hyaluronate in alcoholic liver disease. *Hepatol* **15**: 1054–1059.

Gibson P.R., Fraser J.R.E., Colman J.C., Jones P.A., Jennings G., Dudley F.J. (1994) Serum hyaluronan: a simple index of acute drug-induced changes of liver blood flow in alcoholic liver disease. *Gastroenterol* **105**: 470–474.

Goldberg R.L., Toole B.P. (1984) Pericellular coat of chick embryo chondrocytes: structural role of hyaluronate. *J Cell Biol* **99**: 2114–2122.

Goldberg R.L., Seidman J.D., Chi-Rosso G., Toole B.P. (1984) Endogenous hyaluronate-cell surface interactions in 3T3 and simian virus-transformed 3T3 cells. *J Biol Chem* **259**: 9440–9446.

Goldstein L.A., Zhou D.F.H., Picker L.J., Minty C.N., Bargatze R.F., Ding J.F., Butcher E.C. (1989) A human lymphocyte homing receptor, the Hermes antigen, is related to cartilage proteoglycan core and link proteins. *Cell* **56**: 1063–1072.

Gorham S.D., Olavesen A.H., Dodgson K.S. (1975) Effect of ionic strength and pH on the properties of purified bovine testicular hyaluronidase. *Conn Tissue Res* **3**: 17–25.

Gregory J.D., Laurent T.C., Rodén L. (1964) Enzymatic degradation of chondromucoprotein. *J Biol Chem* **239**: 3312–3320.

Grey A-M., Schor A.M., Rushton G., Ellis I., Schor S.L. (1989) Purification of the migration stimulating factor produced by fetal and breast cancer patient fibroblasts. *Proc Natl Acad Sci USA* **86**: 2438–2442.

Gustafson S., Björkman T., Forsberg N., Lind T., Wikström T., Lidholt K. (1995) Accessible hyaluronan-receptors identical to ICAM-1 in mouse mast-cell tumours. *Glycoconj J* (In press.)

Hall C.L., Turley E.A. (1995) Hyaluronan: RHAMM-mediated locomotion and signaling in tumorigenesis. *J NeuroOncol* In press.

Hall C.L., Wang C., Lange L.A., Turley E.A. (1994) Hyaluronan and the hyaluronan receptor RHAMM promote focal adhesion turnover and transient tyrosine kinase. *J Cell Biol* **126**: 575–588.

Hällgren R., Eklund A., Engström-Laurent A., Schmekel B. (1985) Hyaluronate in bronchoalveolar lavage fluid, a new marker in sarcoidosis reflecting pulmonary disease. *Brit Med J* **290**: 1778–1781.

Hällgren R., Samuelsson T., Laurent T.C., Modig J. (1989) Accumulation of hyaluronic acid in the lung in adult respiratory distress syndrome. *Am Rev Respir Dis* **139**: 682–687.

Hällgren R., Gerdin B., Tengblad A., Tufveson G. (1990a) Accumulation of hyaluronan (hyaluronic acid) in myocardial interstitial tissue parallels development of transplantation dedema in heart allografts in rats. *J Clin Invest* **85**: 668–673.

Hällgren R., Gerdin B., Tufveson G. (1990b) Hyaluronic acid accumulation and redistribution in rejecting rat kidney graft: relationship to the transplantation edema. *J Exp Med* **171**: 2063–2076.

Halliwell B. (1978) Superoxide-dependent formation of hydroxyl radicals in the presence of iron salts. *FEBS Letts* **96**: 238–242.

Hamerman D., Wood D.D. (1984) Interleukin-1 enhances synovial cell hyaluronate synthesis. *Proc Soc Exp Biol Med* **177**: 205–210.

Hardingham T.E., Muir H. (1972) The specific interaction of hyaluronic acid with cartilage proteoglycans. *Biochim Biophys Acta* **279**: 401–405.

Hardwick C., Hoare K., Owens R., Hohn H.P., Hook M., Moore D., Cripps V., Austen L., Nance D.M., Turley E.A. (1992) Molecular cloning of a novel hyaluronan receptor that mediates tumor cell motility. *J Cell Biol* **117**: 1343–1350.

Heatley F., Scott J.E. (1988) A water molecule participates in the secondary structure of hyaluronan. *Biochem J* **254**: 489–493.

Heickendorff L., Ledet T., Rasmussen L.M. (1994) Glycosaminoglycans in the human aorta in diabetes mellitus: a study of tunica media from areas with and without atherosclerotic plaque. *Diabetologia* **37**: 286–292.

Heldin P., Pertoft H. (1993) Synthesis and assembly of the hyaluron-containing coats around normal human mesothelial cells. *Exp Cell Res* **208**: 422–429.

Heldin P., Laurent T.C., Heldin C.H. (1989) Effect of growth factors on hyaluronan synthesis in cultured human fibroblasts. *Biochem J* **258**: 919–922.

Hellsing K. (1969) Immune reactions in polysaccharide media. The effect of hyaluronate, chondroitin sulphate and chondroitin sulphate-protein complex on the precipitin reaction. *Biochem J* **112**: 474–487.

Hellström S., Tengblad A., Hedlund U., Johansson C., Axelsson E. (1990) An improved technique for hyaluronan histochemistry using microwave fixation. *Histochem J* **22**: 677–682.

Hillerdal G., Lindqvist U., Engström-Laurent A. (1991) Hyaluronan in pleural effusions and in serum. *Cancer* **67**: 2410–2414.

Holmes M.W.A., Bayliss M.T., Muir H. (1988) Hyaluronic acid in human articular cartilage. Age-related changes in content and size. *Biochem J* **250**: 435–441.

Honda A., Ohashi Y., Mori Y. (1986) Hyaluronic acid in rabbit pericardial fluid and its production by pericardium. *FEBS Letters* **203**: 273–278.

Honda A., Noguchi N., Takehara H., Ohashi Y., Asuwa N., Mori Y. (1991) Cooperative enhancement of hyaluronic acid synthesis by combined use of IGF-1 and EGF, and inhibition by tyrosine kinase inhibitor genestein, in cultured mesothelial cells from rabbit pericardial cavity. *J Cell Sci* **98**: 91–98.

Hua Q., Knudson C.B., Knudson W. (1993) Internalization of hyaluronan by chondrocytes occurs via receptor-mediated endocytosis. *J Cell Sci* **106**: 365–375.

Huang L., Yoneda M., Kimata K. (1993) A serum-derived hyaluronan-associated protein (SHAP) is the heavy chain of the inter-α-trypsin inhibitor. *J Biol Chem* **268**: 26725–26730.

Hutadilok N., Ghosh P., Brooks P.M. (1988) Binding of haptoglobin, inter-α-trypsin inhibitor, and α-1 proteinase inhibitor to synovial fluid hyaluronate and the influence of these proteins on its degradation by oxygen derived free radicals. *Ann Rheum Dis* **47**: 377–385.

Inoue M., Katakami C. (1993) The effect of hyaluronic acid on corneal epithelial cell proliferation. *Invest Ophthalmol Vis Sci* **34**: 2313–2315.
Jackson D.G., Schenker T., Waibel R., Bell J.I., Stahel R.A.. (1994) Expression of alternatively spliced forms of the CD44 extracellular-matrix receptor on human lung carcinomas. *Int J Cancer Suppl* **8**: 110–115.
Jalkanen S., Bargatze R.F., Herron L.R., Butcher E.C. (1986) Homing receptors and the control of lymphocyte migration. *Immunol Rev* **91**: 39–60.
Jeanloz R.W. (1960) The nomenclature of mucopolysaccharides. *Arthr Rheum* **3**: 233–237.
Juhlin L., Tengblad A., Ortonne J.P., Lacour J.P. (1986) Hyaluronate in suction blisters from patients with scleroderma and various skin disorders. *Acta Dermatol Venereol (Stockh)* **66**: 409–413.
Jeanloz R.W., Forchielli E. (1950) Studies on hyaluronic acid and related substances.I. Preparation of hyaluronic acid and derivatives from human umbilical cord. *J Biol Chem* **186**: 495–511.
King E.S.J. (1935) The Golgi apparatus of synovial cells under normal and pathological conditions and with reference to the formation of synovial fluid. *J Path Bact* **41**: 117–128.
Kishimoto T.K., Larson R.S., Corbi A.L., Dustin M.L., Staunton D.E., Springer T.A. (1989) The leukocyte integrins. *Adv Immunol* **46**: 149–182.
Knudson C.B. (1993) Hyaluronan receptor-directed assembly of chondrocyte pericellular matrix. *J Cell Biol* **120**: 825–834.
Knudson C.B., Toole B.P. (1985a) Fluorescent morphological probe for hyaluronate. *J Cell Biol* **100**: 1753–1758.
Knudson C.B., Toole B.P. (1985b) Changes in the pericellular matrix during differentiation of limb bud mesoderm. *Devel Biol* **112**: 308–318.
Knudson W., Toole B.P. (1988) Membrane association of the hyaluronate stimulatory factor from LX-1 human lung carcinoma cells. *J Cell Biochem* **38**: 165–177.
Knudson W., Biswas C., Toole B.P. (1984) Interactions between human tumor cells and fibroblasts stimulate hyaluronate synthesis. *Proc Natl Acad Sci USA* **81**: 6767–6771.
Knudson W., Biswas C., Li X-Q., Nemec R.E., Toole B.P. (1989) The role and regulation of tumour-associated hyaluronan. **In** *The Biology of Hyaluronan*, eds. D. Evered, J. Whelan. Ciba Foundation Symposium 143, pp. 150–169. Publ. J Wiley and Sons, Chichester.
Knudson W., Bartnik E., Knudson C.B. (1993) Assembly of pericellular matrices by COS-7 cells transfected with CD44 lymphocyte-homing genes. *Proc Natl Acad Sci USA* **90**: 4003–4007.
Kohn L.D., Winand R.J. (1971) Relationship of thyrotropin to exophthalmos-producing substance. *J Biol Chem* **246**: 6570–6575.
Kohn L.D., Winand R.J. (1975) Structure of an exophthalmos-producing factor derived from thyrotropin by partial pepsin digestion. *J Biol Chem* **250**: 6503–6508.
Kornovski B.S., McCoshen J., Kredentser J., Turley E. (1994) The regulation of sperm motility by a novel hyaluronan receptor. *Fertil Steril* **61**: 935–940.
Lamberg S.I., Yuspa S.H., Hascall V.C. (1986) Synthesis of hyaluronic acid is decreased and synthesis of proteoglycans is increased when cultured mouse epidermal cells differentiate. *J Invest Dermatol* **86**: 659–667.
Laurent C., Hellström S., Stenfors L-E. (1986) Hyaluronic acid reduces connective tissue formation in middle ears filled with absorbable gelatin sponge. An experimental study. *Am J Otolaryngol* **7**: 181–186.
Laurent C., Johnson-Wells G., Hellström S., Engström-Laurent A., Wells A. (1991) Localization of hyaluronan in various muscular tissues. A morphological study in the rat. *Cell Tissue Res* **263**: 201–205.
Laurent T.C. (1955) Studies on hyaluronic acid in the vitreous body. *J Biol Chem* **216**: 263–271.
Laurent T.C. (1964) The interaction between polysaccharides and other macromolecules. The exclusion of molecules from hyaluronic acid gels and solutions. *Biochem J* **93**: 106–112.
Laurent T.C. (1970) The structure and function of the intercellular polysaccharides in connective tissue. **In** *Capillary Permeability*, eds. C. Crone, N.A. Larsen. Alfred Benzon Symp II, pp. 261–267. Munksgaard, Copenhagen.
Laurent T.C., Fraser J.R.E. (1986) The properties and turnover of hyaluronan. **In** *Functions of the Proteoglycans*, eds. D. Evered, J. Whelan. Ciba Foundation Symposium 124, pp. 9–29. Publ. J. Wiley and Sons, Chichester, UK.
Laurent T.C., Fraser J.R.E., Pertoft H., Smedsrød B. (1986) Binding of hyaluronate and chondroitin sulphate to liver endothelial cells. *Biochem J* **234**: 653–658.
Laurent T.C., Gergely J. (1955) Light scattering studies on hyaluronic acid. *J Biol Chem* **212**: 325–333.
Laurent T.C, Lilja K., Brunnberg I., Engström-Laurent A., Laurent U.B.G., Lindqvist U. *et al.* (1987) Urinary excretion of hyaluronan in man. *Scand J Clin Lab Invest* **47**: 793–799.
Laurent T.C., Ogston A.G. (1963) The interactions between polysaccharides and other macromolecules. 4. The osmotic pressure of mixtures of serum albumin and hyaluronic acid. *Biochem J* **89**: 249–253.
Laurent T.C., Ryan M., Pietruszkiewicz A. (1960) Fractionation of hyaluronic acid. The polydispersity of hyaluronic acid from the bovine vitreous body. *Biochim Biophys Acta* **42**: 476–485.
Laurent U B G. (1983). Hyaluronate in the human aqueous humour. Acta Ophthalmol **101**: 129–130.

Laurent U.B.G., Granath K.A. (1983) The molecular weight of hyaluronate in the aqueous humour and vitreous body of rabbit and cattle eyes. *Exp Eye Res* **36**: 481–492.

Laurent U.B.G., Laurent T.C. (1981) On the origin of hyaluronate in blood. *Biochem Intl* **2**: 195–199.

Laurent U.B.G., Reed R.K. (1991) Turnover of hyaluronan in the tissues. *Adv Drug Deliv Rev* **7**: 237–256.

Laurent U.B.G., Tengblad A. (1980) Determination of hyaluronate in biological samples by a specific radioassay technique. *Anal Biochem* **109**: 886–894.

LeBaron R.G., Zimmermann D.R., Ruoslahti E. (1992) Hyaluronate binding properties of versican. *J Biol Chem* **267**: 10003–10010.

Lebel L., Smith L., Risberg B., Gerdin B., Laurent T.C. (1988) Effect of increased hydrostatic pressure on lymphatic elimination of hyaluronan from sheep lung. *J Appl Physiol* **64**: 327–332.

Lebel L., Smith L., Risberg B., Gerdin B., Laurent T.C. (1989) Increased lymphatic elimination of interstitial hyaluronan during E. coli sepsis in sheep. *Am J Physiol (Heart Circ. Physiol. 25)* **256**: 1524–1531.

Lee T.H., Wisniewski H-G., Vilcek J. (1992) A novel secretory tumor necrosis factor-inducible protein (TSG-6) is a member of the family of hyaluronate binding proteins, closely related to the adhesion receptor CD44. *J Cell Biol* **116**: 545–557.

Lesley J., Hyman R. (1992) CD44 can be activated to function as an hyaluronic acid receptor in normal murine T cells. *Eur J Immunol* **22**: 2719–2723.

Lesley J., He Q., Miyake K., Hamann A., Hyman R., Kincade P.W. (1992) Requirements for hyaluronic acid binding by CD44: a role for the cytoplasmic domain and activation by antibody. *J Exp Med* **175**: 257–266.

Lesley J., Hyman R., Kincade P.W. (1993) CD44 and its interaction with extracellular matrix. *Adv Immunol* **54**: 271–335.

Lesley J., Howes N., Perschl A., Hyman R. (1994) Hyaluronan binding function of CD44 is transiently activated on T cells during an in vivo immune response. *J Exp Med* **180**: 383–387.

Lesley J., Schulte R., Hyman R. (1990) Binding of hyaluronic acid to lymphoid cell lines is inhibited by monoclonal antibodies against Pgp-1. *Exp Cell Res* **187**: 224–233.

Levick J.R. (1994) An analysis of the interaction between interstitial plasma protein, interstitial flow, and fenestral filtration and its application to synovium. *Microvasc Res* **47**: 68–89.

Linker A., Meyer K., Weissmann B. (1955) Enzymatic formation of monosaccharides from hyaluronate. *J Biol Chem* **213**: 237–248.

Longaker M.T., Adzick N.S., Sadigh D., Hendin B., Stair S.E., Duncan B.W., Harrison M.R., Spendlove R., Stern R. (1990) Hyaluronic acid-stimulating activity in the pathophysiology of Wilms' tumors. *J Natl Cancer Inst* **82**: 135–139.

Longaker M.T., Chiu E.S., Adzick N.S., Stern M., Harrison M.R., Stern R. (1991) Studies in fetal wound healing. V. A prolonged presence of hyaluronic acid characterises fetal wound fluid. *Ann Surg* **213**: 292–296.

Longas M.O., Meyer K. (1981) Sequential hydrolysis of hyaluronate by β-glucuronidase and β-N-acetylhexosaminidase. *Biochem J* **197**: 275–282.

Longas M.A., Bhuyan D.K., Bhuyan K.C., Gutsch C.M., Breitweiser K.O. (1993) Dietary vitamin E reverses the effects of ultraviolet light irradiation on rat skin glycosaminoglycans. *Biochim Biophys Acta* **1156**: 239–244.

Love S.H., Shannon B.T., Myrvik Q.N., Lynn W.S. (1979) Characterization of macrophage agglutinating factor as a hyaluronate-protein complex. *J Reticuloendothel Soc* **25**: 269–282.

Lund P., Hørslev-Petersen K., Helin P., Parving H-H. (1986) The effect of l-thyroxine on skin accumulation of acid glycosaminoglycans in primary myxoedema. *Acta Endocrinol (Copenh)* **113**: 56–58.

Mahadevan P., Larkins R.G., Fraser J.R.E., Fosang A.J., Dunlop M.E. (1995) Increased hyaluronan production in the glomeruli from diabetic rats: a link between glucose-induced prostaglandin production and reduced sulphated proteoglycan. *Diabetologia* **38**: (In Press).

Makgoba M.W., Sanders M.E., Luce G.E.G., Dustin M.L., Springer T.A., Clark E.A., Mannoni P., Shaw S. (1988) ICAM-1: definition by multiple antibodies of a ligand for LFA-1 dependent adhesion of B, T and myeloid cells. *Nature* **331**: 86–88.

Margolis R.U., Margolis R.K., Chang L.B., Preti C. (1975) Glycosaminoglycans of brain during development. Biochemistry 14: 85–88.

Mason R.M., d'Arville C., Kimura J.H., Hascall V.C. (1982) Absence of covalently linked core protein from newly synthesised hyaluronate. *Biochem J* **207**: 445–457.

McBride W.H., Bard J.B.L. (1979) Hyaluronidase-sensitive halos around adherent cells. *J Exp Med* **149**: 507–515.

McCourt P.A.G., Ek B., Forsberg N., Gustafson S. (1995) ICAM-1 is a cell-surface receptor for hyaluronan. *J Biol Chem.* (In press.)

McGary C.T., Raja R.H., Weigel P.H. (1989) Endocytosis of hyaluronic acid by rat liver endothelial cells. Evidence for receptor recycling. *Biochem J* **257**: 875–884.

McNeil J.D., Wiebkin O.W., Cleland L.G., Skosey J.L. (1989) The effects of hyaluronic acid on macrophage Fc receptor binding and phagocytosis are independent of the mode of depolymerization. *Free Rad Res Comms* **6**: 227–233.

Meyer F.A., Koblentz M., Silberberg A. (1977) Structural investigation of loose connective tissue using a series of dextran fractions as non-interacting macromolecular probes. *Biochem J* **161**: 285–291.

Meyer F.A., Yaron I., Yaron M. (1990) Synergistic, additive, and antagonistic effects of interleukin-1β, tumour necrosis factor-α and γ-interferon on prostaglandin E, hyaluronic acid , and collagenase production by cultured synovial fibroblasts. *Arthr Rheum* **33**: 1518–1525.

Meyer K. (1947) The biological significance of hyaluronic acid and hyaluronidase. *Physiol Rev* **27**: 335–357.

Meyer K., Palmer J.W. (1934) The polysaccharide of the vitreous humor. *J Biol Chem*, **107**: 629–634.

Mian N. (1986) Characterization of a high-Mr plasma-membrane-bound protein and assessment of its role as a constituent of hyaluronate synthase. *Biochem J* **237**: 333–342

Mikelsaar R-H., Scott J.E. (1994) Molecular modelling of secondary and tertiary structures of hyaluronan, compared with electron microscopy and NMR data. Possible sheets and tubular structures in aqueous solution. *Glycoconjugate J* **11**: 56–71.

Mikuni-Takagaki Y., Toole B.P. (1981) Hyaluronate-protein complex of Rous sarcoma virus-transformed chick embryo fibroblasts. *J Biol Chem* **256**: 8463–8469.

Miyake K., Underhill C.B., Lesley J., Kincade P.W. (1990) Hyaluronate can function as a cell adhesion molecule and CD44 participates in hyaluronate recognition. *J Exp Med* **172**: 69–75.

Morris E.R., Cutler A.N., Ross-Murphy S.B., Rees D. (1981) Concentration and shear rate dependence of viscosity in random coil polysaccharide solutions. *Carbohydrate Polymers* **1**: 5–21

Mörner C.T. (1889) Chemische studien über den trachealknorpel. *Skand Archiv für Physiol* **1**: 210–243.

Moscatelli D., Rubin H. (1975) Increased hyaluronic acid production on stimulation of DNA synthesis in chick embryo fibroblasts. *Nature* **254**: 65.

Muir H. (1958) The nature of the link between protein and carbohydrate of a chondroitin sulphate complex from hyaline cartilage. *Biochem J* **69**: 195–204.

Munaim S.I., Klagsbrun M., Toole B.P. (1988) Developmental changes in fibroblast growth factor in the chicken embryo limb bud. *Proc Natl Acad Sci USA*. **85**: 8091–8093.

Murakami S., Miyake K., June C.H., Kincade P.W., Hodes R.J. (1990) IL-5 induces a Pgp-1 (CD44) bright B cell subpopulation that is highly enriched in proliferative and Ig secretory activity and binds to hyaluronate. *J Immunol* **145**: 3618–3627.

Murakami S., Miyake K., Abe R., Kincade P.W., Hodes R.J. (1991) Characterization of autoantibody-secreting B cells in mice undergoing stimulatory (chronic) GVH reactions. Identification of a CD44hi population that binds specifically to hyaluronate. *J Immunol* **146**: 1422–1427.

Murata K., Yokoyama Y. (1988) High hyaluronic acid and low dermatan sulphate contents in human pulmonary arteries compared to in the aorta. *Blood Vessels* **25**: 1–11.

Myer L.J., Stern R. (1994) Age-dependent changes of hyaluronan in human skin. *J Invest Dermatol* **102**: 385–389.

Nettelbladt O., Tengblad A., Hällgren R. (1989) Lung accumulation of hyaluronan parallels pulmonary edema in experimental alveolitis. *Am J Physiol* **257** *(Lung Cell Mol Physiol 1)*: L379–L384.

Ng C.K., Handley C.J., Preston B.N., Robinson H.C. (1992) The extracellular processing and catabolism of hyaluronan in cultured adult articular cartilage explants. *Arch Biochem Biophys* **298**: 70–79.

Noble P.W., Lake F.R., Henson P.M., Riches D.W. (1993) Hyaluronate activation of CD44 induces insulin-like growth factor-1 expression by a tumor necrosis factor-alpha-dependent mechanism of murine macrophages. *J Clin Invest* **91**: 2368–2377.

Nyberg A., Engström-Laurent A., Lööf L. (1988) Serum hyaluronate in primary biliary cirrhosis – a biochemical marker for progressive liver damage. *Hepatol* **8**: 142–146.

Ogston A.G., Phelps C.F. (1961) The partition of solutes between buffer solutions and solutions containing hyaluronic acid. *Biochem J* **78**: 827–833.

Ogston A.G., Sherman T.F. (1961) Effects of hyaluronic acid upon diffusion of solutes and flow of solvent. *J Physiol* **17**: 1–8.

Ogston A.G, Stanier J.G. (1950) On the state of hyaluronic acid in synovial fluid. *Biochem J* **46**: 364–376.

Ogston A.G., Stanier J.G. (1951) The dimensions of the particle of hyaluronic acid complex in synovial fluid. *Biochem J* **49**: 585–590.

Ogston A.G., Stanier J.G. (1953) The physiological function of hyaluronic acid in synovial fluid; viscous, elastic and lubricant properties. *J Physiol* **119**: 244–252.

Ogston A.G., Stanier J.G., Toms B.A., Strawbridge D.J. (1950) Elastic properties of ox synovial fluid. *Nature* **165**: 571.

Orkin R.W., Knudson W., Toole B.P. (1985) Loss of hyaluronate-dependent coat during myoblast fusion. *Devel Biol* **107**: 527–530.

Parving H-H., Helin G., Garbarsch C., Johansen A.A., Jensen B.A., Helin P., Lund P., Lyngsøe J. (1982) Acid glycosaminoglycans in myxoedema. *Clin Endocrinol* **16**: 207–210.

Peach R.J., Hollenbaugh D., Stamenkovic I., Aruffo A. (1993) Identification of hyaluronic acid binding sites in the extracellular domain of CD44. *J Cell Biol* **122**: 257–264.

Pearce R.H., Grimmer B.J. (1972) Age and the chemical constitution of normal human dermis. *J Invest Dermatol* **58**: 347–361.

Pessac B., Defendi V. (1972) Cell aggregation: role of acid mucopolysaccharides. *Science* **175**: 898–900.

Philipson L.H., Schwartz N.B. (1984) Subcellular localization of hyaluronate synthetase in oligodendroglioma cells. *J Biol Chem* **259**: 5017–5023.

Pigman W., Rizvi S., Holley H.L. (1961) Depolymerisation of hyaluronic acid by the ORD reaction. *Arthr Rheum* **4**: 240–252.

Pilarski L.M., Miszta H., Turley E. (1993) Regulated expression of a receptor for hyaluronan-mediated motility on human thymocytes and T cells. *J Immunol* **150**: 4292–4302.

Polansky J.R., Toole B.P. (1976) Hyaluronidase activity during thyroxine-induced tadpole metamorphosis. *Devel Biol* **53**: 30–35.

Prehm P. (1983) Synthesis of hyaluronate in differentiated teratocarcinoma cells. Mechanism of chain growth. *Biochem J* **211**: 191–198

Prehm P. (1984) Hyaluronate is synthesized at plasma membranes. *Biochem J* **220**: 597–600.

Prehm P. (1990) Release of hyaluronate from eukaryotic cells. *Biochem J* **267**: 185–189.

Presti D., Scott J.E. (1994) Hyaluronan-mediated protective effect against cell damage caused by enzymatically produced hydroxyl (OH^-) radicals is dependent on hyaluronan molecular mass. *Cell Biochem Funct* **12**: 281–288.

Preston B.P., Davies M., Ogston A.G. (1965) The composition and physicochemical properties of hyaluronic acids prepared from ox synovial fluid and from a case of mesothelioma. *Biochem J* **96**: 449–474.

Radin E.L., Paul I.L., Swann D.A., Schottstaedt E.S. (1971) Lubrication of synovial membrane. *Ann Rheum Dis* **30**: 322–325.

Ragan C.P., Donlan C.P., Coss J.A. Jr., Grubin A.F. (1947) Effects of X-ray irradiation on viscosity of synovial fluid. *Proc Soc Exp Biol Med* **66**: 170–172.

Reed R.K., Lilja K., Laurent T.C. (1988) Hyaluronan in the rat with special reference to the skin. *Acta Physiol Scand* **134**: 405–411.

Reeves and Meyer, (1981) Quoted by Longas and Meyer 1981, loc. cit.

Reid T., Flint M.H. (1974) Changes in glycosaminoglycan content of healing rabbit tendon. *J Embryol Exp Morph* **31**: 489–495.

Ripellino J.A., Klinger M.M., Margolis R.U., Margolis R.K. (1985) The hyaluronic acid binding region as a specific probe for the localization of hyaluronic acid in tissue sections. Application to chick embryo and rat brain. *J Histochem Cytochem* **33**: 1060–1066.

Rodén L., Campbell P., Fraser J.R.E., Laurent T.C., Pertoft H., Thompson J.N. (1989) Enzymic pathways of hyaluronan metabolism. **In** *The Biology of Hyaluronan*, eds. D. Evered, J. Whelan. Ciba Foundation Symposium 143, pp. 60–86. Publ. J .Wiley and Sons, Chichester.

Ropes M.W., Robertson W. von B., Rossmeisl E.C., Peabody R.B., Bauer W. (1947) Synovial fluid mucin. *Acta Med Scandinavica Suppl* **196**: 700–744.

Rosenberg L.C., Buckwalter J.A. (1986) Cartilage proteoglycans. **In** *Articular Cartilage Biochemistry*, pp. 39–57. Eds. K.E. Kluettner, R Schleyerbach, V.C. Hascall. Raven Press, New York.

Ross R. (1975) Connective tissue cells, cell proliferation and synthesis of extracellular matrix — a review. Phil Trans Roy Soc Lond. **B 271**: 247–259.

Ross R. (1986) The pathogenesis of atherosclerosis - an update. *New Engl J Med* **314**: 488–500.

Rydell N.W., Balazs E.A. (1971) Effect of intra-articular injection of hyaluronic acid on the clinical symptoms of osteoarthritis and on granulation tissue formation. *Clin Orthop* **80**: 25–32.

Salmi M., Grön-Virta K., Sointu P., Grenman R., Kalimo H., Jalkanen S. (1993) Regulated expression of exon v6 containing isoforms of CD44 in man: downregulation during malignant transformation of tumors of squamocellular origin. *J Cell Biol* **122**: 431–442.

Salustri A., Ulisse D., Yanagishita M., Hascall V.C. (1990a) Hyaluronic acid synthesis by mural granulosa cells and cumulus cells in vitro is selectively stimulated by a factor produced by oocytes and by transforming growth factor-β. *J Biol Chem* **265**: 19517–19523.

Salustri A., Yanagishita M., Hascall V.C. 1990b Mouse oocytes regulate hyaluronic acid synthesis and mucification by FSH-stimulated cumulus cells.*Devel Biol* **138**: 26–32.

Samuel S.K., Hurta R.A., Spearman M.A., Wright J.A., Turley E.A., Greenberg A.H. (1993) TGF-β1 stimulation of cell locomotion utilizes the hyaluronan receptor RHAMM and hyaluronan. *J Cell Biol* **123**: 749–758.

Sandson J., Hamerman D. (1964) Binding of an alpha globulin to hyaluronateprotein isolated in pathological synovial fluids. *Science* **146**: 70–71.

Sandson J., Hamerman D., Schwick G. (1965) Altered properties of pathological hyaluronate due to a bound inter-alpha trypsin inhibitor. *Trans Assoc Amer Physicians* **78**: 304–313.

Schiller S., Slover G.A., Dorfman A. (1962) Effect of the thyroid gland on metabolism of acid mucopolysaccharides in skin. *Biochim Biophys Acta* **58**: 27–33.

Scott J.E. (1968) Periodate oxidation, pKa and conformation of hexuronic acids in polyuronides and mucopolysaccharides. *Biochim Biophys Acta* **170**: 471–473.

Scott J.E. (1989) Secondary structures in hyaluronan solutions: chemical and biological implications. **In** *The Biology of Hyaluronan*, eds. D. Evered, J. Whelan. Ciba Foundation Symposium 143, pp. 6–15. Publ. J. Wiley and Sons, Chichester.

Scott J.E. (1992) Supramolecular organization of extracellular matrix glycosaminoglycans, in vitro and in the tissues. *FASEB J* **6**: 2639–2645.

Scott J.E., Hughes E.W. (1983) Differential staining of polyanions according to critical electrolyte concentration principles in mixed solvents. *J Microscopy* **129**: 209–219.

Scott J.E., Tigwell M.J. (1978) Periodate oxidation and the shapes of glycosaminoglycuronans in solution. *Biochem J* **173**: 103–114.

Scott J.E., Cummings C., Brass A., Chen Y. (1991) Secondary and tertiary structures of hyaluronan in aqueous solution, investigated by rotary shadowing-electron microscopy and computer simulation. *Biochem J* **274**: 699–705.

Shannon B.T., Love S.H. (1980) Additional evidence for the role of hyaluronic acid in the macrophage disappearance reaction. *Immunol Commun* **9**: 735–746.

Shannon B.T., Love S.H., Myrvik Q.N. (1980) Participation of hyaluronic acid in the macrophage disappearance reaction. *Immunol Commun* **9**: 357–370.

Shannon B.T., Love S.H., Roh B.H., Schroff R.W. (1981) Quantitation of glycosaminoglycans of rabbit lung during delayed-type hypersensitivity reactions and granuloma formation. *Inflammation* **5**: 323–334.

Shatton J., Schubert M. (1954) Isolation of a mucoprotein from cartilage. *J Biol Chem* **211**: 565–573.

Silpananta P., Dunstone J.R., Ogston A.G. (1967) Fractionation of a hyaluronic acid preparation in a density gradient. The isolation and identification of a chondroitin sulphate. *Biochem J* **104**: 404–409.

Silpananta P., Dunstone J.R., Ogston A.G. (1968) Fractionation of a hyaluronic acid preparation in a density gradient. Some properties of the hyaluronic acid. *Biochem J* **109**: 43–50.

Sisson J.C. (1968) Hyaluronic acid in localised myxedema. *J Clin Endocrinol* **28**: 433–436

Sisson J.C., Miles M. (1967) Acid mucopolysaccharide alterations in experimental exophthalmos. *Endocrinol* **80**: 931–937.

Smedsrød B., Pertoft H., Eriksson S., Fraser J.R.E., Laurent T.C. (1984) Studies in vitro on the uptake and degradation of sodium hyaluronate in rat liver endothelial cells. *Biochem J* **223**: 617–626.

Smith T.J., Murata Y., Horwitz A.L., Philipson L., Refetoff S.. (1982) Regulation of glycosaminoglycan synthesis by thyroid hormone in vitro. *J Clin Invest* **70**: 1066–1073.

Smith T.J., Bahn R.S., Gorman C.A., Cheavens M. (1991) Stimulation of glycosaminoglycan accumulation by interferon gamma in cultured human retroocular fibroblasts. *J Clin Endocrinol Metab* **72**: 1169–1171.

Smith T.J., Wang H-S., Evans C.H. (1995) Leukoregulin is a potent inducer of hyaluronan synthesis in cultured human orbital fibroblasts. *Am J Physiol* **268** *(Cell Physiol* **37***)*: C382–C388.

Stamenkovic I., Amiot M., Pesando J.M., Seed B. (1989) A lymphocyte molecule implicated in lymph node homing is a member of the cartilage link protein family. **Cell 56**: 1057–1062.

Stamenkovic I., Aruffo A., Amiot M., Seed B. (1991) The hematopoietic and epithelial forms of CD44 are distinct polypeptides with different adhesion potentials for hyaluronate bearing cells. *EMBO J* **10**: 343–348.

Stephens R.W., Ghosh P., Taylor T.F.K. (1975) The characterisation and function of the polysaccharidases of human synovial fluid in rheumatoid and osteoarthritis. *Biochim Biophys Acta* **399**: 101–112.

St Onge R., Weiss C., Denlinger J.L., Balazs E.A. (1980) A preliminary assessment of Na-hyaluronate injection into "No Man's Land" for primary flexor tendon repair. *Clin Orthop* **146**: 269–275.

Sundblad L. (1965) Glycosaminoglycans and glycoproteins in synovial fluid. **In** *The Amino Sugars. The Chemistry and Biology of Compounds Containing Amino Sugars.* Vol IIA pp. 229–250. Eds. E.A. Balazs, J.W. Jeanloz. Acacemic Press, New York and London.

Sundblad L., Balazs E.A. (1966) Chemical and physical changes of glycosaminoglycans and glycoproteins caused by oxidation reduction systems and radiation. **In** *The Amino Sugars. Metabolism and Interactions.* Vol IIB pp. 229–250. Eds. E.A. Balazs and R.W. Jeanloz. Academic Press, New York and London.

Swann D.A. (1967) The degradation of hyaluronic acid by ascorbic acid. *Biochem J* **102**: 42C–44C.

Swann D.A. (1969) Studies on the structure of hyaluronic acid. Characterization of the product formed when hyaluronic acid is treated with ascorbic acid. *Biochem J* **114:** 819–825.

Tammi R., Ripellino J.A., Margolis R.U., Tammi M. (1988) Localization of epidermal hyaluronic acid using the hyaluronate binding region of cartilage proteoglycan as a specific probe. *J Invest Dermatol* **90**: 412–414.

Tammi R., Säämänen A-M., Maibach, Tammi M. (1991) Degradation of newly synthesized high molecular mass hyaluronan in the epidermal and dermal compartments of human skin in organ culture. *J Invest Dermatol* **97**: 126–130.

Tengblad A., Laurent U.B.G., Lilja K., Cahill R.N.P., Engström-Laurent A., Fraser J.R.E., Hansson H.E., Laurent T.C. (1986) Concentration and relative molecular mass of hyaluronate in lymph and blood. *Biochem J* **236**: 521–525.

Tirtaatmadja V., Boger D.V., Fraser J.R.E. (1984) The dynamic and steady shear properties of synovial fluid and of the components making up synovial fluid. *Rheologica Acta* **23**: 311–321.
Toole B.P. (1972) Hyaluronate turnover during chondrogenesis in the developing chick limb and axial skeleton. *Devel Biol* **29**: 321–329.
Toole B.P. (1976) Morphogenetic role of glycosaminoglycans (acid mucopolysaccharides) in brain and other tissues. **In** *Neuronal Recognition*, pp. 275–329. Ed. S.H. Barondes. Plenum Press, New York.
Toole B.P. (1981) Glycosaminoglycans in morphogenesis. In *Cell Biology of Extracellular Matrix*, pp. 259–294. Ed. E.D. Hay. Plenum Press, New York.
Toole B.P. (1989) Hyaluronate-cell interactions and growth factor regulation of hyaluronate synthesis during limb development. **In** *The Biology of Hyaluronan*, eds. D. Evered, J. Whelan. Ciba Foundation Symposium 143, pp. 138–145. Publ. J. Wiley and Sons, Chichester.
Toole B.P. (1991) Hyaluronan and its binding proteins, the hyaladherins. *Curr Opin Cell Biol* **2**: 839–844.
Toole B.P., Trelstad R.L. (1971) Hyaluronate production and removal during corneal development in the chick. *Devel Biol* **26**: 28–35.
Toole B.P., Biswas C., Gross J. (1979) Hyaluronate and invasiveness of the rabbit V2 carcinoma. *Proc Natl Acad Sci USA* **76**: 6299–6303.
Turley E. (1989) The role of a cell-associated hyaluronan-binding protein in fibroblast behaviour. **In** *The Biology of Hyaluronan*, eds. D. Evered, J. Whelan. Ciba Foundation Symposium 143, pp. 121–137. Publ. J. Wiley and Sons, Chichester.
Turley E., Brassel P., Moore D. (1990) A hyaluronan-binding protein shows a partial and temporally regulated codistribution with actin on locomoting chick heart fibroblasts. *Exp Cell Res* **187**: 243–249.
Turley E.A., Austen L., Moore D., Hoare K. (1993) Ras-transformed cells express both CD44 and RHAMM hyaluronan receptors: only RHAMM is essential for hyaluronan-promoted locomotion. *Exp Cell Res* **207**: 277–282.
Tzaicos C., Fraser J.R.E., Tsotsis E., Kimpton W.G. (1989) Inhibition of hyaluronan uptake in lymphatic tissue by chondroitin sulphate proteoglycan. *Biochem J* **263**: 823–828.
Ueno T., Inuzuka S., Torimura T., Tamaki S., Koh H., Kin M., Minetoma T., Kimura T., Ohira H., Sata M., Yoshida H., Tanikawa K. (1993) Serum hyaluronate reflects hepatic sinusoidal capillarization. *Gastroenterol* **105**: 475–481.
Underhill C.B. (1993) Hyaluronan is inversely correlated with the expression of CD44 in the dermal condensation of the embryonic hair follicle. *J Invest Dermatol* **101**: 820–826.
Underhill C.B., Toole B.P. (1979) Binding of hyaluronate to the surface of cultured cells *J Cell Biol* **82**: 475–484.
Underhill C.B., Toole B.P. (1982) Transformation-dependent loss of the hyaluronate-containing coats of cultured cells. *J Cell Physiol* **110**: 123–128.
Underhill C.B., Toole B.P. (1990) Physical characteristics of hyaluronate binding to the surface of simian virus 40-transformed 3T3 cells. *J Biol Chem* **255**: 4544–4549.
Vaes G. (1965) Studies on bone enzymes. The activation and release of latent acid hydrolases and catalase in bone-tissue homogenates. *Biochem J* **97**: 393–402
Van Praag D., Stone A.L., Richter A.J., Farber S.J. (1972) Composition of glycosaminoglycans (mucopolysaccharides) in rabbit kidney. II. Renal cortex. *Biochim Biophys Acta* **273**: 149–156.
Vuorio E., Einola S., Hakkarainen S., Penttinen R. (1977) Synthesis of underpolymerized hyaluronic acid by fibroblasts cultured from rheumatoid and non-rheumatoid synovitis. *Rheumatol Intl* **2**: 97–102.
Wang C., Tammi M., Tammi R. (1992) Distribution of hyaluronan and its CD44 receptor in the epithelia of human skin appendages. *Histochem* **98**: 105–112.
Wasteson Å., Westermark B., Lindahl U., Pontén J. (1973) Aggregation of feline lymphoma cells by hyaluronic acid. *Int J Cancer* **12**: 169–178.
Watson E.M., Pearce R.H. (1947) The mucopolysaccharide content of the skin in localized pretibial myxedema. *Am J Clin Path* **17**: 507–512.
Waldenström A., Martinussen H.J., Gerdin B., Hällgren R. (1991) Accumulation of hyaluronan and tissue edema in experimental myocardial infarction. *J Clin Invest* **88**: 1622–1628.
Wells A., Larsson E., Tengblad A., Fellström, Tufveson G., Klareskog L., Laurent T.C. (1990a) The localization of hyaluronan in normal and rejected human kidneys. *Transplantation* **50**: 240–243.
Wells A., Lundin Å., Michaëlsson G., Pontén F. (1990b) Hyaluronan in basal cell carcinomas. *Acta Dermatol Venereol (Stockh)* **71**: 274–275.
Wells A., Klareskog L., Lindblad S., Laurent T.C. (1992) Correlation between increased hyaluronan localized in arthritic synovium and the presence of proliferating cells. A role for macrophage-derived factors. *Arthr Rheum* **35**: 391–396.
Westergren-Thorsson G., Särnstrand B., Fransson L-Å., Malmström A. (1990) TGF-β enhances the production of hyaluronan in human lung but not in skin fibroblasts. *Exp Cell Res***186**: 192–195.

Whiteside T.L., Buckingham R.B. (1989) Interactions between cells of the immune system and hyaluronate synthesis by human dermal fibroblasts. **In** *The Biology of Hyaluronan*, eds. D. Evered, J. Whelan. Ciba Foundation Symposium 143, pp. 170–186. Publ. J. Wiley and Sons, Chichester.

Wiebkin O., Muir H. (1973) Influence of the cells on the pericellular environment. The effect of hyaluronic acid on proteoglycan synthesis and secretion by chondrocytes of adult cartilage. *Phil Trans Roy Soc London B* **271**: 283–291.

Wik K.O. (1979) Physicochemical Studies on Hyaluronate. *Acta Univ Upsal. Abs. Uppsala Diss. Fac. Med.* **334**.

Wik K.O., Andersson T., Jacobsson J.O., Granath K.A. (1979) Physicochemical characterization of hyaluronate. ***In*** *Physicochemical Studies on Hyaluronate*, K.O. Wik, *Acta Univ Upsaliensis Abs., Uppsala Diss. Fac. Med.* **334**: 1–29.

Wirth K., Arch R., Somasundaram C., Hofmann M., Weber B., Herrlich P., Matzku S., Zöller M. (1993) Expression of CD44 isoforms carrying metastasis-associated sequences in newborn and adult rats. *Eur J Cancer* **29A**: 1172–1177.

Wisniewski H-G., Burgess W.H., Oppenheim J.D., Vilcek J. (1994) TSG-6, an arthritis-associated protein, forms a stable complex with the serum protein inter-alpha-inhibitor. *Biochemistry* **33**: 7423–7429.

Wolffe E.J., Gause W.C., Pelfrey C.M., Holland S.M., Steinberg A.D., August J.T. (1990) The cDNA sequence of mouse Pgp-1 and homology to human CD44 cell surface antigen and proteoglycan core/link proteins. *J Biol Chem* **265**: 341–347.

Yang B., Yang B.L., Savani R.C., Turley E.A. (1994) Identification of a common hyaluronan binding motif in the hyaluronan binding proteins RHAMM, CD44 and link. *EMBO J* **13**: 286–296.

Yannariello-Brown J., Weigel P. (1992) Detergent solubilization of the endocytic Ca^{2+}-independent hyaluronan receptor from rat liver endothelial cells and separation from a Ca^{2+}-dependent hyaluronan-binding activity. *Biochemistry* **31**: 576–584.

Yoneda M., Suzuki S., Kimata K. (1990) Hyaluronic acid associated with the surfaces of cultured fibroblasts is linked to a serum-derived 85-kDa protein. *J Biol Chem* **265**: 5247–5257.

Yu Q., Banerjee S.D., Toole B.P. (1992) The role of hyaluronan-binding protein in assembly of pericellular matrices. *Devel Dynam* **193**: 145–151.

6 Matrix Proteoglycans

Amanda J. Fosang[1] and Timothy E. Hardingham[2]

[1] *Orthopaedic Molecular Biology Research Unit, Royal Children's Hospital, University of Melbourne, Department of Paediatrics, Parkville, Australia*

[2] *Wellcome Trust Centre for Cell Matrix Research, School of Biological Sciences, University of Manchester, Stopford Building, Oxford Rd, Manchester, UK.*

INTRODUCTION

The feature that distinguishes proteoglycans as a special subset of glycoproteins is the glycosaminoglycan chains, which are covalently attached to the core protein as a post-translational modification. They may be chondroitin sulphate, or its epimerised homologue dermatan sulphate, or keratan sulphate, heparan sulphate or heparin. A large number of protein cores to which glycosaminoglycan chains are attached have now been cloned and sequenced, but they contain no features that distinguish them as a family of closely related proteins. There have emerged amongst them however, some distinctive sub-families that contain protein motifs of related sequence, such as the aggrecan and the leucine-rich proteoglycan families. In this chapter we focus on the proteoglycans with chondroitin sulphate, dermatan sulphate of keratan sulphate chains as these are important components of the extracellular matrix. Those proteoglycans containing heparin/heparan sulphate chains are discussed elsewhere in this book.

PROTEOGLYCAN STRUCTURE

The repeating disaccharide units characteristic of the glycosaminoglycans, and the linkage groups that attach them to their core proteins are shown in Figures 6.1 and 6.2. The chemistry of glycosaminoglycan chain attachment and elongation has been known for some time and excellent reviews can be found elsewhere (Roden and Horowitz, 1978; Roden, 1980; Dorfman, 1981), however the information that dictates the location and the number of chains to be attached to core proteins is still not clear. The first step in addition of chondroitin sulphate chains to the core protein requires the activity of a xylosyltransferase

---

Extracellular Matrix, Volume 2, Molecular Components and Interactions, edited by Wayne D. Comper.

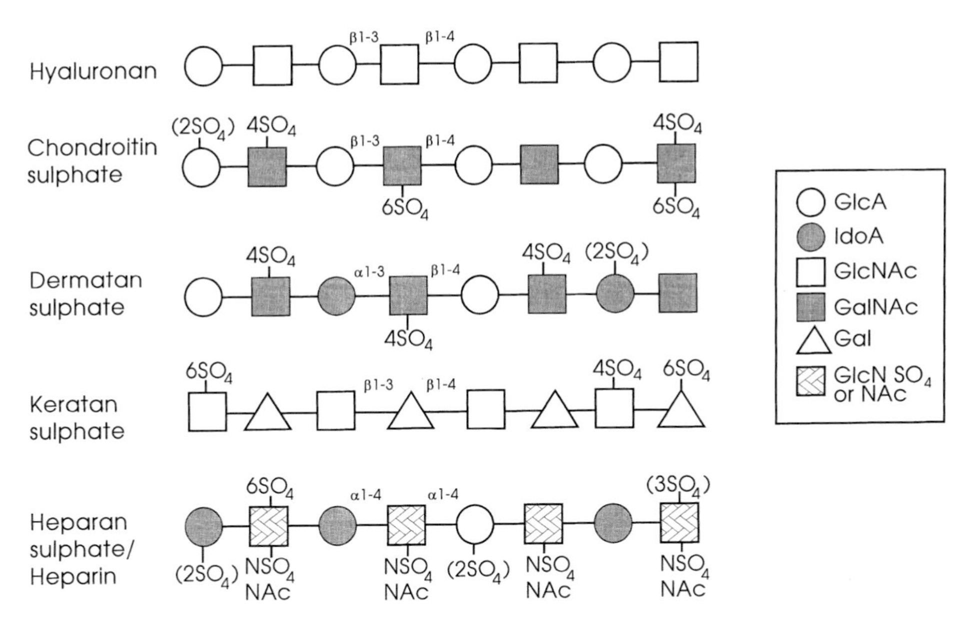

Figure 6.1 Repeating disaccharide structure of glycosaminoglycans. Hyaluronan is synthesised without a covalent link to any protein. Dermatan sulphate is formed from chondroitin sulphate by intracellular epimerisation of glucuronate (GlcA) to iduronate (IdoA). 2-sulphated hexuronate residues are more common in dermatan sulphate than in chondroitin sulphate. Heparin is a more extensively epimerised and sulfated form of heparan sulphate. Heparan sulphate frequently contains some chain segments with little or no epimerisation or sulphation. Heparin is synthesised only on mast cell granule proteoglycan whereas heparan sulphate is found on most cell surfaces and many other proteoglycans.

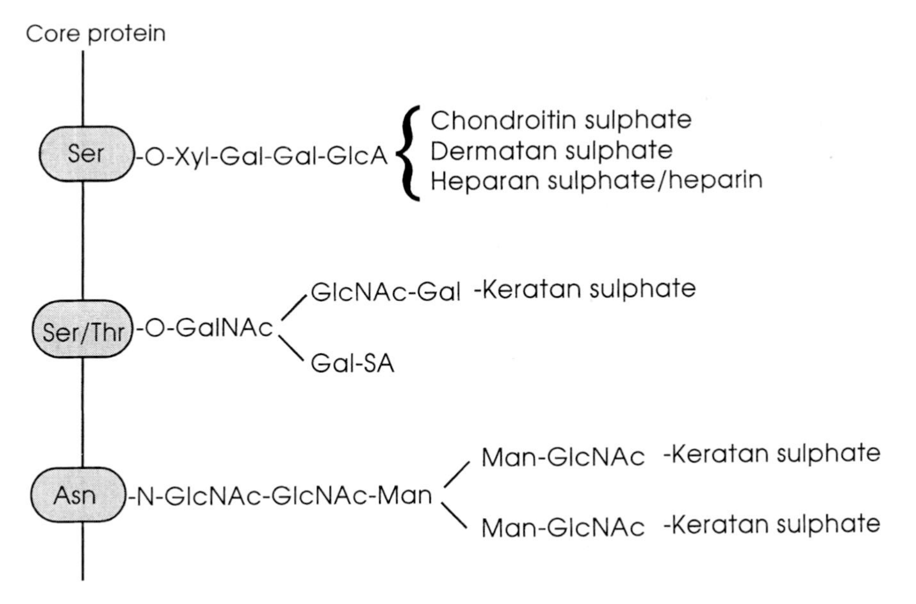

Figure 6.2 Linkages of glycosaminoglycans to protein. Chondroitin sulphate, dermatan sulphate and heparan sulphate (and heparin) all share a common neutral trisaccharide linkage sequence. The xylose may be phosphorylated, but the incidence of this is unclear. Keratan sulphate may be O-linked or N-linked as shown. The O-linked form is most common in cartilage and the N-linked form is present in cornea.

that attaches a xylose residue to an appropriate serine acceptor. The linkage region, -Xyl-Gal-Gal-GlcA-, is the same for chondroitin, dermatan and heparan sulphate chains, but clearly information encoded in the protein core determines which of these chains is extended onto the linkage region. In principle, the number and location of glycosaminoglycan chains attached to the core protein is specified by xylosyltransferase acceptor sequences within the core protein. However there is no single consensus sequence and no absolute rules that allow accurate predictions of which serine residues will be substituted. In proteoglycans many, but not all, Ser-Gly dipeptides are substituted, often in regions containing flanking acidic residues. A comparison of chondroitin sulphate substitution sites in three proteoglycans, decorin (PG40) (Brennan *et al.*, 1984; Krusius and Ruoslahti, 1986; Chopra *et al.*, 1985), rat yolk sac tumour proteoglycan (Oldberg *et al.*, 1981; Bourdon *et al.*, 1985) and the invariant chain of human class II MHC complex molecules (Sant *et al.*, 1985) has been made. The results identified a sequence in each of these proteoglycans consisting of acidic amino acids followed by the tetrapeptide Ser-Gly-Xaa-Gly, where Xaa is any amino acid (Bourdon *et al.*, 1987). Synthetic peptides containing this tetrapeptide sequence served as good substrates for xylosyltransferase, while peptides lacking the sequence were poor acceptors (Bourdon *et al.*, 1987). Thus, Ser-Gly-Xaa-Gly has been proposed as a general consensus for chondroitin sulphate attachment and while it suits many chondroitin sulphate acceptor sites in proteoglycans, the exceptions are numerous. Aggrecan for example appears to have a consensus sequence with a decapeptide repeat containing two Ser-Gly pairs separated by a dipeptide (Krueger *et al.*, 1990, Doege *et al.*, 1991; Upholt *et al.*, 1993). Very few of the serines available for chondroitin sulphate substitution in versican (Zimmermann and Ruoslahti, 1989), PG-M (Shinomura *et al.*, 1993) and PG-Lb proteoglycan (Shinomura and Kimata, 1992) are contained in sequences that are similar to either the tetrapeptide or the decapeptide consensus sequences. Another sequence, acidic-Gly-Ser-Gly-acidic, is prominent in human versican and has also been identified as the chondroitin sulphate attachment site in the $\alpha 2$ chain of chicken type IX collagen (McCormick *et al.*, 1987). Gly-Ser-Gly triplets and Gly-Ser pairs are more common in chick PG-M than Ser-Gly (Shinomura *et al.*, 1993).

CHONDROITIN/DERMATAN SULPHATE AND KERATAN SULPHATE CHAIN STRUCTURE

Repeating Disaccharide Structures

The glycosaminoglycans are long chain unbranched polysaccharides consisting of a repetitive sequence of disaccharide units (Hardingham and Fosang, 1992). In chondroitin sulphate the disaccharide unit contains glucuronate and N-acetylgalactosamine and has usually one sulphate group per disaccharide, which is predominantly either in the 4 or 6 position on N-acetylgalactosamine. A minority (<10%) of disaccharides are non-sulphated and more rarely there are disulphated disaccharides or even trisulphated disaccharides. Dermatan sulphate is a structural isomer of chondroitin sulphate in which some glucuronate is epimerised to iduronate. The epimerisation is enzyme catalysed and occurs in the completed polysaccharide chain during biosynthesis within the cell. The extent of epimerisation varies between none and greater than 90%. The presence of glucuronate and iduronate within the disaccharide repeats of dermatan sulphate gives a larger range of disaccharide structures and thus potentially more unique sequences within the chains than with chondroitin sulphate.

Keratan sulphate contains a different disaccharide repeat to that in chondroitin/dermatan sulphate, although there are some similarities in the structural backbone of the chain. Keratan sulphate contains a lactosamine repeat consisting of N-acetylgalactosamine and galactose. Polylactosamine chains are found attached to other proteins (Feizi, 1989), such as in ovarian mucins, but the distinguishing feature in keratan sulphate is that they are sulphated mainly with 6-0 sulphate on the galactose in monosulphated disaccharide units, and also with some 6-0 sulphate on the N-acetylgalactosamine in disulphated disaccharides. The structural similarity between keratan sulphate and chondroitin/dermatan sulphate derives from the similar linkages in the glucose-galactose units that form the backbone structure for both polysaccharide chains, although the presence of different positions of substitution with amine, uronate and sulphate groups makes their detailed structures quite distinctively different.

Linkages to Protein

Chondroitin/dermatan sulphate chains are linked to protein by a trisaccharide, galactose, galactose and xylose, and the xylose is attached to the hydroxyl of serine in a polypeptide sequence. This linkage is also similar to that of the heparin/heparan sulphate glycosaminoglycan chains. All glycosaminoglycan chains are synthesised by the stepwise addition of single sugars to the growing chain attached to protein, so the first event in chain synthesis is the transfer of xylose to serine in a protein sequence, followed by the transfer of the first and second galactose residues and subsequently the alternating addition of sugars of the disaccharide repeat unit. Evidence from studies of the biosynthesis of aggrecan suggests that the transfer of xylose may occur whilst the protein is within the RER of the cell (Schnyder *et al.*, 1987), but the major part of chain synthesis undoubtedly occurs on proteoglycans within the medial/trans elements of the Golgi (Ratcliffe *et al.*, 1985), where chain synthesis appears to be a very fast and efficient process. The trisaccharide gal-gal-xyl has been shown to contain attached to it both sulphate and phosphate groups on aggrecan synthesised in chondrosarcoma cells, but it is unclear how commonly this occurs in normal chondrocytes, or in the linkage regions of other proteoglycans.

This linkage is characterised as being alkali labile as at high pH there is a β elimination reaction that effectively cleaves the xylose-serine bond converting serine to dehydroalanine and releasing xylose, which remains attached to the chondroitin/dermatan sulphate chain and this survives intact and unmodified by the treatment with alkali.

Keratan sulphate chains are synthesised attached to two different types of linkage to protein one involving an O-linkage to serine or threonine and the other an N-linkage to asparagine. Both types of linkage are also found with some non-sulphated oligosaccharides of glycoproteins. They are distinguished from each other by their sensitivity to alkaline cleavage, as the O-linkage is labile and the N-linkage is stable. The oligosaccharide structure adjacent to the O-linkage contains N-acetylgalactosamine and the keratan sulphate chain extends from a 6-0-galactose attached to it. The linkage sugar also usually has a sialic acid-galactose disaccharide attached at position 3. It is thus very similar to the linkage structure of common O-linked oligosaccharide, such as those found on mucins. The oligosaccharide of the N-linked keratan sulphate is similar to that found in N-linked oligosaccharide of the complex rather than high mannose type. As with the O-linked keratan sulphate the chain extends from a 6-0-galactose attached to N-acetylgalactosamine, which in the corresponding N-linked oligosaccharide is frequently capped with a sialic acid. The structure of the N-linked oligosaccharide linkage region is such that keratan sulphate chains could extend from 2 equivalent branches (biantennary), but evidence so far suggests that only one branch forms keratan sulphate and the other is capped with sialic acid (Stuhlsatz *et al.*, 1989).

Epimerisation of Glucuronate to Iduronate

The presence of iduronate residues in a chain instead of glucuronate has a major effect on the conformation of the chain. Epimerisation of the carboxyl group at C5 of the hexuronate alters the stability of the predominant "chair" conformation ($^{1}C_{4}$) of glucuronate and makes a "skew boat" conformation ($^{2}S_{0}$) an energetically similar alternative (Casu *et al.*, 1993). It also creates a kink in the main axis of the polymer backbone as the linkage between sugars is no longer between only equatorial bonds. The chain may thus gain a kink at an iduronate residue which may flip between alternative conformations. Dermatan sulphate chains of varying glucuronate iduronate content have properties that are not apparent in chondroitin sulphate. In those with about equal proportions of glucuronate and iduronate there are properties of self association that are absent in chains with either very low or very high iduronate content (Fransson *et al.*, 1993). This together with other data implies that self association is favoured where there are mixed segments of alternating glucuronate and iduronate containing disaccharides, and less likely where there are extended blocks of either glucuronate or iduronate disaccharides alone. This property of self association may thus be selected by controlling the extent and the pattern of epimerisation of glucuronate to iduronate during biosynthesis and is another mechanism by which cells may modify the properties of proteoglycans to suit different biological needs. The conversion of glucuronate to iduronate during biosynthesis was found to be more efficient following 4–0 sulphation of the chain (Malmström *et al.*, 1993). Reverse epimerisation was also prevented by 2–0 sulphation of iduronate which may also be a factor contributing to the changes in chain properties.

Sequences in Chondroitin/Dermatan Sulphate Chains

There is now increasing evidence that glycosaminoglycan chains are not randomly sulphated and/or epimerised polymers, but they contain distinct patterns of chain sequence that may include some unique sequences with special functions (Hardingham and Fosang, 1992). So, although their structures are based on repeating disaccharide units, selective sulphation of chondroitin sulphate and keratan sulphate, together with the selective epimerisation of glucuronate to iduronate in dermatan sulphate (and heparan sulphate) produces chains with specific properties.

The identification of the unique sequence within heparin that binds to antithrombin III (Lindahl *et al.*, 1984) was the first example of the controlled expression of a specific sequence within a glycosaminoglycan chain and this has been extended to identify sequences of heparan sulphate that bind to basic fibroblast growth factor (Turnbull *et al.*, 1992; Walker *et al.*, 1994). For chondroitin sulphate, monoclonal antibodies that recognise specific sequences within chains have now revealed how the expression of these structures changes with cell differentiation and with tissue development (Sorrell *et al.*, 1988; Caterson *et al.*, 1990). Changes in the expression of epitopes in chondroitin sulphate were also detected in articular cartilage as part of the cellular responses occurring in models of experimental joint disease. These results suggest that the cellular synthesis of glycosaminoglycan chains is much more closely controlled than has previously been suspected and that much has yet to be understood about the significance of many well-documented changes in glycosaminoglycan compositions during tissue development and ageing and in pathology.

Methods of determining sequences in glycosaminoglycan chains have lagged far behind those for proteins and nucleic acids. However, some attempts have been made to tackle this problem. For dermatan sulphate chains the presence of iduronate residues has been exploited to provide sites for selective enzymatic degradation or chemical cleavage of the chain (Fransson *et al.*, 1993; Yoshida *et al.*, 1993). Chondroitinase ACII fails to cleave at iduronate residues, thus leaving blocks of iduronate disaccharides intact, whereas chondroitinase β only cleaves at iduronate residues, thus leaving glucuronate-rich chain segments. The iduronate residues lacking 2–0-sulphate groups are also sensitive to periodate cleavage, which distinguishes them from those containing 2–0-sulphate and from glucuronate. These selective methods of degradation combined with a procedure that labels the reducing end of the dermatan sulphate chain can be used to reveal the major structural features within dermatan sulphate chains and whether they are proximal or distal to the linkage to protein (Fransson *et al.*, 1993).

For chondroitin sulphate chains on aggrecan, digestion with chondroitinase ACII was used to progressively remove the distal parts of chains and permit the determination of their disaccharide composition separate from the undigested proximal part (Hardingham *et al.*, 1994). This revealed a trend in the distribution of sulphate groups which increased in abundance in the 6-position towards the linkage to protein. This was supported by the analysis with monoclonal antibodies for terminal 6–0 sulphated and 4–0 sulphated disaccharides. These approaches are useful in detecting average changes in composition along chondroitin/dermatan sulphate chains, but more selective methods are needed if unique but rare sequences are also to be detected and sequenced.

For corneal keratan sulphate careful fractionation and chemical analysis revealed a pattern of chain structure in which the region close to the protein linkage contained about 6 monosulphated disaccharides whereas the distal part of chains was mainly disulphated disaccharides (Stuhlsatz *et al.*, 1989). This model of chain structure where only the distal part of chains contained consecutive disulphated disaccharide sequences was also in agreement with the reactivity of chains with monoclonal antibodies. These react only with disulphated disaccharide sequences, but not with monosulphated sequences and only the longer chains were found to be strongly immunoreactive. This general pattern of keratan sulphate structure was also detected in keratan sulphate from cartilage which suggested that a similar pattern of sulphation was present on O-linked and N-linked keratan sulphate (Stuhlsatz *et al.*, 1989).

THE AGGRECAN FAMILY

Proteoglycans of the aggrecan family include the extracellular proteoglycans aggrecan, versican (Zimmermann and Ruoslahti, 1989) and its avian homologue PG-M (Shinomura *et al.*, 1993), neurocan (Rauch *et al.*, 1992), brevican (Yamada *et al.*, 1994) and the cell surface hyaluronan receptor CD44 (Goldstein *et al.*, 1989). They are modular proteoglycans containing combinations of structural motifs, such as epidermal growth factor-like domains, lectin-like domains, complement regulatory protein-like domains, immunoglobulin folds and proteoglycan tandem repeats (Perkins *et al.*, 1989) that are commonly found in other proteins (Figure 6.3). Several other proteins are related to this family of molecules, including link protein (Neame *et al.*, 1986; Deak *et al.*, 1986), BEHAB (Jaworski *et al.*, 1994) and TSG-6 (Lee *et al.*, 1992) and contain some of the same highly conserved structural motifs but lack glycosaminoglycan side chains (Figure 6.3). Aggrecan has

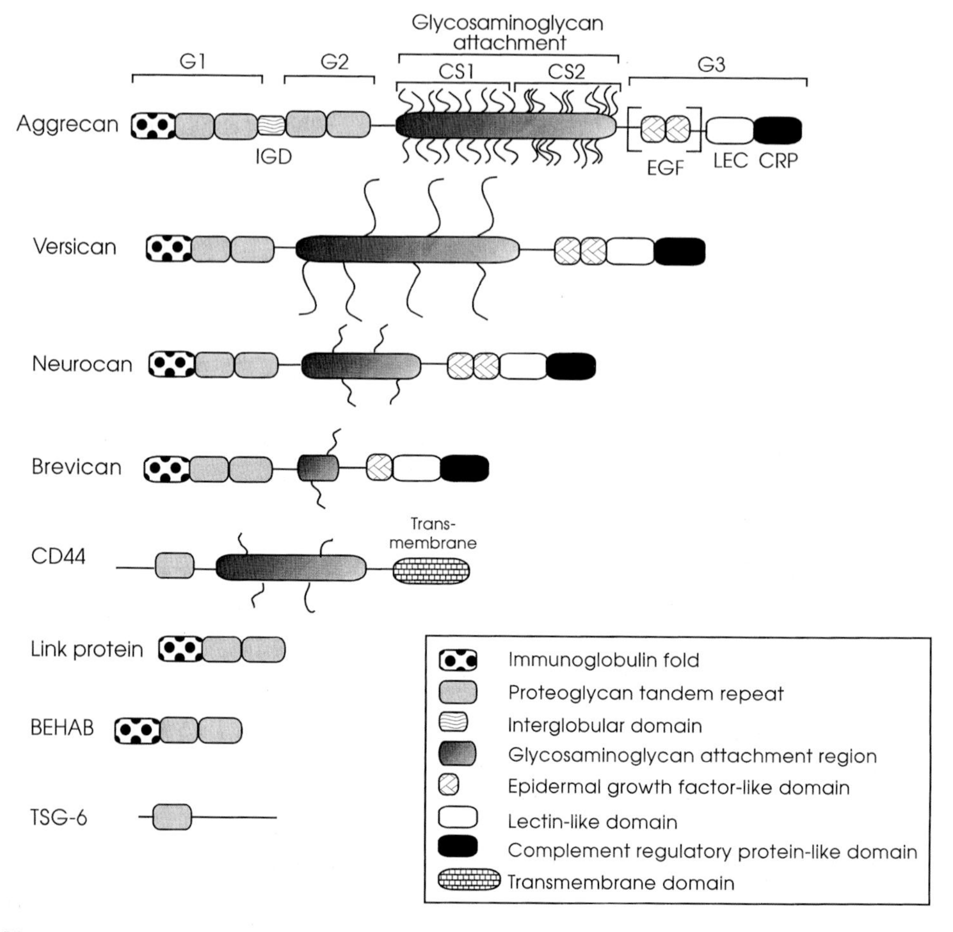

Figure 6.3 Schematic representation of the members of the aggrecan family. The modular arrangement of human aggrecan, versican, neurocan, brevican, CD44, link protein, BEHAB and TSG-6 is shown. The square brackets surround variably spliced domains.

served as the prototype for this family of molecules and its structural and functional properties have been well characterised, in contrast with versican, neurocan and brevican whose functional properties have not been rigorously tested. Aggrecan is expressed in cartilage where it is the most abundantly expressed proteoglycan and there is evidence for aggrecan expression in tendon (Vogel *et al.*, 1994). Neurocan and brevican are expressed exclusively in brain tissue while the distribution of versican and CD44 is more widespread.

AGGRECAN

Aggrecan is found as huge multimolecular aggregates comprising numerous monomers non-covalently bound to hyaluronan (HA). A small glycoprotein, link protein which is homologous with the N-terminal of aggrecan, helps stabilise aggregate formation. Almost 90% of aggrecan mass is substituted glycosaminoglycans which are mostly chondroitin sulphate chains, but also keratan sulphate chains and N- and O-linked oligosaccharides. The complete cDNA sequences of human (Doege *et al.*, 1991), rat (Doege *et al.*, 1987), mouse (Walcz *et al.*, 1994) and chicken (Chandrasekaran and Tanzer, 1992) aggrecan are known and partial sequences for pig (Barry *et al.*, 1992) and bovine (Wiedemann *et al.*, 1984; Oldberg *et al.*, 1987; Antonsson *et al.*, 1989) have also been reported. Human aggrecan has been mapped on chromosome band 15q26 (Korenberg *et al.*, 1993).

Immunoglobulin fold

Aggrecan monomers contain two extended regions which carry the bulk of the glycosylation and three globular domains, G1 and G2 at the N-terminus, and G3 at the C-terminus of the core protein. Based on the N-terminal primary sequences of G1 domain and link protein and compared with the known crystal structures for immunoglobulin constant and variable region domains, secondary structure predictions (Perkins *et al.*, 1989) have identified a pattern of β-sheet structure found in immunoglobulin folds. Both aggrecan and link protein are therefore members of the immunoglobulin superfamily (Bonnet *et al.*, 1986; Perkins *et al.*, 1989). The immunoglobulin-fold contains about 100 amino acids that is predicted to fold into two β sheets in a sandwich conformation stabilised by a conserved disulphide bond. It is encoded in a single exon. Almost all of the proteins which contain immunoglobulin-related structures are cell surface molecules involved in cell recognition, cell adhesion or immune function (Williams, 1985). Their role in cell recognition is emphasised by the fact that many of these molecules interact with other members of the immunoglobulin superfamily, and indeed the interaction between aggrecan G1 domain and link protein is mediated through the immunoglobulin-fold (Grover and Roughley, 1994). It is not surprising to find immunoglobulin-like structures on proteoglycans that have no immune function and are not associated with cell surfaces. The absence of lymphocyte-based immune systems in invertebrates suggests that immunoglobulin-like structures first evolved on cells that had no immune function (Williams, 1985). Thus, immunoglobulin-containing structures present on extracellular proteins such as link protein, aggrecan, and the related proteoglycans versican, neurocan and brevican or cell surface adhesion molecules such as the integrins ICAM-1, ICAM-2 and VCAM-1, may be more primitive than their immunological counterparts with which they are classified.

Proteoglycan tandem repeats

One feature of all the members of the aggrecan family is their ability to bind specifically to hyaluronan. This interaction is non-covalent and is thought to occur through a structural motif called the proteoglycan tandem repeat (PTR) which comprises two homologous loops, or a single loop in the case of CD44 and TSG-6. The C-terminal portion of G1 and link protein, and all of the aggrecan G2 domain comprise the proteoglycan tandem repeat, but while G1 and link protein avidly bind hyaluronan forming trimeric complexes that cannot be dissociated under physiological conditions (Faltz *et al.*, 1979), the G2 domain shows no properties of interaction with hyaluronan (Fosang and Hardingham, 1989). G1 and link protein bind to decasaccharide units on hyaluronan polymers (Hardingham and Muir, 1973; Hascall and Heinegård, 1974) while CD44 with its single loop structure has specificity for a hexasaccharide unit. Thus the two loop arrangement that is found in most members of the aggrecan family is not essential for hyaluronan binding, and indeed recombinant single loops of link protein expressed in a baculovirus system are not deficient in hyaluronan binding (Grover and Roughley, 1994).

The presence of conserved basic sequences in the PTR loops of aggrecan G1 and link protein domains suggested that these regions may interact with the negatively charged hexuronate groups on the hyaluronan polymer. Competitive inhibition experiments using synthetic peptides with sequences homologous with conserved basic regions at the apex of the tandem repeat loops of link protein suggested that these were the sites responsible for hyaluronan binding (Goetinck *et al.*, 1987) however subsequent studies revealed that these interactions were non-specific (Horita *et al.*, 1994). Recent investigations of hyaluronan binding motifs in the hyaluronan receptor RHAMM (Yang *et al.*, 1993) have identified a $B(X_7)B$ motif, where B is either arginine or lysine and X_7 contains no acidic residues and at least one basic amino acid (Yang *et al.*, 1994). This motif is present in the RHAMM protein which has no structural or sequence homologies with members of the aggrecan family, and it is also present within all the proteoglycan tandem repeats that are known to bind hyaluronan. These findings suggest that hyaluronan binding may be dependent upon a charge motif rather than a specific amino acid sequence, and that $B(X_7)B$ may represent a minimum sequence required for hyaluronan binding. Brevican, which contains the same immunoglobulin fold and tandem repeat structures characteristic of this family does not contain a $B(X_7)B$ sequence in its PTR loops, but has two $B(X_7)B$ motifs in its immunoglobulin fold (Yamada *et al.*, 1994). So far brevican has not been tested for its ability to bind hyaluronan. It is also significant that the G2 domain of aggrecan which is unable to bind hyaluronan (Fosang and Hardingham, 1989) lacks the $B(X_7)B$ motif.

Interglobular domain

The short extended region separating the G1 and G2 domains of aggrecan is known as the interglobular domain, or IGD. Rotary shadowing electron microscopy determined that the IGD was of constant length (25nm) and relatively inflexible (Paulsson *et al.*, 1987; Dennis *et al.*, 1990). This apparent stiffness of the IGD may be due to the density of keratan sulphate substitutions (Barry *et al.*, 1992; Fosang *et al.*, 1992) which are greater in the IGD than on the adjoining globular domains. The functional, rather than the structural properties of the IGD make it interesting. Recently it has attracted considerable attention as the site for proteolytic attack on aggrecan during extracellular turnover and cartilage degradation. This is a key site for proteolytic cleavage because it separates the major glycosaminoglycan-bearing portion of aggrecan from the G1 domain that anchors it in the matrix. Two predominant cleavage sites have been identified in the IGD. One is the matrix metalloproteinase site at $N_{341}\downarrow F_{342}$ (Fosang *et al.*, 1991, 1992; Flannery *et al.*, 1992, Fosang *et al.*, 1993) which is cleaved by all members of the matrix metalloproteinase family tested so far. The other is the site at $E_{373}\downarrow A_{374}$ which has been identified in explant culture (Sandy *et al.*, 1991; Ilic *et al.*, 1992; Loulakis *et al.*, 1992) and in human synovial fluids (Sandy *et al.*, 1992; Lohmander *et al.*, 1993). The enzyme responsible for this cleavage has been named "aggrecanase", and although a member of the metalloproteinase family, MMP-8, exhibits aggrecanase activity *in vitro* (Fosang *et al.*, 1994). This enzyme is not expressed by chondrocytes (Dudhia and Hardingham, unpublished results) and the identity of the aggrecanase enzyme present in cartilage remains unknown. The occurrence of C-terminal $DIPEN_{341}$ fragments in cartilage matrix (Flannery *et al.*, 1992; Bayne *et al.*, 1994) and large N-terminal $A_{374}RGSVI$ fragments in joint fluids (Sandy *et al.*, 1992; Lohmander *et al.*, 1993) suggests that cleavage at both the MMP site and the aggrecanase site are involved in aggrecan degradation *in vivo*.

Aggrecanase activity has been investigated in bovine cartilage explant cultures by sequence analysis of aggrecan fragments released into the medium. The results show that four discrete fragments with similar patterns of amino acid sequence around the cleavage sites (Table 6.1) are released; one derived from cleavage in the IGD and three from cleavage in the CS2 domain (Ilic *et al.*, 1992; Loulakis *et al.*, 1992). This similar pattern of amino acids surrounding the cleavage site may represent a consensus sequence recognised by "aggrecanase". Sequences with weak similarity to the aggrecanase consensus sequence can be found in versican, brevican and neurocan at locations comparable to the IGD aggrecanase site (Sandy *et al.*, 1991) (Table 6.1). Specific sites of proteolysis with important roles in matrix turnover may therefore be encoded in this family of proteoglycans.

Table 6.1 Aggrecanase cleavage sites

Proteoglycan	*Cleavage Site*
Aggrecan IGD	$TEGE_{373}$ ARGS
Aggrecan CS2	$KEEE_{1714}$ GLGS
	$TAQE_{1819}$ AGEG
	$VSQE_{1919}$ LGQR
Brevican	$VESE_{400}$ SRGA
Versican	$EAAE_{441}$ ARRG
Neurocan	$DEGE_{377}$ IVSA
	$RLGE_{395}$ QEVI
	$GEQE_{397}$ VITP

Cleavage sites in the bovine aggrecan IGD and CS2 domains have been determined experimentally (Ilic, *et al.*, 1992; Loulakis, *et al.*, 1992). Cleavage sites in human brevican, versican and neurocan are postulated, based on their similarity with the aggrecanase consensus sequence.

G3 Domain

The C-terminal G3 domain of aggrecan comprises another composite of structural motifs from different protein families. These include an alternatively spliced complement regulatory domain at the extreme C-terminal, an adjacent carbohydrate binding domain homologous with type-C animal lectins and an N-terminal epidermal growth factor (EGF)-like domain that is also subject to alternative splicing (Baldwin *et al.*, 1989; Doege *et al.*, 1991; Li *et al.*, 1993). The EGF-like domain (EGF1) that was first reported for human aggrecan (Baldwin *et al.*, 1989) is expressed in approximately 25%–30% of human aggrecan transcripts but is not expressed in other species. In contrast, a second EGF domain, EGF2 is present in human, mouse, rat, cow and dog where the level of expression between species is a constant 4–8% and the sequences are 87–97% identical (Fulop *et al.*, 1993). In humans variable splicing gives rise to about one quarter of aggrecan molecules containing an EGF1 module and very few with EGF2 or both EGF1 and EGF2 modules (Fulop *et al.*, 1993). The majority of aggrecan molecules do not contain any EGF-like domains. This alternative splicing may be unique to aggrecan. There is no evidence for alternative splicing of EGF domains elsewhere despite their wide occurrence in many functionally unrelated proteins, nor is there evidence for alternatively spliced EGF domains in versican (Krusius *et al.*, 1987; Zimmermann and Ruoslahti, 1989) and neurocan (Rauch *et al.*, 1992) which each express two EGF-like modules. It remains to be seen whether brevican (Yamada *et al.*, 1994), which contains a single EGF module and is most closely related to neurocan, may also exhibit splicing variants.

Epidermal growth factor-like sequences present as modular components of larger proteins have been shown to possess growth factor activity. For example, EGF-like repeats of laminin enhance growth of epidermal cells in culture (Panayotou *et al.*, 1989), and similar repeats in the *Notch* protein of *Drosophila melanogaster* (Vassin *et al.*, 1987), the *lin-12* protein of nematode *Caenorhabditis elegans* (Greenwald, 1985), and the sea urchin *Strongylocentrotus purpuratus* (Hursh *et al.*, 1987) are thought to regulate the development of certain cell lineages. Thus, proteoglycan EGF domains may have a role in modulating the proliferative and metabolic activities of cells with the differently spliced products operating in a feedback mechanism. Alternatively, the EGF domain may simply provide an additional site for binding of other matrix proteins (Siegelman *et al.*, 1990). The involvement of EGF sequences in protein-protein interactions is demonstrated not only in EGF binding to its receptor, but also in thrombomodulin where it provides the primary thrombin-binding site (Kurosawa *et al.*, 1988), and in plasminogen activator receptor binding (Appella *et al.*, 1987).

The tridomain structure of aggrecan G3 identifies this molecule as a member of the selectin family (Springer and Lasky, 1991) whose properties are based largely on the lectin component that is thought to mediate cell-cell recognition events occurring in extravasation and inflammation. The selectins such as mel-14 found on the surfaces of leucocytes and endothelial cells, comprise one copy each of the lectin and EGF domains, and specific sequences in both of these domains are involved in cell adhesion (Siegelman *et al.*, 1990). The number of complement regulatory protein-like domains varies among the selectins, ranging from two for the lymphocyte homing receptor to nine for the inducible platelet and endothelial cell surface protein GMP-140 but only one is present in the aggrecan family of proteoglycans. No selectin-like properties of cell adhesion have yet been demonstrated for members of the aggrecan family and there is little information describing interactions of the subdomains of G3 with other cellular or extracellular components. The lectin domain of rat aggrecan translated in a reticulocyte lysate system shows a low affinity selectivity for fucose and galactose ligands (Halberg *et al.*, 1988), raising the possibility that the hydroxylysine-linked galactose substituents of type II collagen may be ligands for G3 in cartilage.

Rotary shadowing electron microscopy of aggrecan showed that more than half of the monomers present in cartilage extracts did not contain G3 domains (Paulsson *et al.*, 1987; Dennis *et al.*, 1990) suggesting that G3 may not have a role at all in the extracellular matrix. Nanomelic chondrocytes in the lethal chicken mutation *nanomelia* produce truncated aggrecan molecules that lack G3 domains and are not secreted from the cell (O'Donnell *et al.*, 1988). Aggrecan mRNA in nanomelic chondrocytes is of normal size and experiments in which ER and Golgi compartments were fused showed that the truncated aggrecan was a competent substrate for xylosylation and glycosaminoglycan chain elongation (Vertel *et al.*, 1994). However the mutation at amino acid 1523 in the CS2 domain introduced a premature stop codon that resulted in synthesis of a shortened aggrecan that was missing the G3 domain and part of the CS domain (Li *et al.*, 1993). The effect of this mutation was to prevent translocation of the truncated aggrecan from the ER to the Golgi (Vertel *et al.*, 1994) and suggests that G3 may have an essential role in intracellular quality control and trafficking.

The lethal chicken mutation *nanomelia* shows a severely defective skeletal phenotype in which extracellular aggrecan is deficient. A similar phenotype, also lethal in the homozygote, is seen in the *cmd* (cartilage matrix deficiency) mouse which expresses normal levels of the cartilage-specific type II collagen but fails to express aggrecan (Kimata *et al.*, 1981). The mutation in the *cmd* mouse is due to a 7bp deletion in the first tandem repeat loop of G1, causing a frame shift which leads to a stop codon further along the G1 domain (Watanabe *et al.*, 1994). Whether the *cmd* transcript is translated and secreted, or whether it is degraded intracellularly because it lacks a G3 domain is not known. The *nanomelia* and *cmd* mutations indicate that aggrecan has a crucial role in matrix formation and cartilage development. Mutations in versican, or neurocan and brevican may also give rise to inheritable defects in mesenchymal, or brain tissues, respectively.

Glycosaminoglycan Attachment Regions

The glycosaminoglycan attachment region between the G2 and G3 domains is the least well conserved part of aggrecan and there are significant differences not only in the size and amino acid sequence, but also the nature of the substitutions in this region; for example rodent aggrecans contain little if any keratan sulphate compared with other species. Adjacent, and C-terminal to G2, is a keratan sulphate-rich domain. This domain comprises hexapeptide repeats of which there are twenty-three in bovine (Antonsson *et al.*, 1989), twelve in human (Doege *et al.*, 1991) and only several poorly conserved repeats in rat and chicken (Upholt *et al.*, 1993) aggrecan. In addition to the keratan sulphate-rich region, keratan sulphate chains are sparsely substituted on the G1 and G2 domains and are more predominant on the extended IGD and CS domains. There is no evidence for glycosaminoglycan attachment in the G3 domain.

The chondroitin sulphate attachment region comprises CS1 and CS2 regions (Figure 6.3) made up of repeating sequences. The CS1 domain repeats contain about twenty amino acids of which there are fifteen in the rat (Doege *et al.*, 1987) and twenty-nine in human aggrecan (Doege *et al.*, 1991). The CS2 domain contains fewer repeats of about 100 amino acids and these are more highly variable. Chicken aggrecan has an additional domain not found in other species containing a different type of twenty amino acid repeat (Chandrasekaran and Tanzer, 1992; Upholt *et al.*, 1993). The glycosaminoglycan attachment regions of versican, neurocan and brevican are completely nonhomologous with aggrecan. Versican and brevican each have an unusual cluster of glutamic acid residues that is also found on a chondroitin sulphate proteoglycan, β-amyloid precursor protein of Alzheimer's disease (Shioi *et al.*, 1992, 1993), but the significance of this acidic cluster is not known.

CD44

The cell surface hyaluronan receptor is a member of the aggrecan family because it has a single loop structure that is equivalent to half a proteoglycan repeat (Stamenkovic *et al.*, 1989). Previously described as Hermes antigen, Pgp-1, H-CAM, extracellular matrix receptor III, Hutch-1 and GP90, the CD44s have emerged as a broad class of integral membrane glycoproteins expressed on the surface of many cell types including haemopoietic and epithelial cells, fibroblasts and brain tissue (Culty *et al.*, 1990). The mature protein has a cytoplasmic tail, a transmembrane domain and an ectodomain that contains the glycosaminoglycan attachment sites and the hyaluronan binding sites. CD44, as well as being expressed in multiple alternatively spliced isoforms, is also expressed in both proteoglycan and non-proteoglycan forms. CD44 proteoglycan may carry chondroitin sulphate, heparan sulphate or both types of glycosaminoglycan side chains in addition to other O- and N-linked sugars, including sulphated N-linked oligosaccharides and O-linked lactosamine chains, and these may be substituted on more than one type of gene transcript. The alternative splicing of CD44 occurs in the cytoplasmic tail (exons 18 and 19) giving rise to short (Goldstein *et al.*, 1989) and long (Stamenkovic *et al.*, 1989) tail forms, and more extensively in the membrane proximal extracellular domain, involving exons 5 to 14 (Screaton *et al.*, 1992). This capacity for alternative splicing (12 of 19 exons) is remarkable and provides enormous potential for functional diversity.

The variations in CD44 isoforms and glycosaminoglycan composition appear to be tissue-specific and may be related to its proposed role in cell recognition and cell adhesion, for example mediating recognition between lymphocytes and high endothelial venules in homing to lymphoid tissues (Jalkanen *et al.*, 1988), in fibroblast adhesion and migration (Jacobson *et al.*, 1984) and T cell activation (Huet *et al.*, 1989; Shimizu *et al.*, 1989; Denning *et al.*, 1990). In addition, CD44 binds the extracellular matrix components collagen (Carter and Wayner, 1988) and fibronectin (Jalkanen and Jalkanen, 1992) and it has been identified as a hyaluronan receptor in several cell systems (Culty *et al.*, 1990; Stamenkovic *et al.*, 1991; Murakami *et al.*, 1991). However not all CD44-expressing cells bind hyaluronan, suggesting that different CD44 isoforms may have different affinities for hyaluronan. Investigations of hyaluronan binding affinity in several epithelial and transformed cells (Baldwin *et al.*, 1989; Stamenkovic *et al.*, 1991; He *et al.*, 1992; Hyman *et al.*, 1994; Dougherty *et al.*, 1994) have lead to studies which have identified regions in

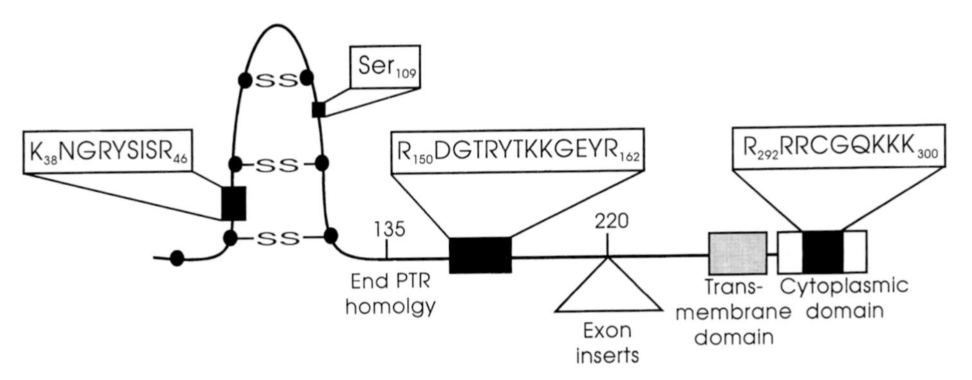

Figure 6.4 Potential sites involved in CD44 interactions with hyaluronan. Schematic diagram showing the approximate locations and the amino acid sequences of regions that may be involved in CD44-hyaluronan interactions (Lesley, *et al.*, 1992; Peach, *et al.*, 1993; Yang, *et al.*, 1994; Dougherty, *et al.*, 1994).

both the extracellular (Peach *et al.*, 1993; Dougherty *et al.*, 1994) and cytoplasmic domains (Lesley *et al.*, 1992) that may be important in hyaluronan binding (Figure 6.4). Two regions are located in the N-terminal domain and conform with the $B(X_7)B$ consensus sequence for hyaluronan binding. A third region which also has the $B(X_7)B$ motif is located on the cytodomain (Yang *et al.*, 1994). A fourth site, Ser_{109} in the PTR loop, is highly conserved and exchange of this residue for Tyr in the CD44R1 leukemia cell line eliminates hyaluronan binding (Dougherty *et al.*, 1994) although the pattern of charge and amino acid sequence surrounding Ser_{109} do not predict that this region would be involved in hyaluronan binding.

LEUCINE-RICH PROTEOGLYCANS

Members of the leucine-rich family of proteoglycans are found in most tissues of the body. They are characterised by containing in their proteins repeating leucine-rich motifs that are similar to a consensus sequence identified in a broad family of proteins ranging from mammalian RNAse inhibitor and FSH receptor, through drosophila Toll and chaoptin proteins, to yeast adenylate cyclase (Kobe and Deisenhofer, 1994). All family members contain multiple leucine rich repeats which vary in number from 3 to 30. The structure of the first member to be crystallised, RNAse inhibitor, showed the leucine-rich motif to form a β sheet/α helix structure, which packs with the β sheets parallel, when several are present, to give a non-globular shape that may facilitate protein-protein interaction (Kobe and Deisenhofer, 1993). Decorin and biglycan were the first leucine-rich proteoglycans to be cloned (Krusius and Ruoslahti, 1986; Fisher *et al.*, 1989), but other members now include fibromodulin (Oldberg *et al.*, 1989) and lumican (Blochberger *et al.*, 1992b) and in the chicken growth plate, proteoglycan Lb (Shinomura and Kimata, 1992) (Table 6.2). With the exception of proteoglycan Lb they fall into a single sub-family of leucine-rich repeat proteins that show some clear general sequence similarities, but proteoglycan Lb contains fewer repeats and has different cysteine positions than the other proteoglycans and it is more closely related to osteoinductive factor. Decorin and biglycan are present in the extracellular matrices of most tissues, although as noted below they have different distribu-

Table 6.2 Leucine-rich proteoglycans and proteins present in cartilage matrix

Proteoglycan or Protein	*Number of CS/DS Attachment Sites*	*Number of Potential N-Linked Glycosylation Sites*	*Sulphated Tyrosine Region*	*Fibrillar Collagen Binding*	*TGF-β Binding*	*Non-glycanated Forms*	*Esitmated Number of Leucine-rich repeats*[a]
Biglycan	2	none	No	No	Yes	Yes	8
Decorin	1	2[d]	No	Yes	Yes	Yes	10
Fibromodulin	none	4[e]	Yes	Yes	Yes	Yes	11
Lumican	none	4[f]	No	Yes	?	Yes	12
Proteoglycan-Lb	1	none	No	No	?	Yes	6
Chondroadherin[b]	none	none	?	?	?	Yes	10
58 kDa cartilage matrix protein[c]	none	none	?	?	?	Yes	?

[a] Taken from (Kobe and Deisenhofer, 1994)
[b] Neame *et al.*, 1994
[c] Bengtsson *et al.*, 1994
[d] Blochberger *et al.*, 1992a
[e] Oldberg *et al.*, 1989; Plaas *et al.*, 1990
[f] Funderburgh *et al.*, 1993

tions. Fibromodulin has a more restricted distribution. It is abundant in cartilage, tendon and sclera and less abundant in skin and bone. Lumican, as its name implies, is found in cornea, but with some expression in muscle and intestine (Blochberger *et al.*, 1992b).

Other non-proteoglycan members of the leucine-rich family have also been identified in cartilage (Table 6.2). These include a 38 kDa protein (chondroadherin) with 10 leucine-rich repeats (Neame *et al.*, 1994), which has been shown to mediate chondrocyte attachment to substratum, although not to facilitate cell spreading. Preliminary sequencing of a 58 kDA cartilage matrix protein has also identified leucine-rich repeat sequences (Bengtsson *et al.*, 1994) which are similar to those of fibromodulin.

Decorin and Biglycan

Decorin and biglycan, known for many years as small DS-PG II and I, or PG-S2 and PG-S1, are commonly found in most connective tissues. Decorin contains a single site for glycosaminoglycan chain attachment at a serine that is 4 amino acids from the N-terminus and biglycan has 2 sites, both also close to the N-terminus. These sites are outside the leucine-rich repeat region. The glycosaminoglycan chains attached to the proteoglycan during biosynthesis are of the chondroitin sulphate/dermatan sulphate type and their structure varies with the tissue and cell type in which the proteoglycan is expressed. For example, in bone the chains are usually free of iduronate (epimerised from glucuronate), but in skin the chains are of high iduronate content. The length of chains also varies between 10 kDa to 40 kDa in different tissues.

Inspite of the close structural relationship between decorin and biglycan there is much to suggest that they fulfil quite different functions. Immunolocalisation of decorin and biglycan during embryonic development (Bianco *et al.*, 1990) shows that although they are frequently expressed in the same tissue they have contrasting distributions. Decorin is most abundant in matrix, whereas biglycan is mainly pericellular; where cells strongly express decorin, they poorly express biglycan, and vice versa. This suggests that decorin and biglycan differ in the intermolecular interactions that determine their tissue distribution and that their expression by cells is also controlled quite independently of each other.

The properties of decorin have been investigated in more detail than biglycan and the other leucine-rich proteoglycans and it is reported to interact with several other matrix proteins through its protein core. Decorin (and fibromodulin) have been shown to inhibit collagen (type I) fibrillogenesis in an in-vitro assay (Vogel *et al.*, 1984; Brown and Vogel, 1989) and as decorin has also been localised bound to specific sites on fibrillar collagen (Scott and Orford, 1981) it has been suggested that it may therefore have a role in regulating collagen fibrillogenesis. This interaction with collagen is a property of the protein core as removal of the glycosaminoglycan chain has no inhibitory effect (Hedbom and Heinegård, 1989, 1993). In contrast biglycan appears to lack binding to fibrillar collagen (Brown and Vogel, 1989), which may go some way to explain their quite different tissue distribution. There is however some evidence to suggest that under physiological conditions, with low phosphate ion concentrations, other proteoglycans including biglycan can bind to collagen, but this is via their glycosaminoglycan chains and not the protein core (Pogány *et al.*, 1994).

Ultrastructural studies using electron dense dyes (Scott and Orford, 1981; Scott, 1988) showed that small proteoglycans were bound at regular sites on fibrillar collagen identified within gap regions near "d" and "e" bands and decorin was identified at this site with specific antibodies (Pringle and Dodd, 1990). Experiments with carefully non-denatured proteoglycans and collagen fibrils showed limited high affinity sites on fibrils which were separate and independent for decorin and fibromodulin (Hedbom and Heinegård, 1993). This supported ultrastructural evidence that keratan sulphate containing proteoglycans bound at "a" and "c" bands in the gap region separately from chondroitin/dermatan sulphate proteoglycans. There is no binding of these proteoglycans to denatured collagen.

Decorin has also been shown to interact via its protein core with type VI collagen (Bidanset *et al.*, 1992). This is a short chain collagen that forms beaded filaments. Under the conditions tested, with collagens coated on plastic in microtitre wells, decorin showed little binding to fibrillar collagens I, II and III, although the binding to type VI was high affinity and quite specific. Decorin was also able to compete with binding between type II collagen and type VI collagen. The complement protein C1q was also shown to be bound by decorin, and this was suggested to be close to, but not within the collagen sequences that form stalk-like segments in the structure of this hexameric protein. Decorin was further shown to inhibit complement activation through C1, but this required the glycosaminoglycan chain in addition to the protein core (Krumdieck *et al.*, 1992).

Decorin has also been shown to bind via its protein core to fibronectin and this binding was partially inhibitable by a decorin derived peptide NKISK (Schmidt *et al.*, 1991). Decorin interacts with the cell-binding domain of fibronectin and it can inhibit fibronectin mediated cell adhesions by fibroblasts (Winnemoller *et al.*, 1991). Decorin is also reported to bind to thrombospondin (Schmidt *et al.*, 1990) and can inhibit cell attachment to a thrombospondin coated surface (Winnemoller *et al.*, 1992). It is also taken up avidly by different mesenchymal cells by a mechanism of receptor-mediated endocytosis and two proteins of 26 and 51 kDa have been identified at the cell surface and in endosomes that bind to the decorin core protein (Hausser *et al.*, 1989). The uptake is inhibited by heparin, but heparin is not itself translocated to lysosomes as is the decorin (Hausser *et al.*, 1993). The effect of heparin appears to result from competition with core protein binding to its receptor, but involvement of cell surface heparan sulphate in receptor activity is also suggested by experiments in which its removal leads to greatly increased endocytosis and cells high in heparan sulphate have lower rates of decorin uptake (Hausser *et al.*, 1993). This specific uptake of decorin also appears to be shared with biglycan and the same 26 kDa and 51 kDa proteins are involved (Hausser *et al.*, 1992).

Decorin has an interesting interaction with transforming growth factor β which has lead to the proposal that it might have an important role in regulating TGFβ action (Yamaguchi *et al.*, 1990). TGFβ is a pleiotropic growth factor (Sporn *et al.*, 1987). Its effect on cells is to selectively up-regulate the expression of some genes and down-regulate others, and these include the genes of extracellular matrix proteins and the proteinases and inhibitors that control matrix turnover. The specific responses to TGFβ differ in different cell types and are also greatly influenced by signals from other hormones, growth factor, cytokines, and also by physical/mechanical signals to the cell from the matrix that surrounds it. The initial observation of the involvement of decorin with the action of TGFβ was from an experiment in which recombinant decorin was expressed at high level in CHO cells (Yamaguchi *et al.*, 1990). This resulted in an inhibition of growth and a change in cell morphology, which was attributed to decorin binding to TGFβ and neutralising its action as a potent growth factor for CHO cells. Experiments supporting this interpretation showed that TGFβ dimer bound tightly to decorin on an affinity matrix and required 6 M urea for its dissociation. Biglycan was also suggested to share this property of interaction with TGFβ. This has now been extended to show that decorin, biglycan and fibromodulin bind all isoforms of TGFβ (β1, β2 and β3), but fibromodulin was less effective than decorin and biglycan in binding to TGFβ3 (Hildebrand *et al.*, 1994). High affinity sites with Kd 1–20 nM were detected as well as sites of lower affinity (Kd 50–200 nM) and decorin and biglycan with glycosaminoglycan chains intact were less effective in binding than deglycosylated or non-glycosylated core proteins. Recombinant TGFβ precursor (the latent form of TGFβ) bound weakly to fibromodulin and not to decorin or biglycan. The main role of these proteoglycans would thus be in binding active TGFβ rather than the inactive latent form. TGFβ is abundantly present in connective tissue matrices, such as bone and cartilage, and it may be held there by interaction with the collagen-bound proteoglycan decorin (and fibromodulin).

As TGFβ largely stimulates matrix production by mesenchymal cells (Sporn *et al.*, 1987) a matrix-store of this factor that is released at times of matrix damage could promote an early cellular response for its repair and renewal. TGFβ is also directly involved in the control of decorin and biglycan expression by cells. In fibroblasts biglycan expression is increased in response to TGFβ (Romaris *et al.*, 1991; Kahari *et al.*, 1991), but in mesangial cells both decorin and biglycan synthesis is greatly increased (Border and Ruoslahti, 1991). This was thus proposed as a potential feedback control of TGFβ action, as increased synthesis and secretion of TGFβ binding-proteoglycans into the matrix could bind the growth factor and limit its further action on cells (Ruoslahti and Yamaguchi, 1991). Local overproduction of TGFβ has been linked to excess tissue fibrosis such as in bleomycin induced lung fibrosis (Westergren-Thorsson *et al.*, 1993) and in experimental glomerulonephritis (Border and Ruoslahti, 1992). Decorin was shown, when given by injection, to block the fibrosis associated with experimental glomerulonephritis in rats *in vivo* (Border *et al.*, 1992) and this suggested that it may have therapeutic potential in controlling TGFβ action. Decorin and biglycan may therefore have an effect on inhibiting cell proliferation by blocking TGFβ action, but it has also been shown that proteoglycans with dermatan sulphate chains may inhibit cell proliferation by binding spermine, which is a polyamine growth promoter (Belting and Fransson, 1993).

The different properties identified so far for decorin suggest a whole range of ways in which it may contribute to matrix organisation and stability and also that it may help regulate mediators that control cell functions. As these studies are extended to other members of the leucine-rich proteoglycan family they may identify common and distinct features that characterise their different functions in the extracellular matrix.

Biglycan shares some properties with decorin such as TGFβ binding, but most evidence suggests that it does not occur in matrices bound to fibrillar collagen. It was also distinguished from decorin in biochemical preparative procedures where it showed self-association into oligomers (Choi *et al.*, 1989) at high ionic strength and much stronger hydrophobic interaction on octyl-Sepharose column chromatography. Thus inspite of similarities in protein sequence and structure there are clear marked differences in its solution properties, which suggests that it has quite different characteristics of interaction with other proteins.

Investigation of the expression of decorin and biglycan in different tissues during embryogenesis by in-situ hybridisation reveals an interesting pattern where frequently both are expressed by the same cells, but the strongest expression of one is often accompanied by weak expression of the other (Bianco *et al.*, 1990). The pattern of expression of biglycan was thus substantially divergent from the expression of decorin. In the developing limb cartilage biglycan was in the outer rim of articular cartilage, which lacked decorin, whereas the central non-articular region of cartilage was more rich in matrix and contained decorin and lacked biglycan. The surrounding soft tissue also contained biglycan and lacked decorin. As development proceeded biglycan was enriched in the territorial capsule of chondrocytes, whereas decorin was in the interterritorial matrix. This contrasting distribution of the two proteoglycans was also observed in some other tissues. In mature articular cartilage it was found that the level of biglycan mRNA decreased with age, whereas decorin mRNA increased, correlating with their tissue content (Roughley *et al.*, 1994). The synthesis of the two proteoglycans was also found to be controlled independently and regulated in opposite ways by TGFβ (Westergren-Thorsson *et al.*, 1991). Biglycan synthesis was increased and decorin decreased in the presence of TGFβ (Roughley *et al.*, 1994). There is thus much evidence that decorin and biglycan are expressed by cells in response to different signals and to fulfil different functions in the extracellular matrix, but there is not yet a complete picture of what these different major functions are.

Fibromodulin

Fibromodulin was initially identified as a 59 kDa protein in cartilage matrix (Heinegård *et al.*, 1986). Although it contains leucine-rich repeat motifs similar to decorin and biglycan it differs from them in having no ser-gly glycosaminoglycan attachment sequence (Oldberg *et al.*, 1989; Antonsson *et al.*, 1993). However it does contain four asparagine sites within the leucine-rich repeat region for N-linked oligosaccharide that are substituted with N-linked keratan sulphate chains (Plaas *et al.*, 1990). Towards the N-terminus there is a tyrosine-rich region which is absent from decorin and biglycan and which appears to be sulphated. It is thus a strongly charged polyanionic matrix component inspite of its lack of a chondroitin/dermatan sulphate chain. Fibromodulin is less widely distributed than decorin and, although it shares some interactive properties with TGFβ, the interaction of fibromodulin and decorin with fibrillar collagen are at separate sites and quite independent (Hedbom and Heinegård, 1993). The role of fibromodulin in matrix organisation may therefore be complementary to that of decorin.

Lumican

Cornea was for long known as the most abundant source of N-linked keratan sulphate. The eventual cloning from chicken cornea of the proteoglycan carrying these keratan sulphate chains identified another member of the leucine-rich family of proteins, lumican (Blochberger *et al.*, 1992a). This was more similar to fibromodulin in sequence than to decorin and biglycan, but lacked a significant N-terminal tyrosine-rich region. The keratan sulphate content suggests that 2 or 3 out of the possible 5 N-linked attachment sites are substituted with keratan sulphate. The localisation of keratan sulphate on fibrillar collagen at the gap zone at the "a" and "c" bands in cornea (Scott, 1988) suggests that lumican also interacts with collagen at a different site from decorin. Lumican is not found exclusively in cornea and immunoreactive protein was found in several tissues, including particularly muscle and intestine (Blochberger *et al.*, 1992a). In adult chicken cornea there is evidence for two decorin isoforms (Blochberger *et al.*, 1992a) one with both chondroitin/dermatan sulphate and keratan sulphate attached and the other with only the chondroitin/dermatan sulphate chain and this differed from the young chick cornea which contained no keratan sulphate. The presence of uniformly thin collagen fibrils with a regular and optimal spacing pattern is essential for corneal transparency. This is lost during corneal scarring (Hassell *et al.*, 1983), which results from changes in collagen and proteoglycan synthesis and in macular corneal dystrophy where keratan sulphate synthesis is deficient (Hassell *et al.*, 1980). Lumican and decorin are thus likely to play an important role in regulating corneal collagen fibre formation and in maintaining the hydration of the cornea and the spacing of fibres into a lattice that gives little dispersion of light.

Proteoglycan Lb

Proteoglycan Lb isolated from embryonic chick epiphysial cartilage is also a member of the leucine-rich family of proteins. However, it is less closely related to decorin, biglycan, fibromodulin and lumican, as it contains only 6 leucine-rich repeats and lacks the same pattern of cysteine residues (Shinomura and Kimata, 1992) and it is more similar to osteoinductive factor. Proteoglycan Lb contains two ser-gly sites for chondroitin/dermatan sulphate chain attachment. It is expressed in growth plate and is found at the lacunae surface of hypertrophic chondrocytes. Other properties of this proteoglycan are not well characterised, but it may have a role in bone morphogenesis.

Leucine-rich Proteoglycans Lacking Glycosaminoglycan Chains

There is an increasing amount of evidence that, as with all proteoglycans, there is considerable variation in the structure and content of chondroitin/dermatan sulphate and keratan sulphate chains synthesised on the leucine-rich family of proteoglycans by different cells and at different stages of development. This includes the presence of protein cores such as that of decorin (Sampaio *et al.*, 1988) or biglycan (Roughley *et al.*, 1993), lacking any glycosaminoglycan chains in cartilage, or non-glycanated forms of decorin and biglycan in intervertebral disc (Johnstone *et al.*, 1993). This might arise as a result of proteinase cleavage within the matrix of the N-terminal peptide sequence with the chondroitin/dermatan sulphate chain attached, or from a failure to initiate chain synthesis, such as with examples of biglycan that contain only a single chondroitin/dermatan sulphate chain rather than two. There is also evidence that lumican is expressed predominantly as a glycoprotein with unsulphated lactosaminoglycan chains rather than as the sulphated form present in cornea (Funderburgh *et al.*, 1991). The functional significance of these variations in structure have yet to be determined.

BIKUNIN PROTEOGLYCAN AND ITS COMPLEXES

The 30 kDa proteoglycan bikunin carries a single chondroitin 4-sulphate chain whose reducing end is linked, at a partial consensus sequence (DEXSG) for glycosaminoglycan attachment, to ser_{10} of the core protein (Enghild *et al.*, 1989; Chirat *et al.*, 1991). It contains two Kunitz type proteinase inhibitor domains (Gebhard *et al.*, 1989) and is found free in serum, urine and in the extracellular matrix of some diseased tissues (Yoshida *et al.*, 1989,

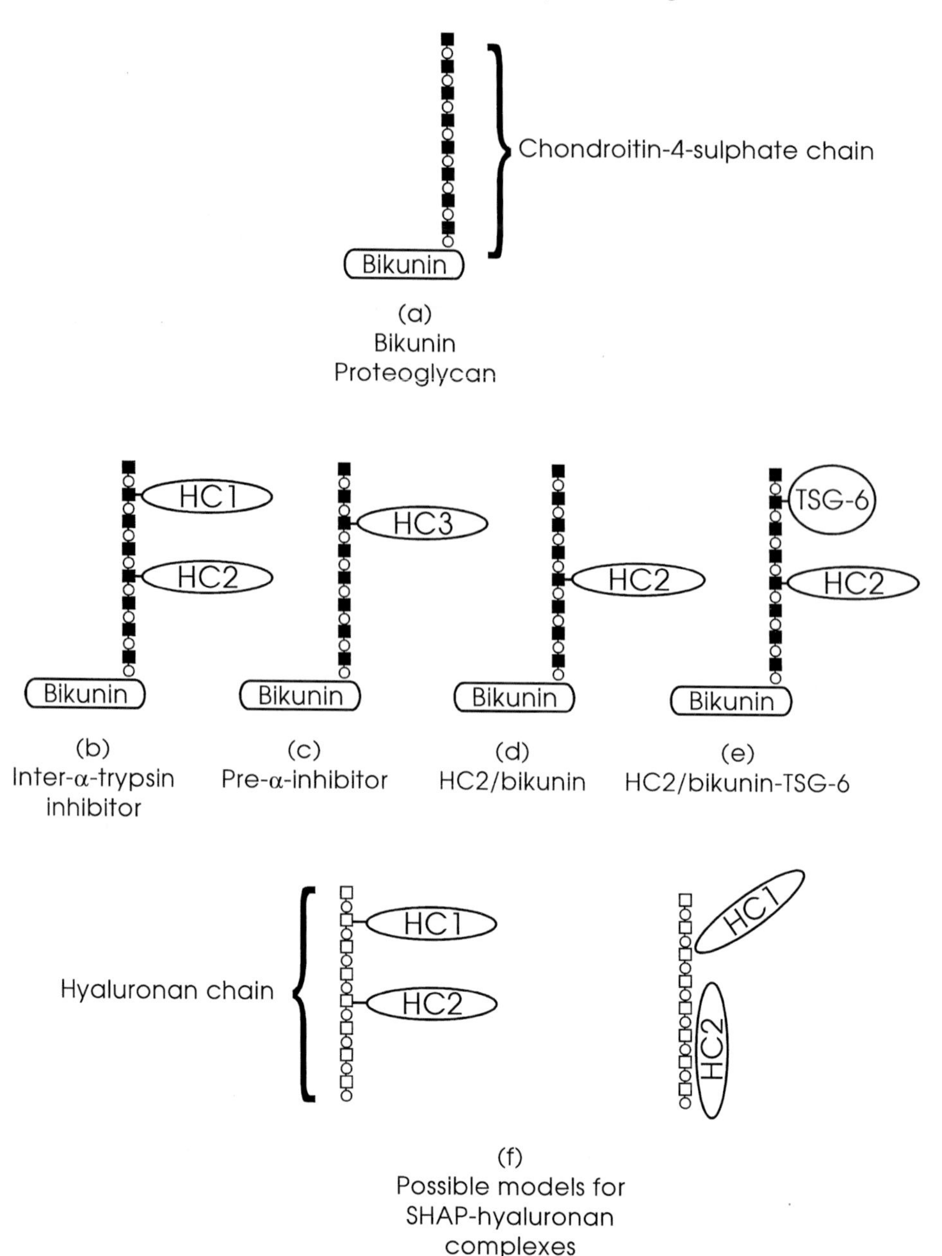

Figure 6.5 Schematic representation of the bikunin proteoglycan and its covalent complexes with heavy chain proteins and TSG-6. (a–e) The chondroitin-4-sulphate chain is shown with its repeating GlcA (○) and GalNac (■) residues and is attached to Ser_{10} of bikunin via a conventional O-linked linkage region for glycosaminoglycans. (b–e) The non-reducing end of chondroitin-4-sulphate is covalently bound to the C-terminal end of the heavy chain proteins, but the position along the polysaccharide of the GalNac residues involved in this linkage are not known. The structures for the trimeric inter-α-trypsin inhibitor and the TSG-6 complex are schematic and have not been confirmed experimentally. (f) Hyaluronan is shown with its repeating GlcA (○) and GlcNac (□) residues. The HC1 and HC2 components of SHAPs may bind hyaluronan via structures which are analogous with inter-α-trypsin inhibitor complexes with chondroitin-4-sulphate (b), or they may bind via a different covalent conformation.

1991). More commonly though, bikunin is found as a covalent complex with one or two heavy chain proteins, HC1, HC2 or HC3, in the form of inter-α-trypsin inhibitor (IαI), pre-α-inhibitor (PαI) or HC2/bikunin (Enghild *et al.*, 1989, 1993)(Figure 6.5). In terms of proteoglycan structure, these complexes are unique because the bikunin chondroitin sulphate chain is covalently linked towards its non-reducing end via carbon-6 of an internal N-acetylgalactosamine residue to an esterified α-carbon on the C-terminal Asp of a heavy chain component. These complexes are unusually stable, resistant to reduction and sensitive to treatment with chondroitinase ABC, however the exact mechanism of how the cross-links are formed is not known.

The cross-link, named a Protein-Glycosaminoglycan-Protein (PGP) cross-link (Enghild *et al.*, 1991), has been well characterised for pre-α-inhibitor and HC2/bikunin (Enghild *et al.*, 1989, 1991, 1993). Recently TSG-6, a member of the aggrecan family of proteins described above has also been identified as part of a highly stable, chondroitinase-sensitive complex with HC2 and bikunin (Wisniewski *et al.*, 1994). The molecular interactions leading to the formation of the TSG-6-HC2/bikunin complex are not known but are likely to involve the same PGP cross-link with chondroitin sulphate. TSG-6 contains a consensus sequence that is conserved in each of the heavy chains structures and which defines the cleavage site of the heavy chain propeptide to generate the new C-terminal aspartic acid residue that forms the ester bond with chondroitin sulphate (Bourguignon *et al.*, 1993).

The bikunin proteoglycan complexes are thought to have roles other than proteinase inhibitors since, for example, IαI accounts for only about 5% of total inhibitory activity despite its relatively high abundance in human serum (Gebhard and Hochstrasser, 1986). Some studies have suggested that IαI serves as a proteinase shuttle that transports proteinases to sites of clearance while others have suggested a role in the assembly and stabilisation of hyaluronan-rich extracellular matrices. An extracellular matrix stabilising factor (ESF) that is essential for the retention of hyaluronan around the expanding cumulus cells of mouse ovarian follicles has been identified as IαI (Chen *et al.*, 1992), and it has been proposed that the unique linkages of the heavy chain subunits to bikunin proteoglycan may present a complex with a spatial arrangement that facilitates a cross-linking function in the matrix (Castillo and Templeton, 1993).

SHAP

Another novel complex involving the HC1 and HC2 subunits of IαI, but without bikunin is SHAP (Serum-derived Hyaluronan Associated Protein) (Huang *et al.*, 1993). SHAP binds hyaluronan to form a stable covalent complex that cannot be dissociated under any conditions tested. SHAP binding is therefore distinct from the other families of hyaluronan-binding proteins discussed above. The interaction of HC1 and HC2 with hyaluronan may occur via structures which are analogous with those that bind chondroitin-4-sulphate in IαI inhibitor or they may bind via some different covalent conformation (Figure 6.5f). The mechanisms involved, the chemical structures, and the biological significance of these unusual glycosaminoglycan-containing complexes are yet to be resolved.

It is interesting that a stable SHAP-hyaluronan complex would probably not be considered a proteoglycan even though it contains protein covalently bound to glycosaminoglycan. Our definition of a proteoglycan therefore should be modified to describe a core protein in which the glycosaminoglycan chain is extended onto N- or O-linked linkage structures by the sequential addition of specific hexoses during proteoglycan biosynthesis.

CONCLUDING COMMENTS

The number of proteoglycans cloned and sequenced is rapidly expanding and proteoglycan forms of known proteins are also being discovered. It is clear that the control of glycosaminoglycan chain addition is part of a cellular mechanism to modify proteins for different functions and further variations in chain structure, including length, sulphation and epimerisation, may provide additional means by which the cells "fine tune" properties to suit particular biological needs. Proteoglycans are important organisers of the extracellular matrix and although for some their major function may be structural, for many there are multiple functions and they are more active participants in matrix events.

References

Antonsson, P., Heinegård, D. and Oldberg, A. (1989) The keratan sulfate-enriched region of bovine cartilage proteolgycan consisits of a consecutively repeated hexapeptide motif. *J. Biol. Chem.*, **264**, 16170–16173.

Antonsson, P., Heinegård, D. and Oldberg, Å. (1993) Structure and deduced amino acid sequence of the human fibromodulin gene. *Biochim. Biophys. Acta,* **1174**, 204–206.

Appella, E., Robinson, E.A., Ullrich, S.J., Stoppelli, M.P., Corti, A., Cassani, G. and Blasi, F. (1987) The receptor-binding sequence of urokinase. A biological fundtion for the growtyh-factor module of proteases. *J. Biol. Chem.*, **262**, 4437–4440.

Baldwin, C.T., Reginato, A.M. and Prockop, D.J. (1989) A new epidermal growth factor-like domain in the human core protein for the large cartilage-specific proteoglycan. *J. Biol. Chem.*, **264**, 15747–15750.

Barry, F.P., Gaw, J.U., Young, C.N. and Neame, P.J. (1992) Hyaluronan-binding region of aggrecan from pig laryngeal cartilage. *Biochem. J.*, **286**, 761–769.

Bayne, E.K., Donatelli, S.A., Singer, I.I., Weidner, J.R., Hutchinson, N.I., Hoerrner, L.A., Williams, H.R., Mumford, R.A., Lohmander, L.S. and Lark, M.W. (1994) Detection of a metalloproteinase-generated aggrecan HABR fragment within human OA and RA cartilage. *Trans. Orthop. Res. Soc*, **40th Annual Meeting**, 308. (Abstract)

Belting, M. and Fransson, L.-Å. (1993) The growth promoter spermine interacts specifically with dermatan sulfate regions that are rich in L-iduronic acid and possess antiproliferative activity. *Glycoconjugate J.*, **10**, 453–460.

Bengtsson, E., Neame, P., Heinegård, D. and Sommarin, Y. (1994) The primary structure of a 58kDa connective tissue protein reveals leucine rich repeats. *Proceedings FECTS XIV*, A5. (Abstract)

Bianco, P., Fisher, L.W., Young, M.F., Termine, J.D. and Robey, P.G. (1990) Expression and localization of the two small proteoglycans biglycan and decorin in developing human skeletal and non-skeletal tissues. *J. Histochem. Cytochem.*, **38**, 1549–1563.

Bidanset, D.J., Guidry, C., Rosenberg, L.C., Choi, H.U., Timpl, R. and Höök, M. (1992) Binding of the proteoglycan decorin to collagen type VI. *J. Biol. Chem.*, **267**, 5250–5256.

Blochberger, T.C., Cornuet, P.K. and Hassell, J.R. (1992a) Isolation and partial characterization of lumican and decorin from adult chicken corneas. A keratan sulfate-containing isoform of decorin is developmentally regulated. *J. Biol. Chem.*, **267**, 20613–20619.

Blochberger, T.C., Vergnes, J-P., Hempel, J. and Hassell, J.R. (1992b) cDNA to chick lumican (corneal keratan sulfate proteoglycan) reveals homology to the small interstitial proteoglycan gene family and expression in muscle and intestine. *J. Biol. Chem.*, **267**, 347–352.

Bonnet, F., Perin, J-P., Lorenzo, F., Jolles, J. and Jolles, P. (1986) An unexpected sequence homology between link proteins of the proteoglycan complex and immunoglobulin-like proteins. *Biochim. Biophys. Acta*, **873**, 152–155.

Border, W.A., Noble, N.A., Yamamoto, T., Harper, J.R., Yamaguchi, Y., Pierschbacher, M.D. and Ruoslahti, E. (1992) Natural inhibitor of transforming growth factor-beta protects against scarring in experimental kidney disease. *Nature*, **360**, 361–364.

Border, W.A. and Ruoslahti, E. (1991) Transforming growth factor-beta 1 induces extracellular matrix formation in glomerulonephritis. *Cell Differ. Dev.*, **32**, 425–431.

Border, W.A. and Ruoslahti, E. (1992) Transforming growth factor-beta in disease: the dark side of tissue repair. *J. Clin. Invest.*, **90**, 1–7.

Bourdon, M.A., Oldberg, Å., Pierschbacher, M.D. and Ruoslahti, E. (1985). *Proc. Natl. Acad. Sci. USA*, **82**, 1321–1325.

Bourdon, M.A., Krusius, T., Campbell, S., Schwartz, N.B. and Ruoslahti, E. (1987) Identification and synthesis of a recognition signal for the attachment of glycosaminoglycans to proteins. *Proc. Natl. Acad. Sci. USA*, **84**, 3194–3198.

Bourguignon, J., Diarra-Mehrpour, M., Thiberville, L., Bost, F., Sesboue, R. and Martin, J.-P. (1993) Human pre-alpha-trypsin inhibitor-precursor heavy chain. cDNA and deduced amino acid sequence. *Eur. J. Biochem.*, **212**, 771–776.

Brennan, M.J., Oldberg, Å., Pierschbacher, M.D. and Ruoslahti, E. (1984) Chondroitin/Dermatan sulfate proteoglycan in human fetal membranes. Demonstration of an antigenically similar proteoglycan in fibroblasts. *J. Biol. Chem.*, **259**, 13742–13750.

Brown, D.C. and Vogel, K.G. (1989) Charactersitics of the *in vitro* interaction of a small proteoglycan (PGII) of bovine tendon with type I collagen. *Matrix*, **9**, 468–478.

Carter, W.G. and Wayner, E. (1988) Characterization of the class III collagen receptor, a phosphorylated transmembrane glycoprotein expressed in nucleated human cells. *J. Biol. Chem.*, **263**, 4193–4201.

Castillo, G.M. and Templeton, D.M. (1993) Subunit structure of bovine ESF (extracellular-matrix stabilizing factor(s)): A chondroitin sulfate proteoglycan with homology to human Iαi (inter-α-trypsin inhibitors). *FEBS Lett.*, **318**, 292–296.

Casu, B., Ferro, D.R., Ragazzi, M. and Torri, G. (1993) Conformation of iduronic acid-containing glycosaminoglycans. In *Dermatan Sulphate Proteoglycans*, J.E. Scott. (ed.) pp. 41–53. London: Portman Press.

Caterson, B., Mahmoodian, F., Sorrell, J.M., Hardingham, T.E., Bayliss, M.T., Carney, S.L., Ratcliffe, A. and Muir, H. (1990) Modulation of native chondroitin sulphate structures in tissue development and in disease. *J. Cell Sci.*, **97**, 411–417.

Chandrasekaran, L. and Tanzer, M.L. (1992) Molecular cloning of chicken aggrecan. Structural analyses. *Biochem. J.*, **288**, 903–910.

Chen, L., Mao, S.J.T. and Larsen, W.J. (1992) Identification of a factor in fetal vvbovine serum that stabilizes the cumulus extracellular matrix. A role for a member of the inter-alpha-trypsin inhibitor family. *J. Biol. Chem.*, **267**, 12380–12386.

Chirat, F., Balduyck, M., Mizon, C., Laroui, S., Sautiere, P. and Mizon, J. (1991) A chondroitin-sulphate chain is located on serine-10 of the urinary trypsin inhibitor. *Int. J. Biochem.*, **23**, 1201–1203.

Choi, H.U., Johnson, T.L., Pal, S., Tang, L.H., Rosenberg, L. and Neame, P.J. (1989) Characterization of the dermatan sulfate proteoglycans, DS-PGI and DS-PGII, form bovine articular cartilage and skin isolated by octyl-sepharose chromatography. *J. Biol. Chem.*, **264**, 2876–2884.

Chopra, R.K., Pearson, C.H., Pringle, G.A., Fackre, D.S. and Scott, P.G. (1985) Dermatan sulphate is located on serine-4 of bovine skin proteodermatan sulphate. *Biochem. J.*, **232**, 277–279.

Culty, M., Miyake, K., Kincade, P.W., Silorski, E., Butcher, E.C. and Underhill, C.B. (1990) The hyaluronate receptor is a member of the CD44 (H-CAM) family of cell surface glycoproteins. *J. Cell Biol.*, **111**, 2765–2774.

Deak, F., Kiss, I., Sparks, K.J., Argraves, W.S., Hampikian, G. and Goetinck, P.F. (1986) Complete amino acid sequence of chicken cartilage link protein deduced from cDNA clones. *Proc. Natl. Acad. Sci. USA*, **83**, 3766–3770.

Denning, S.M., Le, P.T., Singer, K.H. and Haynes, B.F. (1990) Antibodies against the CD44p80, lymphocyte homing receptor molecule augment human peripheral blood T cell activation. *J. Immunol.*, **144**, 7.

Dennis, J.E., Carrino, D.A., Schwartz, N.B. and Caplan, A.I. (1990) Ultrastructural characterization of embryonic chick cartilage proteoglycan core protein and the mapping of a monoclonal antibody epitope. *J. Biol. Chem.*, **265**, 12098–12103.

Doege, K.J., Sasaki, M., Horigan, E., Hassell, J.R. and Yamada, Y. (1987) Complete primary structure of the rat cartilage proteoglycan core protein deduced from cDNA clones. *J. Biol. Chem.*, **262**, 17757–17767.

Doege, K.J., Sasaki, M., Kimura, T. and Yamada, Y. (1991) Complete coding sequence and deduced primary structure of the human cartilage large aggregating proteoglycan, aggrecan. Human specific repeats and additional alternatively spliced forms. *J. Biol. Chem.*, **266**, 894–902.

Dorfman, A. (1981) Proteoglycan biosynthesis. In *Cell Biology of Extracellular Matrix*, E.D. Hay. (ed.) pp. 115–138.New York: Plenum Press.

Dougherty, G.J., Cooper, D.L., Memory, J.F. and Chiu, R.K. (1994) Ligand binding specificity of alternatively spliced CD44 isoforms. Recognition and binding of hyaluronan by CD44R1. *J. Biol. Chem.*, **269**, 9074–9078.

Enghild, J.J., Thogersen, I.B., Pizzo, S.V. and Salvesen, G. (1989) Analysis of inter-alpha-trypsin inhibitor and a novel trypsin inhibitor pre-alpha-trypsin inhibitor, from human plasma. Polypeptide chain stoiciometry and assembly by glycan. *J. Biol. Chem.*, **264**, 15975–15981.

Enghild, J.J., Salvesen, G., Hefta, S.A., Thogersen, I.B., Rutherfurd, S. and Pizzo, S.V. (1991) Chondroitin 4-sulfate covalently cross-links the chains of the human blood protein pre-a-inhibitor. *J. Biol. Chem.*, **266**, 747–751.

Enghild, J.J., Salvesen, G., Thogersen, I.B., Valnickova, Z., Pizzo, S.V. and Hefta, S.A. (1993) Presence of the protein-glycosaminoglycan-protein covalent cross-link in the inter-α-inhibitor-related proteinase inhibitor heavy chain 2/bikunin. *J. Biol. Chem.*, **268**, 8711–8716.

Faltz, L.L., Caputo, C.B., Kimura, J.H., Schrode, J. and Hascall, V.C. (1979) Structure of the complex between hyaluronic acid, the hyaluronic acid binding region, and the link protein of proteoglycan aggregates from the Swarm rat chondrosarcoma. *J. Biol. Chem.*, **254**, 1381–1387.

Feizi, T. (1989) Keratan sulphate oligosaccharide, members of a family of antigens of the poly-N-acetylgalactosamine series. In *Keratan Sulphate Chemistry, Biology, Chemical Pathology*, H. Greiling and J.E. Scott. (ed.) pp. 21–29.London: The Biochemical Society.

Fisher, L.W., Termine, J.D. and Young, M.F. (1989) Deduced protein sequence of bone small proteoglycan I (biglycan) shows homology with proteoglycan II (decorin) and several nonconnective tissue proteins in a variety of species. *J. Biol. Chem.*, **264**, 4571–4576.

Flannery, C.R., Lark, M.W. and Sandy, J.D. (1992) Identification of a stromelysin cleavage site within the interglobular domain of human aggrecan: evidence for proteolysis at this site *in vivo* in human articular cartilage. *J. Biol. Chem.*, **267**, 1008–1014.

Fosang, A.J., Neame, P.J., Hardingham, T.E., Murphy, G. and Hamilton, J.A. (1991) Cleavage of cartilage proteoglycan between G1 and G2 domains by stromelysins. *J. Biol. Chem.*, **266**, 15579–15582.

Fosang, A.J., Neame, P.J., Last, K., Hardingham, T.E., Murphy, G. and Hamilton, J.A. (1992) The interglobular domain of cartilage aggrecan is cleaved by Pump, gelatinases and cathepsin B. *J. Biol. Chem.*, **267**, 19470–19474.

Fosang, A.J., Last, K., Knäuper, V., Neame, P.J., Murphy, G., Hardingham, T.E., Tschesche, H. and Hamilton, J.A. (1993) Fibroblast and neutrophil collagenases cleave at two sites in the cartilage aggrecan interglobular domain. *Biochem. J.*, **295**, 273–276.

Fosang, A.J., Last, K., Neame, P.J., Murphy, G., Knäuper, V., Tschesche, H., Hughes, C.E., Caterson, B. and Hardingham, T.E. (1994) Neutrophil collagenase (MMP-8) cleaves at the aggrecanase site E_{373}–A_{374} in the interglobular domain of cartilage aggrecan. *Biochem. J.*, **304**, 347–351

Fosang, A.J. and Hardingham, T.E. (1989) Isolation of the N-terminal globular domains from cartilage proteoglycans. Identification of G2 domain and its lack of interaction with hyaluronate and link protein. *Biochem. J.*, **261**, 801–809.

Fransson, L.-Å., Chang, F., Yoshida, K., Heinegård, D., Malmström, A. and Schmidtchen, A. (1993) Patterns of epimerization and sulphation in dermatan sulphate chains. In *Dermatan Sulphate Proteoglycans*, J.E. Scott. (ed.) pp. 11–25. London: Portman Press.

Fulop, C., Walcz, E., Valyon, M. and Glant, T.T. (1993) Expression of alternatively spliced epidermal growth factor-like domains in aggrecans of different species. *J. Biol. Chem.*, **268**, 17377–17383.

Funderburgh, J.L., Funderburgh, M.L., Mann, M.M. and Conrad, G.W. (1991) Arterial Lumican. Properties of a corneal-type keratan sulfate proteoglycan from bovine aorta. *J. Biol. Chem.*, **266**, 24773–24777.

Funderburgh, J.L., Funderburgh, M.L., Brown, S.J., Vergnes, J.-P., Hassell, J.R., Mann, M.M. and Conrad, G.W. (1993) Sequence and structural implications of a bovine corneal keratan sulfate proteoglycan core protein. Protein 37B represents bovine lumican and proteins 37A and 25 are unique. *J. Biol. Chem.*, **268**, 11874–11880.

Gebhard, W., Schreitmuller, T., Hochstrasser, K. and Wachter, E. (1989) Two out of the three kinds of inter-alpha-trypsin inhibitor are structurally related. *Eur.J.Biochem.*, **181**, 571–576.

Gebhard, W. and Hochstrasser, K. (1986) Biochemistry of aprotinin and aprotinin-like inhibitors. In *Proteinase Inhibitors*, A.J. Barrett and G. Salvesen. (ed.) pp. 375–388. Amsterdam: Elsevier Science Publishers BV.

Goetinck, P.F., Stirpe, N.S., Tsonis, P.A. and Carlone, D. (1987) The tandemly repeated sequences of cartilage link protein contain the sites for interaction with hyaluronic acid. *J. Cell Biol.*, **105**, 2403–2408.

Goldstein, L.A., Zhou, D.F.H., Picker, L.J., Minty, C.N., Bargatze, R.F., Ding, J.F. and Butcher, E.C. (1989) A human lymphocyte homing receptor, the Hermes antigen, is related to cartilage proteoglycan core and link proteins. *Cell*, **56**, 1063–1072.

Greenwald, I. (1985) lin-21, a nematode homeotic gene, is homologous to a set of mammalian proteins that includes epidermal growth factor. *Cell*, **43**, 583–590.

Grover, J. and Roughley, P.J. (1994) Tej expression of functional link protein in a baculovirus system: analysis of mutants lacking the A, B and B′ domains. *Biochem. J.*, **300**, 317–324.

Halberg, D.F., Proulx, G., Doege, K., Yamada, Y. and Drickamer, K. (1988) A segment of the cartilage proteoglycan core protein has lectin-like activity. *J. Biol. Chem.*, **263**, 9486–9490.

Hardingham, T. and Muir, H. (1973) Binding of oligosaccharides of hyaluronic acid to proteoglycan. *Biochem. J.*, **135**, 905–908.

Hardingham, T.E., Fosang, A.J., Hazell, P., Hey, N.J., Kee, W.J. and Ewins, R.J.F. (1994) The sulphation pattern in chondroitin sulphate chains investigated by chondroitinase ABC and ACII digestion and reactivity with monoclonal antibodies. *Carbohydr. Res.*, **255**, 241–254.

Hardingham, T.E. and Fosang, A.J. (1992) Proteoglycans: Many forms, many functions. *FASEB J.*, **6**,861–870.

Hascall, V.C. and Heinegård, D. (1974) Aggregation of cartilage proteoglycans: oligosaccharide competitors of the proteoglycan-hyaluronic acid interaction. *J. Biol. Chem.*, **249**, 4242–4249.

Hassell, J.R., Newsome, D.A., Krachmar, J.H. and Rodrigues, M.M. (1980) Macular corneal dystrophy: failure to synthesise a mature keratan sulphate proteoglycan. *Proc. Natl. Acad. Sci. USA*, **77**, 3705–3709.

Hassell, J.R., Cintron, C., Kublin, C. and Newsome, D.A. (1983) Proteoglycan changes during restoration of transparency in corneal scars. *Arch. Biochem. Biophys.*, **222**, 362–369.

Hausser, H., Hoppe, W., Rauch, U. and Kresse, H. (1989) Endocytosis of a small dermatan sulphate proteoglycan. Identification of binding proteins. *Biochem. J.*, **263**, 137–142.

Hausser, H., Ober, B., Quentin-Hoffmann, E., Schmidt, B. and Kresse, H. (1992) Endocytosis of different members of the small chondroitin/dermatan sulfate proteoglycan family. *J. Biol. Chem.*, **267**, 11559–11564.

Hausser, H., Witt, O. and Kresse, H. (1993) Influence of membrane-associated heparan sulfate on the internalization of the small proteoglycan decorin. *Exp. Cell Res.*, **208**, 398–406.

He, Q.J., Lesley, J., Hyman, R., Ishihara, K. and Kincade, P.W. (1992) Molecular isoforms of murine CD44 and evidence that the membrane proximal domain is not critical for hyaluronate recognition. *J. Cell Biol.*, **119**, 1711–1719.

Hedbom, E. and Heinegård, D. (1989) Interaction of a 59-kDa connective tissue matrix protein with collagen I and collagen II. *J.Biol.Chem.*, **264**, 6898–6905.

Hedbom, E. and Heinegård, D. (1993) Binding of fibromodulin and decorin to separate sites on fibrillar collagens. *J. Biol. Chem.*, **268**, 27307–27312.

Heinegård, D., Larsson, T., Sommarin, Y., Franzen, A., Paulsson, M. and Hedbom, E. (1986) Two novel matrix proteins isolated from articular cartilage show wide distributions among connective tissues. *J. Biol. Chem.*, **261**, 13866–13872.

Hildebrand, A., Romaris, M., Rasmussen, L.M., Heinegård, D., Twardzik, D.R., Border, W.A. and Ruoslahti, E. (1994) Interaction of the small interstitial proteoglycans biglycan, decorin and fibromodulin with transforming growth factor beta. *Biochem. J.*, **302**, 527–534.

Horita, D.A., Hajduk, P.J., Goetinck, P.F. and Lerner, L.E. (1994) NMR studies of peptides derived from the putative binding regions of cartilage proteins. No evidence for binding to hyaluronan. *J. Biol. Chem.*, **269**, 1699–1704.

Huang, L., Yoneda, M. and Kimata, K. (1993) A serum-derived hyaluronan-associated protein (SHAP) is the heavy chain of the inter α-trypsin inhibitor. *J.Biol.Chem.*, **268**, 26725–26730.

Huet, S.H., Groux, H., Caillou, B., Valentin, H., Prieur, A.M. and Bernard, A. (1989) CD44 contributes to T cell activation. *J.Immunol.*, **143**, 798.

Hursh, D.A., Andrews, M.E. and Raff, R.A. (1987) A sea urchin encodes a polypeptide homologous to epidermal growth factor. *Science*, **237**, 1487–1490.

Hyman, R., Lesley, J. and Schulte, R. (1994) Somatic cell mutants distinguish CD44 expression and hyaluronic acid binding. *Immunogenetics*, **33**, 392–395.

Ilic, M.Z., Handley, C.J., Robinson, H.C. and Mok, M.T. (1992) Mechanism of catabolism of aggrecan by articular cartilage. *Archiv. Biochem. Biophys.*, **294**, 115–122.

Jacobson, K., O'Dell, D., Holifield, B., Murphy, T.L. and August, J.T. (1984) Redistribution of a major cell surface glycoprotein during cell movement. *J. Cell Biol.*, **99**, 1613.

Jalkanen, S., Jalkanen, M., Bargatze, R., Tammi, M. and Butcher, E.C. (1988) Biochemical properties of glycoproteins involved in lymphocyte recognition of high endothelial venules in man. *J. Immunol.*, **141**, 1615–1623.

Jalkanen, S. and Jalkanen, M. (1992) Lymphocyte CD44 binds the COOH-terminal heparin-binding domain of fibronectin. *J. Cell Biol.*, **116**, 817–825.

Jaworski, D.M., Kelly, G.M. and Hockfield, S. (1994) BEHAB, a new member of the proteoglycan tandem repeat family of hyaluronan-binding proteins that is restricted to the brain. *J.Cell Biol.*, **125**, 495–509.

Johnstone, B., Markopoulos, M., Neame, P.J. and Caterson, B. (1993) Identification and characterization of glycanated and non-glycanated forms of biglycan and decorin in the human intervertebral disc. *Biochem. J.*, **292**, 661–666.

Kahari, V-M., Larjava, H. and Uitto, J. (1991) Differential regulation of extracellular matrix proteoglycan (PG) gene expression. Transforming growth factor-beta 1 up-regulates biglycan (PGI) and versican (large fibroblast PG) but down-regulates decorin (PGII) mRNA levels in human fibroblasts in culture. *J. Biol. Chem.*, **266**, 10608–10615.

Kimata, K., Barrach, H-J., Brown, K.S. and Pennypacker, J.P. (1981) Absence of proteoglycan core protein in cartilage from the cmd/cmd (cartilage matrix deficiency) mouse. *J. Biol. Chem.*, **256**,6961–6968.

Kobe, B. and Deisenhofer, J. (1993) Crystal structure of porcine ribonuclease inhibitor, a protein with leucine-rich repeats. *Nature*, **366**, 751–756.

Kobe, B. and Deisenhofer, J. (1994) The leucine-rich repeat: a versatile binding motif. *TIBS*, **19**, 415–421.

Korenberg, J.R., Chen, X.N., Doege, K., Grover, J. and Roughley, P.J. (1993) Assignment of the human aggrecan gene (AGC1) to 15q26 using fluorescence *in situ* hybridization analysis. *Genomics*, **16**, 546–548.

Krueger, R.C., Fields, T.A., Hildreth IV, J. and Schwartz, N.B. (1990) Chick cartilage chondroitin sulfate proteoglycan core protein. *J. Biol. Chem.*, **265**, 12075–12087.

Krumdieck, R., Höök, M., Rosenberg, L.C. and Volanakis, J.E. (1992) The proteoglycan decorin binds C1q and inhibits the activity of the C1 complex. *J. Immunol.*, **149**, 3695–3701.

Krusius, T., Gehlsen, K.R. and Ruoslahti, E. (1987) A fibroblast chondroitin sulfate proteoglycan core protein contains lectin-like and growth factor-like sequences. *J. Biol. Chem.*, **262**, 13120–13125.

Krusius, T. and Ruoslahti, E. (1986) Primary structure of an extracellular matrix proteoglycan core protein deduced from cloned cDNA. *Proc. Natl. Acad. Sci. USA*, **83**, 7683–7687.

Kurosawa, S., Stearns, D., Jackson, K. and Esmon, C. (1988) A 10kDa cyanogen bromide fragment from the epidermal growth factor homology domain of rabbit thrombomodulin contains the primary thrombin binding site. *J. Biol. Chem.*, **263**, 599.–5996.

Lee, T.H., Wisniewski, H-G. and Vilcek, J. (1992) A novel secretory tumor necrosis factor-inducible protein (TSG-6) is a member of the family of hyaluronate binding proteins, closely related to the adhesion receptor CD44. *J. Cell Biol.*, **116**, 545–557.

Lesley, J., He, Q., Miyake, K., Hafmann, A., Hyman, R. and Kincade, P.W. (1992) Requirements for hyaluronic acid binding by CD44: A role for the cytoplasmic domain and activation by antibody. *J.Exp.Medicine*, **175**, 257–266.

Li, H., Schwartz, N.B. and Vertel, B.M. (1993) cDNA cloning of chick cartilage chondroitin sulfate (aggrecan) core protein and identification of a stop codon in the aggrecan gene associated with the chondrodystrophy, nanomelia. *J. Biol. Chem.*, **268**, 23504–23511.

Lindahl, U., Thunberg, L., Backstrom, G., Riesenfeld, J., Nordling, D. and Bjork, I. (1984) Extension and structural variability of the antithrombin-binding sequence in heparin. *J. Biol. Chem.*, **259**, 12368–12376.

Lohmander, L.S., Neame, P.J. and Sandy, J.D. (1993) The structure of aggrecan fragments in human synovial fluid: Evidence that aggrecanase mediates cartilage degradation in inflammatory joint disease, joint injury, and osteoarthritis. *Arthritis Rheum.*, **36**, 1214–1222.

Loulakis, P., Shrikhande, A., Davis, G. and Maniglia, C.A. (1992) N-Terminal sequence of proteoglycan fragments isolated from medium of interleukin-1-treated articular-cartilage cultures. *Biochem. J.*, **284**,589–593.

Malmström, A., Coster, L., Fransson, L.-Å., Hagner-McWhirter, A. and Westergren-Thorsson, G. (1993) Formation of L-iduronic acid during biosynthesis of dermatan sulphate. In *Dermatan Sulphate Proteoglycans*, J.E. Scott. (ed.) pp. 129–137. London: Portman Press.

McCormick, D., van der Rest, M., Goodship, J., Lozano, G., Ninomiya, Y. and Olsen, B.R. (1987) Structure of the glycosaminoglycan domain in the type IX collagen-proteoglycan. *Proc. Natl. Acad. Sci. USA*, **84**, 4044–4048.

Murakami, S., Miyake, K., Abe, R., Kincade, P.W. and Hodes, R.J. (1991) Characterization of autoantibody-secreting B cells in mice undergoing stimulatory (chronic) graft-versus-host reactions. Identification of a CD44 hi population that binds specifically to hyaluronate. *J. Immunol.*, **146**, 1422–1427.

Neame, P., Christner, J. and Baker, J. (1986) The primary structure of link protein from rat chondrosarcoma proteoglycan aggregate. *J. Biol. Chem.*, **261**, 3519–3535.

Neame, P.J., Sommarin, Y., Boynton, R.E. and Heinegård, D. (1994) The structure of a 38-kDaleucine-rich protein (chondroadherin) isolated from bovine cartilage. *J. Biol. Chem.*, **269**, 21547–21554.

O'Donnell, C.M., Kaczman-Daniel, K., Goetinck, P.F. and Vertel, B.M. (1988) Nanomelic chondrocytes synthesize a glycoprotein related to chondroitin sulfate proteoglycan core protein. *J. Biol. Chem.*, **263**, 17749–17754.

Oldberg, A., Antonsson, P. and Heinegård, D. (1987) The partial amino acid sequence of bovine cartilage proteoglycan, deduced from a cDNA clone, contains numerous Ser-Gly sequences arranged in homologous repeats. *Biochem. J.*, **243**, 255–259.

Oldberg, Å., Hayman, E.G. and Ruoslahti, E. (1981) Isolation of a chondroitin sulfate proteoglycan from a rat yolk sac tumor and immunochemical demonstration of its cell surface localization. *J. Biol. Chem.*, **256**, 10847–10852.

Oldberg, Å., Antonsson, P., Lindblom, K. and Heinegård, D. (1989) A collagen-binding 59-kd protien (fibromodulin) is structurally related to the small interstitial proteoglycans PG-S1 and PG-S2 (decorin). *EMBO*, **8**, 2601–2604.

Panayotou, G., End, P., Aumailley, M., Timpl, R. and Engel, J. (1989) Domains of laminin with growth-factor activity. *Cell*, **56**, 93–101.

Paulsson, M., Morgelin, M., Wiedemann, H., Beardmore-Gray, M., Dunham, D.G., Hardingham, T.E., Heinegård, D., Timpl, R. and Engel, J. (1987) Extended and globular protein domains in cartilage proteoglycans.

Biochem. J., **245**, 763–772.

Peach, R.J., Hollenbaugh, D., Stamenkovic, I. and Aruffo, A. (1993) Identificaction of hyaluronic acid binding siltes in the extracellular domain of CD44. *J. Cell Biol.*, **122**, 257–264.

Perkins, S.J., Nealis, A.S., Dudhia, J. and Hardingham, T.E. (1989) Immunoglobulin fold and tandem repeat structures in proteoglycan N-terminal domains and link protein. *J. Mol. Biol.*, **206**, 737–754.

Plaas, A.H.K., Neame, P.J., Nivens, C.M. and Reiss, L. (1990) Identification of the keratan sulphate attachment sites on bovine fibromodulin. *J. Biol. Chem.*, **265**, 20634–20640.

Pogány, G., Hernandez, D.J. and Vogel, K.G. (1994) The *in vitro* interaction of proteoglycans with type I collagen is modulated by phosphate. *Arch. Biochem. Biophys.*, **313**, 102–111.

Pringle, G.A. and Dodd, C.M. (1990) Immunoelectron microscopic localization of the core protein of decorin near the d and e bands of tendon collagen fibrils by the use of monoclonal antibodies. *J. Histochem. Cytochem.*, **38**, 1045–1411.

Ratcliffe, A., Fryer, P.R. and Hardingham, T.E. (1985) Proteoglycan biosynthesis in chondrocytes: protein A-gold localisation of proteoglycan protein core and chondroitin sulphate within golgi subcompartments. *J. Cell Biol.*, **101**, 2355–2365.

Rauch, U., Karthikeyan, L., Maurel, P., Margolis, R.U. and Margolis, R.K. (1992) Cloning and primary structure of neurocan, a developmentally regulated, aggregating chondroitin sulfate proteoglycan of brain. *J. Biol. Chem.*, **267**, 19536–19547.

Roden, L. (1980). In *The Biochemistry of Glycoproteins and Proteoglycans*, W.J. Lennarz. (ed.) pp. 267–371. New York: Plenum Publishing Corp.

Roden, L. and Horowitz, M.I. (1978) . In *The Glycoconjugates*, M.I. Horowitz and W. Pigman. (ed.) pp. 3–71. Academic Press.

Romaris, M., Heedia, A., Molist, A. and Bassols, A. (1991) Differnetial effect of transforming growth factor beta on proteoglycan synthesis in human embryonic lung fibroblasts. *Biochim. Biophys. Acta*, **1093**, 229–233.

Roughley, P.J., White, R.J., Magny, M-C., Liu, J., Pearce, R.H. and Mort, J.S. (1993) Non proteoglycan forms of biglycan increase with age in human articular cartilage. *Biochem. J.*, **295**, 421–426.

Roughley, P.J., Melching, L.I. and Recklies, A.D. (1994) Changes in the expression of decorin and biglycan in human articular cartilage with age and regulation by TGF-β. *Matrix*, **14**, 51–59.

Ruoslahti, E. and Yamaguchi, Y. (1991) Proteoglycans as modulators of growth factor activities. *Cell*, **64**, 867–869.

Sampaio, L., Bayliss, M.T., Hardingham, T.E. and Muir, H. (1988) Dermatan sulphate proteoglycan from human articular cartilage. Variation in its content with age and its structural comparison with a small chondroitin sulphate proteoglycan from pig laryngeal cartilage. *Biochem. J.*, **254**, 757–764.

Sandy, J.D., Neame, P.J., Boynton, R.E. and Flannery, C.R. (1991) Catabolism of aggrecan in cartilage explants. Identification of a major cleavage site within the interglobular domain. *J. Biol. Chem.*, **266**, 8683–8685.

Sandy, J.D., Flannery, C.R., Neame, P.J. and Lohmander, L.S. (1992) The structure of aggrecan fragments in human synovial fluid. Evidence for the involvement in osteoarthritis of a novel proteinase which cleaves the Glu 373 — Ala 374 bond of the interglobular domain. *J. Clin. Invest.*, **89**, 1512–1516.

Sant, A.J., Cullen, S.E. and Schwartz, B.D. (1985) Biosynthetic relationships of the chondroitin sulfate proteoglycan with Ia and invariant chain glycoproteins. *J. Immunol.*, **135**, 416–422.

Schmidt, G., Hausser, H. and Kresse, H. (1990) Extracellular accumulation of small dermatan sulphate proteoglycan II by interference with the secretion-recapture pathway. *Biochem. J.*, **266**, 591–595.

Schmidt, G., Hausser, H. and Kresse, H. (1991) Interaction of the small proteoglycan decorin with fibronectin. Involvement of the sequence NKISK of the core protein. *Biochem. J.*, **280**, 411–414.

Schnyder, J., Payne, T. and Dinarello, C.A. (1987) Human monocyte or recombinant interleukin 1's are specific for nthe secretion of a metalloproteinase from chondrocytes. *J. Immunol.*, **138**, 496–503.

Scott, J.E. (1988) Proteoglycan-fibrillar collagen interactions. *Biochem. J.*, **252**, 313–323.

Scott, J.E. and Orford, C.R. (1981) Dermatan sulphate-rich proteoglycan associates with rat tail tendon collagen at teh d band in the gap region. *Biochem. J.*, **197**, 213–216.

Screaton, G.R., Bell, M.V., Jackson, D.G., Cornelius, F.B., Gerth, U. and Bell, J.I. (1992) Genomic structure of DNA encoding the lymphocyte homing receptor CD44 reveals at least 12 alternatively spliced exons. *Proc. Natl. Acad. Sci. USA*, **89**, 12160–12164.

Shimizu, Y., Seventer, G.A., Siraganian, R., Wahl, L. and Shaw, S. (1989) Dual role of the CD44 molecule in T cell adhesion and activation. *J. Immunol.*, **143**, 2457.

Shinomura, T., Nishida, Y., Ito, K. and Kimata, K. (1993) cDNA cloning of PG-M, a large chondroitin sulfate proteoglycan expressed during chondrogenesis in chick limb buds. Alternative spliced multiforms of PG-M and their relationships to versican. *J. Biol. Chem.*, **268**, 14461–14469.

Shinomura, T. and Kimata, K. (1992) Proteoglycan-Lb, a small dermatan sulfate proteoglycan expressed in embryonic chick epiphyseal cartilage, is structurally related to osteoinductive factor. *J. Biol. Chem.*, **267**,

1265–1270.
Shioi, J., Anderson, J.P., Ripellino, J.A. and Robakis, N.K. (1992) Chondroitin sulfate proteoglycan form of the alzheimer's beta-amyloid precursor. *J. Biol. Chem.*, **267**, 13819–13822.
Shioi, J., Refolo, L.M., Efthimiopoulos, S. and Robakis, N.K. (1993) Chondroitin sulfate proteoglycan form of cellular and cell-surface Alzheimer amyloid precursor. *Neurosci. Lett.*, **154**, 121–124.
Siegelman, M.H., Cheng, I.C., Weissman, I.L. and Wakeland, E.K. (1990) The mouse lymphnode homing receptor is identical with the lymphocyte cell surface marker Ly-22: Role of the EGF domain in endothelial binding. *Cell*, **61**, 611–622.
Sorrell, J.M., Lintala, A.M., Mahmoodian, F. and Caterson, B. (1988) Epitope-specific changes in chondroitin sulphate/dermatan sulphate proteoglycan as markers of lymphopoietic and granulopoietic compartments of developing bursae of fabricius. *J. Immunol.*, **140**, 4263–4270.
Sporn, M.B., Roberts, A.B., Wakefield, L.M. and De Crombrugghe, B. (1987) Some recent advnaces in the chemistry and biology of transforming growth factor-beta. *J. Cell Biol.*, **105**, 1039–1045.
Springer, T.A. and Lasky, L.A. (1991) Sticky sugars for selectins. *Nature*, **349**, 196–197.
Stamenkovic, I., Amiot, M., Pesando, J.M. and Seed, B. (1989) A lymphocyte molecule implicated in lymph node homing is a member of the cartilage link protein family. *Cell*, **56**, 1057–1062.
Stamenkovic, I., Aruffo, A., Amiot, M. and Seed, B. (1991) The hemopoietic and epithelial forms of CD44 are distinct polypeptides with differnet adhesion potentials for hyluronate-bearing cells. *EMBO J.*, **10**, 343–348.
Stuhlsatz, H.W., Keller, R., Becker, G., Oeben, M., Lennartz, L., Fischer, D.C. and Greiling, H. (1989) Structure of keratan sulphate proteoglycans: core proteins, linkage regions, carbohydrate chains. In *Keratan Sulphate Chemistry, Biology, Chemical Pathology*, H. Greiling and J.E. Scott. (ed.) London: The Biochemical Society.
Turnbull, J.E., Fernig, D.G., Ke, Y., Wilkinson, M.C. and Gallagher, J.T. (1992) Identification of the basic fibroblast growth factor binding sequence in fibroblast heparan sulfate. *J. Biol. Chem.*, **267**,10337–10341.
Upholt, W.B., Chandrasekaran, L. and Tanzer, M.L. (1993) Molecular cloning and analysis of the protein modules of aggrecans. *Experientia*, **49**, 384–392.
Vassin, H., Bremer, K.A., Knust, E. and Campos-Ortega, J.A. (1987) The neurogenic gene Delta of Drosophila melanogaster is expressed in neurogenic territories and encodes a putitive transmembrane protein with EGF-like repeats. *EMBO*, **6**, 3431–3440.
Vertel, B.M., Grier, B.L., Li, H. and Schwartz, N.B. (1994) The chondrodystrophy, nanomelia: biosynthesis and processing of the defective aggrecan precursor. *Biochem. J.*, **301**, 211–216.
Vogel, K.G., Paulsson, M. and Heinegård, D. (1984) Specific inhibition of type I and type II collagen fibrillogensis by the small proteoglycans from tendon. *Biochem. J.*, **223**, 587–597.
Vogel, K.G., Sandy, J.D., Pogany, G., Robbins, J.R. (1994) Aggrecan in bovine tendon. *Matrix Biology.* **14**, 171–179
Walcz, E., Deak, F., Erhardt, P., Coulter, S.N., Fulop, C., Horvath, P., Doege, K.J. and Glant, T.T. (1994) Complete coding sequence, deduced primary structure, chromosomal localization, and structural analysis of murine aggrecan. *Genomics*, **22**, 364–371.
Walker, A., Turnbull, J.E. and Gallagher, J.T. (1994) Specific heparan sulfate saccharides mediate the activity of basic fibroblast growth factor. *J. Biol. Chem.*, **269**, 931–935.
Watanabe, H., Kimata, K., Line, S., Strong, D., Gao, L., Kozak, C.A. and Yamada, Y. (1994) Mouse cartilage matrix deficiency (*cmd*) caused by a 7 bp deletion in the aggrecan gene. *Nature Genet.*, **7**, 154–157.
Westergren-Thorsson, G., Antonsson, P., Malmstrom, A., Heinegård, D. and Oldberg, A. (1991) The synthesis of a family of structurally related proteoglycans in fibroblasts is differently regulated by TGF-β. *Matrix*, **11**, 177–183.
Westergren-Thorsson, G., Hernnäs, J., Särnstrand, B., Oldberg, Å., Heinegård, D. and Malmström, A. (1993) Altered expression of small proteoglycans, collagen, and transforming growth factor-β_1 in developing bleomycin-induced pulmonary fibrosis in rats. *J. Clin. Invest.*, **92**, 632–637.
Wiedemann, H., Paulsson, M., Timpl, R., Engel, J. and Heinegård, D. (1984) Domain structure of cartilage proteoglycans revealed by rotary shadowing of intact and fragmented molecules. *Biochem. J.*, **224**, 331–333.
Williams, A.F. (1985) Immunoglobulin-related domains for cell surface recognition. *Nature*, **314**, 579–580.
Winnemoller, M., Schmidt, G. and Kresse, H. (1991) Influence of decorin on fibroblast adhesion to fibronectin. *Eur. J. Cell Biol.*, **54**, 10–17.
Winnemoller, M., Schon, P., Vischer, P. and Kresse, H. (1992) Interactions between thrombospondin and the small proteoglycan decorin: interference with cell attachment. *Eur. J. Cell Biol.*, **59**, 47–55.
Wisniewski, H.-G., Burgess, W.H., Oppenheim, J.D. and Vilcek, J. (1994) TSG-β, an arthritis-associated hyaluronan binding protein, forms a stable complex with the serum protein inter-α-inhibitor. *Biochem*, **33**, 7423–7429.
Yamada, H., Watanabe, K., Shimonaka, M. and Yamaguchi, Y. (1994) Molecular cloning of brevican, a novel brain proteoglycan of the aggrecan/versican family. *J. Biol. Chem.*, **269**, 10119–10126.
Yamaguchi, Y., Mann, D.M. and Ruoslahti, E. (1990) Negative regulation of transforming growth factor-beta by

the proteoglycan decorin. *Nature*, **346**, 281–284.
Yang, B., Zhang, L. and Turley, E.A. (1993) Identification of two hyaluronan-binding domains in the hyaluronan receptor RHAMM. *J. Biol. Chem.*, **268**, 8617–8623.
Yang, B., Yang, B.L., Savani, R.C. and Turley, E.A. (1994) Identification of a common hyaluronan binding motif in the hyaluronan binding proteins RHAMM, CD44 and link protein. *EMBO J.*, **13**, 286–296.
Yoshida, E., Sumi, H., Maruyama, M., Tsushima, H., Matsuoka, Y., Sugiki, M. and Mihara, H. (1989) Distribution of acid stable trypsin inhibitor immunoreactivity in normal and malignant human tissues. *Cancer*, **64**, 860–869.
Yoshida, E., Yoshimura, M., Ito, Y. and Mihara, H. (1991) Demonstration of an active component of inter-alpha-trypsin inhibitor in the brains of alzheimer type dementia. *Biochem. Biophys. Res. Commun.*, **174**, 1015–1021.
Yoshida, K., Arai, M., Kohno, Y., Maeyama, K., Miyazono, H., Kikuchi, H., Morikawa, K., Tawada, A. and Suzuki, S. (1993) Activity of bacterial eliminases towards dermatan sulphates and dermatan sulphate proteoglycan. In *Dermatan Sulphate Proteoglycan*, J.E. Scott. (ed.) pp. 55–70. London: Portman Press.
Zimmermann, D.R. and Ruoslahti, E. (1989) Multiple domains of the large fibrobalst proteoglycan versican. *EMBO J.*, **8**, 2975–2981.

7 Heparan Sulphate Proteoglycans: The Control of Cell Growth

John T. Gallagher

CRC Department of Medical Oncology, Paterson Institute, Christie Hospital Wilmslow Road, Manchester, UK

INTRODUCTION

Organisms control the growth, development and function of their cellular constituents by means of both soluble effectors such as growth factors and hormones, and insoluble cell-binding proteins of the extracellular matrix (ECM). These external regulatory molecules act on their target cells through specific receptors in the plasma membrane which generate chemical messengers inside the cell. The heparan sulphate proteoglycans (HSPG's) are essential components of cellular response systems because although in most cases they have no known intrinsic signalling capability they function as co-factors or co-receptors due to the recognition properties of their HS chains which enable them to bind and activate growth factors released from cells in a latent or inactive form (Ruoslahti, 1991, Klagsbrun and Baird, 1991). The activated proteins can then be recognised as ligands by their cognate receptors. The mechanisms of activation are unclear although conformational changes and dimerisation of protein effectors may occur (Gallagher and Turnbull, 1992). It is also possible that proximity-based effects are involved whereby the HS chain binds to both the growth factor and its receptor (Kan *et al.*, 1993). Such fundamental properties of HS were not appreciated when they were first detected on cell surfaces over 20 years ago (Gallagher *et al.*, 1986, for review) and it is only in the past 3 years or so that we have begun to recognise the complexity and scale of the regulatory functions of HS (Lindahl *et al.*, 1994; Gallagher, 1994; Couchman and Woods, 1993). Paramount amongst the HS-dependent growth factors is the so-called basic fibroblast growth factor (bFGF) and this review will describe the structure of the cell surface HSPGs and their role in dual receptor systems for bFGF and other growth factors.

Extracellular Matrix, Volume 2, Molecular Components and Interactions, edited by Wayne D. Comper.

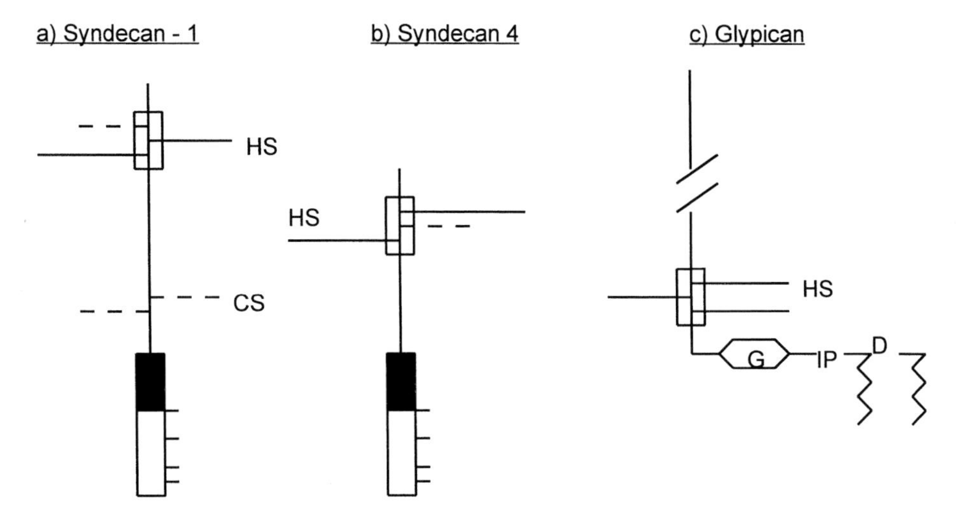

Figure 7.1 Syndecan and Glypican Proteoglycans. The syndecans are transmembrane PGs with the HS chains positioned close to the amino terminus of the ectodomains. The Figure illustrates two members of the syndecan family, syn-1 which is expressed mainly on epithelial cells in adult tissues and syn-4 (amphiglycan or ryudocan) which is synthesised by most cell types. The core protein molecular weights of syn-1 and syn-4 are 32 KDa and 20 KDa respectively. Both occur as hybrid PGs bearing HS (—) and short CS (---) chains. Four conserved tyrosines in the cytoplasmic domain are indicated (-). The glypican family are GPI-anchored PGs bearing 2–3 HS chains. Glypican itself (63 kDa protein) is shown in the Figure. This was the first member of the family to be cloned and it is widely expressed whereas the other GPI-anchored HSPG named cerebroglycan is confined to the developing brain (G = glycan, IP = inositol phosphate, D = diacyl glycerol)

CELL SURFACE HEPARAN SULPHATE PROTEOGLYCANS

Nine cell membrane PGs have been identified that contain HS chains, the two major groups being the members of the syndecan and glypican families (Figure 7.1). The syndecans are type I transmembrane proteins with molecular weights of their protein components as predicted from cDNA sequences in the range 20–35 kDa (Bernfield *et al.,* 1992). There are four PGs in the syndecan family whereas only two glypicans have been formally described as PGs with core protein molecular weights in the region of 60–64 kDa (David, 1993; Stipp *et al.,* 1994). The glypicans are implanted in the membrane by a covalently-linked glycosylphosphatidyl inositol (GPI) anchor at the carboxyl terminus of the protein (Figure 7.1).

The Syndecans And Their Glycanation Sites

The members of the syndecan family are numbered 1-to-4 (syn-1 to syn-4) in chronological order of their discovery. The name syndecan was given to the first of these PGs to be cloned and sequenced (Saunders *et al.,* 1989) and was eventually adopted as the family name (Bernfield *et al.,* 1992). (The Greek root of syndecan apparently implies a function of binding cells to the ECM). Trivial names reflecting with modest accuracy cell lineage associations or functions are often used to describe syn-2, also known as fibroglycan, and syn-4 which has alternative rather arcane names of ryudocan and amphiglycan (David 1993). The syndecans contain 2–3 HS chains positioned distally in the ectodomains of the core proteins towards the amino terminus (Figure 7.1). Preparations of syn-1 and syn-4 also contain, on average, about one chondroitin sulphate (CS) chain in this distal region and, uniquely among the syndecans, syn-1 is further modified by addition of 1 or 2 CS chains located closer to the cell surface (Kokenyesi and Bernfield, 1994; Figure 7.1). The existence of such hybrid PGs has raised the question of what determines whether HS or CS is added to glycanation sites in core proteins. Both GAGs are attached by O-glycosidic linkages to serine (ser) residues at ser-gly sequences and both are assembled on an identical protein linkage tetrasaccharide sequence of (ser)-Xyl-Gal_2-GlcA (Kjellèn and Lindahl, 1991; Gallagher, 1989). Commitment to HS or CS is made by addition of the first amino sugar, either GlcNAcα 1-4 for HS or GalNAcB1-4 for CS, to the GlcA at the non-reducing end of this sequence. Since there is a strong degree of selectivity for HS or CS at glycanation sites located in different regions of the PG core proteins the glycosyl transferase enzymes that add the first amino sugars must recognise additional properties of the protein including local conformational features and peptide domains in the vicinity of the ser-gly residues.

Data from studies on syn-1 and syn-4 indicate that HS addition is strongly favoured where there is a ser-gly repeat (SGSG) and a nearby sequence of acidic residues. In syn-1 two HS chains are added to each ser in this double acceptor site (Kokenyesi and Bernfield, 1994). The two CS sites near to the cell membrane in syn-1 (Figure 7.1) are flanked by acidic amino acids (E G S G E and E T S G E nb E = glutamate); and a similar sequence is also found around the most distal ser-gly nearest the N-terminus (D G S G D nb D = aspartate). Interestingly this site also contains CS (Figure 7.1 and 7.2). A fully glycanated syn-1 will therefore contain two HS chains and three CS chains but in reality the PG is mainly an HS-bearing structure because the CS sites are usually occupied with low efficiency and the chains are relatively short (Rapraeger *et al.*, 1985; Sanderson and Bernfield, 1988). The average number and length of CS chains in syn-1 of mouse mammary epithelial cells is increased by treatment with TGF-β (Rapraeger, 1989). In contrast to syn-1 the three distal glycanation sites in syn-4 (which are in broadly similar sequences to those in syn-1 (Figure 7.2) can be substituted with HS or CS but HS is the predominant GAG at each site (Shworak *et al.*, 1994). Thus although some properties of the core protein sequence can be equated with selective glycanation, one must be cautious about predicting the exact glycan structure of a PG solely on the basis of amino acid sequence information. The influence of the variable quantities of CS on syndecan function is unclear. It is not known whether syn -2 and syn-3 are also synthesised as hybrid HS/CS PGs.

Apart from the amino acid sequences in the regions of HS glycanation the ectodomains of the syndecans show little similarity in primary structure. Close sequence homologies which unite the syndecans in a single family are readily identified in the transmembrane and cytoplasmic domains (Bernfield *et al.*, 1992) suggesting that these regions plus the HS chains are mainly responsible for the regulation of PG function. The spatial and temporal variations in expression of the syndecans in the developing embryo and the striking differences in patterns of synthesis in cultured cells derived from different tissues provide strong, though indirect evidence for distinct biological roles of these PGs (Kim *et al.*, 1994; David, 1993; Bernfield *et al.*, 1992). Some indications for specificity are beginning to emerge: syn-4 for example is the only HSPG that is localised in focal adhesions of cultured cells and it is likely to have a significant influence on cell adhesion and migration on ECM sultrates such as fibronectin (Woods and Couchman, 1994). Syn-1, which is mainly

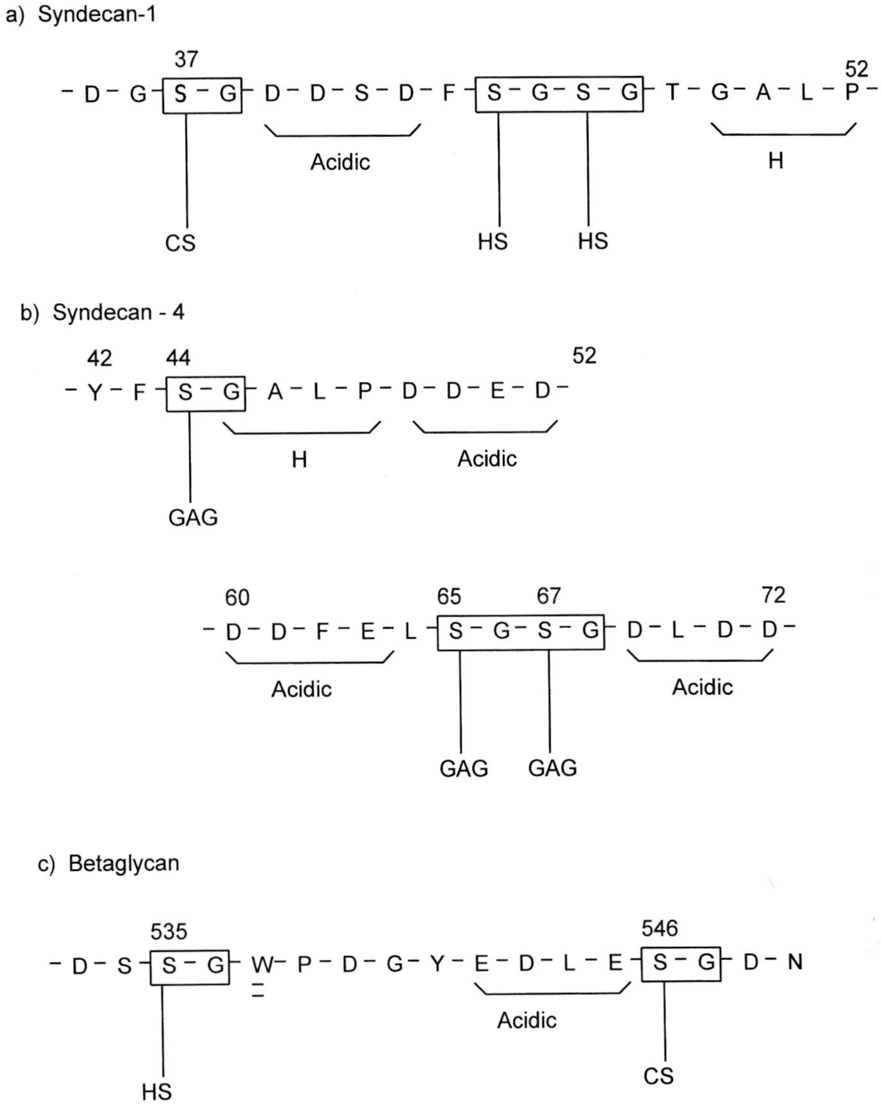

a) Syndecan-1

37 52
- D - G - [S - G] - D - D - S - D - F - [S - G - S - G] - T - G - A - L - P -
Acidic H
CS HS HS

b) Syndecan - 4

42 44 52
- Y - F - [S - G] - A - L - P - D - D - E - D -
H Acidic
GAG

60 65 67 72
- D - D - F - E - L - [S - G - S - G] - D - L - D - D -
Acidic Acidic
GAG GAG

c) Betaglycan

535 546
- D - S - [S - G] - W - P - D - G - Y - E - D - L - E - [S - G] - D - N
Acidic
HS CS

Figure 7.2 Amino acid sequences that promote the synthesis of HS and CS in syndecan and *b*etaglycan. Consensus sequences for synthesis of GAGs at ser residues in protein acceptors are SGXG (Bourdon *et al.*, 1987) or an SG flanked by acidic residues. HS synthesis is strongly favoured at SGSG sequences in close proximity to a short stretch of acidic amino acids. Some of these features are illustrated in the known GAG attachment sites in syn-1 (a), syn-4 (b) and βetaglycan (c). The sequences shown for the two syndecans correspond to the glycanation sites close to the N-terminus of the ectodomain (Figure 7.1). The S-G-S-G sequence in syn-1 has an HS on both serines and CS is present on ser 37 where the SG is flanked by acidic residues. In syn-4 ser residues at positions 37, 65 and 67 are mainly substituted by HS but CS can occur as an alternative GAG at each site. Syn-1 and syn-4 both contain a hydrophobic (H) GALP sequence which may be important for glycanation. βetaglycan (c) contains both HS and CS, with HS addition at ser 535 favoured by the acidic domain and an adjacent tryptophan (W). The ser-gly where CS is attached (ser 546) is flanked by acidic amino acids. Glycanation of βetaglycan is an optional modification and the protein can be expressed on the cell surface in a non-glycanated form. The lack of a double GAG acceptor sequence (S-G-S-G) may cause the protein to be a less efficient primer of HS synthesis than the syndecans. SG glycantation sites are boxed. Because of its extended length the N-terminal glycanation region of syn-4 is shown as two separate sequences from amino acid residues 42–52 and 60–72. Numbering of peptide sequences is from the N-terminal amino acid of the complete protein sequence.

produced by epithelial cells, is localised to baso-lateral regions of simple epithelium and over-expression of the syn-1 gene suppresses the transformed phenotype of cultured cells (Leppa *et al.*, 1992). Surprisingly this suppressive effect can be produced by transfection with the syn-1 ectodomain (Mali *et al.,* 1995).

Glypicans

The core proteins of the two GPI-anchored HSPGs are homologous over their entire length and contain 2-to-3 ser-gly repeats in similar sequences to the HS-glycanation sites in syn-1 (David *et al.,* 1990; Stipp *et al.,* 1994). The first of these proteins to be sequenced was named glypican and this was then used as the family name following the tradition established with the syndecans. Glypican is synthesised by many types of cell culture and it is normally co-expressed with one or more of the syndecans (David, 1994). The other known GPI-anchored HSPG has been named cerebroglycan because it appears to be produced only in the embryonic brain (Stipp *et al.,* 1994). A developmentally-regulated transcript OCI-5 cloned from a rat intestinal cell line has a deduced amino acid sequence which is similar to glypican and represents a third member of this family. To date OCI-5 has not been identified as an HSPG. Functional differences between the glypicans have not been demonstrated. Glypican appears to lack any CS chains being present on cells exclusively as an HS-bearing species (Shworak *et al.*, 1994).

Glycanated Variants Of Membrane Proteins

Glycanation by HS is an optional modification step for three other transmembrane proteins —(i) β etaglycan (Massague, 1992), (ii) a receptor for fibroblast growth factors (Takagi *et al.*, 1994) and (iii) the CD44/Hermes antigen (Brown *et al.*, 1991).

βetaglycan

The PG form of βetaglycan contains only one HS chain and it is often present on cell surfaces as a hybrid (HS and CS) PG. It was originally known as the Type 3 TGF-β receptor because it binds TGF-β by its protein core; it is not a signalling protein but may facilitate ligand transfer to the TGF-β receptors 1 and 2 which together form the signal transducing complex (Lopez-Casillas *et al.,* 1994). Synthesis of βetaglycan is widespread in cultured cells but it is a low abundance HSPG by comparison with the syndecans and glypicans. The HS glycanation site contains only a single ser-gly and a proximal acidic sequence that is essential for efficient initiation of HS synthesis (Figure 7.2; Zhang and Esko, 1994).

Fibroblast Growth Factor Receptor

Rat parathyroid cells synthesise an HSPG variant of FGFR (Takagi *et al.,* 1994). The glycanated receptor is a two or three loop form of FGFR-2 (III b variant, see later) that binds acidic FGF (aFGF) and KGF (keratinocyte growth factor or FGF7) but not bFGF (Fantl *et al.,* 1993; Fernig and Gallagher, 1995). Only a minor fraction of the receptors contain an HS chain which is located in the Ig—loop (loop III) nearest the membrane. The parathyroid cells thus appear to have merged the dual receptor mechanism for the FGFs (ie HSPG plus FGFR) into a single component. It will be interesting to examine other cells and tissues for the presence of HS-bearing FGFRs although FGFR-2 is not modified by HS when expressed in CHO cells (Takagi *et al.*, 1994). It is assumed that this novel receptor PG has the normal signalling properties of the FGFRs.

CD44

CD44 was originally identified as the lymphocyte homing receptor (Jalkanen *et al.*, 1988) but it is now known to be a widely expressed protein involved in the control of cell adhesion and migration (Lesley *et al.*, 1994). It is one of the cell surface molecules that binds to hyaluronic acid (HA; Stamenkovic *et al.*, 1991). More recent work has shown that alternative splicing generates many protein isoforms of CD44 and one of these (CD44 E variant or epican) synthesised in keratinocytes is efficiently glycanated by HS (Kugelman *et al.*, 1992). CD44 E incorporates the V3 exon that contains ser-gly repeats in a suitable context to serve as sites for HS synthesis (Jackson *et al.*, 1995). Keratinocytes also synthesise other HSPGs (e.g. syndecan 1, Sanderson *et al.*, 1992) and the particular role fulfilled by HS-CD44E is unknown; it does not bind to HA (Jackson *et al.*, 1995).

Perlecan—On The Membrane Periphery

The only other HSPG to be identified is present in the pericellular matrix and basement membranes rather than being integrated in the plasma membrane. This near-ubiquitous and relatively abundant matrix - associated component is called perlecan (due to its beads-on-a-string appearance in the electron microscope) and though it contains a very large core protein core (470 KDa) the HS chains, 2–3 in number, are grouped at one end of the protein (Noonan *et al.*, 1993). The close proximity of the HS chains in perlecan and in the membrane HSPGs reflects the main peptide sequence requirements for initiation of HS synthesis and may indicate that inter-chain co-operative effects are significant in HS function.

STRUCTURE OF HEPARAN SULPHATE

Disaccharides and Domains

The heparan sulphates are glycosaminoglycans (GAGs) and in common with other members of the GAG family they are characterised by a high negative charge density and the presence of a repeating disaccharide unit (Kjellèn and Lindahl, 1991; Gallagher, 1989). The HS-disaccharide consists of α, β1-4 linked glucosamine (GlcN) and uronic acid (UA) in polymer chains that vary in length from 50–150 disaccharide units. Although the repeat unit is a recognisable feature of HS the polymer is actually far more complex than is suggested by a repeating unit structure. The glucosamine residue is substituted at the primary amine by an acetyl (Ac) or sulphate group and the uronic acid is present as glucuronic acid (GlcA) when linked to C-1 of GlcNAc but when the GlcN is N-sulphated the uronate is more commonly in the form of iduronate (IdoA) rather than GlcA (Figure 7.3). Structural complexity and variability in HS are further and significantly amplified by addition of ester-linked (O) sulphate groups at different positions around the sugar rings. In broad quantitative terms, the majority of HS species contain about equal amounts of N-acetylated and N-sulphated disaccharides but they are not uniformly distributed in the GAG chain.

a) N-acetylated

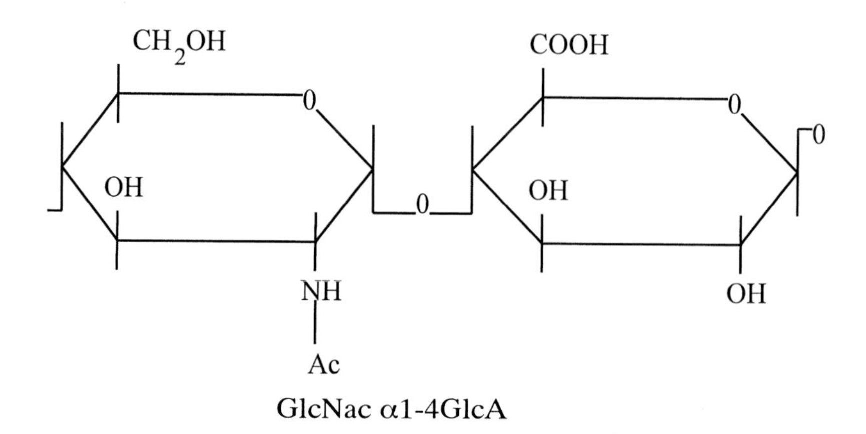

b) N-sulphated

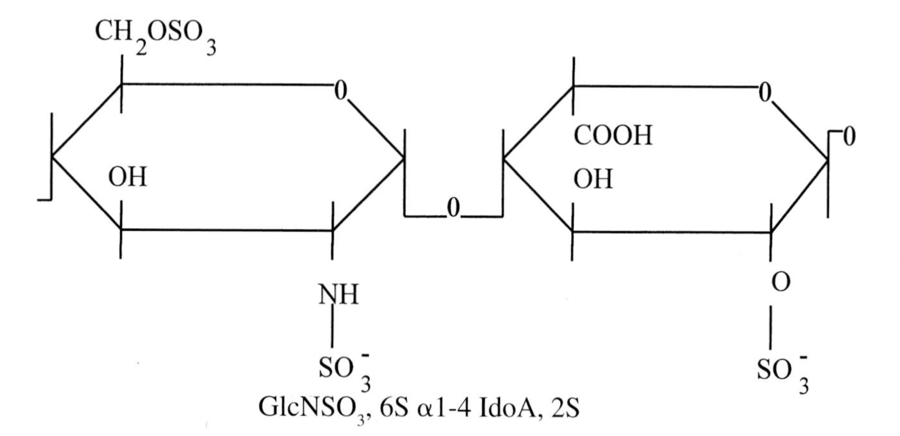

Figure 7.3 Disaccharides in HS. HS contains N-acetylated and N-sulphated disaccharides in roughly equal proportions. The N-acetylated units (a) contain GlcA and are rarely sulphated but may contain a sulphate group at C-6 when they are adjacent to N-sulphated units. N-sulphated disaccharides often contain IdoA rather than GlcA and when extensively modified O-sulphates are present at C-2 of IdoA and C-6 of $GlcNSO_3$ (as illustrated); occasionally, as in the AT-III binding site, the $GlcNSO_3$ is sulphated at C-3. Variations in degree and position of O-sulphation in the N-sulphated disaccharides give rise to considerable structural diversity in HS.

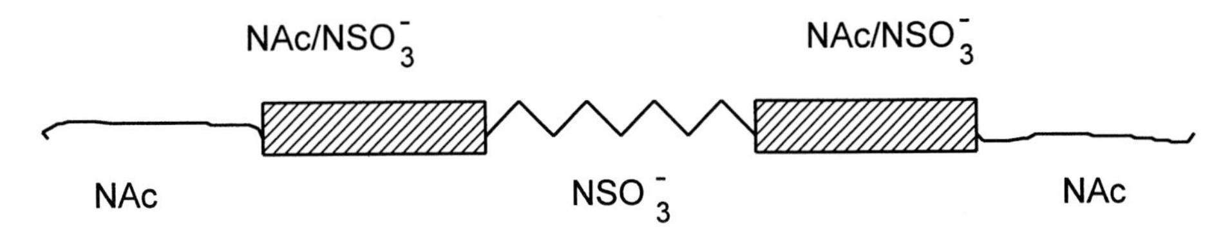

Figure 7.4 Sulphated region of HS. In the HS chain N-acetylated and N-sulphated disaccharide units tend to occur in separate sequences. The N-sulphated domains (^^^) contain the majority of the O-sulphate groups and IdoA residues whereas the N-acetylated regions (—) are largely unsulphated. Though not formally proven it is likely that there are transition zones (▨) between the regions of high and low sulphation comprised of alternate sequences of N-acetylated and N-sulphated units with some C-6 sulphation of GlcNAc and $GlcNSO_3$, and containing small amounts of non-sulphated IdoA.

The characteristic feature of the polymeric structure of HS, that distinguishes it from other GAGS including the chemically-related heparin, is that the N-and O-sulphated sugars tend to occur in clusters from 2–8/9 disaccharides in length and these clusters are separated by regions containing mainly N-acetylated units with few sulphate groups (Turnbull and Gallagher, 1991a, Figure 7.4).

In heparin, a product of connective tissue mast cells over 80% of disaccharides are N-sulphated and the polymer is heavily and quite uniformly substituted with O-sulphate groups (Lindahl, 1989).

Biosynthesis

The clustering of sulphation in HS arises through complex mechanisms that operate during biosynthesis when a series of structural modifications are imposed on a non-sulphated nascent polymer chain (Lindahl *et al.,* 1994). As described earlier the HS chain is assembled on protein acceptors primed with the GAG-linkage sequence of Xyl-$(Gal)_2$-GlcA. Synthesis begins with the sequential, alternate addition of GlcNAc and GlcA to form the HS precursor heparan, and as the chain length grows modifications are initiated by an N-sulphotransferase which converts GlcNAc residues to $GlcNSO_3$ (Petterson *et al.,* 1994; Wei *et al.,* 1993). Additional modifications then occur in regions of N-sulphation and are most extensive where N-sulphated disaccharides run in sequences. An epimerase converts GlcA to IdoA and O-sulphotransferases then transfer sulphate groups to C-2 of IdoA, C-6 and C-3 of $GlcNSO_3$ and C-6 of GlcNAc (Lindahl, 1989). Those modifications occur to different degrees in different cell types (see for example Turnbull and Gallagher, 1991), the main variables being the concentration and patterns of O-sulphation (Gallagher and Walker, 1985; Lyon *et al.,* 1994). A consequence of these variations in polymer sulphation is that the HS family is highly polymorphic with a corresponding diversity in protein recognition properties and cellular functions (Kjellèn and Lindahl, 1991).

Targeted Modifications

The action of the N-sulphotransferase, the enzyme of primary modification, is critical to the final molecular organisation of HS. The enzyme is regulated not only with respect to the degree of modification it achieves but also in relation to the sites of modification. The lengths of the N-sulphated clusters rarely exceed 9 disaccharide units and sequence analysis of HS in the inner regions of the chain indicates that the disaccharide units near to the protein are not modified, the first N-sulphate group being about ten units downstream from the protein linkage region. The most proximal (to the protein core) cluster of N-sulphates is a further 6–10 disaccharides from the protein (for review see Gallagher *et al.*, 1992; Lyon *et al.*, 1994). This region is identified by the enzyme heparinase which attacks HS (and heparin) at $GlcNSO_3$ (±6S)-IdoA,2S units (Linhardt *et al.*, 1991). IdoA,2S is most commonly found in the relatively long N-sulphated domains and heparinase susceptibility of a region of HS is indicative of a sulphation cluster (Turnbull & Gallagher, 1991b). The relatively precise positioning of the proximal $GlcNO_3$ and IdoA, 2S is consistent with the view that the polymer modifying enzymes that convert the heparan precursor to HS are targeted to distinct regions of the chain. The significance of this is presently unknown.

Transition Zones

Despite the predominant occurrence of N-sulphates in contiguous sequence they are also found as solitary units (i.e. flanked by sequences of N-acetylated disaccharides) and in so-called mixed sequences where N-acetylated and N-sulphated units occur in an alternate arrangement. These mixed sequences are common in HS and they are the only regions where GlcNAc residues are sulphated at C-6 (Gallagher *et al.*, 1986). Though not formally established it is likely that mixed sequences represent a transition zone between the sulphate-free and highly-sulphated domains of HS (Figure 7.4). These regions are important areas of structural diversity. They are deficient in 0-sulphates in transformed and tumour cells (Winterbourne and Mora, 1981; Pejler and David, 1987) and they define the most common regions in which the antithrombin III binding site occurs in HS and heparin (Lindahl *et al.*, 1994). This sequence is a pentasaccharide of structure:

[$GlcNSO_3$-IdoA] - GlcNAc (6S) - GlcA - $GlcNSO_3$(3S,6S) - IdoA (2S) - $GlcNSO_3$ (6S).

The unit in square brackets is not part of the AT-III site but is the usual adjacent disaccharide. It is shown to illustrate that the site partially overlaps a mixed sequence of N-acetylated and N-sulphated units. The sulphation pattern and sugar sequence of the AT-III binding region is highly distinctive and the hallmark of the sequence is the 3-O-sulphate on the central $GlcNSO_3$, a rare substituent in HS which occurs mainly, though not exclusively in this sequence. The AT-III site is vital for the anticoagulant properties of both therapeutic heparins and HS on endothelial cells. It acts by inducing a conformational change in AT-III which greatly accelerates the rate at which this protease inhibitor binds and inactivates the pro-coagulant proteases thrombin and factor Xa (Evans *et al.*, 1992).

INTERACTION OF HS WITH bFGF

Dual Receptor Mechanism For The FGFs

bFGF is a member of the fibroblast growth factor (FGF) family comprised of nine members of regulatory molecules, which stimulate the growth, migration and differentiation of various cell types (Fernig and Gallagher, 1995). The FGFs bind in a selective manner to four structurally-related tyrosine kinase receptors, each encoded by a separate gene and for at least three of these receptors, the products of the flg (FGFR-1), bek (FGFR-2) and FGFR-3

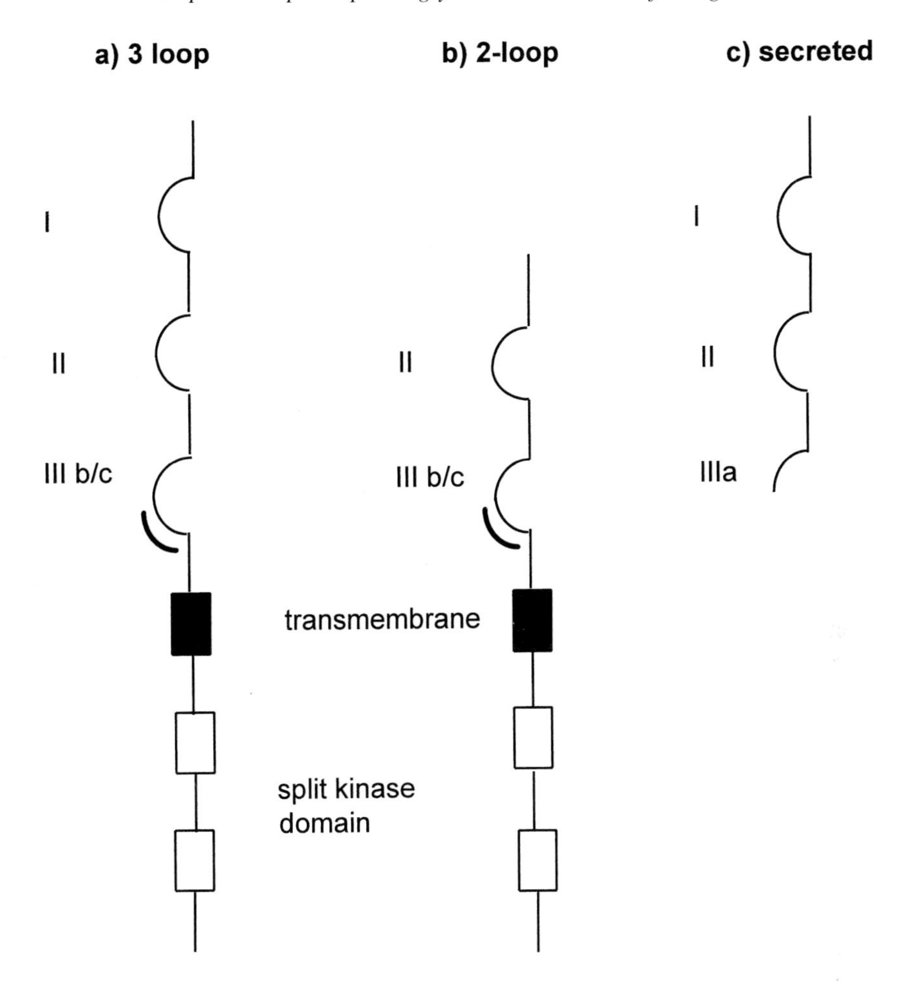

Figure 7.5 Structural characteristics of the tyrosine kinase receptors for the FGFs, the FGFRs. The receptors are encoded by four structurally-related genes and isoforms are generated from three of these genes (FGFR-1 or flg, FGFR-2 or bek and FGFR-3) by alternative splicing. The ectodomains typically contain two or three Ig loops and differential use of three exons that encode the region of Ig loop III nearest the membrane (heavy curved line) gives rise to IIIb or IIIc transmembrane forms and the IIIa secreted form. The IIIb and IIIc variants of both the two loop and three loop bek receptors have distinct binding properties for the FGFs. bFGF and aFGF bind to the IIIc variant whereas the IIIb counterpart binds aFGF and FGF-7 (keratinocyte growth factor) but not bFGF. In common with other receptors of the tyrosine kinase type, the FGFRs must dimerise in order to elicit intracellular signals

genes alternative splicing generates a number of protein receptor isoforms, with distinct ligand binding properties (Figure 7.5; for reviews see McKeehan *et al.,* 1994; Givol and Yayon, 1992).

It has been recognised for many years that the FGFs bind to heparin, a property that is widely exploited in purification of these proteins, and that FGFs are bound to HS in the pericellular matrix. It was only in 1991 with the publication of work first by Yayon and colleagues and then by Rapraeger and co-workers (Yayon *et al.,* 1991; Rapraeger *et al.,*

1991; Olwin and Rapraeger, 1992) that it become clear that HS was an essential co-receptor for expression of biological activity of bFGF. Subsequently it was shown that acidic FGF (FGF-1) and FGF-4 also required HS for activity (Guimond *et al.*, 1993). These findings led to the search for the structures of growth factor binding sites in HS. To date most studies have concentrated on the binding site for bFGF.

The High Affinity Binding Site In HS For bFGF. A Site For Growth Factor Activation

It is clear from several investigations of HS and heparin, in their native form or after selective desulphation and specific scission, that N-sulphate groups and IdoA, 2S residues are essential for binding to bFGF (Maccarana *et al.*, 1993; Ishihara *et al.*, 1993 b, c; Habuchi *et al.*, 1992; Turnbull *et al.*, 1992).

The natural GAG co-receptor or co-factor for bFGF is HS rather than heparin and the shortest HS structure with minimal sulphation that binds to bFGF is a pentasaccharide with a single IdoA, 2S of general sequence:

$$\text{UA-GlcNSO}_3\text{-UA-GlcNSO}_3\text{-IdoA, 2S}$$

in which the UA can be GlcA or IdoA (Maccarana *et al.*, 1993). Sequences of this length bind to bFGF with low affinity and are unable to induce an appropriate configuration in the growth factor compatible with mitogenic activity (Guimond *et al.*, 1993).

Other investigations of structure-activity relationships in HS have exploited the specificity of the enzyme heparitinase which only cleaves HS at hexosaminidic linkages to GlcA (linkage cleaved GlcN.R (6S) *a*1-4 GlcA where R = Ac or SO_3^-, Lindhardt *et al.*, 1991). This specificity means that the enzyme mainly attacks the N-acetylated and, to a lesser extent, the alternate sequences of HS (Figure 7.4) and releases the IdoA-rich N-sulphated domains which can be fractionated and tested for binding activity. Heparitinase scission of cultured fibroblast HS led to the affinity isolation and characterisation of a saccharide sequence named Oligo-H which bound to bFGF as strongly as the intact HS (Turnbull *et al.*, 1992). The sequence contained seven disaccharide units (i.e. dp14 fragments, dp = degree of polymerisation) and it had the following structure:

$$\text{GlcA-GlcNSO}_3^-\ [\text{–IdoA, 2S – GlcNSO}_3]_5\text{–IdoA – GlcNAc}$$

The striking feature of this sequence is the internal repeat of five identical disaccharides (IdoA,2S-GlcNSO$_3$) which is similar to a typical heparin sequence (basic unit IdoA, 2S-GlcNSO$_3$,6S) but without any 6-O-sulphate groups. A slight reduction in bFGF affinity was noted in HS-saccharides that contained four rather than five internal repeat units.

Heparitinase-released fragments from porcine mucosal HS of similar size (dp12–14) and structure to Oligo-H were able to induce mitogenic activity of bFGF in cell cultures treated with sodium chlorate which inhibits GAG sulphation (Walker *et al.*, 1994). These cells are unresponsive to bFGF unless saccharides of the appropriate structure are also added to the culture medium (Rapraeger *et al.*, 1991). Heparin fragments of size $dp_{10\text{-}12}$ are also able to activate bFGF in HS-deficient cells but shorter sequences are inactive (Guimond *et al.*, 1993; Ishihara *et al.*, 1993a). If we assume that the HS binding site in bFGF accommodates a pentasaccharide (Macccarana *et al.*, 1993) then active $dp_{10\text{-}12}$ heparin or $dp_{12\text{-}14}$ HS saccharides could, in principle, bind two bFGF molecules to form a dimeric ligand that would in turn facilitate dimerisation and signal transduction by the FGFRs (Figure 7.6; Ornitz *et al.*, 1992; Hsu *et al.*, 1993). An alternative possibility suggested from findings that FGFRs contain a heparin/HS binding site (Kan *et al.*, 1993; McKeehan *et al.*, 1994) is that active saccharides promote bFGF action because they have two protein binding regions, one for bFGF and the other for FGFR. In this model the HS saccharide is working through a proximity based mechanism by bringing ligand and receptor into close or direct contact.

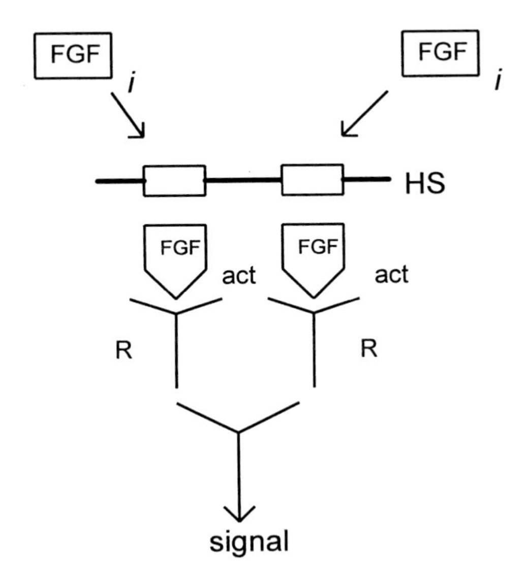

Figure 7.6 Activation of bFGF by HS : dimerisation model. HS is an essential co-receptor or co-factor for bFGF but its mechanism of action is unclear. One possibility is that the active site in HS, which is an extended saccharide sequence, can accommodate two bFGF molecules. Latent or inactive forms of the growth factor (FGFi) bind in close proximity to each other in the active site (▭—▭) The bFGFs undergo a conformational change (FGFact) exposing binding sites for FGFRs (R). These FGFs positioned close together on the HS chain act as a dimeric ligand, facilitating dimerisation and signal transduction by the FGFRs.

An HS-mediated conformational change in bFGF may be a necessary accompaniment to any activation process (Figure 7.6).

At the present time it is not clear whether the high affinity Oligo-H sequence itself, devoid of 6-sulphate groups, encompasses all the structural requirements of the bFGF active site. Adequate quantities of such material have not been prepared for testing in bioassays. One study has suggested that Oligo-H is an incomplete structure and that one or more 6-sulphate groups must be present in the active site with the added suggestion that they are involved in binding to FGFR rather than to bFGF (Guimond *et al.*, 1993). However, data from other investigations are not consistent with a role for 6-sulphates in bFGF activation (Walker *et al.*, 1994; Ishihara *et al.*, 1994). Thus, whilst many of the essential features of the active site for bFGF in HS have been elucidated some points of fine detail need to be resolved and the activation mechanism remains as open question.

It should be stressed that bFGF binding sequences are not restricted to any specific type of HSPG and it is not known whether a particular HSPG species has the major role to play on the cell surface in facilitating the action of an individual growth factor. For instance, is bFGF activity promoted more effectively by syn-1 than syn-4? Can the actions of HS chains on the glypicans be distinguished in any way from those on the syndecans or CD44E? Does the nature of the FGFR isoform influence the effectiveness of different HSPG co-receptors? In general the HSPG core proteins do not determine the sulphation pattern of their HS chains. HS fine structure is controlled by the cell or tissue of origin (Gallagher and Walker, 1985). This has been more clearly shown recently in studies on syn-1 which is modified with HS chains of variable length, sulphation and binding properties when synthesised by cultured epithelial, fibroblastic and lymphoid cells (Sanderson *et al.*, 1992; Sanderson *et al.*, 1994; Kato *et al.*, 1994; Lyon *et al.*, 1994). Furthermore clear differences in the HS (and CS) content of syn-1 were present when the PG was isolated from simple or stratified epithelial tissues (Sanderson and Bernfield, 1988).

HS BINDING PROPERTIES OF OTHER GROWTH FACTORS AND CYTOKINES

Acidic FGF (aFGF) a close relative of bFGF within the FGF family, shares a similar dependence with bFGF on HS for its biological activity (Guimond *et al.*, 1993), requiring N-sulphated saccharides of at least 10–12 sugar residues in length for cell signalling in bioassays with HS-deficient cells (Ishihara, 1994). Comparative studies with partially desulphated heparins have demonstrated that aFGF recognises a different sulphation pattern from bFGF (Guimond *et al.*, 1993) and a requirement for some 6-O-sulphate groups as well as N-sulphates and IdoA, 2S has been established (Ishihara 1994). It would be too simplistic however to suggest that the minimal sulphation of the active site for a aFGF corresponds to an Oligo-H type structure with an overlay of several 6-sulphates. This is because it has been shown that at a specific stage in development of the murine neuroepithelium an HS species is produced that binds and activates aFGF but not bFGF (Nurcombe *et al.*, 1993). If the active site for aFGF was Oligo-H plus 6-sulphate, it would also activate bFGF because this growth factor is "blind" to most, if not all, 6-sulphates.

Investigations of the saccharide-binding properties of other growth factors and chemokines that are regulated by HS have been few in number but from the limited information available it seems likely that FGF-4 (Rapraeger *et al.*, 1993, Ishihara, 1994), the chemokine IL-8 (Witt and Lander, 1994) and hepatocyte growth factor (Lyon *et al.*, 1994) all have their own special requirements for high affinity binding to HS. We can safely anticipate more important developments in this general area during the next few years as awareness grows of the importance of HS as a central player in the control of cell behaviour and as methods for the purification and sequencing of HS saccharides continue to improve.

References

Bernfield, M., Kokenyesi, R., Kato, M., Hinkes, M.T., Spring, J., Gallo, R.L. and Lose, E. (1992) Biology of the syndecans: A family of transmembrane heparan sulphate proteoglycans. *Ann. Rev. Cell Biol,* **8**, 365–393.

Bourdon, M.A., Krusius, T., Campbell, S., Schwartz, N.B. and Ruoslahti, E. (1987) Identification and synthesis of a recognistion signal for the attachment of glycosaminoglycans to proteins. *Proc. Natl. Acad. Sci. USA,* **84,** 3194–3198.

Brown, T.A., Boachard, T., St John, T., Wayner, E. and Carter, W.G. (1991) Human keratinocytes express a new CD44 core protein (CD44E) as a heparan sulphate intrinsic membrane proteoglycan with additional exons. *J. Cell. Biol.* **113**, 207–221.

Couchman, J.R. and Woods, A. (1993) Structure and biology of pericellular proteoglycans. In *Cell Surfaces and Extracellular Glycoconjugates: Structure and Function* ed by D.D. Roberts & R.P. Mechem, New York: *Academic Press*, 33–82.

David, G. (1993) Integral membrane heparan sulphate proteoglycans. *FASEB J.* **7**, 1023–1030.

Evans, D.L., Marshall, C.J., Christey, P.B. and Carrell, R.W. (1992) Heparan binding site, conformational change and activation of antithrombin. *Biochemistry,* **31**, 12629–12642.

Fantl, W.J., Johnson, D.E. and Williams, C.T. (1993) Signalling by receptor tyrosine kinases. *Annu. Rev. Biochem.,* **62,** 453–481.

Fernig, D. and Gallagher, J.T. (1995) Fibroblast growth factors and their receptors: an information network controlling tissue growth, morphognesis and repair. *Progress in Growth Factor Research,* **5**, 353–377.

Gallagher, J.T. and Walker, A. (1985) Molecular distinctions between heparan sulphate and heparin. Analysis of sulphation patterns indicates tht heparan sulphate and heparin are separate families of N-sulphated polysaccharides. *Biochem. J.* **230**, 665–674.

Gallagher, J.T., Lyon, M. and Steward, W.P. (1986) Structure and function of heparan sulphate proteoglycans. *Biochem. J.* **236,** 313–325.

Gallagher, J.T. (1989) The extended family of proteoglycans: social members of the pericellular zone. *Curr. Op. Cell Biol.,* **1**, 1201–1218.

Gallagher, J.T. and Turnbull, J.E. (1992) Heparan sulphate in the binding and activation of basic fibroblast growth factor. *Glycobiology,* **2**, 523–528.

Gallagher, J.T., Turnbull, J.E. and Lyon, M. (1992) Patterns of sulphation in heparan sulphate polymorphism based on a common structural theme. *Int. J. Biochem,* **24**, 553–560.

Gallagher, J.T. (1994) Heparan sulphates as membrane receptors for the firbroblast growth factors. *Wur. J.Clin. Chem. Clin. Biochem,* **32**, 234–247.

Givol, D. and Yayon, A. (1992) Complexity of FGF receptors: genetic basis for structural diversity and functional specificity. *FASEB J,* **6**, 3362–3369.

Guimond, S., Maccarana, M., Olwin, B.B., Lindahl, U. and Rapraeger, A.C. (1993) Activity and inhibitory heparin sequences for FGF - 2 (basic FGF). *J.Biol. Chem,* **268**, 23906–23914.

Habuchi, H., Suzuki, S., Saito, T., Tamura, T., Harada, T., Yoshida, K. and Kimata, K. (1992) Structure of a heparan sulphate oligosaccharide that binds to basic fibroblast growth factor. *Biochem. J.,* **285**, 805–813.

Ishihara, M. (1994) Structural requirements in heparin for binding and activation of FGF - 1 and FGF - 4 are different from that for FGF - 2. *Glycobiology,* **4**, 817–824.

Ishihara, I., Shaklee, P.N., Yang, Z., Liang, W., Wei, Z., Stack, R.J. and Holme, K. (1994) Structural features in heparin which modulate specific biological activities mediated by fibroblast grwoth factor. *Glycobiology*, **4**, 451–458.

Ishihara, M., Tyrell, D.T., Strauber, G.B., Cousens, L.S. and Stack, R.J. (1993a) Preparation of affinity-fractionated, heparin-derived oligisaccharides and their effects on selected biological activities mediated by basic fibroblast growth factor. *J.Biol. Chem,* **268**, 4675–4683.

Ishihara, M., Guo, Y., Wei, Z., Swiedler, S.J., Orellana, A.Z., Hirschberg, C.B. (1993b) Regulation of biosynthesis of basic fibroblast growth factor binding domains of heparan sulphate by heparan sulphate N-deacetylase/n-sulphotransferase supression. *J. Biol. Chem.,* **268**, 20091–20095.

Ishihara, M., Guo,Y. and Swiedler, S.J. (1993c) Selective impairment of the synthesis of basic fibroblast growth factor binding domains of heparan sulphate in a COS cell mutant defective in N-sulphotransferase. *Glycobiology,* **3**, 83–88.

Jackson, D., Bell, J.I., Dickinson, R., Timans, J., Sheilds, J. and Whittle, N. (1995) Proteoglycan forms of lymphocyte homing receptor CD44 are alternatively spliced variants containing the v3 exon. *J. Cell. Biol,* **128**, 673–685.

Kan, M., Wang, F., Xu, J., Crabb, J.W., Hon, J. and Mc Keehan, W.L. (1993) An essential heparin-binding domain is in the fibroblast growth factor receptor kinase. *Science,* **259**, 1918–1921.

Kato, M., Wang, H., Bernfield, M., Gallagher, J.T. and Turnbull, J.E. (1994) Cell surface syndecan 1 on distinct cell types differs in fine chain structure and ligand binding of its heparan sulphate chains. *J. Biol. Chem.,* **269**, 18881–18890.

Kim, C.W., Goldberger, O.A., Gallo, R.L. and Bernfield, M. (1994) Members of the syndecan family of heparan sulphate proteoglycans are expressed in distinct cell, tissue and development-specific patterns. *Molecular Biology of the Cell,* **5**, 797–805.

Kjellèn, L. and Lindahl, U. (1991) Proteoglycans – Structures and interactions. *Annu. Rev. biochem,* **60**, 443–476.

Klagsbrun, M. and David, A. (1991) A dual receptor system is required for basic fibroblast growth factor activity. *Cell,* **67**, 229–231.

Kokenyesi, R. and Bernfield, M. (1994) Core protein structure and sequence determine the site and presence of heparan sulphate and chondroitin sulphate on syndecan - 1. *J. Biol. Chem.,* **269**, 12304–12309.

Kugelman, L.C., Ganguly, S., Haggerty, J.G., Weismann, M. and Milstone, L.M. (1992) The core protein of epican, a heparan sulphate proteoglycan on keratinocytes, is an alternative form of CD44. *J Invest. Dermatol.*, **99**, 381–385.

Leppa, S., Mali, M., Miettinen, H. and Jalkanen, M. (1992) Syndecan expression regulates cell morphology and growth of mouse mammary epithelial tumour cells. *Proc. Natl, Acad. Sci. USA*, **89**, 932–936.

Lindahl, U. (1989) Biosynthesis of heparin and related polysaccharides. *Heparin* (ed by Lane, D.A. and Lindahl, U.), 159–189 Edward Arnold (London).

Linhardt, R.J., Turnbull, J.E., Wang, H., Loanathan, D and Gallagher, J.T. (1991) Examination of the substrate specifications of heparin and heparan sulphate lyases. *Biochemistry*, **29**, 2611–2617.

Lopez-Casillas, F., Paune, H., Andres, J.C. and Massague, J. (1994). β etaglycan can act as a dual modulator of TFG-B access to signalling receptors: mapping of ligand binding and GAG attachment sites. *J. Cell. Biol.*,**124**, 557–568.

Lyon, M., Deakin, J and Gallagher, J.T. (1994) Liver heparan sulphate structure: a novel molecular design. *J. Biol. Chem.*, **269**, 11208–11215.

Maccarana, M., Casu, B. and Lindahl, U. (1993) Minimal sequence in heparin/heparan sulphate required for binding of basic fibroblast growth factor. *J. Biol. Chem.*, **268**, 23898–23905.

Mali, M., Andtfold, H., Miettinen, H.M and Jalkanen, M. (1994) Suppression of tumour cell growth by syndecan 1 ectodomain. *J. Biol. Chem.*, **269**, 27795–27798.

Massague, J. (1992) Receptors for the TGFB family. *Cell*, **69**, 1067–1070.

McKeehan, W.L. and Kan, M. (1994) Heparan sulphate fibroblast growth factor receptor complex: structure-function relationships. *Molecular Reprod. Dev.*, **39**, 69–82.

Noonan, D.M. and Hassell, J.R. (1993) Perlecan, the large low-density proteoglycan of basement membranes: structure and variant forms. *Kidney Int.*, **43**, 53–60.

Nurcombe, V., Ford, M.D., Wildschut, J.A. and Bartlett, P.F. (1993) Developmental regulation of neural response to FGF-1 and FGF-2 by heparan sulphate proteoglycan. *Science*, **260**, 103–106.

Olwin, B.B. and Rapraeger, A. (1992) Repression of myogenic differentiation by aFGF, bFGF and K-FGF in dependent on cellular heparan sulphate. *J. Cell. Biol.*, **118**, 631–639.

Ornitz, D.M., Yayon, A., Flanagan, J.G. Suahn, C.M., Levi, E. and Leder, P. (1992) Heparin is required for cell free binding of basic fibroblast growth factor to a soluble receptor and for mitogenesis in whole cells. *Mol. Cell. Biol.*, **12**, 240–247.

Patterson, I., Kushe, M., Unger, E., Wlad, H., Nyland, I., Lindahl, U. and Kjellen, L. (1991) Biosynthesis of heparin. Purification of a 110 kDa mouse mastocytoma protein required for both glucosaminyl N-deacetylation and N-sulphation. *J. Biol. Chem.*, **266**, 8044–8049.

Pejler, G. and David, G. (1987) Basement membrane heparan sulphate with high affinity for antethrombin synthesised by normal and transformed mouse mammary epithelial cells. *Biochem. J.*, **248**, 69–77.

Rapraeger, A., Jalkanen, M., Endo, E., Koda, J and Bernfield, M. (1985) The cell surface proteoglycan from mouse mammary epithelial cells bears chondroitin sulphate and heparan sulphate glycosaminoglycans. *J. Biol. Chem.*, **260**, 11046–11052.

Rapraeger, A. (1989) Transforming growth factor (type beta) promotes the addition of chondroitin sulphate to cell surface proteoglycan syndecan of mouse mammary epithelia. *J Cell. Biol.*, **109**, 2509–2518.

Rapraeger, A.C., Krufla, A and Olwin, B.B. (1991) Requirement of heparan sulphate of bFGF mediated fibroblast growth and myoblast differentiation. *Science*, **252**, 1705–1708.

Sanderson, R.A., Turnbull, J.E., Gallagher, J.T. and Lander, A.D. (1994) Fine structure of heparan sulphate regulates syndecan 1 function and cell behaviour. *J. Biol. Chem.*, **269**, 13100–13106.

Sanderson, R.D., Hinkes, M.T. and Bernfield, M. (1992) Syndecan 1, a cell surface proteoglycan, changes in size and abundance when keratinocytes stratify. *J. Invest. Dermatol.*, **99**, 390–396.

Shworak, N.W., Shirakawa, M., Mulligan, R.C. and Rosenberg R.D. (1994) Characterisation of ryudocan glycosaminoglycan acceptor sites. *J. Biol. Chem.*, **269**, 21204–21214.

Stamenkovic, I., Aruffo, A., Amiot, M. and Seed, B. (1991) The haemotopoietic and epithelial forms of CD44 are distinct polypeptides with different adhesion potentials for hyaluronate-bearing cells. *EMBO J.*, **10**, 343–348.

Stipp, C.S., Litwack, E.D. and Lander, A.D. (1994) Cerebroglycan: an integral membrane heparan sulphate proteoglycan that is unique to the developing nervous system and expressed speicifically during neural differentiation. *J. Cell. Biol.*, **124**, 149–160.

Takagi, Y., Shrivastav, S., Miki, T. and Sakaguchi, K. (1994) Molecular cloning and expression of the acidic fibroblast growth facto receptors in a rat parathyroid cell line (PFr). *J. Biol. Chem.*, **269**, 23743–23749.

Turnbull, J.E. and Gallagher, J.T. (1991a) Distribution of iduronate-2-sulphate residues in heparan sulphate. Evidence for an ordered polymeric structure. *Biochem. J.*, **273**, 553–559.

Turnbull, J.E. and Gallagher, J.T. (1991b) Sequence analysis of heparan sulphate indicates defined locations of N-suphated glucosamine and iduronate-2 - sulphate residues proximal to the protein - linkage region. *Biochem. J.*, **277**, 297–303.

Turnbull, J.E., Fernig, D.E., Ke, Y., Wilkinson. M.C. and Gallagher, J.T. (1992) Identification of the basic fibroblast growth factor binding sequence in fibroblast heparan sulphate. *J. Biol. Chem.,* **267**, 10337–10341.

Walker, A., Turnbull, J.E. and Gallagher, J.T. (1994) Specific heparan sulphate saccharides mediate the activity of basic fibroblast growth factor. *J. Biol. Chem.,* **269**, 931–935.

Woods, A. and Couchman, J.R. (1994) Syndecan 4 heparan sulphate proteoglycan is a selectively enriched and widespread focal adhesion component. *Mol. Biol. Cell.,* **5**, 183–192.

Wei, Z., Swiedler, S.J., Ishihara, M., Orellana, A. and Hirschberg, C.B. (1993) A single protein catalyses N-deacetylation and N-sulphation during the biosynthesis of heparan sulphate. *Proc, Natl, Acad. Sci. USA.,* **90**, 3885–3888.

Winterbourne, D. and Mora, P. (1981) Cells selected for high tumourgenicity or transformed by simian virus 40 synthesise heparan sulphate with reduced degree of sulphation. *J. Biol. Chem.,* **256**, 4310–4320.

Witt, D.P. and Lander, A. (1994) Differential binding of chemokines to glycosaminoglycan subpopulations. *Curr. Biol.,* **4**, 394–400.

Woods, A. and Couchman, J.R. (1992) Heparan sulphate proteoglycans and signalling in cell adhesion. *Adv. Exp. Med. Biol.,* **313**, 87–96.

Yayon, A., Klagsbran, M., Esko, J.D., Leder, P. and Prnitz, D.M. (1991) Cell surface heparin-like molecules are required for binding of basic fibroblast growth factor to its high affinity receptor. *Cell.,* **64**, 841–848.

Zhang, L. and Esko, J.D. (1994) Amino acid determinants that drive heparan sulphate assembly in a proteoglycan. *J. Biol. Chem.,* **269**, 19295–19299.

Zhu, X., Hsu, B.T., Rees, D.C. (1993) Structural studies of the binding of the anti-ulcer drug sucrose octasulphate to acidic fibroblast growth factor. *Structure,* **1**, 27–34.

8 Regulation of Gene Expression by the Extracellular Matrix

Nancy Boudreau and Mina J. Bissell

Life Sciences Division, Lawrence Berkeley Laboratory, University of California, Berkeley, USA

INTRODUCTION

The extracellular matrix (ECM) profoundly influences the major programs of cell function including growth, development and differentiation. The composition of the ECM is often cell-type specific but the major protein components are shared by many tissues and include fibronectin, laminin, collagens and proteoglycans. These components not only provide structural support and compartmentalization within tissue structures, but also directly impart instructive information to individual cells. Many comprehensive reviews have detailed the evidence supporting the role for ECM in mediating cellular development and differentiation (Adams and Watt, 1993; Hay, 1993; Lin and Bissell, 1993). Much less information is available detailing the mechanism by which the ECM can control these complex phenomena. Recent work has provided clues as to how ECM evokes signals carrying information to the nucleus to elicit tissue-specific gene expression. We discuss some of the recent evidence and provide working models for how ECM may stop proliferation, prevent apoptosis and influence tissue-specific gene expression.

ECM and Receptor-Mediated Signaling

The discovery of distinct ECM binding proteins, the integrins, on the surface of cells provided initial proof for a link between the ECM and the cytoplasm. Integrin receptors are heterodimers comprised of alpha and beta subunits which span the plasma membrane (see next chapter). The cytoplasmic domain of the alpha subunits intimately associates with cytoskeletal proteins such as vinculin and talin (Burridge *et al.*, 1988; Hynes, 1992). Up to 19 different heterodimeric combinations are possible, with the particular combination of

Extracellular Matrix, Volume 2, Molecular Components and Interactions, edited by Wayne D. Comper.

alpha and beta subunits showing different specificities for binding to individual ECM components (Hynes, 1992). The overlapping binding affinities and the presence of multiple integrin receptors for a given ECM component on individual cell types perhaps suggests specialized functions of the receptors.

Despite the extensive biochemical characterization of integrins, the nature of the cytoplasmic signals generated from ECM components binding to their respective integrin receptors are still poorly defined. Strong evidence linking changes in tyrosine phosphorylation (reviewed by Juliano and Haskill, 1993; Roskelley *et al.*, 1995) and changes in intracellular pH (Ingber *et al.*, 1990a), have been associated with ECM-integrin interactions. The most direct evidence to indicate that these signals are involved in ECM-directed gene activity is that treatment with function-blocking anti-integrin antibodies often alters or prohibits ECM-induced gene expression (Streuli *et al.*, 1991, Streuli *et al.*, 1995, Howlett *et al.*, 1995). In addition, exposure of cells to oncogenic agents such as RSV which often abrogates differentiated gene expression, profoundly alters the pattern of integrin phosphorylation observed (Hirst *et al.*, 1986; Anesiekvich *et al.*, 1991).

Whereas the ability to elicit distinct biochemical signals in response to ECM has been demonstrated in numerous cell types, a direct link between these intracellular signals and subsequent changes in gene expression remain to be established. For example, unlike steroid hormone receptor binding, and the subsequent translocation of this complex to the nucleus which directly interact with elements in the promoters of various genes, such a link has not firmly been established for ECM-mediated transcriptional activation. It has, however, been possible to gain an understanding of some of the transcriptional regulatory mechanisms by directly examining the genes which are upregulated in response to ECM.

Two genes which are strongly upregulated in the presence of ECM include β-casein in mammary epithelial cells (Emermen and Pitelka, 1977; Barcellos-Hoff *et al.*, 1989) and albumin in hepatocytes (Bissell *et al.*, 1990, Caron, 1990; Lui *et al.*, 1991). The expression of β-casein or albumin is characteristic of the functional differentiated phenotype displayed by the mammary gland or the liver *in vivo,* and these tissues have provided physiologically relevant models to examine the mechanisms employed by ECM in regulating specialized gene expression.

Maintenance of Tissue-Specific Gene Expression

Genes such as β-casein and albumin, comprise only a fraction of the cellular genes whose expression is modified when cells are induced to undergo differentiation in the presence of the ECM. As postulated by Bissell (1981) and Blau (1992), normal cells maintain their inherent plasticity, and as such the differentiated state is not fixed and requires continuous active regulation to be maintained. In addition to the requirement for continuous signals to maintain expression of β-casein or albumin, the genes which contribute to the cell's microenvironment would be prime targets for continual active maintenance, since the failure to maintain a suitable environment is reflected by a change or loss of function of the tissue. Evidence suggests that the nature of the ECM is directly responsible for maintaining this proper environment by regulating the expression of ECM components, growth factors and the ability of cells to respond to hormones.

FEEDBACK MODULATION OF ECM SYNTHESIS AND COMPOSITION BY THE ECM

Previous studies from this laboratory have demonstrated that in the presence of an intact basement membrane (BM), mammary epithelial cells express very low levels of the components of the BM including laminin, fibronectin, HSPG and type IV collagen (Parry *et al.*, 1985, Streuli and Bissell, 1990). However when these cells are cultured on conventional tissue culture plastic, the mRNA and protein levels of the individual ECM components are markedly increased, apparently in a frustrated attempt to recreate the proper extracellular environment (Streuli and Bissell, 1990).

Whether or not the presence of intact ECM components directly feeds back to the cells to modulate their own expression has not been explored further. For example, it is not known whether integrins participate in relaying the status of the extracellular environment to the cell nucleus and subsequently modulate expression of ECM components accordingly. Furthermore, the regions of the ECM genes which might respond to such cues have not been defined.

Alternatively, fragments of ECM components may be responsible for inducing the synthesis of ECM components. For example, when fibroblasts are embedded into type I collagen gels, they repress synthesis of type I collagen (Eckes *et al.*, 1993). whereas addition of collagen hexapeptides or proteolytic fragments of collagen induces expression of collagen genes (Katayama *et al.*, 1993). Again, whether integrins are involved and whether they can distinguish between intact ECM components or their fragments or whether distinct integrins are employed have not been addressed.

Another possibility is that maintenance of the correct ECM microenvironment may arise indirectly through mediators such as TGFß, which has been shown to markedly increase expression of a number of ECM components including fibronectin, collagen (Ignotz and Massague, 1986; Wrana *et al.*, 1988) and glycosaminoglycans (Chen *et al.*, 1987; Boudreau *et al.*, 1992). Although generally expressed at low levels *in vivo*, or in cells cultured on ECM, expression of TGFβ is markedly elevated in cells cultured in the absence of ECM (Streuli *et al.*, 1993). Such factors could potentially contribute to the high levels of expression of ECM components observed under these conditions. A closer examination of the mechanisms underlying this feedback regulation is critical to understanding the maintenance of the differentiated phenotype.

NEGATIVE REGULATION OF GENE EXPRESSION BY THE ECM IS REQUIRED FOR DIFFERENTIATION

In addition to monitoring the synthesis and integrity of the ECM, the differentiated cell must also ensure that gene products with the potential to compromise this environment be actively suppressed. The inappropriate expression of matrix degrading proteinases in mammary epithelium can effectively destroy the surrounding ECM resulting in a loss of function of the tissue (Sympson *et al.*, 1994). Thus the functionally differentiated cells must also ensure expression of inhibitors of matrix-degrading proteases and/or directly suppress expression of genes encoding matrix-degrading proteases. Similar to the feedback modulation of expression of ECM components, evidence exists that the cells can infact distinguish between intact and fragmented ECM components and adjust the expression of proteases such as stromelysin accordingly (Werb *et al.*, 1989; Tremble *et al.*, 1994).

Because different components of the ECM have distinct influences on the pattern of gene expression in a particular cell, maintaining the differentiated state also requires that cells suppress the expression of those ECM components which could potentially alter this phenotype. A dramatic example of the consequences of introducing an inappropriate ECM component is found in mammary epithelial cells. Tenascin is a large, multi-domain glycoprotein that is not expressed in the functional milk-producing mammary gland. Expression of tenascin begins during the process of involution, a time when cells are destroying the laminin rich basement membranes and consequently losing their capacity to express milk proteins (Chammas *et al.*, 1994; Jones *et al.*, 1995). Recent work from our laboratory has shown that the addition of exogenous tenascin to fully differentiated mammary epithelial cells directly inhibits transcription of the β-casein gene (Jones *et al.*, 1995). Thus in order to maintain mammary gland function during pregnancy and lactation, expression of tenascin must be suppressed. Interestingly, mammary epithelial cells cultured in the absence of an ECM express high levels of tenascin mRNA and protein, whereas those cultured on an intact, laminin rich basement membrane express little or no tenascin (Jones *et al.,* 1995).

The presence of laminin is required to maintain endothelial cells in their differentiated, quiescent state (Pauly *et al.*, 1992) while a fibronectin-rich matrix will induce endothelial cells to proliferate (Ingber *et al.*, 1990a). Therefore, the differentiated state of endothelial cells requires not only that the presence of laminin be maintained but also that expression of fibronectin be suppressed.

These examples illustrate that the ECM plays a critical role not only in directing expression of tissue-specific genes, but also by suppressing the expression of genes which would compromise the proper microenvironment and provide direct support for the concept of 'dynamic reciprocity' postulated by Bissell *et al.*, (1982). The failure to maintain the appropriate extracellular environment not only results in the loss of tissue-specific gene expression, but the cell now becomes vulnerable to various other fates ranging from neoplasia, transdifferentation or apoptotic death (see Figure 8.1).

ECM, DIFFERENTIATION AND SUPPRESSION OF CELL CYCLE

It has long been observed that the processes of proliferation and differentiation are mutually exclusive. Cells are often required to exit the cell cycle in order to express differentiated properties and re-entry into the cell cycle often results in the loss of the differentiated gene expression (Stein *et al.*, 1992). The relationship between proliferation, ECM synthesis and deposition and differentiation has been well studied in osteoblasts. Stein and colleagues have demonstrated that osteoblast progenitor cells will exit the cell cycle once they begin to synthesize and deposit ECM. Once the ECM is in place, osteoblasts display differentiated properties, including expression of the osteoblast-specific gene, osteocalcin. Evidence that the organization and maturation of the ECM are required for suppression of proliferation comes from the finding that removal of differentiated cells from their matrices results in a reentry into the cell cycle with a subsequent loss of capacity to express genes such as osteocalcin (Stein *et al.*, 1990). Similarly in hepatocytes, albumin production begins after

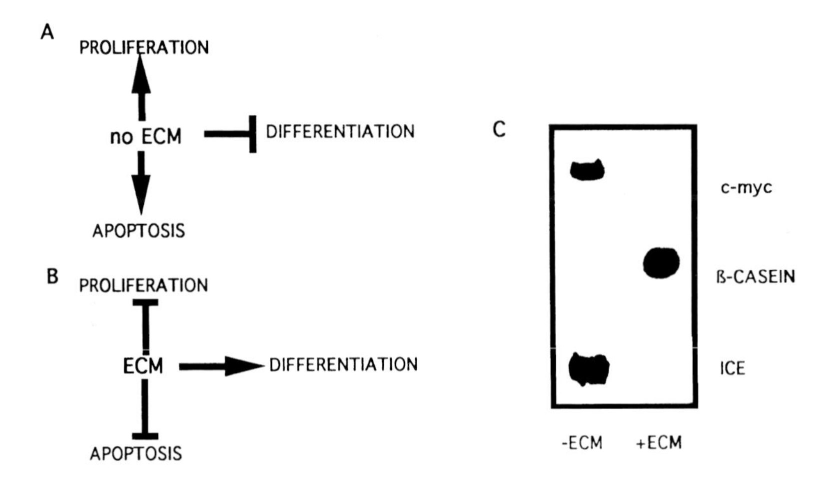

Figure 8.1 ECM regulates the cellular programs of growth, apoptosis and differentiation

A) In the absence of ECM cells proliferate or undergo apoptosis and are unable to express the differentiated phenotype.

B) When the appropriate ECM is present, cells will shut off proliferation and apoptosis and undergo differentiation

C) mRNA expression in mammary epithelial cells illustrates that in the absence of ECM (–) high levels of c-myc and ICE, but not *b*-casein are expressed. However, when ECM is present (+) the genes associated with proliferation (myc) and apoptosis (ICE) are not expressed while the *b*-casein gene, characteristic of the differentiated phenotype, is expressed at high levels.

the program of growth is switched off (Mooney *et al.,* 1991) and re-activating proliferation will lead to a loss of albumin expression (Padgham *et al.*, 1993). In mammary epithelial cells, expression of milk proteins occurs only in the absence of proliferation (Desprez *et al.*, 1993) which can be induced by plating cells onto exogenous basement membrane ECM (Petersen *et al.*, 1992). Additional studies have indicated that ECM-integrin interactions contribute to this cessation of proliferation (Howlett *et al.*, 1995). In contrast, malignant mammary epithelial cells do not cease proliferating in response to ECM (Petersen *et al.*, 1992) and studies by Giancotti and Ruoslahti (1989) have shown that restoring functional integrin receptors to transformed cells leads to a rapid decline in the rate of proliferation.

A similar reciprocal relationship between proliferation, ECM synthesis and differentiation is observed during the process of angiogenesis. When in contact with an intact laminin-rich basement membrane, endothelial cells are quiescent and express genes associated with the differentiated state (Pauly *et al.,* 1992). In order for new vessels to develop, the cells must first degrade the existing ECM allowing them to migrate into the adjacent connective stroma, which is also permissive for proliferation. The proliferating cells begin to resynthesize and deposit a basement membrane and subsequently cease proliferating and resume their differentiated function once the proper basement membrane is reestablished (Ausprunk and Folkman, 1977; Kalebic *et al.*, 1983; Schnaper *et al.*, 1993). Similarly, vascular smooth muscle cells normally exist in a quiescent differentiated state but are induced to proliferate and lose their differentiated characteristics when the existing ECM is disrupted through the activity of matrix-degrading proteinases (Pauly *et al.*, 1994).

REGULATION OF GROWTH FACTORS, CYTOKINES AND TRANSCRIPTION FACTORS BY THE ECM

Cells cultured in the absence of ECM synthesize many factors that are not produced by cells *in vivo*, or when cultured on ECM. As previously mentioned, culturing mammary epithelial cells on ECM reduces transcription of tenascin (Jones *et al.*, 1995) and TGFβ1 (Streuli *et al.*, 1993). In osteoblasts, the expression of TGFβ also drops dramatically as the cells accumulate a fibronectin and type I collagen matrix (Owen *et al.*, 1990). We have also demonstrated that expression of TGFα is high in mammary epithelial cells cultured on plastic but is suppressed in the presence of ECM (Lin *et al.*, 1995). In addition, physical damage to the existing ECM will increase expression of TNFα (Hershovitz *et al.*, 1993) or allow release of ECM-associated bFGF (Flaumenhaft *et al.*, 1989; Vlodavsky *et al.*, 1990). Depending upon the cell type studied, many of these cytokines are capable of supporting proliferation and the ability of the ECM to suppress expression of these factors may contribute to the ability of the ECM to inhibit proliferation.

In addition, down regulating expression of these cytokines is also essential for maintenance of the differentiated state. For example, direct application of TGFα to mammary epithelial cells cultured on ECM will impair expression of the whey acidic milk protein (Lin *et al.*, 1995). Addition of high doses of TNFα will also impair differentiation of mouse mammary epithelial cells (Ip *et al.*, 1992) and induce expression of matrix-degrading proteinases (Niedbala, 1993). As well, the osteocalcin gene contains a TNFα negative response element which contributes to its shutdown when TNF is present (Nanes *et al.*, 1994). Whether these factors interfere with tissue-specific gene expression directly or through their ability to stimulate proliferation is not clear. However, TGFβ, which acts as a growth inhibitor in epithelial cells, also inhibits β-casein expression in mammary epithelium (Robinson *et al.*, 1993) or production of albumin in hepatocytes (Busso *et al.*, 1990).

Despite the strong correlation between ECM and inhibition of proliferation, direct evidence linking ECM to suppression of cell-cycle related genes is scarce. Hepatocytes cultured on exogenous basement membranes, show a rapid down-regulation of a number of early response genes associated with proliferation including *c-myc, c-jun and jun* B (Rana *et al.*, 1994). One consequence of reduced *jun* expression is the subsequent loss of expression of the transcription factor AP-1, a product of the jun and fos genes. Interestingly, the ability of ECM to reduce AP-1 activity appears to function independently of ECM's ability to inhibit proliferation (Rana *et al.*, 1994). Such an ECM-mediated reduction in AP-1 likely supports expression of the differentiated phenotype, as high levels of AP-1 induced by expression of a variety of oncogenes have been shown to inhibit expression of the β-casein gene in mammary epithelium (Jehn *et al.*, 1992). The lack of AP-1 binding to β-casein promoter sequences indicates that inhibition of mammary-specific gene expression by this factor may be indirect. One possibility is that AP-1 expression may induce transcription of cytokines such as TGFβ or TNFα which in turn interfere with expression of the differentiated phenotype.

Nonetheless, the ability of the ECM to suppress expression of these factors is critical in maintaining the differentiated phenotype and the conditions of culturing cells without an appropriate ECM can be considered to be analogous to a response to injury whereby a marked increase in synthesis of cytokines ensues (Sieweke and Bissell, 1994).

Perhaps a more direct means by which the ECM directly suppresses proliferation and concomitantly supports the differentiated phenotype is through modulating factors which may be both inhibitory for certain genes and stimulatory for others. A good example is ID-1, a helix-loop-helix factor which in addition to supporting proliferation, also prevents differentiation-associated factors from interacting with tissue-specific genes. For example, overexpression of ID-1 in osteoblasts not only induces proliferation, but directly impairs expression of the differentiation-associated gene, osteocalcin. Furthermore, even when proliferation is blocked, high levels of ID-1 will prevent binding of essential transcription factors to the OCE element E box of the osteocalcin promoter and as a result, block transcription of the osteocalcin gene (Tamura and Noda, 1994).

Direct evidence for a role of the ECM in modulating expression of ID-1 come from studies in our laboratory. When mammary epithelial cells are cultured in the presence of ECM, ID-1 is down-regulated, proliferation decreases and expression of β-casein ensues. Overexpression of ID-1 in these cultures will prevent the expression of β-casein whereas treatment of cells with anti-sense oligonucleotides to ID-1 leads to enhanced transcription of the β-casein gene (Desprez *et al.*, 1995). Thus the down regulation of ID-1 in presence of ECM not only results in a loss of proliferative stimulus but simultaneously 'de-represses' β-casein gene expression.

Another factor with the potential to reciprocally modulate differentiation and proliferation is the appropriately named yin-yang-1 or YY1. YY1 can function either as a repressor or activator depending on the promoter context (Tijan and Maniatis, 1994) and in proliferating myoblasts has been shown to simultaneously activate expression of c-myc while directly inhibiting expression of the alpha actin gene associated with the differentiated myocyte phenotype (Lee *et al.*, 1994). In mammary epithelial cells, YY1 interacts with and suppresses transcription of the β-casein gene (Raught and Rosen, 1994; Meier and Groner, 1994). Whether YY1 concomitantly induces proliferation or is down-regulated by the ECM in these cells is not known.

These examples again illustrate the cells's dependence on the appropriate extracellular environment to express genes associated with differentiated functions, but additionally suggest how the lack of appropriate ECM may result in the expression of genes with the potential to deregulate growth of the cells.

TRANSDIFFERENTIATION

Transdifferentiation is defined as the process, whereby a differentiated cell begins to express genes associated with another phenotype; for example, epithelium becomes mesenchymal or muscle cells acquire neurogenic properties. As cells maintain their inherent plasticity, it follows that genes not associated with a particular differentiated phenotype must also be actively suppressed. Increasing evidence demonstrates that contact with the appropriate ECM is required to suppress the potential for transdifferentiation. Work on lens epithelium by Hay *et al.* (1993) have elegantly demonstrated the potential for differentiated cells to alter their phenotype depending on the extracellular environment. When epithelial basement membranes are destroyed by enzyme treatment, the cells begin to express genes such as vimentin which are generally associated with the fibroblastic phenotype. In addition the cells begin to adopt a fibroblastic morphology. Addition of retinoic acid stimulates cells to resynthesize their basement membrane and as a result cells suppress vimentin expression and begin to express the keratin genes characteristic of differentiated epithelial cells.

Whether the loss of one phenotype arises directly from the lack of ECM-directed expression of tissue-specific genes or arises through generation of factors produced in the absence of ECM has not been explored. Studies by Bachem *et al.*, (1993) have shown that both TGFβ and TNFα, two factors produced at high levels in the absence of ECM, can directly induce fat storing cells to acquire a myofibroblastic phenotype.

More recent evidence indicates that both the composition and physical properties of the BM can regulate transdifferentiation of retinal pigment epithelium to neural retina, i.e. maintaining the same composition of ECM components but altering the mechanical properties can also lead to transdifferentiation into neural retina (Opas and Dziak, 1994).

In addition, the process of transdifferentiation appears to alter the expression of cell adhesion molecules which contribute to the differentiated state. When keratinocytes are induced to assume a myogenic phenotype, they accordingly lose expression of adhesion molecules characteristic of keratinocyctes, such as desmosomal proteins and integrins, and begin to express integrin receptors generally associated with a fibroblastic or myogenic phenotype (Boukamp and Fusenig, 1993).

These examples not only demonstrate that the inherent phenotypic plasticity is preserved in differentiated tissues but also demonstrate that an ECM which supports the expression of genes contributing to a particular phenotype must simultaneously suppress the expression of genes which would specify another phenotype and function.

APOPTOSIS

Apoptosis or programmed cell death (PCD) is most often associated with cells whose function has been completed or with those which cannot carry out their designated activities properly (Ellis *et al.*, 1991). Post-lactating mammary epithelial cells provide a good example of cells whose function has been completed and thus subsequently undergo apoptosis (Walker *et al.*, 1988; Strange *et al.*, 1992). Also, many cells which escape normal growth regulation are highly susceptible to apoptosis (White, 1994). It is not surprising then that the presence of an ECM which supports differentiated function and suppresses growth will protect cells from undergoing PCD.

Studies by Meredith *et al.*, (1993) and Frisch and Francis (1994) have indeed shown that adhesion to ECM is required to maintain survival of endothelial and epithelial cells, respectively. Recently we and others (Pullan *et al.*, 1995) have shown that an intact three-dimensional ECM is required to prevent apoptosis in cultured mammary epithelial cells. Destruction of this ECM by overexpression of proteolytic enzymes in cultured cells or in transgenic animals or disrupting cell-ECM interactions with anti-β1 integrin antibodies results in apoptosis and indicates that both adhesion to and subsequent signals generated by the ECM are necessary to prevent cells from undergoing the program of cell death (Boudreau *et al.*, 1995).

The ability of the ECM to guarantee survival of cells again relates to the ECM's ability to actively suppress expression of genes which in this case will induce PCD (Figure 8.1). Unlike necrotic cell death, apoptosis involves active *de novo* expression of new genes to complete the nuclear disintegration and DNA fragmentation. Recently several genes have been identified which appear to induce apoptosis, including interleukin-1β converting enzyme (ICE) (Muira *et al.*, 1993), which is the mammalian homologue of the *C.elegans* death gene *ced-3* (Yuan *et al.*, 1993); functionally related cysteine proteinases, Nedd-2 or Ich-1(Kumar *et al.*, 1994; Wang *et al.*, 1994); bax and bclx, which are homologues of the apoptosis suppresser gene bcl-2 (Oltavi *et al.*, 1993; Boise *et al.*, 1993) and FAS, a TNFα related extracellular molecule (Suda *et al.*, 1993). In addition many cell-cycle associated genes including the tumor suppresser p53 (Miyashita *et al.*, 1994; Reed,1994) and c-myc (Evan *et al.*, 1991) have been linked to apoptosis, suggesting that regulation of cell-cycle and apoptosis may be intimately linked.

In mammary epithelial cells, the loss of ECM correlates with the induction of expression of ICE, bax and c-myc (Boudreau *et al.*, 1995; Pullan *et al.*, 1995: Strange *et al.*, 1992; Boudreau *et al.*, 1996 in press). Whether the ECM directly suppresses transcription of these genes or suppresses expression of other factors which in turn induce expression of effectors of apoptosis is not known. For example, addition of TGFβ, or TNFα, factors whose expression is induced in the absence of ECM, induce apoptosis in uterine epithelium, hepatocytes, and endothelial cells (Rotello *et al.*, 1991; Oberhammer *et al.*, 1992; Shinagawa *et al.*, 1991; Robaye *et al.*, 1991). However, the ability of these cytokines to influence the pattern of expression of genes involved in PCD has not yet been directly investigated.

As interference with ECM-integrin interactions activates genes which induce proliferation (Giancotti and Ruoslahti, 1990) or apoptosis (Boudreau *et al.*, 1995) it follows that the ECM must be transmitting signals to repress these cellular programs. How these signals differ from ECM-derived signals which promote the program of differentiation remains to be determined. However, by studying genes whose transcription is directly activated in response to ECM, it may be possible to identify regulatory processes which may also be involved in suppressing gene transcription.

ECM-RESPONSIVE ELEMENTS WITHIN DNA

Genes such as β-casein or albumin, which are actively transcribed in the presence of ECM contain elements within their promoters known as ECM-response elements. The first such element, identified in the promoter of the β-casein gene of mammary epithelial cells, is a 161 bp sequence which confers ECM- and prolactin-responsiveness when attached to the otherwise inactive 121 bp promoter of the β-casein gene. This element also appears to function in a mammary cell-specific manner (Schmidhauser *et al.*, 1990, 1992). In the albumin gene, a considerably larger 1 kb enhancer also displays similar ECM-responsiveness in hepatocytes (DiPersio *et al.*, 1990).

More recently, an ECM/collagen-response element has been located in the LpS1 gene of the sea urchin, *Lychtenius Pyctus*. Expression of the LpS1 gene is lost in the presence of agents such as β-APN which disrupt the collagenous matrix (Benson *et al.*, 1991) and a 120 bp region within the promoter confers ECM-inducible transcriptional activation to this and heterologous promoters (Craig Tomlinson, personal communication).

The magnitude of the ECM's ability to influence gene transcription is exemplified by its ability to regulate the activity of some of the most potent transcriptional activators in eukaryotic cells, the viral enhancers. Schmidhauser *et al.*, (1994) and Romagnolo *et al.*, (1993) have shown the activity of the mouse mammary tumor virus (MMTV) enhancer is also activated in the presence of ECM.

How the ECM activates transcription in such a diverse set of genes is not clear, particularly when considering there is no significant sequence homology within these 'response elements' identified to date.

ECM DIRECTED SYNTHESIS OF TRANSCRIPTION FACTORS

The isolation of such ECM-response elements has provided a great opportunity to further study the mechanisms by which ECM can directly stimulate gene transcription. One obvious means by which the ECM could activate gene transcription would be to generate tissue-specific transcription factors which interact with these elements. Studies in hepatocytes cultured on extracellular matrices, show increased expression of HNF3 and eH-TF (Liu *et al.*, 1991) and C/EBP (Rana *et al.*, 1994), liver-enriched transcription factors which interact with the albumin enhancer.

In the mammary gland, a number of tissue-specific transcriptional activators have been identified including MGF (Schmitt-Ney *et al.*, 1991; Watson *et al.*, 1991) and more recently POU-domain factors (Jehn *et al.*, 1994). Although the expression of these factors correlates with transcriptional activation of the β-casein gene, their regulation by ECM or their ability to interact with the ECM-response element of the β-casein gene have not been determined. It will also be of interest to determine whether any of these factors which support tissue-specific gene expression can simultaneously act to suppress the expression of genes which interfere with differentiated function.

ECM DIRECTED CHROMATIN ORGANIZATION

In addition to the requirement for specific transcription factors, transcription of many genes requires that histone protein complexes covering the corresponding DNA consensus sequences must be arranged in a manner which allow these factors access to their appropriate regulatory elements (Wolfe, 1994). Recent studies by MacPherson *et al.*, (1993) have provided evidence that histones covering the albumin enhancer must be reorganized to permit ECM-induced activation of transcription. The shift in histone phasing appears to be mediated by HNF3, the same factor whose expression is dependent upon the presence of ECM (Liu *et al.*, 1991). Precisely how HNF3 carries out this task is not yet known. One means of altering histone conformation occurs through the activity of specific deacetylase enzymes within the nucleus (Helzden *et al.*, 1991). In mammary epithelial cells cultured without ECM, acetylation of histones by addition of sodium butyrate increases transcriptional activity of the β-casein enhancer to levels similar to those which occur in response to ECM (Schmidhauser *et al.*, 1994). Together, these data suggest that histone acetylation may be a common means by which ECM regulates transcriptional activity of genes.

ECM AS A MORPHOREGULATOR

The task of coordinating the activation and suppression of genes which collectively comprise the differentiated phenotype is enormous and how the ECM can coordinate these activities is an equally daunting question. Nonetheless, a key to this may lie in the earliest observations, which noted that the ECM had a profound influence on the morphology of the cells with which it was in contact (Grobstein, 1975; Wessells, 1977). Ironically, in attempting to establish that ECM directly influences gene expression, investigators sought to segregate these structural aspects from the biochemical and molecular events. Although the demonstration of different biochemical signals and responses generated in isolated cells adhering to various ECM contributed immensely to establishing the ECM as a *bona fide* regulator of gene expression, more sophisticated experimental approaches have indicated that the morphology bestowed by the ECM is a significant component in its ability to regulate gene expression.

Previous work in our and other laboratories have noted the requirement for a malleable ECM substratum in order for cells to respond appropriately. For example, when mid-pregnant mammary epithelial cells are plated onto tissue culture plastic or a one-dimensional fixed collagen substratum, they assume a flattened cobblestone morphology and fail to express β-casein. When the collagen gels are released from the culture dishes and floated into the media, the cells acquire a polarized, cuboidal morphology (Emerman and Pitelka, 1977) and respond by expressing high levels of β-casein (Barcellos-Hoff *et al.*, 1989). Microscopic observation of these cultures revealed that the cells utilized the flexible substratum to synthesize and deposit an endogenous basement membrane rich in laminin (Streuli and Bissell, 1990). When gels are fixed the cells are unable to deposit their own matrix and thus cannot express genes associated with the differentiated phenotype. Similarly when cultured on flexible, porous membranes, mammary epithelial cells can also synthesize and deposit an endogenous ECM and acquire the ability to make casein without the addition of exogenous matrix components (Parry *et al.*, 1987; Reichmann *et al.*, 1989; Boudreau *et al.*, 1995).

Although mammary epithelial cells absolutely require the presence of the component laminin and its subsequent interaction with $\beta 1$ integrins to activate transcription of the β-casein gene, the morphology induced through laminin-integrin interactions are also essential for the appropriate response. Roskelley *et al.*, (1994) have shown that laminin is not capable of evoking synthesis of β-casein until the cells have acquired the proper rounded morphology. Pre-treatment with the phorbol ester, TPA prevented the morphological changes induced by laminin and subsequent expression of β-casein, whereas addition of TPA to pre-rounded cells had no effect on laminin's ability to activate β-casein transcription.

Studies in endothelial cells have indicated that the flattened cellular morphology induced by fibronectin directly contributes to the ability of these cells to proliferate (Ingber 1990b). Tensegrity model postulated by Ingber stress the requirement for proper cell shape in order to achieve expression of differentiation-associated genes. For example, hepatocytes plated on laminin normally express albumin, but when cell morphology is distorted, despite maintaining proper laminin-integrin contacts, the ability to express albumin was impaired (Singhvi *et al.*, 1994). Thus although ECM-integrin connections remain intact, the altered cytoskeleton and nuclear morphology were incapable of properly transducing the signals required to activate transcription of the albumin gene.

These results convincingly demonstrate that tissue-specific gene expression evoked by the ECM requires more than simple adhesion of cell to various ECM components or simple ligation of the corresponding integrin receptor. That is, although ECM binding to an integrin can initiate a biochemical signal, it must also ensure that the proper cytoskeletal and nuclear configuration are present to propagate and translate these incoming signals in a meaningful way. Although the precise contribution of the cellular architecture to signaling events is not clear, it is conceivable that organizing physically linked substrates and effectors will promote interactions and generate subsequent mediators that would not be possible over greater physical distances. Thus the role of the ECM is that of both initiating and directing traffic along this 'information superhighway' which transports essential information from the extracellular environment to the nucleus to establish and maintain the differentiated phenotype.

ECM AS A MASTER REGULATORY SWITCH

The ability of the ECM to coordinate expression and suppression of genes collectively comprising the differentiated phenotype lends support to the notion of the ECM as a master regulatory switch. Recently the homeotic genes have been assigned the role of master regulatory genes which are responsible for the development of specialized tissue structure and function. Many of the HOX gene products have been shown to act as transcription factors, although little is known with respect to their downstream target genes (Botas, 1993). However their profound ability to modulate tissue morphology and acquisition of differentiated, specialized functions, suggest that the ECM and homeotic genes may be intimately associated.

Acknowledgement

This work was supported by the U.S. Department of Energy, Office of Health and Environmental Research (contracts DE-ACO3-76-SF01012 and DE-ACO3-76-SF0098) and by the National Institutes of Health (NCI CA 57621). Nancy Boudreau was supported by a research fellowship from the Canadian Medical Research Council.

References

Adams, J.C. and Watt, F.M. (1993) Regulation of development and differentiation by the extracellular matrix. *Development* **117**, 1183–1198

Aneskievich, B.J., Haimovich, B. and Boettiger, D. (1991) Phosphorylation of integrin in differentiating ts-Rous sarcoma virus-infected myogenic cells. *Oncogene* **6**, 1381–1390

Ashkenas, J., Damsky, C.H., Bissell, M.J. and Werb, Z. (1994) Integrins signaling and the remodeling of the extracellular matrix. In *Integrins* pp. 79–109, D. Cheresh and R. Mecham eds. Academic Press

Ausprunk, D.H. and Folkman, J. (1977) Migration and proliferation of endothelial cells in preformed and newly formed blood vessels during angiogenesis. *Microvas. Res.* **14**, 53–65

Bachem, M.G., Sell, K.M., Melchior, R., Kropf, J., Eller, T. and Gressener, A.M. (1993) Tumor necrosis factor alpha and transforming growth factor beta 1 stimulate fibronectin synthesis and the transdifferentiation of fat-storing cell in the rat liver into myofibroblasts. *Virchows Archiv. B. Cell Pathology* **63**, 123–130

Barcellos-Hoff, M.H., Aggeler, J., Ram, T.G. and Bissell, M.J. (1989) Functional differentiation and alveolar morphogenesis of primary mammary cultures on reconstituted basement membrane. *Development*, **105**, 223–235

Benson, S., Rawson, R., Killian, C. and Wilt, F. (1991) Role of the extracellular matrix in tissue-specific gene expression in the sea urchin embryo. *Molecular Reproduction and Development* **29**, 220–226

Bissell, D.M., Caron, J.M., Babiss, L .E. and Friedman, J.M. (1990) Transcriptional regulation of the albumin gene in cultured rat hepatocytes. Role of basement-membrane matrix. *Mol. Biol. Med.* **7**, 187–197

Bissell, M.J. (1981) The differentiated state of normal and malignant cells or how to define a normal cell in culture. *Int. Rev. Cytology.* **70**, 27–100

Bissell, M.J., Hall, H.G. and Parry, G. (1982) How does the extracellular matrix direct gene expression? *J. Theor. Biol.* **99**, 31–68

Blau, H.M. (1992) Differentiation requires continuous active control. *Ann. Rev. Biochem.* **61**, 1213–1230

Boise, L.H., Gonzalez-Garcia, M., Postema, C.E., Ding, L., Lindsten, T., Turka, L.A., Mao, X., Nunez, G. and Thompson, C.B. (1993) bcl-x, a bcl-2 related gene that functions as a dominant regulator of apoptotic cell death. *Cell* **74**, 597–608.

Botas, J. (1993) Control of morphogenesis and differentiation by HOM/Hox genes. *Curr. Op. Cell Biol.* **5**, 1015–1022

Boudreau, N., Clausell, N., Boyle, J. and Rabinovitch, M. (1993) Transforming growth factor beta regulates increased ductus arteriosus endothelial glycosaminoglycan synthesis and a post-transcriptional mechanism controls increased smooth muscle fibronectin. *Lab. Invest.* **67**, 350–359

Boudreau, N., Sympson, C.J., Werb, Z. and Bissell, M.J. (1995) Extracellular matrix suppresses expression of interleukin-1β converting enzyme and apoptosis in mammary epithelial cells. *Science* **287**, 891–893

Boudreau, N., Werb, Z., Bissell, M.J. (1995 in press) Suppression of apoptosis by basement membrane requires three dimensional tissue organization and withdrawal from the cell cycle. PNAS

Boukamp, P. and Fusenig, N. E. (1993) "Trans-differentiation" from epidermal to mesenchymal/myogenic phenotype is associated with a drastic change in cell-cell and cell-matrix adhesion molecules. *J. Cell Biol.* **120**, 981–993

Burridge, K., Fath, K., Kelly, G., Nuckolls, G. and Turner, C. (1988) Focal adhesions; transmembrane junctions between the extracellular matrix and the cytoskeleton. *Annu. Rev. Cell Biol.* **4**, 487–525

Busso, N., Chesne, C., Delers, F., Morel, F. and Guillouzo, A. (1990) Transforming growth factor beta inhibits albumin synthesis in normal human hepatocytes and in hepatoma HepG2 cells. *Biochem. Biophys. Res. Comm.* **171**, 647–654

Caron, J.M. (1990) Induction of albumin gene transcription in hepatocytes by extracellular matrix proteins. *Molec. Cell. Biol.* **10**, 1239–1243

Chammas, R., Taverna, D., Cella, N., Santos, C. and Hynes, N.E. (1994) Laminin and tenascin assembly and expression regulate HC11 mouse mammary cell differentiation. *J. Cell Sci.* **107**, 1031–1040

Chen, J.K., Hoshi, H. and McKeehan, W.L. (1987) TGFß specifically stimulates synthesis of proteoglycan in human adult arterial smooth muscle cells. *Proc. Natl. Acad. Sci.* **84**, 5287–5291

Chen, D., Magnuson, V., Hill S., Arnaud, C., Steffensen, B. and Klebe, R.J. (1992) Regulation of integrin gene expression by substrate adherence. *J. Biol. Chem.* **267**, 23502–23506

Desprez, P.Y., Roskelley, C.D., Campisi, J. and Bissell, M.J. (1993) Isolation of functional cell lines from a mouse mammary epithelial cell strain; the importance of basement membrane and cell-cell interaction. *Mol. Cell Diff.* **1**, 99–110

Desprez, P.Y., Hara, E., Campisi, J. and Bissell, M.J. (1995) ID overexpression inhibits extracellular matrix dependent expression of β-casein in mammary epithelial cells. *Molec. Cell Biol.* **15**, 3398–3404

Dipersio, C.M., Jackson, D.A. and Zaret, K.S. (1991) The extracellular matrix coordinately modulates liver transcription factors and hepatocyte morphology *Molec. Cell Biol.* **11**, 4405–4414

Eckes, B., Mauch, C., Huppe, G. and Kreig, T. (1993) Downregulation of collagen synthesis in fibroblasts within three-dimensional collagen lattices involves transcriptional and posttranscriptional mechanisms. *Febs Lett.* **318**, 129–133.

Edelman, G.M. and Jones, F.S. (1993) Outside and downstream the homeobox. *J. Biol. Chem.* **268**, 20683–20686

Ellis, R.E. ,Yuan, J. and Horvitz R. H. (1991) Mechanisms and functions of cell death. *Annu. Rev. Cell Biol.* **7**, 663–698

Emerman, J.T. and Pitelka, D.R. (1977) Maintenance and induction of morphological differentiation in dissociated mammary epithelium on floating collagen membranes. *In Vitro* **13**, 316–328

Evan, G.I., Wyllie, A.H., Gilbert, C.S., Litttlewood, T.D., Land, H., Brooks, M., Waters, C.M., Penn, L.Z. and Hancock, D.C. (1992) Induction of apoptosis in fibroblasts by c-myc protein. *Cell*, **69**, 119–128

Flaumenhaft, R., Moscatelli, D., Saskela, O. and Rifkin, D.B. (1989) Role of extracellular matrix in the action of basic fibroblast growth factor; matrix as a source of growth factor for long-term stimulation of plasminogen activator and DNA synthesis. *J. Cell Physiol.* **140**, 75–81

Frisch, S.M. and Francis, H. (1994) Disruption of epithelial cell-matrix interactions induces apoptosis. *J. Cell Biol.* **124**, 619–626

Giancotti, F.G. and Ruoslahti, E. (1990) Elevated levels of the $\alpha5\beta1$ fibronectin receptor suppresses the transformed phenotype of CHO cells. *Cell*, **60**, 849–859

Grobstein, C. (1975) In *Extracellular Matrix Influences on Gene Expression*, H.C. Slavin and R.C. Gruelich eds. pp. 9–19 Academic Press, New York.

Hay, E. (1993) Extracellular matrix alters epithelial differentiation. *Curr. Op. Cell Biol.* **5**, 1029–1035

Henzdel, M.J., Delcuve, G.P. and Davie, J.R. (1991) Histone deacetylase is a component of the internal nuclear matrix. *J. Biol. Chem.* **266**, 21936–21942

Hershovitz, R., Cahalon, L., Gilat, D., Miron, S., Miller, A. and Lider, O. (1993) Physically damaged extracellular matrix induces TNF-α secretion by interacting resting CD4+ T cells and macrophages. *Scand. J. Immunol.* **37**, 111–115

Hirst, R., Horwitz, C., Buck, C. and Rohrschneider, L. (1986) Phosphorylation of the fibronectin receptor complex in cells transformed by oncogenes that encode tyrosine kinases. *Proc. Natl. Acad. Sci.* **83**, 6470–6474

Howlett, A.R., Bailey, N., Damsky, C., Petersen, D.W. and Bissell, M.J. (1995) Cellular growth and survival are mediated by $\beta1$ integrins in normal human breast epithelium but not in breast carcinoma. *J. Cell Sci.* **108**, 1945–1957

Hynes, R.O. (1992) Integrins; versatility, modulation and signaling in cell adhesion. *Cell*, **69**, 11–25

Ignotz, R. A., Endo, T., Massague, J. (1987) Regulation of fibronectin and type I collagen mRNA by TGFß. *J. Biol. Chem.* **262**, 6443–6446

Ingber, D.E. and Folkman, J. (1989) Mechanochemical switching between growth and differentiation during FGF-stimulated angiogenesis *in vitro*; role of the extracellular matrix. *J. Cell. Biol.* **109**, 317–330

Ingber, D.E., Prusty, D., Frangioni, J.V., Cragoe, E.J. Jr., Lechene, C. and Schwartz, M.A. (1990a) Control of intracellular pH and growth by fibronectin in capillary endothelial cells. *J. Cell Biol.* **110**, 1803–1811

Ingber, D.E. (1990b) Fibronectin controls capillary endothelial cell growth by modulating cell shape. *Proc. Natl. Acad. Sci.* **87**, 3579–3583

Ip, M.M., Shoemaker, S.F. and Darcy, K.M. (1992) Regulation of rat mammary epithelial cell proliferation and differentiation by tumor necrosis factor-alpha. *Endocrinology* **130**, 2833–2844

Jehn, B., Costello, E., Marti, A., Keon, N., Deane, R., Li, F. ,Friis, R.R., Burri, P.H., Martin, F. and Jaggi, R. (1992) Overexpression of mos, ras, src and fos inhibits mouse mammary epithelial cell differentiation. *Molec. Cell. Biol.* **12**, 3890–3902.

Jehn, B., Chicaiza, G., Martin, F. and Jaggi, R. (1994) Isolation of three POU-domain containing cDNA clones from lactating mouse mammary gland. *Biochem. Biophys. Res. Commun.* **200**, 156–162

Jones, P.L., Boudreau, N., Myers, C.A., Erickson, H. and Bissell, M.J. (1995) Tenascin-C inhibits extracellular matrix-dependent gene expression in mammary epithelial cells: localization of active regions using recombinant tenascin fragments. *J. Cell Sci.* **108**, 519–527

Juliano, R.L. and Haskill, S. (1993) Signal transduction from the extracellular matrix. *J. Cell Biol.* **120**, 577–585

Kalebic, T., Garbisa, S., Glaser, B. and Liotta, L.A. (1983) Basement membrane collagen; degradation by migrating endothelial cells. *Science* **221**, 281–283

Katayama, K., Armendariz-Borunda, J., Raghow, R., Kang, A. H. and Seyer, J. M. (1993) A pentapeptide from type I procollagen promotes extracellular matrix production. *J. Biol. Chem.* **268**, 9941–9944

Kumar, S., Kinoshita, M., Noda, M., Copeland, N.G. and Jenkins, N.A. (1994) Induction of apoptosis by the mouse Nedd2 gene, which encodes a protein similar to the product of the *C. elegans* cell death gene *ced-3* and the mammalian IL-1ß converting enzyme. *Genes and Development* **8**, 1613–1626.

Lin, C.Q. and Bissell, M.J. (1993) Multi-faceted regulation of cell differentiation by extracellular matrix. *FASEB J.* **7**, 737–744

Liu, J-K., DiPersio, M.C. and Zaret, K.S. (1991) Extracellular signals that regulate liver transcription factors during hepatic development. *Molec. Cell. Biol.* **11**, 773–784

Lee, T.C., Zhang, Y. and Schwartz, R.J. (1994) Bifunctional transcriptional properties of YY1 in regulating muscle actin and c-myc gene expression during myogenesis. *Oncogene* **9**, 1047–1052

McPherson, C.E., Shim, E-Y., Freidman, D.S. and Zaret, K.S. (1993) An active tissue-specific enhancer and bound transcription factors existing in a precisely positioned nucleosomal array. *Cell* **75**, 387–398

Meier, V.S. and Groner, B. (1994) The nuclear factor YY1 participates in the repression of the β-casein gene promoter in mammary epithelial cells and is counteracted by mammary gland factor during lactogenic hormone induction. *Molec. and Cell. Biol.* **14**, 128–137

Meredith, J.B., Fazeli, B. and Schwartz, M. A. (1993) The extracellular matrix as a cell survival factor. *Molec. Biol. Cell* **4**, 953–961

Miyashita, T., Krajewski, S., Krajewska, M., Wang, H.G. ,Lin, H.K., Liebermann, D.A., Hoffman, B. and Reed, J.C. (1994) Tumor suppresser p53 is a regulator of bcl-2 and bax gene expression *in vitro* and *in vivo*. *Oncogene* **9**, 1799–1805

Mooney, D., Hansen, L., Vacanti, J., Langer, R., Farmer, S. and Ingber, D.E. (1992) Switching from differentiation to growth in hepatocytes: control by extracellular matrix. *J. Cell Physiol.* **151**, 497–505

Montesano, R., Pepper, M.S., Mohle-Steinlein, U., Risau, W., Wagner, E.F. and Orci, L. (1990) Increased proteolytic activity is responsible for the aberrant morphogenetic behavior of endothelial cells expressing middle T oncogene. *Cell* **62**, 435–445

Nanes, M.S., Kuno, H., Demay, M.B., Hendy, G.N., DeLuca, H.F., Titus, L. and Rubin, J. (1994) A single element confers responsiveness to 1, 25-dihydroxyvitamin D3 and TNF alpha in the rat osteocalcin gene. *Endocrinology* **134**, 1113–1120

Niedbala, M.J. (1993) Cytokine regulation of endothelial cell extracellular proteolysis. *Agents and Actions* **42**, 179–193

Oberhammer, F.A., Pavelka, M., Sharma, S., Tiefenbacher, R., Purchio, A.F., Bursch, W and Schulte-Jermann, R. (1992) Induction of apoptosis in cultured hepatocytes and in regressing liver by TGFβ. *Proc. Natl. Acad. Sci.* **89**, 5408–5412

Oltavi, Z.N., Milliman, C.L. and Korsmyer, S.J. (1993) Bcl-2 heterodimerizes *in vivo* with a conserved homologue, bax, that accelerates cell death. *Cell* **74**, 609–619

Opas, M. and Dziak, E. (1994) bFGF induced transdifferentiation of RPE to neuronal progenitors is regulated by the mechanical properties of the substratum. *Developmental Biology* **161**, 440–454

Owen, T.A., Holthuis, J., Shalboub, V., Barone, L.M., Wilming, L., Tassinari, M.S., Kennedy, M.B., Pockwinse, S., Lian, J.B. and Stein, G.S. (1990) Progressive development of the rat osteoblast phenotype in vitro; reciprocal relationships in expression of genes associated with osteoblast proliferation and differentiation during formation of the bone extracellular matrix. *J. Cell Physiol.* **143**, 420–430

Padgham, C.R., Boyle, C.C., Wang, X.J., Raleigh, S.M., Wright, M.C. and Paine, A.J. (1993) Alteration of transcription factors mRNAs during isolation and culture suggests the activation of a proliferative mode underlies their de-differentiation. *Biochem. Biophys. Res. Comm.* **197**, 599–605

Parry, G., Lee, E.Y., Farson, D., Koval, M. and Bissell, M.J. (1985) Collagenous substrata regulate the nature and distribution of glycosaminoglycans produced by differentiated cultures of mouse mammary epithelial cells. *Exp. Cell Res.* **156**, 487–499

Parry, G., Cullen, B., Kaetzel, C.S., Kramer, R. and Moss, L.J. (1987) Regulation of differentiation and polarized secretion in mammary epithelial cells maintained in culture; extracellular matrix and membrane polarity influences *J. Cell. Biol.* **105**, 2043–2050

Pauly, R.R., Passaniti, A., Crow, M., Kinsella, J.L., Papadopoulos, N., Montecone, R., Lakatta, E.G. and Martin, G. R. (1992) Experimental models that mimic the differentiation and dedifferentiation of vascular cells. *Circulation* **86**, 68–73

Pauly, R.R., Passaniti, A., Bilato, C., Monticone, R., Cheng, L., Papadopoulos, N., Gluzband, Y.A., Smith, L., Weinstein, C., Lakatta, E.G. and Martin, G.R. (1994) Migration of cultured vascular smooth muscle cells through a basement membrane barrier requires type IV collagenase activity and is inhibited by cellular differentiation. *Circ. Res.* **75**, 41–54

Petersen, O.W., Ronnov-Jessen, L., Howlett, A.R. and Bissell, M.J. (1992) Interaction with basement membrane serves to rapidly distinguish normal and malignant human breast epithelial cells. *Proc. Natl. Acad. Sci.* **89**, 9064–9068

Pullan, S., Wilson, J., Tilly, J., Hickman, J.A., Dive. C. and Streuli, C.H. (1995) Requirement of basement membrane for the suppression of programmed cell death in mammary epithelium. *J. Cell. Biol.* in press

Rana, B., Mischoulon, D., Xie, Y., Bucher, N. and Farmer, S.R. (1994) Cell-extracellular matrix interactions can regulate the switch between growth and differentiation in rat hepatocytes; Reciprocal expression of C/EBP alpha and immediate-early growth response transcription factors. *Molec. Cell Biol.* **14**, 5858–5869

Raught, B., Khursheed, B., Kazansky, A. and Rosen, J. (1994) YY1 represses β-casein gene expression by preventing the formation of a lactation-associated complex. *Molec. Cell. Biol.* **14**, 1752–1763

Reichmann, E., Ball, R., Groner, B. and Friis, R.R. (1989) New mammary epithelial and fibroblastic cell clones in coculture from structures competent to differentiate functionally. *J. Cell. Biol.* **108**, 1127–1138

Reed, J.C. (1994) Bcl-2 and the regulation of programmed cell death. *J. Cell Biol.* **124**, 1–6

Robaye, B., Mosselmans, R., Fiers, W., Dumont, J.E. and Galand, P. (1991) Tumor necrosis factor induces apoptosis in normal endothelial cells *in vitro*. *Am. J. Pathol.* **138**, 447–453

Robinson, S.D., Roberts, A.B. and Daniel, C.W. (1993) TGF beta suppresses casein synthesis in mouse mammary explants and may play a role in controlling milk levels during pregnancy. *J. Cell Biol.* **120**, 245–251

Romagnolo, D., Akers, R.M., Wong, E.A., Boyle, P.L., McFadden, T.B., Byatt, J.C. and Turner, J. D. (1993) Lactogenic hormones and extracellular matrix regulate expression of IGF-1 linked to MMTV-LTR in mammary epithelial cells. *Molec. Cell Endocrinology* **96**, 147–157

Roskelley, C.D., Desprez, P.Y. and Bissell, M.J. (1995) Extracellular matrix-dependent tissue-specific gene expression in mammary epithelial cells requires both physical and biochemical signal transduction *Proc. Natl Acad. Sci.* **91**,12378–12382

Rotello, R.J., Lieberman, R.C., Purchio, A.F. and Gerschenson, L.E. (1991) Coordinated regulation of apoptosis and cell proliferation by TGFß1 in cultured uterine epithelial cells. *Proc. Natl. Acad. Sci.* **88**, 3412–3415

Schmid, V., Baader, C., Bucciarelli, A. and Reber-Muller, S. (1993) Mechanochemical interactions between striated muscle cells of jellyfish and grafted extracellular matrix can induce and inhibit DNA replication and transdifferentiation in vitro. *Developmental Biology* **155**, 483–496

Schmidhauser, C., Bissell, M.J., Myers, C.A. and Casperson, G.F. (1990) Extracellular matrix and hormones transcriptionally regulate bovine 5' sequences in stably transfected mouse mammary cells. *Proc. Natl. Acad. Sci.* **87**, 9118–9122

Schmidhauser, C., Casperson, G.F., Myers, C.A., Sanzo, K.T., Bolton, S. and Bissell, M.J. (1992) A novel transcriptional enhancer is involved in the prolactin and extracellular matrix regulation of casein gene expression. *Mol. Biol. Cell* **3**, 699–709

Schmidhauser, C., Casperson, G.F. and Bissell, M.J. (1994) Transcriptional activation by viral enhancers: Critical dependence on extracellular matrix-cell interactions in mammary epithelialcells. *Molecular Carcinogenesis* **10**, 1–6

Schmidhauser, C., Myers, C.A., Mossil, R., Casperson, G.F. and Bissell, M.J. (1994) Extracellular matrix dependent gene regulation in mammary epithelial cells. Hanahan *Symposia Proceedings* – in press

Schmitt-Ney, M., Doppler, R.K. and Groner, B. (1991) β-casein gene promoter activity is regulated by the hormone-mediated relief of transcriptional repression and a mammary-gland-specific nuclear factor. *Mol. Cell. Biol.* **11**, 3745–3755

Schnaper, H.W., Kleinman, H.K. and Grant, D.S. (1993) Role of laminin in endothelial cell recognition and differentiation. *Kidney International*, **43**, 20–25

Shinagawa, T., Yoshioka, K., Kakumu, S., Wakita, T., Ishikawa, T., Itoh, Y. and Takayanagi, M. (1991) Apoptosis in cultured rat hepatocytes; the effects of TNFα and interferon gamma. *J. Pathol.* **165**, 247–253

Sieweke, M.H. and Bissell, M.J. (1995) The tumor-promoting effect of wounding: a possible role for TGFβ-induced stromal alterations. *Crit. Rev. Oncogenesis*. pp. 1–17

Singhvi, R., Kuman, A., Lopez, G.P., Stephanopoulos, G.N., Wang, D.I., Whitesides, G.M. and Ingber, D.E. (1994) Engineering cell shape and function. *Science* **264**, 696–698.

Stein, G.S., Lian, J.B. and Owen, T. A. (1990) Relationship of cell growth to the regulation of tissue-specific gene expression during osteoblast differentiation. *FASEB J.* **4**, 3111–3123

Stein, G.S., Lian, J.B., Owen, T.A., Holthuis, J., Bortell, R. van Wijnen, A.J. (1992) Mechanisms that mediate a functional relationship between proliferation and differentiation. In *Molecular and Cellular Approaches to the Control of Proliferation and Differentiation*, San Diego Academic Press, pp. 299–341.

Strange, R., Li, F., Saurer, S., Burkhardt, A. and Friis, R.R. (1992) Apoptotic cell death and tissue remodeling during mouse mammary gland involution. *Development* **115**, 49–58

Streuli, C.H. and Bissell, M.J. (1990) Expression of extracellular matrix components is regulated by substratum. *J. Cell Biol.* **110**, 1405–1415

Streuli, C.H., Bailey, N. and Bissell, M.J. (1991) Control of mammary epithelial cell differentiation; basement membrane induces tissue-specific gene expression in the absence of cell-cell interaction and morphological polarity. *J. Cell Biol.* **115**, 1383–1395

Streuli, C.H., Schmidhauser, C., Kobrin, M., Bissell, M.J. and Derynck, R. (1993) Extracellular matrix regulates expression of the TGFß1 gene. *J. Cell Biol.* **120**, 253–260

Suda, T., Takahashi, T., Golstein, P. and Nagata, S. (1993) Molecular cloning and expression of the Fas ligand, a novel member of the tumor necrosis factor family. *Cell* **75**, 1169–1178

Sympson, C.J., Talhouk, R.S., Alexander, C.M., Chin, J.R., Clift, S.M., Bissell, M.J. and Werb, Z. (1994) Targeted expression of stromelysin-1 in mammary gland provides evidence for a role of proteinases in branching morphogenesis and the requirement for an intact basement membrane for tissue-specific gene expression. *J. Cell Biol.* **125**, 681–693

Tamura, M. and Noda, M. (1994) Identification of a DNA sequence involved in osteoblast-specific gene expression via interaction with helix-loop-helix type transcription factors. *J. Cell Biol.* **126**, 773–773

Tijan, R. and Maniatis, T. (1994) Transcriptional activation; a complex puzzle with few easy pieces. *Cell* **77**, 5–8

Tremble, P., Chiquet-Ehrismann, R. and Werb, Z. (1994) The extracellular matrix ligands fibronectin and tenascin collaborate in regulating collagenase gene expression in fibroblasts. *Mol. Biol. Cell* **5**, 439–453

Vlodavsky, I., Korner, G., Ishai-Michaeli, R., Bashkin, P. and Fuks, Z. (1990) Extracellular matrix-resident growth factors and enzymes: possible involvement in tumor metastisis and angiogenesis. *Cancer Met. Rev.* **9**, 203–226

Walker, N.I., Bennett, R.E. and Kerr, J.F. (1989) Cell death by apoptosis during involution of the lactating breast in mice and rats. *Am. J. Anat.* **185**, 19–32

Wang, L., Muira, M., Bergeron, L., Zhu, H. and Yuan, J. (1994) Ich-1, an ICE-related gene encodes both positive and negative regulators of programmed cell death. *Cell* **78**, 739–750

Watson, C.J., Gordon, K.E., Robertson, M. and Clark, J.A. (1991) Interaction of DNA-binding proteins with a milk protein gene promoter *in vitro*; identification of a mammary specific factor. *Nucleic Acids Res.* **19**, 6603–6610

White, E. (1994) Tumor biology. p53, guardian of Rb. *Nature* **371**, 21–22

Werb, Z., Tremble, P.M., Behrendtsen, O., Crowley, E. and Damsky, C. H. (1989) Signal transduction through the fibronectin receptor induces collagenase and stromelysin gene expression *J. Cell. Biol.* **109**, 877–889

Wessells, N.K. (1977) In '*Tissue Interaction and Development*' W. A. Beryamin ed. pp. 213–231, New York.

Wolfe, A.P. (1994) Transcription: In tune with the histones. *Cell* **77**, 13–16

Wrana, J.L., Maeno, M., Hawrylshyn, B., Yao, K-L., Dommenicucci, C. and Sodek, J. (1988) Differential effects of TGFβ on the synthesis of extracellular matrix proteins by normal fetal rat calvarial bone cell population. *J. Cell. Biol.* **106**, 915–924

Yuan, J., Shaman, S., LeDoux, S., Ellis, H.M. and Horvitz, R.H. (1993) The *C. elegans* cell death gene *ced-3* encodes a protein similar to mammalian interleukin-1β-converting enzyme. *Cell* **75** 641–652

Zutter, M.M., Santoro, S.A., Painter, A.S., Tsung, Y.L. and Gaford, A. (1994) The human alpha 2 integrin gene promoter. Identification of positive and negative regulatory elements for cell-type and developmentally restricted gene expression. *J. Biol. Chem.* **269**, 463–469

9 Molecular Recognition of the Extracellular Matrix by Cell Surface Receptors

Kristofer Rubin[1], Donald Gullberg[2], Bianca Tomasini-Johansson[1], Rolf K. Reed[3], Cecilia Rydén[1] and Thomas K. Borg[4]

[1]*Department of Medical and Physiological Chemistry, University of Uppsala Biomedical Center, Sweden*
[2]*Department of Animal Physiology, University of Uppsala Biomedical Center, Sweden*
[3]*Department of Physiology ,University of Bergen, Norway*
[4]*Department of Developmental Biology and Anatomy, University of South Carolina, Columbia, South Carolina, USA*

INTRODUCTION

Cellular Binding of Extracellular Matrix (ECM) molecules

One major function of the extracellular matrix (ECM) is to provide a solid support for cells. This enables the complex microarchitecture of tissues to be formed, thus allowing all body cells - with the exception of the red blood cells — to fulfill their differentiated functions when interacting with ECM structures. The ECM may vary from a relatively rigid framework that forms a scaffold on which the cells are arranged to a complex fluid medium through which the cells flow. As cells respond to the extracellular environment, they change their phenotype. Furthermore, cellular interactions with ECM structures play important roles in many pathophysiological processes, such as inflammatory responses, tumor invasion, and metastasis. Such effects could be indirect, resulting from the three-dimensional structure of the ECM influencing cell shape and/or cytoskeletal organization. Much experimental data have been provided that favor the hypothesis that cell shape can modulate many important cell-physiological functions such as: proliferation, differentiation, and expression of various gene products (Folkman and Moscona, 1978; Ben-Ze'ev *et al.*, 1980; Bissell *et al.*, 1982; Nakagawa *et al.*, 1989; Ingber, 1991). Several processes have been shown — at the physiological level — to be controlled by cellular interactions with ECM components; examples include embryonal development, as well as repair mechanisms during adult life. It has become increasingly clear that ECM molecules also exert direct influence on cellular phenotype via specific signal-transducing receptors (Damsky and Werb, 1992; Hynes, 1992; Juliano and Haskill, 1993). Many physiological processes require that the cell-ECM interactions are dynamic. Examples include such diverse cellular processes as adherence of platelets to subendothelial structures, maturation and migration

---

Extracellular Matrix, Volume 2, Molecular Components and Interactions, edited by Wayne D. Comper.

of intestinal cells, and the control of interstitial fluid pressure and thereby of fluid balances in tissues.

Special ECM receptors involved in clearing debris in the blood — originating from ECM turnover in the body — are present in the liver of mammals. These types of receptors are particularly expressed by the distinct type of endothelial cells that line liver sinusoids — the liver endothelial cells (LEC). LEC are pluripotent with regard to their ability to interact with extracellular matrix components and have the capacity to specifically bind, endocytose, and degrade a number of ECM molecules, such as denatured collagens, glycosaminoglycans, laminin, and nidogen (Smedsrød *et al.*, 1990). This function of LEC will not be discussed further in the present chapter.

Both Gram-negative and Gram-positive bacteria interact specifically with ECM molecules. Recent investigations with animal models have demonstrated that these interactions are of relevance for infectious processes.

Cell Adhesion

Much of our knowledge of cell-ECM contacts originate from studies of cell adhesion *in vitro*, *i.e.* how cells adhere to immobilized ECM components. Early work revealed that different cell types were able to, directly, specifically and in the absence of serum, adhere to various immobilized ECM glycoproteins such as: fibronectin (Pearlstein, 1976; Grinnell *et al.*, 1977; Höök *et al.*, 1977; Ruoslahti and Hayman, 1979), laminin (Terranova *et al.*, 1980; Carlson *et al.*, 1981; Johansson *et al.*, 1981; Couchman *et al.*, 1983), and collagens (Grinnell and Minter, 1978; Rubin *et al.*, 1979; Schor and Court, 1979; Rubin *et al.*, 1981). Although a large proportion of the characterized ECM-molecules support cell adhesion, some ECM molecules such as SPARC, thrombospondin, and tenascin, are able to inhibit cell adhesion. (Chiquet-Ehrismann *et al.*, 1988; Murphy-Ullrich *et al.*, 1991; Sage and Bornstein, 1991). Tenascin seems to be dualistic with regard to adhesion-promoting activity, as this ECM glycoprotein has also been reported to support specific cell adhesion (Prieto *et al.*, 1993; Sriramarao *et al.*, 1993; Wehrle-Haller and Chiquet, 1993).

It has been suggested that the mechanisms for cell adhesion involve weak, non-specific physical forces — according to the theory of the interactions of lyophobic colloids (Curtis, 1966). Basically, this theory postulates that binding strength in cell adhesion is generated by London dispersion forces. At certain distances, these forces are able to override the repulsive electrostatic charge interactions between two negative surfaces, e.g., between two cells or cell and ECM structures. Eukaryotic cell surfaces are, however, highly dynamic and it is therefore, unlikely that low-affinity and unspecific charge interactions would be allowed to develop in sufficient numbers to support cell adhesion. The relative importance of these unspecific forces may be larger in bacterial adhesion since diffusion of bacterial surface components is sterically hindered by the rigid bacterial cell wall.

The prevailing concept on the nature of animal cell adhesion holds that adhesive processes are mediated by specific cell surface molecules — the adhesion receptors. Such receptors started to be isolated and characterized during the mid-seventies (Hausman and Moscona, 1975; Thiery *et al.*, 1977; Müller and Gerisch, 1978). Since then, our knowledge of adhesion receptors has expanded dramatically. Four families of adhesion receptors have been characterized: 1) the immunoglobulin superfamily (Williams and Barclay, 1988; Edelman and Crassin, 1991; Buck, 1992); 2) the cadherins (Takeichi, 1988; Geiger and

Ayalon, 1992)]; 3) the integrins (Buck and Horwitz, 1987; Hynes, 1987; Ruoslahti and Pierschbacher, 1987; Ginsberg *et al.*, 1988; Hemler, 1990; Springer, 1990; Akiyama *et al.*, 1990); and 4) the selectins (Bevilacqua and Nelson, 1993; Vestweber, 1993). Several cell surface proteoglycans have been found to have features in common with the adhesion receptors, and it can be argued that they should be placed among the families of adhesion receptors (Jalkanen *et al.*, 1992). Adhesion receptors do not merely provide physical links between intra- and extracellular structural elements, some also act as true receptors in the sense that ligand-engaged receptors transduce signals to the cell interior. Analogous to animal cells, bacterial cells have been shown to possess surface proteins that mediate specific adhesion to ECM molecules. Some of these bacterial proteins have been characterized, and their biological properties are being elucidated as discussed later in this chapter.

Integrins have special importance for eukaryotic cellular adhesion to ECM structures. The pivotal roles of these adhesion receptors in cellular adhesion, to ECM structures of cultured cells, are indicated by the fact that antibodies directed to integrins are able to completely inhibit a large number of such adhesion processes. It should be emphasized, however, that this does not exclude the presence of other classes of adhesion receptors, playing a similarly important role in adhesion. Thus, some cell-ECM adhesion processes depend upon the participation and activation of different classes of adhesion receptors necessary for functional adhesion to take place. An example is the reported co-operation between intercalated proteoglycans and integrins in complex adhesion processes; some cultured cells require the presence of both integrin and proteoglycan interaction with the ECM-substrate (Woods *et al.*, 1986; Jalkanen *et al.*, 1992). Other intercellular adhesion processes also require a co-operation between several classes of adhesion receptors. Functional adhesion of leukocytes to activated endothelial cells, one leading to diapedesis, requires that selectins, activated integrins, and members of the Ig-superfamily of adhesion receptors cooperate in a complex fashion (Albelda and Buck, 1990; Springer, 1990; Butcher, 1991; Bevilacqua and Nelson, 1993).

Morphology of Vertebrate Cell-ECM Contacts, *in Vivo* and *in Vitro*

Specialized contacts between the ECM and the cell surface have been well documented in cultured cells. Such contacts do also occur *in vivo*, although there are relatively few well described examples. An example of specialized cell-ECM contacts *in vivo* is the contact between ECM-components and striated muscle. The ECM, especially the collagenous structures associated with striated muscle, occurs in a precise three-dimensional arrangement (Figure 9.1) the arrangement of the epimysium, perimysium, and endomysium forming a three-dimensional network that is intimately associated with muscle function (Caulfield and Borg, 1979; Borg and Caulfield, 1980). This arrangement forms an elastic, stress-tolerant network that is essential to muscle compliance and efficiency (Borg *et al.*, 1981). The elasticity is dependent upon two principal factors: 1) the arrangement of the collagen fibers into a weave-like network; and 2) the anchoring of the collagen fibers at specific sites on the muscle.

Both cardiac and skeletal muscle have a precise, parallel arrangement of the myofibers (Caulfield and Borg, 1979; Borg and Caulfield, 1980). It is functionally important to maintain this alignment during contraction in order to generate maximum force. The lateral alignment of the muscle fibers is held together by bundles of collagen, termed struts.

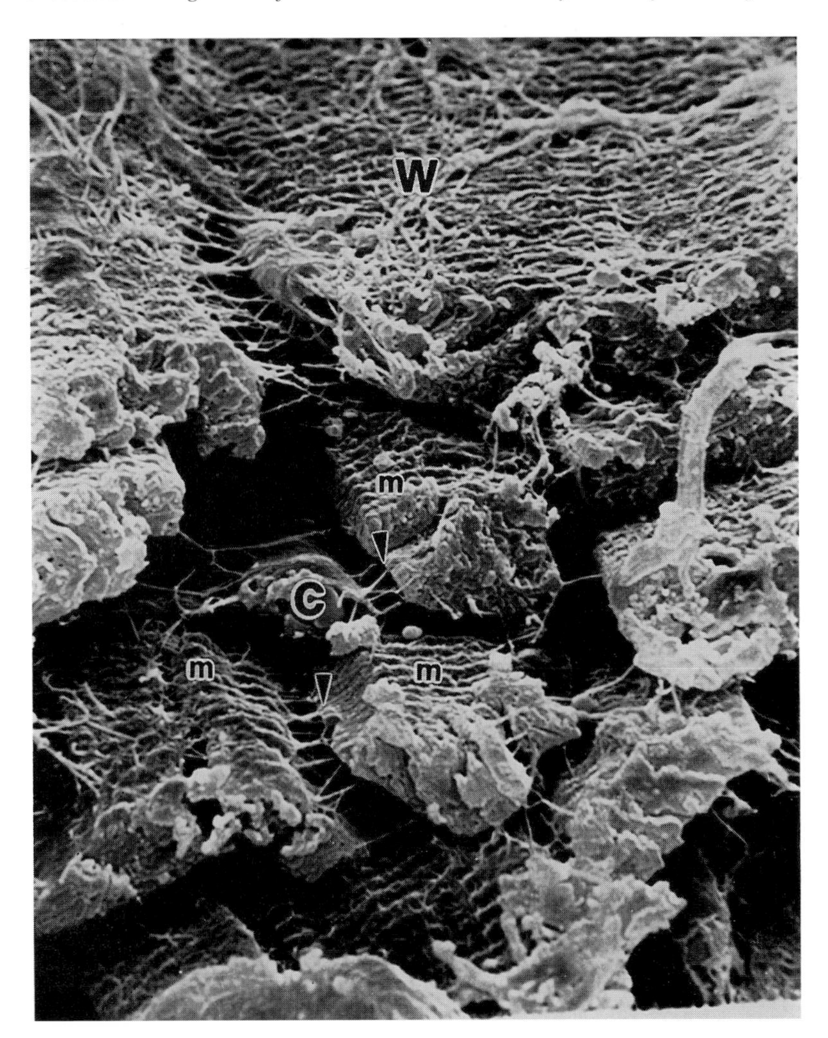

Figure 9.1 Morphological arrangement of cell-collagen contacts in rat heart. Scanning electron micrograph showing the morphological arrangement of collagen in association with rat cardiac muscle. The endomysial weave (W) network surrounds the group of myocytes while individual struts (arrows) connect adjacent myocytes, as well as myocytes to capillaries (C). Reprinted from Borg *et al.*, (1981) with permission from Williams & Williams.

Electron microscopic observations have revealed that these collagen struts are associated with the myocardial surface at sites just lateral to the Z-band, and connect adjacent myocytes (Caulfield and Borg, 1979; Borg and Caulfield, 1980; Borg *et al.*, 1981). Similar struts connect myocytes to capillaries. This arrangement of endomysial collagen in the heart provides for capillary patency in the presence of high intraventricular pressure.

Both *in vivo* and *in vitro* studies of cardiac myocytes have shown that these sites of attachment of collagen to the cell surface involve a cytoplasmic connection with certain adhesion receptors (Terracio *et al.*, 1991). The basement membrane components laminin and collagen type IV — and internally — the cytoskeletal proteins α-actinin and vinculin have been localized at this site. The staining pattern for integrins (see below) has shown a precise distribution at the Z-line and co-localize with the cytoskeletal staining pattern (Carver *et al.*, 1994).

Specialized attachment sites between collagen fiber bundles and tissue cells have also been demonstrated in the lymphatic system (Leak, 1970). Collagen fiber bundles interconnect the lymphatic vessels with the surrounding parenchyma or connective tissue in a loose arrangement. When edema forms, e.g., in inflammation, the collagen fibers become taut and pull in a three-dimensional manner to maintain patency of the lymphatic vessel. While immuno-localization studies of integrin distribution have not been performed in this system, it is likely that integrins will be involved in attachment of the collagen to both the vascular and connective tissue cells. In order to provide vascular patency, the attachment sites must be precisely located. Little is known of how these sites are determined developmentally during angiogenesis or during repair following injury.

Epithelial cells of stratified squamous, transitional, and pseudostratified tissues form organelles, where cells make close contact with the underlying basement membrane. These sites, termed hemidesmosomes, were originally detected by electron microscopy and shown to interact with the epithelial intermediate filaments — or tonofilaments — at the cytoplasmic side (Kelly, 1966). Our knowledge of the molecular composition of these specialized organelles mediating close contact between cells and the underlying ECM, has substantially increased and several macromolecules, particular for this structure, have been isolated as reviewed in Jones *et al.* (1994).

CELL-SURFACE RECEPTORS FOR ECM MOLECULES

Integrins — A Large Family of Adhesion Receptors Mediating Adhesion to ECM Structures

Introduction

The discovery of a family of adhesion receptors mediating specific cellular adhesion to ECM structures allowed for rapid progress in the field of cell-ECM interactions. These receptors — the integrins (Buck and Horwitz, 1987; Hynes, 1987; Ruoslahti and Pierschbacher, 1987) — comprise a large family of membrane-spanning glycoproteins that mediate cell adhesion to extracellular matrix (ECM) proteins, as well as to other cells. (For reviews see: Albelda and Buck, 1990; Akiyama *et al.*, 1990; Hemler, 1990; Springer, 1990; Ruoslahti, 1991; Reichardt and Tomaselli, 1991; Hynes, 1992; Loftus *et al.*, 1994). The name *integrin* was introduced in 1986 in a paper describing molecular cloning and the complete amino acid sequence of the avian integrin β_1-subunit (Tamkun *et al.*, 1986).

Two main strategies were used to isolate receptors that mediate adhesion to ECM glycoproteins. One took advantage of the monoclonal antibody technique. Monoclonal antibodies with the ability to inhibit cell adhesion to ECM structures were isolated, and the corresponding antigens characterized. These monoclonal antibodies reacted with polypeptide complexes containing components with M_r:s in the range 110,000 – 160,000 (Greve and Gottlieb, 1982; Neff *et al.*, 1982; Chen *et al.*, 1985; Knudsen *et al.*, 1985; Brown and Juliano, 1986). In the second strategy, membrane proteins with the ability to bind ECM components were purified by affinity chromatography. Instrumental for the sucess of this approach was the determination of the amino acid sequence of the cell-binding site in fibronectin (Pierschbacher and Ruoslahti, 1984). Synthetic peptides encompassing this sequence, i.e. peptides containing L-arginyl-glycyl- L-aspartyl-L-serine (RGDS), allowed specific elution of fibronectin receptor complexes from affinity columns containing immobilized fibronectin (Pytela *et al.*, 1985; Pytela *et al.*, 1985). Later studies revealed that the adhesion receptors isolated by the two methods were similar, and belonged to a family of related adhesion receptors. It was also found that these adhesion receptors were closely related to a group of cell-surface molecules that had been isolated as lymphocyte antigens. These proteins had been named VLA-proteins (very late antigens of activation) (Hemler *et al.*, 1985). Furthermore, a group of surface glycoproteins that are involved in leukocyte adhesion and diapedesis, and with the molecular characteristic of being α/β heterodimers, were shown to belong the expanding integrin family related to integrins (Kishimoto *et al.*, 1987; Hemler, 1990; Springer, 1990). Several purified platelet glycoproteins, notably the fibrinogen receptor gpIIb/IIIa, and also the platelet receptors for collagen and fibronectin were soon shown to belong to the family of integrins (Ginsberg *et al.*, 1987; Ginsberg *et al.*, 1988; Kieffer and Phillips, 1990).

Besides mediating cellular adhesion, integrins also function as signalling molecules conveying information from the extracellular environment to the cell interior. This signalling function of integrins is often referred to as "outside - in" signalling by integrins (Hynes, 1992). Studies with cultured cells have shown that a large number of diverse cell physiological processes are regulated, or modulated by integrins. Examples include cellular growth, motility, and gene expression (Schiro *et al.*, 1991; Damsky and Werb, 1992b; Schwartz, 1992; Hynes, 1992; Juliano and Haskill, 1993; Leavesley *et al.*, 1993). Clustering of certain integrins at the cell surface with anti-integrin antibodies, or by ligand-binding, results in a rapid increase in intracellular free Ca^{2+}, alkalinization of the cytosol, and phosphorylation of intracellular proteins (Hynes, 1992; Schwartz, 1992; Juliano and Haskill, 1993). Data have recently been provided showing that signals mediated from ligand-occupied, or clustered, integrins regulate cell survival and inhibit the induction of apoptosis in a variety of cell types (Meredith *et al.*, 1993; Montgomery *et al.*, 1994).

Examples have been provided showing that integrin-mediated adhesion reactions can influence signal transduction processes evoked by growth factor stimulation. Taking platelet-derived growth factor (PDGF) as an example, it has been shown that integrin-mediated adhesion is necessary for PDGF-induced increases in free cytoplasmic Ca^{2+} (McNamee *et al.*, 1993). PDGF stimulation activates phospholipase C-γ in attached and suspended cells, however, only in the former did this lead to an increase in the concentration of free intracellular Ca^{2+}. This was explained by the fact that adhesion of cells is necessary for the formation of phosphatidylinositol bisphosphate (PIP2), a substrate for activated phospholipase C-γ (PLC-γ) (McNamee *et al.*, 1993). Activated PLC-γ generates free inositol triphosphate (IP3) from PIP2. IP3 is a potent regulator of free cytoplasmic Ca^{2+}.

Less information is available concerning possible *in vivo* functions of integrins. Integrins have, however, been shown to be involved in such diverse processes as normal embryonal development, inflammatory processes and malignancies (Albelda and Buck, 1990; Hemler, 1990; Springer, 1990; Ruoslahti, 1991; Hynes, 1992; Reed *et al.*, 1992; Elliott *et al.*, 1994). By using the techniques of homologous recombination in embryonic stem cells and the establishment of transgenic mice, the importance of integrin in embryonal development has been demonstrated. Thus, embryos deficient in one integrin-subunit (α_5) die by day 10–11 of gestation (Yang *et al.*, 1993) (see also page 283).

Several integrins can act as receptors for any one particular ECM glycoprotein. Examples are: laminin which is recognized by at least six different integrins, collagens by four, and fibronectins by at least eight. This suggests that ligand occupation of different integrins may give different functional consequences. This view is supported by studies in which functional aspects of chimeric integrin α-subunits were analyzed (Chan *et al.*, 1992). It was demonstrated that the cytoplasmic domains of certain α-subunits had distinct and different functions in cellular migration and contraction (Chan *et al.*, 1992).

Structure of Integrins

Integrins are non-covalently associated heterodimers, composed of from one α- and one β-subunit (Albelda and Buck, 1990; Hemler, 1990; Springer, 1990; Ruoslahti, 1991; Hynes, 1992; Palmer *et al.*, 1993). At present, 8 β-, and 15 α-subunits, all representing distinct gene products, have been characterized. Integrin subunits are type I transmembrane glycoproteins, with the carboxy terminal in the cytoplasm and the amino terminal located extracellularly. The extent of similarity at the amino acid level is 40 and 50% between different α- and β-subunits, respectively. These integrin-subunits associate to form around 20 different integrin α/β heterodimers, as described to date. Both subunits contribute amino acids to the ligand-binding domain of the heterodimeric receptors; isolated integrin subunits possess no ligand binding capacity. Available data suggest that the ligand-binding sites of integrins are formed by amino acid sequences present in both subunits. Both the α- and β-subunits have short cytoplasmic domains, containing around 50 or fewer amino acids (Sastry and Horwitz, 1993). The exception is the ~1000 amino acids large cytoplasmic domain of the β_4-subunit. Integrin β-subunits migrate as polypeptides with M_r:s between 90,000 – 115,000 – except the β_4-subunit that migrates as an M_r 205,000 peptide. The α-subunits migrate as polypeptides ranging in M_r between 120,000 and 205,000. Electron microscope studies of integrin heterodimer structures revealed a globular head with two stalks or tails, partly intercalated in the lipid bilayer (Carrell *et al.*, 1985; Nermut *et al.*, 1988) with a total molecular length of 280 Å. The globular head has a dimension of around 80 by 120 Å, and the two tails are around 180 – 200 Å long.

A molecular characteristic shared by all integrin β-subunits, is the presence of cystine-containing repeats in the extracellular part of the molecule, close to the transmembrane segment. With the exception of the β_4 and β_8 integrin subunits the cytoplasmic domains of the integrin β-subunits show many similarities (Sastry and Horwitz, 1993).

Integrin α-subunits are characterized by a set of well conserved cysteine residues and three or four aspartic acid-rich metal-binding domains of the general structure DXDXDGXXD, where X could be any amino acid. Some α-chains possess a ~190 amino acid divalent cation-binding domain, the I (inserted, interactive) domain that participates in ligand-binding (Michishita *et al.*, 1993). These domains show amino acid similarity with similar sized domains present in a number of other proteins, such as the vonWillebrand factor, cartilage matrix protein and complement factors C2 and B. Several α-subunits are proteolytically processed after synthesis, and appear on the cell surface as a low-M_r transmembrane chain disulfide-bonded to the high-M_r extracellular part of the molecule. The amino acid sequences of the cytoplasmic domains of the integrin α-subunits show variation with few similarities evident. In several integrin α-subunits, however, the sequence KXGFFKR (the amino acid at the X position varies) is present in the junction region between the transmembrane and cytoplasmic domains.

Many members of the integrin family of adhesion receptors arise from alternative splicing of integrin subunit precursor mRNA:s, thus giving rise to the presence of integrin isoforms. Among the α-subunits, α_3, α_6, and α_7 are known to occur in isoforms differing as to primary structure of the cytoplasmic domains. (For a review see: Sastry and Horwitz, 1993). Isoforms of the α_6 and α_7-subunits, differing in the putative ligand-binding domains, have also been described (Ziober *et al.*, 1993). Three different β_1 integrin isoforms have been demonstrated (Altruda *et al.*, 1990; Languino and Ruoslahti, 1992), and seem to have different functions (Balzac *et al.*, 1993).

Biosynthesis of Integrins

Isolated α- or β-subunits do not appear on cell surfaces; α/β heterodimers are assembled intracellularly and transported to the cell surface after synthesis. Available data suggest that the β_1 integrin-subunit is synthesized in excess over the α-subunits. Lenter and Vestweber have presented data suggesting that part of the newly synthesized β_1-subunits seem to reside in an underglycosylated form, associated with the chaperon calnexin in the endoplasmatic reticulum for prolonged periods of time (Lenter and Vestweber, 1994). As much as 80–90% of the total pool of cellular integrin β_1-subunits is present in this underglycosylated immature form. This large pool of immature β_1-subunits most likely assures a constant supply of β-subunits that could associate with newly synthesized α-subunits. Regulation of integrin synthesis seems to be exerted at the level of α-subunits in the systems studied (Heino *et al.*, 1989; Zhang *et al.*, 1993; Lenter and Vestweber, 1994; Åhlén and Rubin, 1994). Thus, increases in integrin expression are achieved via increases in the synthesis of particular α-subunits. At least one integrin α-subunit, the α_6-subunit, seems to associate with calnexin after synthesis, but is processed and released to the mature form at a faster rate than the β_1-subunit. Promoter regions for the β_1, α_2, α_4, and α_5 integrin genes have been characterized (Birkenmeier *et al.*, 1991; Cervella *et al.*, 1993; Rosen *et al.*, 1994; Zutter *et al.*, 1994). These promoters lack TATA boxes but contain binding sites for different transcription factors. The β_1 gene has two promoter regions, one that is ubiquitously active and another that seems to be subjected to more precise regulations (Cervella *et al.*, 1993).

Specificity and Classification of Integrins

The published information concerning integrin specificity is very large and rapidly expanding. A comprehensive review of this field is not feasible in this chapter. Instead, we refer to the chapters on 'Basement membranes' (Volume 1, Chapter 6) and Non-collagenous matrix proteins (Volume 2, Chapter 3) in this series for additional data on integrin binding-sites in ECM molecules. Reported binding specificities of ECM-binding integrins are summarized in Figure 9.2. Characteristics of integrin ligand-binding sites, especially those of the platelet integrin α_{IIb}/β_3 have been reviewed recently (Loftus *et al.*, 1994).

Integrins have been divided in subfamilies based on the particular β-subunit shared by members within each subfamily (Albelda and Buck, 1990; Hemler, 1990; Springer, 1990;

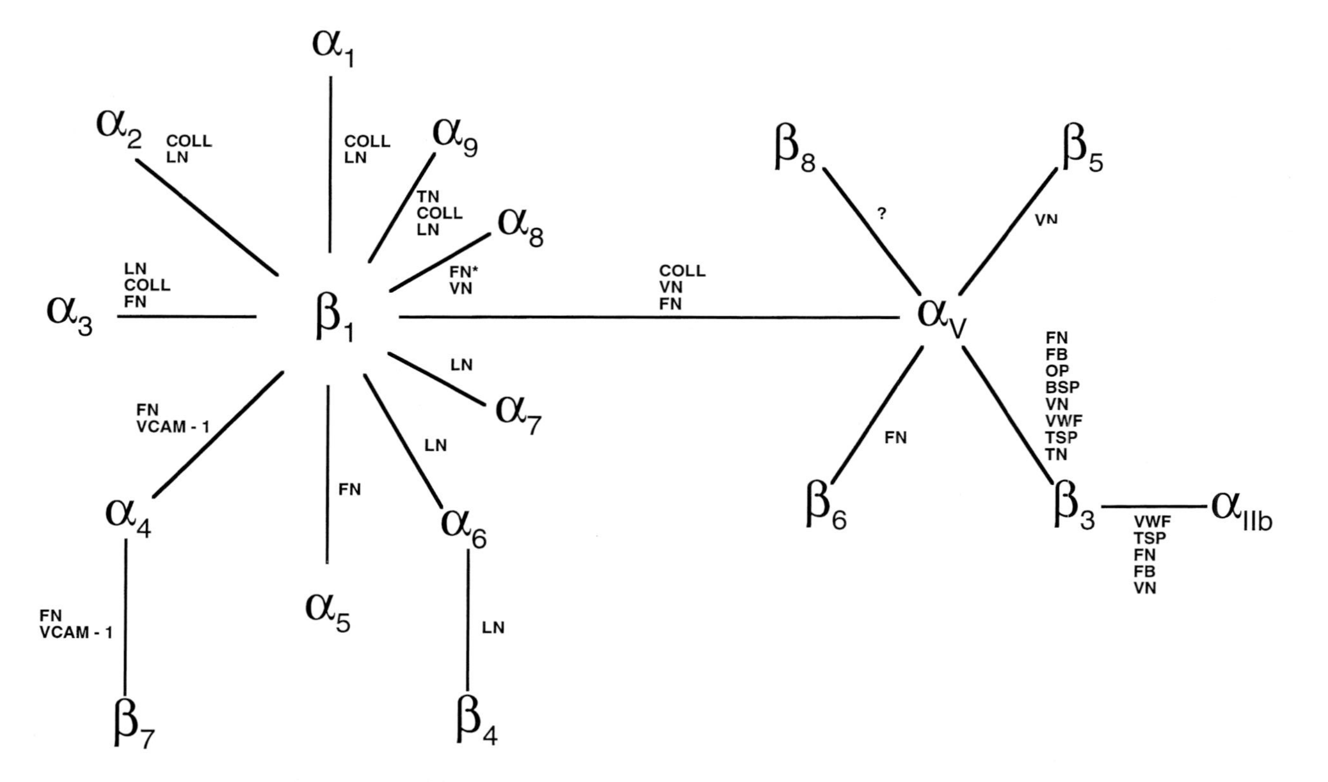

Figure 9.2 Reported specificities of integrins binding to ECM molecules. Data are from references given in the text. Bone sialoprotein (BSP); Different collagen types (COLL); Fibrinogen (FB); Fibronectins (FN); Osteopontin (OP); Tenascin (TN); Thrombospondin (TSP); Vascular cell adhesion molecule-1 (VCAM-1); Vitronectin (VN); von Willebrand factor (VWF). Asterisks indicate recently reported ligand specificy of a_8 [Müller, U., Bossy, K., Venstrom, K. and Reichardt, L.F. (1994) Molecular Biology of the Cell, Supplement volume 5, Abstract #2467].

Ruoslahti, 1991; Hynes, 1992). Three integrin subfamilies were originally reported, the β_1-, β_2-, and β_3-subfamilies. Of these, β_1 and β_3 integrins are of importance for cellular interactions with the ECM; β_2 integrins encompass leukocyte adhesion receptors that mediate intercellular adhesion. Later studies have demonstrated the presence of several additional integrin β-subunits, as well as the fact that certain α-chains can associate with more than one β-subunit. These findings make the original division of integrins into three subfamilies, based on a common β-subunit, less feasible.

Ten β_1 integrins have been characterized. These are the $\alpha_1\beta_1$ through $\alpha_9\beta_1$, and $\alpha_V\beta_1$. One feature, common to many β_1 integrins, is their specific binding of more than one ligand. This can be exemplified with the $\alpha_1\beta_1$ and $\alpha_2\beta_1$ integrins that have been purified as collagen receptors (Kunicki *et al.*, 1988; Kramer and Marks, 1989; Gullberg *et al.*, 1992; Hynes, 1992). It has been shown that $\alpha_3\beta_1$ can also act as a collagen receptor in some cell types (Wayner *et al.*, 1988; Elices *et al.*, 1991). In addition to being collagen receptors, these integrins function as laminin receptors, and $\alpha_3\beta_1$ as a fibronectin receptor (Wayner *et al.*, 1988; Elices and Hemler, 1989; Languino *et al.*, 1989). Furthermore, evidence that $\alpha_2\beta_1$ and $\alpha_3\beta_1$ mediate intercellular adhesion has been presented (Carter *et al.*, 1990; Larjava *et al.*, 1990; Sriramarao *et al.*, 1993; Symington *et al.*, 1993). It is not known if these diverse ligands contain a common three-dimensional structure that is recognized by all three integrins. In the case of $\alpha_1\beta_1$ and $a_2\beta_1$ data have been presented suggesting that these integrins recognize distinct sites in collagen, both dependent on the triple-helical conformation(Gullberg *et al.*, 1992; Kern *et al.*, 1993). The binding site for $\alpha_1\beta_1$ in collagen type IV contains an R and a D residue that are present in different collagen α-chains of the triple helix, thus indicating that this integrin may recognize a site that is dependent on the rigid triple-helical conformation (Eble *et al.*, 1993). This finding is in agreement with the view that $\alpha_1\beta_1$ only recognizes native collagen. Other β_1 integrins include the laminin receptors $\alpha_6\beta_1$ and $\alpha_7\beta_1$, and the fibronectin receptors $\alpha_4\beta_1$ and $\alpha_5\beta_1$. Some β_1 integrins, such as the fibronectin-binding integrins $\alpha_5\beta_1$ and $\alpha_V\beta_1$, recognize the amino acid sequence RGD (Pierschbacher *et al.*, 1983). Rat $\alpha_9\beta_1$ has been suggested to be a laminin/collagen receptor (Forsberg *et al.*, 1994), as well as a tenascin receptor (Yokosaki *et al.*, 1994).

The α_4-subunit that associates with two β-subunits, the β_1- and β_7-subunits, forms adhesion receptors that are of importance for lymphocyte homing. In addition, both these α_4-containing integrins mediate adhesion to fibronectin (Ruegg *et al.*, 1992; Chan and Aruffo, 1993). Similar to the α_4-subunit, the α_6-subunit is also known to associate with two different β-subunits, the β_1- and β_4-subunits. The $\alpha_6\beta_4$ and $\alpha_6\beta_1$ integrins act as laminin receptors in cultured cells; in tissues, $\alpha_6\beta_4$ is localized to hemidesmosomes of epithelia and is also present in endothelia. Although at their extracellular part both integrins bind laminins, their cytoplasmic domains interact with two different cytoskeletal systems. Intracellularly, the $\alpha_6\beta_4$ is associated with the intermediate filament system, whereas $\alpha_6\beta_1$ interacts with the microfilament (actin) system.

The $\alpha_V\beta_3$ integrin was originally purified as a vitronectin receptor; later studies have shown that this integrin binds a number of additional ECM molecules, including osteopontin, bone sialoprotein (BSP) (Oldberg *et al.*, 1986; Oldberg *et al.*, 1988), denatured collagen (Pfaff *et al.*, 1993), and tenascin (Prieto *et al.*, 1993). The α_V-chain can associate with several β-subunits, namely the β_1-, β_3-, β_5-, β_6, and β_8 -subunits. Ligand specificities varies among these α_V integrins, as can be exemplified by $\alpha_V\beta_6$ that functions as a fibronectin receptor (Busk *et al.*, 1992) , whereas $\alpha_v\beta_5$ is a vitronectin receptor, and $\alpha_v\beta_3$ binds a large number of ECM molecules. Most α_v integrins seem to be dependent on the RGD.

Integrin specificity has also been found to depend on which cell type that express a particular integrin. This can be exemplified by the $\alpha_2\beta_1$ integrin; $\alpha_2\beta_1$ isolated from platelets, melanoma cells, and fibroblasts functions as a collagen receptor (Takada *et al.*, 1988; Kramer and Marks, 1989; Staatz *et al.*, 1989; Gullberg, 1992), whereas, $\alpha_2\beta_1$ expressed in endothelial cells functions as a receptor for both collagen and laminin (Elices and Hemler, 1989; Languino *et al.*, 1989). Transfection of cells that do not express $\alpha_2\beta_1$ with α_2 cDNA revealed that these differences in specificity depend on the cellular environment, and are not an inherited property of the α_2-subunit; and furthermore that the α_2 specificities could be generated in a single cell type (Bosco *et al.*, 1993).

Integrins Link Intracellular Fibers with the Extracellular Matrix at Focal Adhesions — Structural and Regulatory Implications

Focal adhesions i.e. sites where the cell membrane comes in close contact with the underlying substrate have been well described *in vitro*. Focal adhesions were originally defined by reflection interference microscopy, a technique which made a dark image where there was close contact between a cell and the substrate (Abercrombie *et al.*, 1971; Izzard and Lochner, 1976; Burridge *et al.*, 1988; Turner and Burridge, 1991; Schwartz, 1992; Turner, 1994). Early morphological studies described this junction in fibroblasts as a "fibronexus" where the external ECM components contacted the cell surface, and the internal cytoskeletal components were localized (Singer, 1979). The *in vivo* significance of the focal adhesions is, presently, only hypothetical. However, cells cultured on ECM substrates change the number and pattern of focal adhesions concomitant with changes in their behavioral properties, such as migration, adhesion, and spreading. These findings clearly indicate a physiological significance of the focal adhesions.

Integrins concentrate and co-localize with several cytoskeletal components at focal adhesion sites (Burridge *et al.*, 1988; Turner and Burridge, 1991; Schwartz 1992; Turner, 1994). Some of these cytoskeletal proteins are able to interact directly with the integrin β_1-subunit. Thus, in early studies Horwitz *et al.*, (1986) demonstrated a low affinity interaction between talin and avian β_1-integrins, using equilibrium gel filtration. α-Actinin interacts directly with a synthetic peptide corresponding to the integrin β_1-subunit cytoplasmic domain with, an apparent of $K_d \sim 10^{-8}$. The affinity for purified β_1- and β_3-integrins, as determined in solid-phase assays, had lower apparent affinities (K_d:s $\sim 10^{-6}$) (Burridge *et al.*, 1990; Otey *et al.*, 1990). Recent evidence suggests that the protein tyrosine kinase-focal adhesion kinase- ($pp125^{FAK}$) (Hanks *et al.*, 1992; Kornberg *et al.*, 1992; Schaller *et al.*, 1994), interacts directly with β_1-, β_2- and β_3-integrin subunits, and — in addition — with the cytoskeletal protein paxillin (Schaller and Parsons, 1994). Protein kinase C (Jaken *et al.*, 1989) and, in transformed cells, $p60^{v\text{-}src}$ (Rohrschneider, 1980), have been found to be concentrated at focal adhesions. Other intracellular proteins that appear to be localized to focal adhesions are vinculin, paxillin, zyxin and tensin (Burridge *et al.*, 1988; Turner and Burridge, 1991; Schwartz, 1992; Turner, 1994). The exact function of the latter proteins has not been elucidated in detail, but evidence exists that they participate in the formation and regulation of membrane-cytoskeletal interactions.

Integrins organize into focal adhesions only after having bound to their proper ligand. Examples being $\alpha_5\beta_1$ and $\alpha_v\beta_3$ that only assemble into focal adhesions in cells cultured on fibronectin and vitronectin, respectively (Fath *et al.*, 1989). The structural information necessary to assemble β_1 integrins into focal adhesions seems to reside in the β_1 subunit. This concept is based on the fact that chimeric proteins containing only the β_1 cytoplasmic domain assemble in focal adhesions spontaneously (Geiger *et al.*, 1992; LaFlamme *et al.*, 1992). Available data suggest that the α-subunit masks the site in the β-subunit that possesses a default affinity for focal adhesion components. This theory is validated by the fact that removal of the cytoplasmic domain from the α_1-subunit leads to the formation of an $\alpha_1\beta_1$ that assembles into focal adhesions on cells cultured on fibronectin (Briesewitz *et al.*, 1993).

Protein phosphorylation events seem to be important for adhesion of cells to ECM glycoproteins and for the formation of focal adhesions. This can be illustrated by the fact that $pp125^{FAK}$ and paxillin become phosphorylated on protein tyrosine residues during attachment and spreading of cells (Burridge *et al.*, 1992; Guan and Shalloway, 1992; Hanks *et al.*, 1992; Kornberg *et al.*, 1992). Tyrosine phosphate is concentrated at focal adhesion sites in fibroblasts spreading on fibronectin, as detected by immunofluorescence (Burridge *et al.*, 1992), and treatment of fibroblasts with the protein tyrosine kinase inhibitor-herbimycin A-inhibits the formation of focal adhesions and the assembly of stress fibers in fibroblasts (Burridge *et al.*, 1992). Similar results are seen in human dermal fibroblasts that are spreading on collagen type I (Figure 9.3). These findings demonstrate that tyrosine kinase activity is important in cell-ECM adhesion and are in line with that some tyrosine kinase-equipped growth factor receptors, e.g. the PDGF β-receptors, affect processes that are dependent on cell-ECM adhesion, such as cell migration (Ferns *et al.*, 1990; Siegbahn *et al.*, 1990; Eriksson *et al.*, 1992). Vinculin is transiently translocated from focal adhesions in fibroblasts after PDGF-stimulation (Herman and Pledger, 1985). PDGF-stimulation also leads to phosphorylation on and/or redistribution of other focal adhesion proteins, including $pp125^{FAK}$ (Tidball and Spencer, 1993; Rankin and Rozengurt, 1994). Furthermore, PDGF-BB leads to a transient redistribution of β_1-integrins from focal adhesions to the cell circumference (K. Åhlén and K. Rubin, unpublished observation). Members of the ras-related family of GTP-binding proteins are important for growth factor-induced assembly of focal adhesions and actin stress fibers (Ridley and Hall, 1992), as well as for growth factor-induced membrane ruffling (Ridley *et al.*, 1992). These growth factor activities are, at least, indirectly coupled to integrin function, thus suggesting that ras-related GTP-binding proteins are important for integrin function. These examples clearly indicate that tyrosine kinase-equipped growth factor receptors influence focal adhesion assembly and function, as well as integrin function.

In conclusion, integrin β-subunits interact with both structural and regulatory proteins, and participate in the formation of cytoskeletal complexes at focal adhesion sites. Some enzymatic activities of importance for the control of cell behavior are concentrated and regulated at these sites. It is, therefore, possible that integrins, by regulating the assembly of focal adhesion sites, modulate cellular activities, including the response to growth factor stimulation. Such a model offers an attractive explanation for the outside-inside signalling function of integrins, as to how ECM molecules could influence cell behavior.

Regulation of Integrin Activity

One prominent feature of many integrin-mediated adhesion reactions is their dependence on activation, i.e. cells that are able to regulate the activity of integrins expressed at the cell

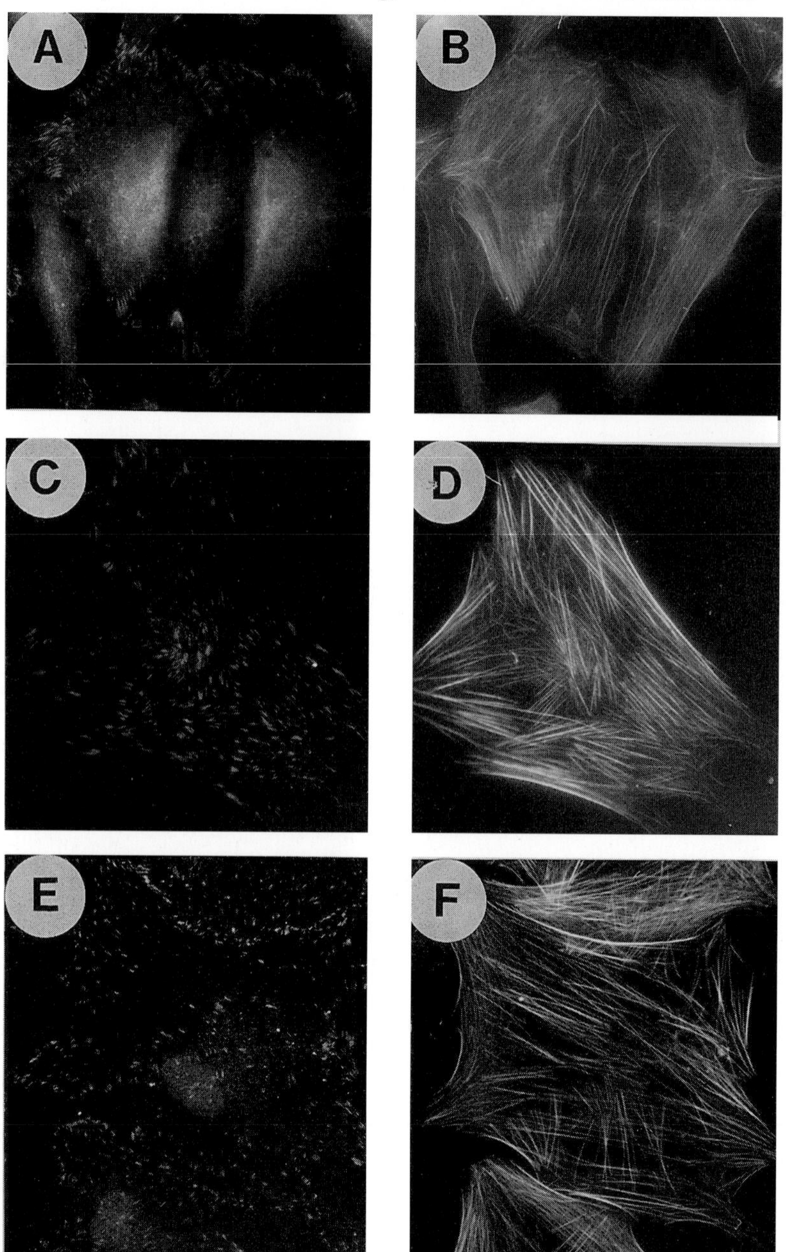

Figure 9.3 Organization of focal adhesions and actin-containing stress fibers in diploid human fibroblasts cultured on collagen type I. Fibroblasts (AG 1518) were seeded on dishes coated with bovine collagen type I and cultured in serum-free MCDB medium for 2 hours. Cells were subjected to immunofluorescence staining with monoclonal anti-human integrin a_2-subunit antibodies (A); anti-paxillin (C); anti-phosphotyrosine (E) and bound mouse IgG was detected with biotinylated anti-mouse IgG followed by Texas Red-conjugated Streptavidin. The actin-containing stress fibre networks in the same cells were visualized using FITC-conjugated phallacidin (B, D and F). Focal adhesions in A, C and E are seen as white streaks situated both at the margins and centrally of the cell bodies, typically at tips of actin-containing stress-fibers.

surface. Integrins participate in many adhesion reactions that are dynamic. Examples that have been particularly well studied are platelet and leukocyte integrins. In non-activated platelets, the integrin α_{IIb}/β_3 has a low affinity for fibrinogen, however, as a consequence of platelet activation, α_{IIb}/β_3 gains a high affinity for fibrinogen. This change in affinity of α_{IIb}/β_3 is instrumental for fibrinogen-induced aggregation of activated platelets. Similarly, β_2-integrins on circulating leukocytes have relatively low affinities for their ligands, whereas β_2-integrins on chemoattractant-stimulated leukocytes acquire a high affinity for their counter-receptors present on activated endothelia. Both these types of integrins can undergo intramolecular affinity modulation (Shattil *et al.*, 1985; Dustin and Springer, 1989; Du *et al.*, 1991) due to processes that have been referred to as inside-outside signalling (Hynes 1992). This property of integrins seems to be unique among the families of adhesion receptors. Determinations of integrin affinity have, in many cases, revealed relatively low affinities with K_d:s in the μM range. K_d:s for $\alpha_L\beta_2$ (LFA-1) present on T cells could undergo a 200-fold increase in affinity after T cell activation, although only a portion of the LFA-1 molecules acquired the high affinity state (Lollo *et al.*, 1993).

The molecular background to these activations of integrins is not completely understood, but available data favors the idea that transitions between the two affinity stages result from changes in conformation of the integrin heterodimers (Sims *et al.*, 1991; Humphries *et al.*, 1993; Calvete *et al.*, 1994; Williams *et al.*, 1994). This is underscored by the presence of monoclonal anti-integrin β_3-antibodies that only recognize activated β_3 integrins (Shattil *et al.*, 1985; Frelinger *et al.*, 1990). It has been shown that recombinant α_{IIb}/β_3 heterodimers, in which the cytoplasmic domain of the α_{IIb}-subunit has been removed, are constitutively in the high affinity-state (O'Toole *et al.*, 1991; Loftus *et al.*, 1994; O'Toole *et al.*, 1994). The latter finding indicates that the α-cytoplasmic domain exerts restriction on the integrin heterodimer, thus hampering transition between the affinity states.

Evidence that β_1 integrins may exist in high and low affinity stages is indicated by the fact that certain monoclonal anti-β_1 integrin antibodies are able to increase the affinity of avian (Neugebauer and Reichardt, 1991) and human (Kovach *et al.*, 1992; van de Wiel-van Kemenade *et al.*, 1992; Arroyo *et al.*, 1993; Faull *et al.*, 1993) β_1 integrins. These antibodies increase β_1 integrin-mediated adhesion reactions, possibly by shifting the equilibrium between the active and inactive states in favor of the active state.

Divalent cations can affect the affinity of integrins and are necessary for cell-matrix adhesion. Example of this are the ability of Mn^{2+} to increase the affinity of $\alpha_5\beta_1$ for fibronectin (Gailit and Ruoslahti, 1988) and that divalent cation sites play an essential role in the regulation of activity and ligand specificity of $\alpha_4\beta_1$ (Masumoto and Hemler, 1993). The collagen-binding integrins $\alpha_1\beta_1$ and $\alpha_2\beta_1$ both are dependent on Mg^{2+} for their function (Staatz *et al.*, 1989; Gullberg *et al.*, 1992; Kern *et al.*, 1993)

Protein kinase C (PkC) activity has also been shown to be of importance in cell-ECM adhesion. This is based on studies which show that stimulation of CHO cells with phorbol esters increases attachment of these cells to fibronectin (Danilov and Juliano, 1989), and that PkC inhibitors block the formation of focal adhesions and inhibit cell spreading (Woods and Couchman, 1992; Vuori and Ruoslahti, 1993). Membrane-bound PkC becomes activated when CHO cells spread on immobilized fibronectin, and phorbol esters enhance both processes (Vuori and Ruoslahti 1993). It is possible that the adhesion-stimulating effect of PkC in the above examples was due to a direct phosphorylation of the integrins by PkC, however, when investigated no such phosphorylation was detected

(Danilov and Juliano, 1989). In MG-63 osteosarcoma cells, and in AG1523 diploid human fibroblasts, activation of PkC by phorbol esters led to phosphorylation of the β_5 but not of the β_3-subunit (Freed *et al.*, 1989). Integrin β-subunits can be phosphorylated as shown by the phosphorylation of the β_1-subunit by pp60^{src} in RSV transformed cells. This phosphorylation leads to a diminished affinity for both talin and fibronectin (Tapley *et al.*, 1989). Recent data have shown that tyrosine phosphorylated β_1 integrin subunits have a different distribution than their unphosphorylated counterparts (Johansson *et al.*, 1994). Furthermore, these authors showed that the p85 subunit of phosphatidylinositol 3'-kinase binds specifically to tyrosine phosphorylated β_1-subunits, suggesting that the phosphorylation is part of an integrin-mediated signalling cascade — and/or conversely — is involved in the regulation of integrin affinity or activity.

Available data strongly suggest that ECM molecule-binding integrins exist in high and low avidity stages. Thus, the intrinsic property of integrins — to be able to undergo intramolecular avidity modulation — suggests that these adhesion receptors participate in dynamic adhesion reactions. Much data have been gathered showing that the avidity state of integrins is regulated by various kinase avidities in cells, some of which are related to growth factor activities. It is likely that growth factor effects on adhesion phenomena can, in part, be executed via changes in integrin avidities. A tentative graphical description, based on available data, of the high and low avidity transition stages of integrins is presented in Figure 9.4.

Regulation of Integrin Expression *in Vitro*

Integrin expression can be regulated by several growth factors and cytokines in a variety of cell types. Several examples have been provided on how growth factor- and hormone-induced increases in integrin expression can influence cell behavior. Retinoic acid up-regulates the expression of $\alpha_1\beta_1$ on neuroblastoma cells, leading to an increased outgrowth of neurites on laminin (Rossino *et al.*, 1991). Similarly, $\alpha_1\beta_1$ is upregulated by nerve growth factor in PC12 cells, parallel to the neurite outgrowth-stimulating activity of this growth factor (Zhang *et al.*, 1993). Epidermal growth factor stimulates the expression of $\alpha_2\beta_1$ on human keratinocytes, as detected by flow cytometry analysis (Chen *et al.*, 1993). This increase promoted keratinocyte locomotion on collagen, a process that is believed to be of importance for re-epithelialization and wound-closure during wound healing.

The angiogenic growth factor, basic fibroblast growth factor (bFGF), stimulates synthesis of several integrins on microvascular endothelial cells (Enenstein *et al.*, 1992; Klein *et al.*, 1993). To what extent the angiogenic capability of bFGF is coupled to such changes in integrin expression is not known at present. In a comparison of the effects of transforming growth factor-β (TGF-β) and PDGF-BB on integrin expression by vascular smooth muscle cells it was found that these factors differ as to which of the integrin heterodimer they stimulate (Janat *et al.*, 1992). TGF-β stimulates the synthesis and expression of integrins in a large number of different cell types (Ignotz and Massague, 1987; Heino *et al.*, 1989; Heino and Massague, 1989; Ignotz *et al.*, 1989; Janat *et al.*, 1992). TGF-β increases steady state levels of β_1-subunit mRNA, as well as synthesis and post-translational processing of this subunit (Heino *et al.*, 1989; Ignotz *et al.*, 1989). TGF-β also stimulates the synthesis of α-subunits. Available data indicate that the synthesis of α-subunits is the rate-limiting step in the synthesis of integrin heterodimers.

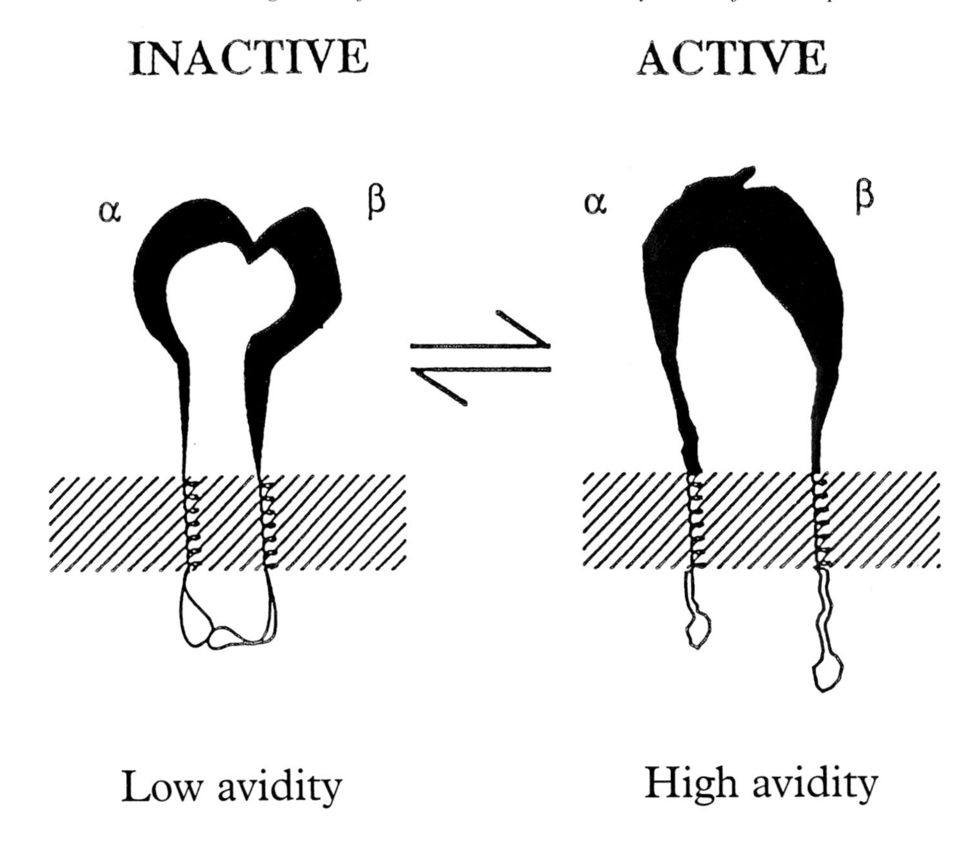

Figure 9.4 Tentative schematic illustration of the two avidity states of integrin heterodimers. The figure is based on studies referenced in the text. The two states differ with respect to conformation; the high-avidity state exposes sites for cytoskeletal proteins present on the β-subunit cytoplasmic domain, as well as high avidity ligand binding sites. [Reproduced with permission from: Rubin, K., Sundberg, C., Åhlén, K. and Reed, R.K. (1995) Integrins: transmembrane links between the extracellular matrix and the cell interior. In: *Interstitium, Connective Tissue and Lymphatics* Reed, McHale, Bert, Winlove and Laine Eds. pp 29–40. Portland Press. London]

PDGF, TGF-β, interleukin-1 and tumor necrosis factor-α stimulate integrin expression in cultured fibroblasts, although specificity can be discerned in as much as the different factors stimulate the expression of different integrins. Furthermore, in WI-38 fibroblasts, TGF-β stimulates the synthesis of several α-subunits, including the α_1-, α_2- and α_3-subunits — leading to a marked increase in cell surface expression of these integrins. Interleukin-1 (IL-1) and tumor necrosis factor-α (TNF-α) stimulate the expression of $\alpha_1\beta_1$ on human foreskin fibroblasts but not the synthesis of $\alpha_2\beta_1$ and $\alpha_3\beta_1$ (Santala and Heino, 1991). In contrast, PDGF-BB stimulates the synthesis of α_2, but not α_1 or α_3 in such cells (Åhlén and Rubin, 1994).

Contraction of Three-Dimensional Collagen Lattices *in Vitro*

Fibroblasts are able to contract three-dimensional lattices of reconstituted collagen fibers, leading to a decrease in lattice volume. This process is referred to as fibroblast-mediated collagen gel contraction (Bell *et al.*, 1979; Stopak and Harris, 1982; Buttle and Ehrlich, 1983; Grinnell and Lamke, 1984). A visible decrease in lattice volume can usually be observed within hours after the initiation of the contraction process. Collagen gel contraction has been suggested to result from fibroblast migration through the lattice and rearrangement of collagen fibers into bundles thereby compacting the structure. Reagents disrupting cytoskeletal fibers — such as cytochalasin — effectively block collagen gel contraction (Tomasek and Hay, 1984). Because fibroblasts undergo changes in phenotypic characteristics during and after contraction, their morphology and synthesis of certain ECM molecules are changed (Elsdale and Bard, 1972; Nusgens *et al.*, 1984; Tomasek and Hay, 1984; Mauch *et al.*, 1988; Greve *et al.*, 1990). Furthermore, the cellular sensitivity for growth factor stimulation ceases (Nakagawa *et al.*, 1989), data have been published suggesting that cell-surface expression of PDGF β-receptors decreases, or that cell surface-expressed receptors become refractory (Tingström *et al.*, 1992; Lin and Grinnell, 1993).

A role for β_1 integrins has been implicated in fibroblast-mediated collagen gel contraction (Gullberg *et al.*, 1990). Particularly, the collagen-binding integrin $\alpha_2\beta_1$ seems to play an important role (Klein *et al.*, 1991; Schiro *et al.*, 1991; Hunt *et al.*, 1994). Stimulation of fibroblasts with certain growth factors and cytokines influences their ability to contract collagen gels. TGF-β (Montesano and Orci, 1988), PDGF (Clark *et al.*, 1989; Gullberg *et al.*, 1990), endothelins (Guidry and Höök, 1991), and angiotensin II (Burgess *et al.*, 1994) stimulate collagen gel contraction. A direct comparison of the effects of TGF-β and PDGF, using the same assay system, revealed that the effects of PDGF are evident after a few hours, whereas the stimulatory effect of TGF-β had a slower onset (Tingström *et al.*, 1992). These findings indicate that TGF-β and PDGF, respectively, exert their stimulatory effects via different intracellular signalling pathways. The inhibitory effects of anti-β_1 integrin antibodies on fibroblast-mediated collagen gel contraction could in part be overcome by increasing the concentration of PDGF-BB (Gullberg *et al.*, 1990). This finding suggests that PDGF-BB induces an increase in collagen-binding β_1 integrin expression and/or activity.

Both PDGF and TGF-β can be produced by activated macrophages (Ross *et al.*, 1990). These cells also produce cytokines that are able to inhibit fibroblast-mediated collagen gel contraction. These cytokines include IL-1 and TNF-α (Gillery *et al.*, 1989; Tingström *et al.*, 1992). Furthermore, prostaglandin E_2 (Ehrlich and Wyler, 1983) inhibits collagen gel contraction, an effect that may be due to changes in intracellular levels of cAMP, since increases in intracellular cAMP inhibit collagen gel contraction (Ehrlich and Wyler, 1983; Van Bockxmeer *et al.*, 1984). This inhibition seems to be mediated by changes in phosphorylation of myosin light chain kinase (Ehrlich *et al.*, 1991). A direct role of PKC for collagen gel contraction has also been demonstrated (Guidry, 1993).

Thus, collagen gel contraction provides an example of a cellular process that is dependent on an interplay between growth factors/cytokines and integrins. It is interesting to note that contraction of fibroblast-populated collagen gels is regulated by a number of growth factors and cytokines that are produced locally during wound healing. Collagen gel contraction has, in fact, been suggested to be an *in vitro* model for wound contraction (Bell *et al.*, 1979; Grinnell, 1994). Furthermore, the strong and rapid modulation of collagen gel contraction *in vitro*, by inflammatory factors such as PDGF and prostaglandin E_2, suggests that the *in vivo* counterpart to collagen gel contraction is also of importance in some aspect(s) of inflammatory reactions. One possibility, discussed further in the present chapter, is that these regulations may be of importance in the control of fluid balance in tissues.

Integrins in Embryonic Development

Introduction

Examples presented in this section have been taken from a few well characterized systems where cell-matrix interactions have been studied during differentiation and development in vertebrates and invertebrates. (For further reading we refer to excellent reviews on this subject: Sanes 1989; Reichardt and Tomaselli, 1991; Adams and Watt, 1993; Brown, 1993; Ekblom, 1993).

A strong emphasis is placed on the importance of integrins in developmental processes. There are, however, indications that also non-integrin receptors can act to mediate the effects of the ECM on morphogenesis and differentiation during development. Mutants lacking integrin β chains exist in *Drosophila melanogaster* (MacKrell *et al.*, 1988). In this species a number of developmental processes proceed normally in the absence of the affected integrins. In vertebrates, the identification of dystroglycan as a laminin binding protein of the dystrophin associated membrane protein complex, indicates a possible role for this non-integrin in transmembranal linkages.

Cell-ECM Interactions During Invertebrate Development

ECM molecules in Drosophila. The *Drosophila* embryo lacks an interstitial matrix. Instead, all internal organs and glands are surrounded by basement membranes. Invertebrate homologues to vertebrate laminins, collagen type IV, tenascin, and syndecan, as well as other novel ECM components, have been characterized in *Drosophila* (Fessler and Fessler, 1989; Baumgartner *et al.*, 1994; Fogerty *et al.*, 1994; Nelson *et al.*, 1994; Spring *et al.*, 1994). Except for the recently described embryonic lethal mutants in the laminin α_1 chain, no other mutations in *Drosophila* ECM genes have been described (Henchcliffe *et al.*, 1993). The inability to detect an effect on neuronal or muscle development, up to the time of death, during late embryogenesis in the *lamA* null mutant indicate that redundant mechanisms are operative in *Drosophila*. Except for the recently characterized laminin α_1 chain no other α chain has so far been described in *Drosophila* (Kusche-Gullberg *et al.*, 1992).

Drosophila integrins. It was discovered in the late 1980s that the position specific (PS) antigens in *Drosophila melanogaster* are homologs of vertebrate integrins. The β_{PS} integrin which, with regard to the amino acid sequence, is 43% identical with vertebrate β_1 subunits, was originally found to be associated with two α-chains: α_{PS1}, which during embryogenesis is expressed on ectodermally and endodermally derived cells, and α_{PS2} which is expressed on mesodermal derivatives such as somatic and visceral muscles. More recently, a 90 kD band associated with β_{PS} integrins, but distinct from α_{PS1} and α_{PS2}, has been identified as a third α-chain. This α-chain has been named α_{PS3} (Gotwals *et al.*, 1994). A second *Drosophila* β chain (nu) has been described (Yee and Hynes, 1993). The $\beta\nu$ chain is 33% identical with vertebrate and with other *Drosophila* β chains and shows a remarkable tissue-specific distribution during embryogenesis in that it is only expressed in endodermal cells around the midgut. The α-chain associated with $\beta\nu$ has not been identified.

Alternative splicing of Drosophila integrins. A variant of the β_{PS}-subunit that exhibits alternative splicing in the extracellular domain has been characterized using PCR (Zusman *et al.*, 1993). Analysis of the function of the two splice forms of β_{PS} show that both forms are needed during embryogenesis but that either form can function during postembryonic β_{PS}-mediated events. The location of the alternative exons suggests that these two forms of the β_{PS} integrin may interact with alternative α subunits or different ligands. It could be important to determine if this splice variant will also be found in vertebrate integrins where the hitherto described β-chain splice-variants are restricted to sites either close to the membrane spanning domain or to the regions coding for the cytoplasmic domains. One β_1 integrin splice variant in vertebrates has been shown to be expressed in a tissue specific manner (Balzac *et al.*, 1993).

The α_{PS2} was the first integrin chain described to show developmentally regulated splicing in the postulated ligand binding extracellular domain (Brown *et al.*, 1989). This was hypothesized to generate α_{PS2} integrin heterodimers with different affinities for their ligand. This theory is supported by recent experiments with cells transfected with the two α_{PS2} splice variants that were allowed to interact with the first described endogenous *Drosophila* integrin ligand (see below). Now that alternative splicing in the postulated ligand binding domain has been shown to occur for the vertebrate α_6 and α_7 integrin chains (Ziober *et al.*, 1993), it will be interesting to see whether the affinity modulation that appears to exist in *Drosophila* can also be extended to these vertebrate integrins. The reported splicing of α_6 integrin will have implications for a number of biological systems where α_6 has been shown to play a role. It has been suggested that the alternative splicing of the α_7 integrin explains the diverse functions of α_7 during different stages of myogenic differentiation.

Ligands for Drosophila integrins. Early findings on the effect of RGD peptides on gastrulation, and the effect of anti-fibronectin antibodies indicated that a *Drosophila* homologue of fibronectin might exist (Naidet *et al.*, 1987). The expression pattern of the known integrins was, however, not reconcilable with a role for integrins during this event. Later studies showed that the experiments with RGD peptides could not be repeated (Leptin *et al.*, 1992). That the α_{PS2} integrin during later developmental events interacts with an RGD containing ligand was first shown *in vitro*. Analyses with cultured cells — transfected with α_{PS2} integrins — demonstrated vertebrate vitronectin and fibronectin to be RGD dependent ligands for the $\alpha_{PS2}\beta_{PS}$ integrin (Hirano *et al.*, 1991; Bunch and Brower, 1992). Vitronectin could also promote the differentiation of myoblasts into myotubes *in vitro* (Gullberg *et al.*, 1994). Using cells transfected with the two splice variants of α_{PS2}, the first *Drosophila* integrin ligand — which has been named tiggrin — has been identified (Fogerty *et al.*, 1994). Tiggrin contains an RGD sequence near the C-terminus, and the $\alpha_{PS2}\beta_{PS}$ interaction with tiggrin is inhibited by RGD peptides. In the developing embryo, tiggrin shows a remarkable co-localization with α_{PS2} at muscle attachments. The α_{PS2} splice form, which is expressed at muscle attachments, also displays the highest affinity for tiggrin *in vitro*. *Drosophila* tenascin also contains an RGD-sequence, opening the possibility that tenascin interacts with the $\alpha_{PS2}\beta_{PS}$ integrin. Although a number of analyses indicated that laminin is not a ligand for β_{PS} integrins (Volk *et al.*, 1991; Fristrom *et al.*, 1993; Gullberg *et al.*, 1994) recent transfection experiments with $\alpha_{PS1}\beta_{PS}$ integrins have shown that laminin is a ligand for this integrin heterodimer in *in vitro* assays (Gotwals *et al.*, 1994).

Function of Drosophila *integrins.* Two mutations exist that alter *Drosophila* integrin chains: *lethal myospheroid ((l)mys)*, and inflated *(if)*, affecting the β_{PS}- and the β_{PS2}-chains, respectively. *In vivo* analyses of these mutations have indicated a role for *Drosophila* β_{PS} integrins in muscle attachments and in the attachment of epithelial cells to basement membranes, in the adult wing and eye (Brabant and Brower, 1993). Gastrulation and cell differentiation appears normal in the *mys* embryos. This is consistent with the lack of effect of RGD-containing peptides on *Drososphila* gastrulation (Leptin *et al.*, 1992). During embryogenesis in *mys* embryos, muscle specific genes are expressed but, the morphogenesis of muscle attachments is abnormal. Integrins in *Drosophila* thus serve as a link between the cytoskeleton and the basement membrane.

In summary, genetic analyses of integrins in *Drosophila* show that integrins are primarily needed at sites of strong cell adhesion. *Drosophila* should continue to be an interesting system in studies of integrin functions in development. The absence of yet identified integrins on early migratory and circulatory cells indicates that novel integrins are likely be identified in *Drosophila*.

Cell-ECM Interactions During Vertebrate Development

To illustrate the importance of cell-ECM interactions during vertebrate development we will discuss some widely studied systems: differentiation of skeletal muscle and epithelial polarization in the kidney and the salivary gland.

Myogenic differentiation. The adult skeletal muscle is a highly specialized tissue composed of differentiated muscles of different fiber types. During mouse development, skeletal muscles initially form in the myotome during embryonic days 9–10 of gestation (E9–E10). Embryonic myoblasts form primary fibers at mouse E13–E14. As fetal development begins E16–E17, secondary muscle fibers form, parallel to the primary fibers. After birth, existing primary and secondary fibers grow by fusion with satellite myoblasts (Miller, 1992). It is not known what precise role the cellular interactions with the extracellular matrix play during these events, although some new insights have been gained during the few past years. Relatively little is known about the pattern of integrin-expression, as well as function, during development. This is partly due to the limited availability of immunological reagents to different integrin α-chains reactive with murine and avian integrins. With the available antibodies, immunohistochemistry and antibody injection experiments have generated some information, and new *in vivo* knowledge will be collected with the availability of transgenic mice lacking different integrin α-chains.

No naturally occurring β_1 integrin mutants have, so far, been described in higher organisms; β_1 deficient cells have, however, been described. The recently described targeted deletion of β_1 integrins in F9 cells affects morphogenesis. In the *$\beta 1$, null* F9 cells, endoderm specific genes are expressed upon *in vitro* differentiation to endoderm, but the morphological differentiation is abnormal (Stephens *et al.*, 1993). Furthermore, the generation of β_1 integrin-deficient embryonic stem cells has recently been described (Fässler *et al.*, 1994). Whereas the undifferentiated stem cells show changed adhesive properties, the exact role of β_1 integrins for the differentiation potential of these cells is now the subject of intensive studies.

The importance of integrins for muscle integrity has not yet been evaluated in vertebrates. However, two other components involved in the transmission of force — the cytoskeletal protein dystrophin and the ECM-protein laminin-2, both cause muscle dystrophies in mice (Hoffman *et al.*, 1987; Xu *et al.*, 1994).

In vivo antibody injection experiments have indicated a role for integrins in muscle formation. Studies with avian embryos have shown that injection of antibodies to β_1, integrin after myotome formation, disrupt the subsequent muscle formation (Jaffredo *et al.*, 1988; Drake *et al.*, 1991). Depending on the time and site of injection different muscles are affected. The results from antibody injection experiments implicate a role for integrins in myoblast migration. This is in contrast to integrin β-chain mutants in *D. melanogaster* and *C. elegans* where myoblasts migration is unaffected (Volk *et al.*, 1991; Hresko *et al.*, 1994). The inability of RGD-peptides to affect abdominal muscle formation indicates that RGD-dependent receptors such as a_v containing receptors and the $\alpha_5\beta_1$ integrin might not be the major receptors involved in the early migratory events from the somite leading to the formation of these muscles.

Immunohistochemical analyses have indicated the presence of α_1, α_5, α_6 and α_V on myogenic cells early during development. At the time of birth α_1, α_V, α_5 and α_6 are largely downregulated in skeletal muscle (Muschler and Horwitz, 1991; Bronner-Fraser *et al.*, 1992; Duband *et al.*, 1992; Hirsch *et al.*, 1994). The theory that separate cell surface molecules are important during different developmental stages of myogenesis is thus supported by the tightly regulated expression of integrin a-chains. A prominent example is the $\alpha_7\beta_1$ integrin is developmentally regulated *in vivo* and is induced as a laminin-containing basement membrane forms around the primary myotubes (Song *et al.*, 1992). Secondary myoblasts also express α_7, and the expression of α_7 increases on secondary myotubes. Furthermore, recent data show that $\alpha_7\beta_1$ is also subject to a developmentally regulated alternative splicing in both the cytoplasmic, as well as the putative ligand binding domains (Song *et al.*, 1992; Collo *et al.*, 1993; Ziober *et al.*, 1993). In addition to the $\alpha_7\beta_1$ integrin, adult skeletal muscle expresses the $\alpha_9\beta_1$ integrin (Song *et al.*, 1992; Palmer *et al.*, 1993).

In vitro and *in vivo* analyses have shown that $\alpha_4\beta_1$ integrin is important during secondary myogenesis, where it interacts with VCAM-1 (Rosen *et al.*, 1992). Based on the expression pattern *in vivo*, and functional analyses *in vitro*, it has been suggested that $\alpha_4\beta_1$ integrin on primary myotubes acts to align secondary embryonic myoblasts and that it thereby facilitates secondary myogenesis. *In vitro* analyses show a specific upregulation of $\alpha_4\beta_1$ integrin during myotube formation and antibodies to $\alpha_4\beta_1$ can inhibit myotube formation. It is possible that the inhibitory effect of anti-β_1 integrin antibodies on myoblast fusion *in vitro*, initially observed in chick myogenic cultures was in fact due to the inhibition of the $\alpha_4\beta_1$ VCAM-1 interaction (Menko and Boettiger, 1987).

In *Drosophila* the PS2 integrin is involved in the stability of muscle attachments and is alternatively spliced in the ligand-binding domain during embryogenesis. Another type of regulation seems to occur during vertebrate development. In mice fetuses α_v integrins are present at myotendinous junctions (MTJ) (Hirsch *et al.*, 1994). The β integrin chain associated with α_v at these sites is unknown. Prior to birth, α_v is downregulated and is no longer present on myotubes. Studies of avian development have indicated that $\alpha_7\beta_1$ appears at MTJs during fetal development but, unlike α_v, persists at myotendinous junctions in the adult animal (Bao *et al.*, 1993). The role of integrins at the MTJ is also interesting in relation to the recent finding of PDGF receptor expression at the myotendinous junction of skeletal muscles (Tidball and Spencer, 1993). *In vitro* studies have shown that PDGF causes phosphorylation of talin. PDGF-mediated phosphorylation of talin at the MTJ would allow for breakages of cytoskeletal linkages and allow for myofibril growth.

In transgenic mice, in which the genes for fibronectin and the a_5 chain, respectively, have been inactivated, it has been shown that both targeted mutations — *FN null* and a_5 *null* — are recessively lethal, with the *FN null* embryos showing more pronounced defects (George *et al.*, 1993; Yang *et al.*, 1993). *FN null* embryos implant, undergo gastrulation, and erythropoiesis takes place. In contrast, neural tube, mesoderm, and vascular development is defective in these embryos. Because no normal somites develop, no myotubes forms at the time of lethality, around E10. In contrast, cardiac myocytes do develop, since beating hearts were observed.The a_5 *null* mutant embryos start to show defects in posterior mesodermally derived structures around day 9 and subsequently, die around E11. Somites, and head and heart structures develop normally, whereas mesodermally derived structures in the posterior parts do not. Mesodermal cells, isolated from a_5 *null* embryos *in vitro*, formed focal contacts, assembled fibrils, and migrated in response to fibronectin, indicating the use of other fibronectin receptors for these processes. It will be important to determine which receptors are active during these events in the a_5 *null* mutant cells. Likewise, it will be essential to determine what extracellular matrix molecules are used to direct the initial mesoderm migration in *FN null* mutant embryos and if the lack of fibronectin leads to compensatory upregulation of other matrix molecules in early events, such as gastrulation. The experience from the two null mutants indicates the existence of potentially complex compensatory mechanisms, and, in addition, show that antibody injection experiments provide valuable complementary information when studying *in vivo* functions.

Epithelial polarization. The use of organ cultures has been very helpful in following epithelial polarization for kidney and salivary gland development. In the kidney, the formation of polarized epithelial cells from mesenchymal stem cells has been shown to be mediated by laminin-1 domains E3 and E8 (Sorokin *et al.*, 1990; Sorokin *et al.*, 1992). The effect of the E8 fragment seems to be mediated by the $\alpha_6\beta_1$ integrin. *In situ* hybridization data indicating a temporal and spatial co-expression of laminin-1 and dystroglycan in the developing kidney, as well as antibody perturbation experiments in kidney organ cultures, strongly suggest that dystroglycan is a receptor for the E3 fragment of laminin-1 (Durbej, M. and Ekblom, P., unpublished observation). Recent studies of epithelial polarization during the formation of epithelial sheets in the developing submandibular gland, likewise indicate a role for the E8 and E3 regions of laminin (Kadoya *et al.*, 1994). When the effect of two antibodies that inhibited branching morphogenesis — namely antibodies to the α_6 integrin and laminin E3 domain — were compared at the EM level, quite different effects were observed. Whereas antibodies to E3 seemed to disrupt the basement membrane, antibodies to α_6 integrin inhibited branching without disrupting epithelial attachment to it. Based on this finding the authors suggest that the E3 region is involved in basement membrane assembly and the E8 region in the binding to cells via the α_6 integrin.

Expressions and Functions of Integrins *in Vivo*

The availability of specific monoclonal antibodies against individual integrin-subunits has allowed investigation of the localization of integrins in tissues and of changes in their expression during development and in specific disease conditions (Buck *et al.*, 1990; Defilippi *et al.*, 1991; Koukoulis *et al.*, 1991; Pignatelli *et al.*, 1991; Albelda 1992; Damjanovich *et al.*, 1992; Korhonen *et al.*, 1992; Korhonen *et al.*, 1992; Pignatelli *et al.*, 1992; Pignatelli *et al.*, 1992; Stallmach *et al.*, 1992; Gipson *et al.*, 1993; Johnson *et al.*, 1993; Koukoulis *et al.*, 1993; Koukoulis *et al.*, 1993; Lindmark *et al.*, 1993; Mette *et al.*, 1993; van den Berg *et al.*, 1993). These investigations, which are primarily correlative, have documented the distribution of α-chains in relation to individual ECM components, and documented how this distribution changes with altered expression of ECM components. In general, most normal tissues express a restricted number of integrins that are most easily detected in epithelial, smooth muscle and endothelial cells, i.e. cells that are in intimate contact with basement membranes. However, resident cells that are not associated with a basement membrane have also been shown to express integrins *in vivo*. One example is the $\alpha_2\beta_1$ integrin, that is expressed in dermal fibroblasts and various dendritic cells in lymph nodes, and in the thymus (Wayner *et al.*, 1988; Zutter and Santoro, 1990).

Immunohistochemical evidence showing the anchoring function of integrins has been clearly demonstrated with the presence of $\alpha_6\beta_4$ in epidermal desmosomes (Jones *et al.*, 1991; Hormia *et al.*, 1992; Koukoulis *et al.*, 1993). The pattern showed that the distribution was confined to the basal surface and was associated with the hemidesmosomes in a variety of tissues. Antisera directed at both the extracellular and intracellular cytoplasmic domains indicate the specific localization. *In vitro* studies with the rat bladder carcinoma cell line 804G indicated that $\alpha_6\beta_4$ is important in the formation of hemidesmosomes, as well as its role in the maintenance of that structure. It appears that this integrin is specific to the epithelium, as attempts to localize the $\alpha_6\beta_4$ in the hemidesmosomes of the intercalated disc of the heart have been negative (Kennel *et al.*, 1992).

The pattern of integrins, as detected by immunohistochemistry, is associated with altered adhesiveness during development, such as in the endometrium (Damsky *et al.*, 1992a; Isemura *et al.*, 1993), and testis (Schaller *et al.*, 1993). The cyclic expression of ECM components during ovulation was followed by the cyclic expression of specific integrins. As the profile of fibronectin and laminin changed, so did the expression of α_2, α_5, and α_6 (Bischof *et al.*, 1993). These changes were correlated with the migration of the epithelium and repair of the endometrial layer.

Expression of Integrins in Tumors — Observations in Situ

Because of the importance of adhesion and migration mechanisms in tumor growth, invasion, and metastasis, the integrin profiles of many tumor types have been investigated and compared to those of their normal tissue counterparts. Most tumors display changes in at least some of their integrins that can be correlated with malignancy. However, it is not possible to establish patterns that apply to all tumors since the changes in integrin profiles are characteristic for each tumor type.

Nevertheless, it can be stated, in general, that with an increase in tumorigenicity there is a decrease in integrin expression, as detected by immunohistochemical methods. Particularly, decreases in $\alpha_2(\beta_1)$, $\alpha_5(\beta_1)$, and/or $\alpha_6(\beta_1)$, and β_4 have been consistently noted in certain kinds of tumors. The disappearance of these integrins from the tumor cell surface suggests a decrease in cell-matrix and cell-cell binding that would favor uncontrolled growth and migration. This pattern is exemplified in carcinoma of the breast (Koukoulis *et al.*, 1991; Natali *et al.*, 1992; Pignatelli *et al.*, 1992a; Arihiro *et al.*, 1993), lung (Damjanovich *et al.*, 1992), colon (Koretz *et al.*, 1991; Pignatelli *et al.*, 1992b; Stallmach *et al.*, 1992; Koukoulis *et al.*, 1993), pancreas (Weinel *et al.*, 1992), kidney (Korhonen *et al.*, 1992; Korhonen *et al.*, 1992), bladder (Liebert *et al.*, 1993), and in basal cell carcinoma of the skin (Savoia *et al.*, 1993).

In some tumors, the observed decrease in integrin expression is also characterized by a loss in polarization. This is particularly the case for $\alpha_6\beta_4$, which is normally expressed at the basolateral surface of epithelia. Loss of expression and polarization of this integrin, coincident with a decrease in certain basement membrane components, has been observed in carcinoma of the breast (Natali *et al.*, 1992) and bladder (Liebert *et al.*, 1993). Similarly, in colon carcinoma, the loss of continuity of $\alpha_2\beta_1$ and $\alpha_3\beta_1$ expression at the basolateral surface of tumor glands, correlates with tumor malignancy (Lindmark *et al.*, 1993).

In breast carcinoma, the decrease in $\alpha_2\beta_1$ and $\alpha_5\beta_1$ expression has been detected at the RNA level by *in situ* hybridization (Zutter *et al.*, 1993). The decrease in $\alpha_6\beta_4$ expression is also at the RNA level, since it was found missing in tumors in a differential display screening for tumor suppressors of breast carcinoma (Sager *et al.*, 1993). Thus, it appears that with tumor development, the mechanisms controlling the synthesis and/or localization of certain integrins are affected.

A second pattern that is observed in certain tumors is an increase in expression of the integrins $(\alpha_v)\beta_3$ and $\alpha_3(\beta_1)$. This is the case in glioblastoma (Gladson and Cheresh, 1991), astrocytoma (Paulus *et al.*, 1993), renal cell carcinoma (Korhonen *et al.*, 1992a, b), and melanoma (Albelda 1993; Natali *et al.*, 1993). $\alpha_6\beta_4$ is overexpressed in squamous cell carcinoma of the head and neck (Van Waes and Carey, 1992).

Expression of Integrins in Tumors — Observations *in Vivo*

The importance of $\alpha_v\beta_3$ in melanoma is further supported by experiments in which the *in vivo* tumorigenicity of a melanoma cell variant lacking α_v — thus lacking cell surface expression of $\alpha_v\beta_3$ — was restored by transfection of α_v (Felding-Habermann *et al.*, 1992). Tumor cell growth was increased when the $\alpha_v\beta_3$-expressing cells were injected subcutaneously into nude mice but, interestingly, the *in vitro* growth properties of these cells were not affected.

Similar kinds of experiments have been carried out to assess the changes brought about by a particular integrin in the tumorigenicity and invasive potential of various kinds of tumor cells. The overexpression of $\alpha_5(\beta_1)$ in Chinese hamster ovary cells resulted in a decrease in tumor growth when the transfected cells were injected subcutaneously into nude mice (Giancotti and Ruoslahti, 1990). The decrease in tumor growth was associated with increased fibronectin matrix production and decreased migration. Similar gene transfer experiments, as described above, have been carried out in a colon carcinoma cell line. The result was also a decrease in tumorigenicity, with increased $\alpha_5\beta_1$ expression and concomitant fibronectin matrix assembly (Stallmach *et al.*, 1994). This is in agreement with the decrease in α_5, observed in colon tumors with increasing degree of malignancy (Stallmach *et al.*, 1992). A diminished fibronectin-containing extracellular matrix (Hynes and Yamada, 1982), as well as a decrease in $\alpha_5\beta_1$ expression, are often characteristics of virus-transformed cells in culture (Plantefaber and Hynes, 1989). Thus, the association, at least in part, of $\alpha_5\beta_1$ expression with fibronectin matrix assembly (Fogerty and Mosher,

1990) is probably related to decreased growth and migratory or invasive potential of some tumor cells (Giancotti and Ruoslahti, 1990).

Transfection of the integrin subunit α_2 into human rhabdomyosarcoma cells increased their metastatic potential when injected both subcutaneously and intravenously, whereas their growth properties were not affected (Chan *et al.*, 1991). Increased adhesion to both laminin and collagen was associated with expression of $\alpha_2\beta_1$. Transfection of the integrin α_4 into a highly metastatic melanoma cell line did not affect growth properties but did inhibit metastasis when the cells were injected subcutaneously, and not when injected intravenously. Thus, the initial stage of metastasis, that is, invasion prior to intravasation was blocked by the expression of $\alpha_4\beta_1$ (Qian *et al.*, 1994).

It is clear that different integrins affect tumor cell growth and invasiveness in a manner particular to that integrin, and perhaps particular to each cell type and *in vivo* microenvironment. Important information can be derived from these types of experiments, since tumors might undergo changes in their integrin profile that accommodate their adhesion and migration requirements during growth, invasion and metastasis. However, it is difficult to derive any clinically relevant conclusions based only on the type of gene transfer experiments listed above. For instance, contrary to the association between α_4 expression and the decreased invasiveness in melanoma cells (Qian *et al.*, 1994), others have found a slight increase in α_4 expression with increased melanoma malignancy, as observed *in situ* (Albelda, 1993; Schadendorf *et al.*, 1993).

A collection of data supporting *in vivo* observations is obviously important in assessing the relevance of a particular integrin for a particular tumor type. Such a collection is perhaps beginning to emerge for the integrin $\alpha_v\beta_3$ in melanoma. In addition to the gene transfer experiments described above, a monoclonal antibody against the platelet glycoprotein $\alpha_{IIb}\beta_3$ has been shown to inhibit the growth of subcutaneously injected melanoma cells in nude mice, presumably by blocking the activity of the similar integrin, $\alpha_v\beta_3$, on melanoma cells (Boukerche *et al.*, 1989). In addition, an RGD-containing peptide that can inhibit the interaction between $\alpha_v\beta_3$ and its ligands, blocks metastasis of intravenously injected melanoma cells (Humphries *et al.*, 1988). It also blocks the adhesion of metastatic melanoma lines to lymph node vitronectin *in situ* (Nip *et al.*, 1992).

Thus, $\alpha_v\beta_3$ apparently plays a role both in tumor growth and in the later stages in metastasis. The molecular mechanisms involved and the identity of its ligand(s) are yet to be clarified. Interestingly, a recent study showed that ligation of $\alpha_v\beta_3$ on melanoma cells seeded in three-dimensional collagen, conferred growth advantage and, in the absence of serum, prevented apoptotic cell death (Montgomery *et al.*, 1994). Partial proteolysis by the melanoma cells caused denaturation of the collagen thus exposing sites, most probably RGD-sequences, recognized by $\alpha_v\beta_3$ (Montgomery *et al.*, 1994). Whether other $\alpha_v\beta_3$ ligands, presented either in one or three-dimensional matrices, can provoke similar effects is still unclear. Another study showed that perturbation of $\alpha_v\beta_3$ by binding to certain antibodies or to vitronectin, caused the upregulation of collagenase IV in a melanoma cell line (Seftor *et al.*, 1992). Even though this finding was apparently not duplicated with a different, more aggressive tumor cell line (Seftor *et al.*, 1993), it suggests a potential mechanism via which $\alpha_v\beta_3$ might promote the expression of proteolytic activity needed for growth sustenance and/or invasion by melanoma cells. Another important, albeit indirect, function for the $\alpha_v\beta_3$ in tumorigenesis has been identified in that its expression is confined to tissues undergoing neovascularization and antibodies to $\alpha_v\beta_3$ inhibit the angiogenic process (Brooks *et al.*, 1994).

In addition to its classical ligand, vitronectin, $\alpha_v\beta_3$ recognizes most other RGD-containing adhesion-promoting proteins. These include fibrinogen, von Willebrand factor, and vitronectin — ligands not shared by most β_1 integrins. Binding to fibrinogen might be advantageous for tumor cells in the circulation since it might provide a means to interact with platelets and monocytes which could aid in escaping immune surveillance. Binding to vonWillebrand factor present at the endothelial cell basal surface, and exposed with injury in blood vessels, might provide a point of attachment for circulating tumor cells that would then be followed by extravasation. Vitronectin appears to be present mostly in inflamed or diseased tissues, sites that might favor tumor growth and invasion. It is also interesting to note that $\alpha_v\beta_3$ is present in podosomes (Zambonin-Zallone *et al.*, 1989). These are focal contact-like sites found in transformed fibroblasts and normally invasive cells that concentrate proteolytic activity and are thus thought to be involved in localized matrix degradation (Chen *et al.*, 1984; Tarone *et al.*, 1985). The interaction between $\alpha_v\beta_3$ and its ligand(s) might therefore be protease resistant. In the case where vitronectin would be the ligand, local protease resistance could come about via vitronectin binding to plasminogen activator inhibitor-1 and keeping it in an active state (Declerck *et al.*, 1988; Mimuro and Loskutoff, 1989), thus promoting inhibition of urokinase and plasminogen activators at the cell surface (Ciambrone and McKeown-Longo, 1990). This would allow tumor cells to maintain interactions with the supporting matrix needed for growth and/or migration.

As exemplified here for $\alpha_v\beta_3$ in melanoma, the molecular mechanisms involved in the recognition of the extracellular matrix by tumor cell integrins, and the signals this recognition can elicit, are beginning to be elucidated. This knowledge, together with that obtained for other molecules (cadherins, CAMs, selectins), should allow for a better understanding of the adhesion processes involved in tumor growth and dissemination.

Control of Interstitial Fluid Pressure in Dermis — a Possible Role for β_1 Integrins via *in Vivo* Contraction of Extracellular Matrix Structure

Recent experiments suggest that β_1 integrins in dermis may be involved in a phenomenon analogous to their role in the *in vitro* collagen gel contraction, described earlier in this chapter. In several acute inflammatory reactions, the initial edema formation is enhanced by the interstitial fluid pressure becoming more negative, *i.e.* the loose connective tissue will provide a "suction" on the fluid in the capillaries (Reed, 1994). This phenomenon was first observed in thermal skin injury (Lund *et al.*, 1989) and has later been observed in dextran-induced anaphylaxis in skin and trachea in rats, as well as in rat trachea following mast cell degranulation induced by C48/80 and Polymyxin B. Furthermore, the phenomenon is also seen in neurogenic inflammation in trachea, as well as following inflammation in skin induced by carrageenan and xylene (Reed, 1994).

Increased negativity of interstitial fluid pressure has also been observed following subdermal injection of anti-β_1 integrin IgG, concomitant with edema formation, and with a time response and magnitude similar to that observed in the inflammatory reactions cited above. The effect was dose-dependent, and there was no effect of pre-immune IgG, anti-fibronectin IgG, or synthetic peptides containing the sequence RGD (Figure 9.5). The anti-β_1 integrin IgG was used based on the reasoning that the interaction between cells and extracellular matrix components of loose connective tissue can be described as analogous to the collagen gel contraction (see above). First, experiments have shown that loose connective tissue which is allowed free access to saline will swell due to its content of glycosaminoglycans and hyaluronan (Meyer, 1983). Furthermore, the swelling of the hyaluronan and proteoglycan gel was described to be physically restrained by a microfibrillar network. Second, the reasoning behind injecting the anti-β_1 integrin IgG was that if the fibroblasts were able to mediate tension in the connective tissue fibers, it should be possible to reverse, or diminish, this tension by blocking with anti-β_1 integrin IgG. Subsequently, the tissue should be able to swell, and since time is required for fluid to enter the tissue, interstitial fluid pressure would initially become more negative until the increased negativity of the interstitial pressure would balance the expanding glycosaminoglycan/ hyaluronan gel of the tissue. To strengthen the concept of an analogy with collagen gel contraction, increased levels of intracellular cAMP have been shown to slow, or inhibit collagen gel contraction, and also to cause an increased negativity of the interstitial fluid pressure (Rodt *et al.*, 1994). Furthermore, recent data in our laboratories show that PDGF can normalize a dextran-induced decrease in interstitial fluid pressure (Rodt *et al.*, In Press). PDGF also strongly stimulates collagen gel contraction (Gullberg *et al.*, 1990). The concept is shown schematically in Figure 9.6.

The commonly accepted role of interstitial fluid pressure is to counteract alterations in interstitial fluid volume, since increased fluid volume, induced by increased capillary filtration pressure, would raise interstitial fluid pressure. The rise in interstitial fluid pressure will then act across the capillary wall to directly counteract further filtration of fluid, and at the same time enhance lymph flow; both changes will maintain a constant interstitial fluid volume, rather than enhance the fluid filtration. The observed decrease of interstitial fluid pressure from –1 to between –5 to –10 mm Hg provides a substantial increase in the net filtration pressure across the capillary which is normally 0.5 to 1 mm Hg. Thus, the decrease of interstitial fluid pressure represents between 5 and 20 times an increase in the pressure which generates the fluid filtration. In burn injury, pressures as low as –150 mm Hg have been observed (Lund *et al.*, 1989), but in this situation, the decrease in pressure is partly caused by collagen denaturation. However, more important than the absolute magnitude of the increased negativity of interstitial fluid pressure, is the fact that these observations provide evidence for loose connective tissue being able to "actively" influence capillary filtration and generate edema. This "active" control is likely mediated via collagen-binding β_1 integrins and, therefore, points to the possibility of a wide participation of the connective tissue cells in control of the stresses in the extracellular fiber networks of the loose connective tissue. This opens up the possibility that a wide range of physiological and inflammatory mediators can influence the stress on fiber networks via modulating the activity of β_1 integrin function, which in turn will influence e.g. transcapillary transport and fluid balance of the loose connective tissue.

Non-Integrin Receptors for Extracellular Matrix Proteins

While this review has focused primarily on the role of integrins as mediators of the interaction between the eukaryotic cell surface and the ECM, other components on the cell surface have also been suggested to regulate cell-behavior via the ECM. Sugar transferases, especially galactosyltransferase (galtase), have been shown to influence cell migration (Loeber and Runyan, 1990; Shur, 1993). Most investigations focus on a particular type of ECM receptor; however, Loeber and Runyan (1990) selectively blocked integrins and/or galtase to show that multiple adhesion mechanisms were involved in the regulation of cell

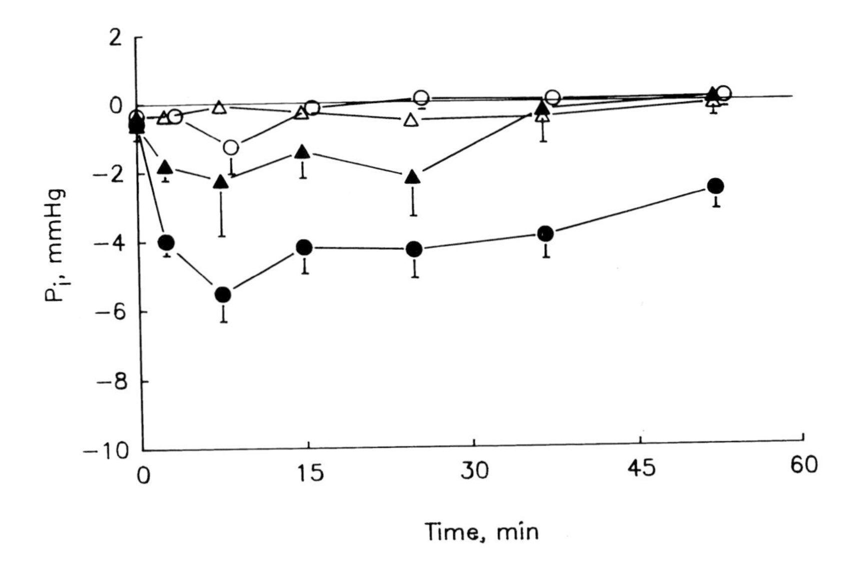

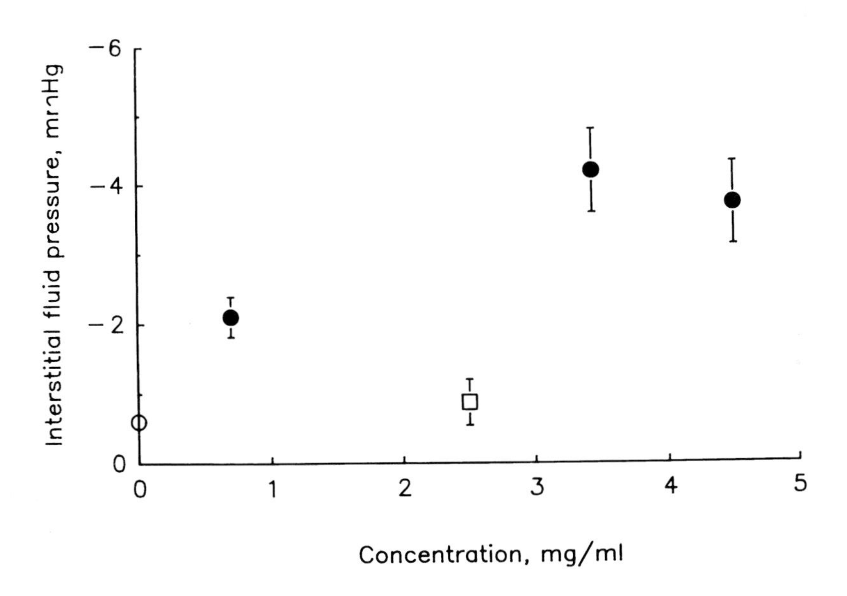

Figure 9.5 Effects of anti-rat β_1 integrin IgG on interstial fluid pressure.

A. Graph showing the effects of anti-rat β_1 integrin immunoglobulin G (●) (3.4–4.5 mg/ml) and preimmune immunoglobulin G (❍) (2.5 mg/ml) from the same rabbit on dermal interstitial fluid pressure. Also shown is rabbit anti-rat plasma fibronectin immunoglobulin G (△) (2.5 mg/ml) and the synthetic peptide Gly-Arg-Gly-Asp-Thr-Pro (▲) (6 mg/ml).

B. Graph showing interstitial fluid pressure as a function of the concentration of anti-β_1 integrin immunoglobulin G (●) during the first 5 minutes after subdermal injection. Also shown is interstitial fluid pressure at preinjection control (grand mean, ❍) and preimmune immunoglobulin G (❑).

Values are mean ±1 SEM (the error bar is contained within some of the data points). [Reproduced with permission from: Reed, R.K., Rubin, K., Wiig, H. and Rodt, Å. (1992) Blockade of β_1 -integrins in skin causes edema through lowering of interstitial fluid pressure. Reprinted from *Circulation Res.* **71**: 978–983 copyright The American Heat Association]

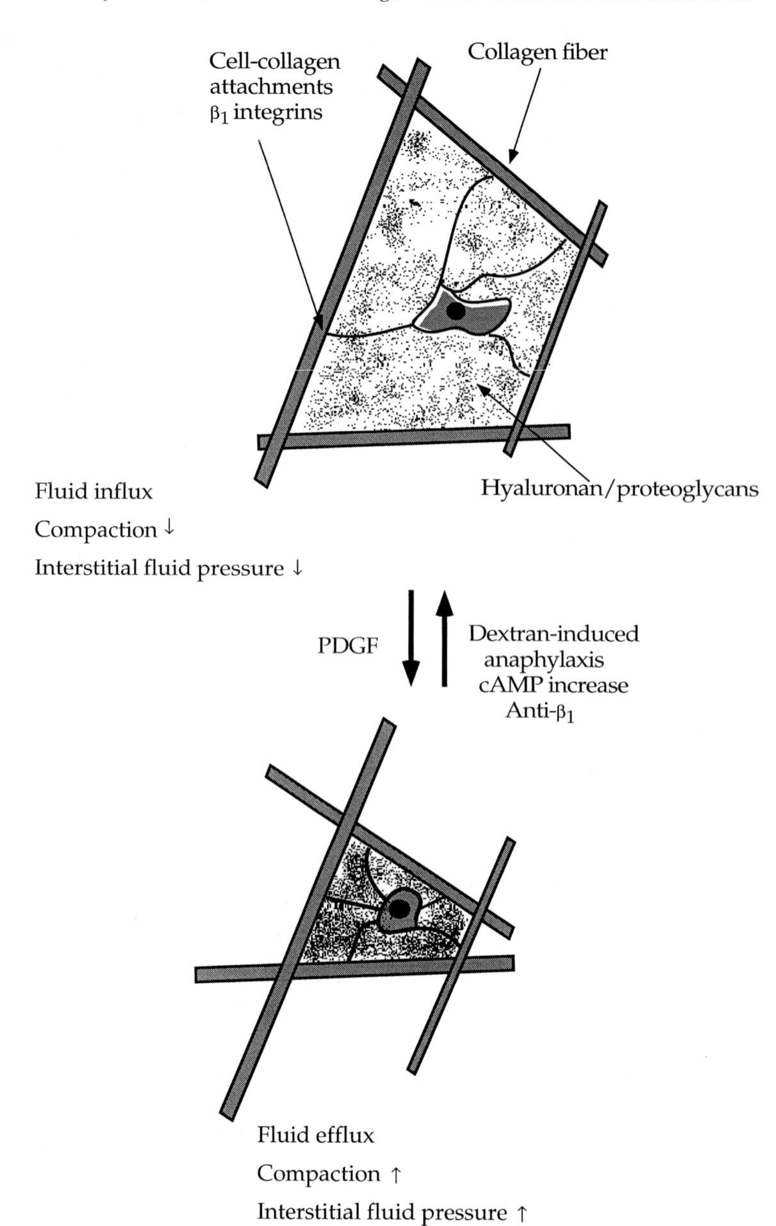

Figure 9.6 Hypothetical role of collagen-binding β_1-integrins in the regulation of interstitial fluid pressure. A graphical presentation of a hypothesis postulating a role for connective tissue cells in the control of fluid balance in loose connective tissues. The figure based on data from studies that are referenced in the text. A hyaluronan/proteoglycan gel is restrained from taking up fluid and expand by a collagen network that, in turn, is kept under tension by connective tissue cells. Cells exert their tension on the collagen network via collagen-binding β_1-integrins, and are able to modulate this tension by regulating the activity of the integrins. The binding activity of the latter is stimulated by PDGF and is inhibited by increases in intracellular cAMP levels. This model of *in vivo* contractility of dermal connective tissue cells shares many properties with the fibroblast-mediated contraction of three-dimensional collagen gels *in vitro*.

migration on laminin, collagen, and fibronectin. These data suggested that there was "cross-talk" among different types of receptors as occupancy of specific ECM receptors modulates the activity of the other.

A non-integrin 67/69 kD laminin-binding protein, with some properties of an adhesion-receptor for laminin, has been found in a variety of tissues and cell lines including tumor, skeletal muscle, and placental cells (Mecham, 1991). It is most likely, identical to a separately described elastin-binding protein that recognizes a hydrophobic sequence in elastin. This protein has been ascribed a role in tumor metastasis. Its exact role as an adhesion receptor has not been clarified, nor has its molecular structure been clearly elucidated; available sequence information suggests a 33 kD non-glycosylated protein lacking a typical transmembrane domain (Yow *et al.*, 1988; Rao *et al.*, 1989). It has also been suggested that the 67/69 laminin-binding protein is possibly a chaperone.

An M_r 31,000 non-integrin collagen binding-protein, anchorin CII, was purified from avian chondrocytes by affinity chromatography (Mollenhauer and von der Mark, 1983). Antibodies to this protein partly inhibited adhesion of chondrocytes to collagen. Further characterizations and protein sequencing of anchorin CII revealed that it belongs to the calpactin family and that it lacks a typical transmembrane domain (Pfäffle *et al.*, 1988). Whereas other members of the calpactin family are located at the inner side of the plasma membrane, anchorin CII seems to be partly released and bound to the outer surface of cells. Its role in the adhesion of cells to collagen is, however, not clear. It may be that anchorin CII takes part in cellular interactions with collagen as part of an interactive multiprotein complex at the cell surface.

The CD44 family of non-integrin ECM receptors encompasses polymorphic glycoproteins which are encoded by one single gene that is alternatively spliced, and the core proteins are subjected to cell lineage-specific glycosylation (Haynes *et al.*, 1989; Jackson *et al.*, 1992). CD44 molecules appear as membrane intercalated proteoglycans, with chondroitin sulfate side chains. They play a physiological role in leukocyte trafficking (Haynes *et al.*, 1989; Jackson *et al.*, 1992). Certain variants of these glycoproteins have affinity for hyaluronan (Thomas *et al.*, 1992) and others can act as collagen receptors (Carter and Wayner, 1988; Faassen *et al.*, 1993). They have also been implicated in important functions in tumor metastasis (Gunthert *et al.*, 1991; Sy *et al.*, 1992)

INTERACTIONS OF BACTERIAL CELLS WITH ECM STRUCTURES

Introduction

A large number of specific interactions between bacterial cells, both Gram-positive and Gram-negative, and isolated ECM components have been described. Since the initial report by Kuusela in 1978 of an interaction between fibronectin and staphylococcal cells (Kuusela, 1978) many different bacteria have been shown to interact with ECM components. In addition, cell wall components binding to ECM components have been purified from bacteria. *In vivo* experiments have confirmed the hypothesis that specific interactions between ECM components and bacteria are of importance for the development of certain infectious processes (Nelson *et al.*, 1991; Schennings *et al.*, 1993; Patti *et al.*, 1994).

Gram-negative bacteria, such as *Escherichia coli*, interact mainly with carbohydrate moieties on eukaryotic cells (Leffler and Svanborg-Edén 1981; Svanborg-Edén *et al.*,

1982). However, *E. coli* also possesses protein binding with ECM components, as exemplified by fibronectin and collagen (Korhonen *et al.*, 1984; Hoschutzky *et al.*, 1989; Westerlund *et al.*, 1989a, b). The majority of interactions described between gram-positive bacteria — such as staphylococci and streptococci — and ECM components, are based on protein-protein interactions (Espersen and Clemmensen, 1982; Rydén *et al.*, 1983; Switalski *et al.*, 1993). Lectin-mediated interactions between grampositive bacteria and ECM components have also been reported (McGavin *et al.*, 1993). The latter interactions are unspecific in the sense that the staphylococcal lectin like components recognize a large number of ECM molecules.

Only a few examples of the many bacterial/ECM interactions that have been described will be discussed in this review. These interactions provide examples of how microbial cells and ECM components can interact — and of their possible *in vivo* significance.

Interactions between Staphylococcal Cells and Fibronectin

Fibronectin binds with high affinity to both live and heat-inactivated staphylococcal cells. Dissociation constants in the range of 10^{-8}–10^{-9} M have been reported (Espersen and Clemmensen, 1982; Proctor *et al.*, 1982; Rydén *et al.*, 1983). The major staphylococcal binding domain is located in a 29 kDa amino-terminal fragment of the fibronectin molecule (Mosher and Proctor, 1980; Proctor *et al.*, 1982; Kuusela *et al.*, 1984). Binding of bacteria occurs in a part of fibronectin that is distant from the eukaryotic cell and collagen binding sites. This makes it possible for fibronectin to anchor staphylococci to the host cells, or to fibronectin, in the ECM. Since fibronectin is present in wound fluids and is immobilized in the granuloma tissue, the interaction with staphylococcal cells may contribute to the common finding of staphylococcal contamination and also to the infection of wounds.

The interaction between fibronectin and staphylococcal cells is mediated by a specific cell wall protein that is not involved in other ECM interactions. A fibronectin-binding protein (FNBP) has been isolated from staphylococcal cell walls, and has been identified in several different strains. The molecular weight of the isolated FNBP varies in these reports, probably due to difficulties to purify the native protein without degradation (Espersen and Clemmensen, 1982; Proctor *et al.*, 1982; Rydén *et al.*, 1983).

Genes encoding two fibronectin binding proteins — FNBP A and B — have been cloned and sequenced from the staphylococcal genome (Signäs *et al.*, 1989; Jönsson *et al.*, 1991). These proteins show substantial amino acid similarity with each other, as well as similarity with other cell wall proteins isolated from staphylococcal cells. The nucleotide sequences, as well as amino acid sequences of the gene-products, closely resemble fibronectin-binding proteins isolated from streptococcal cells, as well as other cell wall proteins from gram-positive cocci (Lindgren *et al.*, 1992; Schneewind *et al.*, 1992), indicating a common predecessor gene for these proteins.

Staphylococcal mastitis is a great problem in veterinary medicine. Staphylococci isolated from bovine mastitis cases have been shown to bind fibronectin, and a mouse model for mastitis has been established (Jonsson *et al.*, 1985). Recombinant FNBP has been used as an antigen in vaccination trials, aiming at elucidating the potential of this treatment for diminishing the rate of mastitis and the severity of the disease. So far, it has been shown that such vaccination is effective in the mouse-mastitis model and in cows (Nelson *et al.*, 1991). In the latter, vaccination with recombinant FNBP completely prevents mastitis and drastically reduces morbidity in this disease. These experiments strongly suggest that the ability of staphylococci to bind fibronectin is a virulence factor in mastitis. The interaction between fibronectin and staphylococci has been shown to be important for disease development in an experimental model of endocarditis (Kuypers and Proctor, 1989). The ability to bind fibronectin, in combination with the collagen binding capacity of such staphylococci, is important in the development of experimental rat endocarditis (see below). Rats vaccinated with recombinant FNBP are protected from the development of experimental endocarditis (Schennings *et al.*, 1993).

Interactions Between Staphylococcal Cells and Collagen

Several coagulase-negative and coagulase-positive strains of staphylococci bind collagen in specific reactions that also mediate staphylococcal attachment to immobilized collagen (Holderbaum *et al.*, 1986; Maxe *et al.*, 1986; Voytek *et al.*, 1988). The affinity for staphylococcal binding of collagen, $K_d \sim 10^{-7}$, is lower than that of the fibronectin-staphylococcal interaction. Collagen types I-VI are all recognized by the staphylococcal collagen-binding component(s) (Speziale *et al.*, 1986). Despite the availability of increasingly more effective antibiotics, staphylococcal infections of bone and joints persist as a significant clinical problem. Staphylococcal infections account for 60–70 % of infections at these locations. It has been hypothesized that the interactions between staphylococci and, especially, collagen types I and II may have a pathogenetic importance in these infections. Collagen types I and II are the main organic constituents in bone and cartilage, respectively. The capacity to bind collagen is not restricted to staphylococci isolated from bone infections, but seems to occur among many strains of staphylococci, regardless of tissue specificity. On the other hand, *S.aureus* strains — isolated from patients suffering from septic arthritis — are more prone to adhere to cartilage *in vitro* than strains obtained from infections at other sites (Switalski *et al.*, 1993). A correlation between the ability to interact with the cartilage collagen type II and arthritogenicity, has been reported (Switalski *et al.*, 1993).

The staphylococcal surface protein responsible for collagen binding has been purified (Switalski *et al.*, 1989). The gene encoding this M_r 110,000 collagen adhesin has also been cloned and sequenced (Patti *et al.*, 1992). The structure of this protein resembles that of protein A, as well as that of the FNBP:s. The purified collagen adhesin inhibits the binding of collagen type II to staphylococci, and also inhibits the adhesion of staphylococci to cartilage slices *in vitro*.

Because the gene for the collagen adhesin in *S. aureus* strain FDA 574 has been identified, it has been possible to identify this gene in other staphylococcal strains. A clinical isolate of *S. aureus* — strain Phillips — has the gene for the collagen adhesin, and expresses a surface protein that binds collagen. A mutant of this strain was constructed in such a way that the gene for the collagen adhesin was destroyed without destroying any of the other functions of the staphylococcal genome (Patti *et al.*, 1994). This mutant strain — Ph 100 — was not capable of binding collagen. In a second experiment, a mutant was constructed by inserting the intact collagen adhesin gene into a *S. aureus* strain naturally lacking the gene (Patti *et al.*, 1994). These different strains have been utilized in two *in vivo* models of infection — a septic arthritis model in mice and a model for endocarditis in rats. In brief, in the arthritis model after injecting staphylococci i.v., the strains harboring the collagen adhesin gene gave rise to septic arthritis in 70% of the animals; the mutant, devoid of the collagen adhesin gene and the naturally negative strain, were much less virulent, giving rise to arthritis in only 30% of injected animals — and a much milder clinical and histological picture (Patti *et al.*, 1994).

Interactions Between Staphylococcal Cells and Bone Sialoprotein

A glycoprotein with high affinity for hydroxyapatite — bone sialoprotein (BSP) — comprises about 12% of the organic bone ECM. BSP has only been detected in bone and dentin, and the glycoprotein is predominantly expressed in osteoid close to the interface between bone and cartilage. Staphylococcal cells, isolated from cases of septic arthritis and osteomyelitis, bind BSP (Rydén *et al.*, 1987). Since BSP is selectively expressed in bone tissue, this interaction may be important for targeting staphylococcal cells to bone tissue in bone infection. A staphylococcal surface protein binding BSP has been isolated and shows an M_r of 97,000 (Yacoub *et al.*, 1994). The binding site for staphylococci in BSP has been localized in the aminoterminal end of the molecule (Rydén *et al.*, 1989), separate from the eukaryotic binding RGD-containing sequence, which is located in the C-terminus.

All of the staphylococcal strains inducing septic arthritis in mice have been found to be capable of binding BSP in significant amounts, and this interaction may add to the virulence of these bacteria in this experimental model (Bremell *et al.*, 1991). It has been shown histologically that not only the joint cartilage, but also the bone tissue, is invaded by staphylococci early on in the disease. Most likely, a combination of interactions between ECM components of the host and bacterial surface structures are important for disease development.

Evolution of Bacterial Cell-Surface Proteins

Several cell-surface proteins have been isolated from Gram-positive cocci and the genes encoding these proteins have been cloned, such as the streptococcal proteins M-protein, Arp4 protein, protein H, and protein G (Sjöbring and Björck, 1992), and the staphylococcal protein A (Uhlén *et al.*, 1984), all binding immunoglobulins. Genes encoding fibronectin-binding proteins have been cloned and sequenced — both from different streptococci (Lindgren *et al.*, 1992) and from staphylococci (Signäs *et al.*, 1989) — as well as a collagen adhesin from staphylococcal cells (Patti *et al.*, 1992). When comparing these gene-products, it is striking that they share many features like secretion signal peptides, cell-wall spanning regions and membrane anchoring domains, as well as repeated domains of various functions (Goward *et al.*, 1993). One such well defined protein is protein A of *S. aureus*, which has a typical functional domain structure with Ig-binding capacity. Several cell-surface proteins with Ig-binding capacity have been isolated from the cell walls of groups A, C, and G streptococci; these have also shown repeated domain structures, which may have evolved through stepwise duplication of a DNA sequence, thus pointing to common synthesis procedures. Motifs, important for binding, seem to have been transferred both within and across groups. The secretion signal peptide sequence shows high similarity among these proteins, and so does the carboxyterminal, particularly a motif (LPXTGE) in the membrane anchoring region common to both streptococcal and staphylococcal proteins. The LPXTGE motif has been shown to be important for the attachment of the proteins to the cell membrane (Schneewind *et al.*, 1992).

SUMMARY

The establishment and regulation of cellular phenotype of eukaryotic cells is a dynamic interaction between various external stimuli. The ECM constitutes an important contributor of such stimuli. The mediation of signals from the ECM is dependent upon specific receptors. Several classes of such receptors have been isolated, the most thoroughly studied being the integrin family of adhesion receptors. These adhesion receptors form a transmembrane link between organized extracellular structures, i.e. ECM components, and cytoskeletal structures. Integrins are true receptors in the sense that they transduce signals from the cell exterior to the inside. Thus, not only do integrins establish physical contact between outside and inside structural elements but they also convey regulatory signals across the membrane.

Integrins are unique among known adhesion receptors in that they have the capability to undergo intramolecular affinity modulation, and to respond to signals generated from stimulation of the integrin-carrying cell by growth factors, cytokines and prostaglandins. Our understanding of the specific signalling pathways is still in its infancy, and clarification of these pathways at the molecular level will be essential to our understanding of the fundamental processes of cell-ECM interactions. However, it is clear that integrins transmodulate cellular responses to growth factor stimulation, and are themselves affected by such stimulation.

It is clear that the components of the ECM are not static and that they are constantly changing and turning over. The rate of these changes influences the cellular phenotype during development and during physiological and patho-physiological processes in the adult. The mechanisms whereby the ECM exerts its control of these processes can now be more readily investigated, with some of the main components identified, i.e. individual ECM molecules and receptors for the ECM. Such investigations will not only increase our understanding of fundamental biological phenomena, but will also, undoubtedly, find practical applications for new strategies to control certain diseases, such as inflammatory conditions and malignancies.

Many genera of bacteria are able to interact directly and specifically with several ECM components. In many cases, specific bacterial ECM-binding proteins have been characterized. Experimental evidence has been provided showing that bacterial-ECM interactions have a patho-physiological importance for certain infectious diseases. Further characterization of such interactions will, most likely, lead to the introduction of new therapeutic principles not based on conventional antibiotics.

Acknowledgements

Karina Åhlén is gratefully acknowledged for help in the production of Figure 9.3, and Raquel Tomasini for valuable editorial support. Original research presented in this chapter was supported by grants from the Swedish Cancer Foundation and the Norwegian Council for Science and the Humanities.

References

Abercrombie, M., Heaysman, J.E.M. and Pegrum, S.M. (1971). The locomotion of fibroblasts in culture. IV. Electron microscopy of the leading lamella. *Exp Cell Res* **67**: 359–367.

Adams, J. and Watt, F.M. (1993). Regulation of development and differentiation by the extracellular matrix. *Development* **117**: 1183–1198.

Åhlén, K. and Rubin, K. (1994). Platelet-derived growth factor-BB stimulates synthesis of the integrin α_2-subunit in human diploid fibroblasts. *Exp Cell Res* **215**: 347–353

Akiyama, S.K., Nagata, K. and Yamada, K.M. (1990) Cell surface receptors for extracellular matrix components. *Biochimica et Biophysica Acta* **1031**: 91–110

Albelda, S.M. (1992). Differential expression of integrin cell-substratum adhesion receptors on endothelium. *Experientia* **61**: 188–192.

Albelda, S.M. (1993). Role of integrins and other cell adhesion molecules in tumor progression and metastasis. *Lab Invest* **68**: 4–17.

Albelda, S.M. and Buck, C.A. (1990). Integrins and other cell adhesion molecules. *FASEB J* **4**: 2868 – 2880.

Altruda, F., Cervella, P., Tarone, G., Botta, C., Balzac, F., Stefanuto, G. and Silengo, L. (1990). A human integrin $\beta 1$ subunit with a unique cytoplasmic domain generated by alternative mRNA processing. *Gene* **95**: 261– 266.

Arihiro, K., Inai, K., Kurihara, K., Takeda, S., Khatun, N., Kuroi, K. and Toge, T. (1993). A role of VLA-6 laminin receptor in invasion of breast carcinoma. *Acta Pathol Jpn* **43**: 662–669.

Arroyo, A.G., Garcia-Pardo, A. and Sanchez-Madrid, F. (1993). A high affinity conformational state on VLA integrin heterodimers induced by an anti-β_1 chain monoclonal antibody. *J Biol Chem* **268**: 9863–9868.

Balzac, F., Belkin, A.M., Koteliansky, V.E., Balabanov, Y.V., Altruda, F., Silengo, L. and Tarone, G. (1993). Expression and functional analysis of a cytoplasmic domain variant of the $\beta 1$ integrin subunit. *J. Cell. Biol* **121**: 171–178.

Bao, Z.Z., Lakonishok, M., Kaufman, S. and Horwitz, A.F. (1993). $\alpha_7\beta_1$ integrin is a component of the myotendinous junction on skeletal muscle. *J. Cell Sci.* **106**: 579–590.

Baumgartner, S., Martin, D., Hagios, C. and Chiquet-Ehrismann, R. (1994). Ten^m, a *Drosophila* gene related to tenascin, is a new pair-rule gene. *EMBO J* **13**: 3728–3740.

Bell, E., Ivarsson, B. and Merrill, C. (1979). Production of a tissue-like structure by contraction of collagen lattices by human fibroblasts of different proliferative potential *in vitro*. *Proc. Natl. Acad. Sci. USA* **76**: 1274–1278.

Ben-Ze'ev, A., Farmer, S.R. and Penman, S. (1980). Protein synthesis requires cell-surface contact while nuclear events respond to cell shape in anchorage-dependent fibroblasts. *Cell* **21**: 365–372.

Bevilacqua, M.P. and Nelson, R.M. (1993). Selectins. *J Clin Invest* **91**: 379–387.

Birkenmeier, T.M., McQuillan, J.J., Boedeker, E.D., Argraves, W.S., Ruoslahti, E. and Dean, D.C. (1991). The $\alpha 5\beta 1$ fibronectin receptor. Characterization of the a5 gene promoter. *J Biol Chem* **266**: 20544–20549.

Bischof, P., Redard, M., Gindre, P., Vassilakos, P. and Campana, A. (1993). Localization of α_2, α_5 and α_6 integrin subunits in human endometrium, decidua and trophoblast. *Eur J Obstet Gynecol Reprod Biol* **51**: 217–226.

Bissell, M.J., Hall, H.G. and Parry, G. (1982). How does extracellular matrix direct gene expression? *J Theor Biol* **99**: 31–68.

Borg, T.K. and Caulfield, J.B. (1980). Morphology of connective tissue in skeletal muscle. *Tissue and Cell* **12**: 197–207.

Borg, T.K., Ranson, W.F., Moslehy, F.A. and Caulfield, J.B. (1981). Structural basis of ventricular stiffness. *Lab Invest* **44**: 49–54.

Bosco, M., Chan, C. and Hemler, M.E. (1993). Multiple functional forms of the integrin VLA-2 can be derived from a single α_2 cDNA clone: interconversion of forms induced by an anti-$\beta 1$ antibody. *J Cell Biol* **120**: 537–543.

Boukerche, H., Berthier-Vergnes, O., Bailly, M., Dore, J.F., Leung, L.L. and McGregor, J.L. (1989). A monoclonal antibody (LYP18) directed against the blood platelet glycoprotein IIb/IIIa complex inhibits human melanoma growth in vivo. *Blood* **74**: 909–912.

Brabant, M.C. and Brower, D.L. (1993). PS2 integrin requirements in *Drosophila* embryo and wing morphogenesis. *Dev Biol* **157**: 49–59.

Bremell, T., Lange, S., Yacoub, A.I., Rydén, C. and Tarkowski, A. (1991). Experimental *Staphylococcus aureus* arthritis in mice. *Infect Immun* **59**: 2615–2623.

Briesewitz, R., Kern, A. and Marcantonio, E.E. (1993). Ligand-dependent and -independent integrin focal contact localization: the role of the α chain cytoplasmic domain. *Mol Biol Cell* **4**: 593–604.

Bronner-Fraser, M., Artinger, M., Muschler, J. and Horwitz, A.F. (1992). Developmentally regulated expression of $\alpha 6$ integrin in avian embryos. *Development* **115**: 197–211.

Brooks, P.C., Clark, R.A.F. and Cheresh, D.A. (1994) Requirement of vascular integrin $\alpha_v\beta_3$ for angiogenesis. *Science* **264**: 569–571.

Brown, N.H. (1993). Integrins hold *Drosophila* together. *Bioessays* **15**: 383–390.

Brown, N.H., King, D.L., Wilcox, M. and Kafatos, F.C. (1989). Developmentally regulated alternative splicing of *Drosophila* integrin PS2 a transcripts. *Cell* **59**: 185–195.

Brown, P.J. and Juliano, R.L. (1986). Expression and function of a putative cell surface receptor for fibronectin in hamster and human cell lines. *J Cell Biol* **103**: 1595–1603.

Buck, C., Albelda, S., Damjanovich, L., Edelman, J., Shih, D.T. and Solowska, J. (1990). Immunohistochemical and molecular analysis of β_1 and β_3 integrins. *Cell Differ Dev* **32**: 189–202.
Buck, C.A. (1992). Immunoglobulin superfamily: structure, function and relationship to other receptor molecules. *Semin Cell Biol* **3**: 179–188.
Buck, C.A. and Horwitz, A.F. (1987). Integrin, a transmembrane glycoprotein complex mediating cell-substratum adhesion. *J Cell Sci Suppl* **8**: 231–250.
Bunch, T.A. and Brower, D.L. (1992). *Drosophila* PS2 integrin mediates RGD-dependent cell-matrix interactions. *Development* **116**: 239–247.
Burgess, M.L., Carver, W.E., Terracio, L., Wilson, S.P., Wilson, M.A. and Borg, T.K. (1994). Integrin-mediated collagen gel contraction by cardiac fibroblasts. Effects of angiotensin II. *Circ Res* **74**: 291– 298.
Burridge, K., Fath, K., Kelly, T., Nuckolls, G. and Turner, C. (1988). Focal adhesions: transmembrane junctions between the extracellular matrix and the cytoskeleton. *Annu Rev Cell Biol* **4**: 487–525.
Burridge, K., Nuckolls, G., Otey, C., Pavalko, F., Simon, K. and Turner, C. (1990). Actin-membrane interaction in focal adhesions. *Cell Differ Dev* **32**: 337–342.
Burridge, K., Turner, C.E. and Romer, L.H. (1992). Tyrosine phosphorylation of paxillin and pp125FAK accompanies cell adhesion to extracellular matrix: a role in cytoskeletal assembly. *J Cell Biol* **119**: 893–903.
Busk, M., Pytela, R. and Sheppard, D. (1992). Characterization of the integrin $\alpha v\beta 6$ as a fibronectin-binding protein. *J Biol Chem* **267**: 5790–5796.
Butcher, E. C. (1991). Leukocyte-endothelial cell recognition: three (or more) steps to specificity and diversity. *Cell* **67**: 1033–1036.
Buttle, D.J. and Ehrlich, H.P. (1983). Comparative studies of collagen lattice contraction utilizing a normal and a transformed cell line. *J Cell Physiol* **116**: 159–166.
Calvete, J.J., Mann, K., Schafer, W., Fernandez-Lafuente, R. and Guisan, J.M. (1994). Proteolytic degradation of the RGD-binding and non-RGD-binding conformers of human platelet integrin glycoprotein IIb/IIIa: clues for identification of regions involved in the receptor's activation.*Biochem J* **298**: 1–7.
Carlson, R., Engvall, E., Freeman, A. and Ruoslahti, E. (1981). Laminin and fibronectin in cell adhesion: enhanced adhesion of cells from regenerating liver to laminin. *Proc Natl Acad Sci (USA)* **78**: 2403–2406.
Carrell, N. A., Fitzgerald, L. A., Steiner, B., Erickson, H. P. and Phillips, D. R. (1985). Structure of human platelet membrane glycoproteins IIb and IIIa as determined by electron microscopy. *J Biol Chem* **260**: 1743–1749.
Carter, W.G. and Wayner, E.A. (1988). Characterization of the class III collagen receptor, a phosphorylated, transmembrane glycoprotein expressed in nucleated human cells. *J Biol Chem* **263**: 4193–4201.
Carter, W.G., Wayner, E.A., Bouchard, T.S. and Kaur, P. (1990). The role of integrins $\alpha 2\beta 1$ and $\alpha 3\beta 1$ in cell-cell and cell-substrate adhesion of human epidermal cells. *J Cell Biol* **110**: 1387–1404.
Carver, W., Price, R., Raso, D., Terracio, L. and Borg, T.K. (1994). Distribution of $\beta 1$ integrin in developing rat heart. *J Histochem Cytochem* **42**: 167–175.
Caulfield, J.B. and Borg, T.K. (1979). The collagen network of the heart. *Lab Invest* **40**: 364–372.
Cervella, P., Silengo, L., Pastore, C. and Altruda, F. (1993). Human β_1-integrin gene expression is regulated by two promoter regions. *J Biol Chem* **268**: 5148–5155.
Chan, B.M., Kassner, P.D., Schiro, J.A., Byers, H.R., Kupper, T.S. and Hemler, M.E. (1992). Distinct cellular functions mediated by different VLA integrin a subunit cytoplasmic domains. *Cell* **68**: 1051–1060.
Chan, B. M., Matsuura, N., Takada, Y., Zetter, B.R. and Hemler, M.E. (1991). In vitro and in vivo consequences of VLA-2 expression on rhabdomyosarcoma cells. *Science* **251**: 1600–1602.
Chan, P.Y. and Aruffo, A. (1993). VLA-4 integrin mediates lymphocyte migration on the inducible endothelial cell ligand VCAM-1 and the extracellular matrix ligand fibronectin. *J Biol Chem* **268**: 24655–24664.
Chen, J.D., Kim, J.P., Zhang, K., Sarret, Y., Wynn, K.C., Kramer, R.H. and Woodley, D.T. (1993). Epidermal growth factor (EGF) promotes human keratinocyte locomotion on collagen by increasing the α_2 integrin subunit. *Exp Cell Res* **209**: 216–223.
Chen, W.-T., Hasegawa, E., Hasegawa, T., Weinstock, C. and Yamada, K. M. (1985). Development of cell surface linkage complexes in cultured fibroblasts. *J Cell Biol* **100**: 1103–1114.
Chen, W. T., Olden, K., Bernard, B.A. and Chu, F.F. (1984). Expression of transformation-associated protease(s) that degrade fibronectin at cell contact sites. *J Cell Biol* **98**: 1546–1555.
Chiquet-Ehrismann, R., Kalla, P., Pearson, C.A., Beck, K. and Chiquet, M. (1988). Tenascin interferes with fibronectin action. *Cell* **53**: 383–390.
Ciambrone, G.J. and McKeown-Longo, P.J. (1990). Plasminogen activator inhibitor type I stabilizes vitronectin-dependent adhesions in HT-1080 cells. *J Cell Biol* **111**: 2183–2195.
Clark, R.A., Folkvord, J.M., Hart, C.E., Murray, M.J. and McPherson, J.M. (1989). Platelet isoforms of platelet-derived growth factor stimulate fibroblasts to contract collagen matrices. *J Clin Invest* **84**: 1036–1040.
Collo, G., Starr, L. and Quaranta, V. (1993). A new isoform of the laminin receptor $\alpha 7\beta 1$ is developmentally regulated in skeletal muscle. *J. Biol. Chem.* **268**: 19019–19024.
Couchman, J.R., Höök, M., Rees, D.A. and Timpl, R. (1983). Adhesion, growth, and matrix production by fibroblasts on laminin substrates. *J Cell Biol* **96**: 177–183.

Curtis, A.S.G. (1966). Cell adhesion. *Sci. Prog., Oxf.* **54**: 61–86.
Damjanovich, L., Albelda, S.M., Mette, S.A. and Buck, C.A. (1992). Distribution of integrin cell adhesion receptors in normal and malignant lung tissue. *Am J Respir Cell Mol Biol* **6**: 197–206.
Damsky, C.H., Fitzgerald, M.L. and Fisher, S.J. (1992a). Distribution patterns of extracellular matrix components and adhesion receptors are intricately modulated during first trimester cytotrophoblast differentiation along the invasive pathway, in vivo. *J Clin Invest* **89**: 210–222.
Damsky, C.H. and Werb, Z. (1992b). Signal transduction by integrin receptors for extracellular matrix: cooperative processing of extracellular information. *Curr Opin Cell Biol* **4**: 772–781.
Danilov, Y.N. and Juliano, R.L. (1989). Phorbol ester modulation of integrin-mediated cell adhesion: a post-receptor event. *J Cell Biol* **108**: 1925–1933.
Declerck, P.J., De Mol, M., Alessi, M.C., Baudner, S., Paques, E.P., Preissner, K.T., Muller- Berghaus, G. and Collen, D. (1988). Purification and characterization of a plasminogen activator inhibitor 1 binding protein from human plasma. Identification as a multimeric form of S protein (vitronectin). *J Biol Chem* **263**: 15454 –15461.
Defilippi, P., van Hinsbergh, V., Bertolotto, A., Rossino, P., Silengo, L. and Tarone, G. (1991). Differential distribution and modulation of expression of alpha1/beta1 integrin on human endothelial cells. *J Cell Biol* **114**: 855–863.
Drake, C.J., Davis, L.A., Hungerford, J.E. and Little, C.D. (1991). Perturbation of ß1 integrin-mediated adhesions results in altered somite cell shape and behavior. *Dev. Biol.* **149**: 327–338.
Du, X.P., Plow, E.F., Frelinger, A. 3., O'Toole, T. E., Loftus, J. C. and Ginsberg, M. H. (1991). Ligands "activate" integrin $\alpha_{IIb}\beta_3$ (platelet GPIIb-IIIa). *Cell* **65**: 409–416.
Duband, J.-L., Belkin, A.M., Syfrig, J., Thiery, J.P. and Koteliansky, V.E. (1992). Expression of α_1 integrin, a laminin-collagen receptor, during myogenesis and neurogenesis in the avian embryo. *Development* **116**: 585–600.
Dustin, M.L. and Springer, T.A. (1989). T-cell receptor cross-linking transiently stimulates adhesiveness through LFA-1. *Nature* **341**: 619–624.
Eble, J.A., Golbik, R., Mann, K. and Kuhn, K. (1993). The $\alpha_1\beta_1$ integrin recognition site of the basement membrane collagen molecule [α1(IV)]2 α2(IV). *EMBO J* **12**: 4795–4802.
Edelman. G.M. and Crossin, K.L. (1991) Cell adhesion molecules: Implications for a molecular histology. *Annu Rev Biochem* **60**: 155–190
Ehrlich, H.P., Rockwell, W.B., Cornwell, T.L. and Rajaratnam, J.B. (1991). Demonstration of a direct role for myosin light chain kinase in fibroblast-populated collagen lattice contraction. *J Cell Physiol* **146**: 1–7.
Ehrlich, H.P. and Wyler, D.J. (1983). Fibroblast contraction of collagen lattices in vitro: inhibition by chronic inflammatory cell mediators. *J Cell Physiol* **116**: 345–351.
Ekblom, P. (1993). Basement membranes in development. In: *Molecular and Cellular Aspects of Basement Membrane* DH Rohrbach and R Timpl (eds) pp 359–383. San Diego: Academic Press Inc.
Elices, M.J. and Hemler, M.E. (1989). The human integrin VLA-2 is a collagen receptor on some cells and a collagen/laminin receptor on others. *Proc Natl Acad Sci USA* **86**: 9906–9910.
Elices, M.J., Urry, L.A. and Hemler, M.E. (1991). Receptor functions for the integrin VLA-3: fibronectin, collagen, and laminin binding are differentially influenced by Arg-Gly-Asp peptide and by divalent cations. *J Cell Biol* **112**: 169–181.
Elliott, B.E., Ekblom, P., Pross, H., Niemann, A. and Rubin, K. (1994). Anti-β_1 integrin IgG inhibits pulmonary macrometastasis and the size of micrometastasis from a murine mammary carcinoma. *Cell Adhesion and Communication* **1**: 319–332.
Elsdale, T.R. and Bard, J.B.L. (1972). Collagen substrata for studies on cell behavior. *J Cell Biol* **54**: 626–637.
Enenstein, J., Waleh, N. S. and Kramer, R. H. (1992). Basic FGF and TGF-b differentially modulate integrin expression of human microvascular endothelial cells. *Exp Cell Res* **203**: 499–503.
Eriksson, A., Siegbahn, A., Westermark, B., Heldin, C.H. and Claesson-Welsh, L. (1992). PDGF α- and β-receptors activate unique and common signal transduction pathways. *EMBO J* **11**: 543–550.
Espersen, F. and Clemmensen, I. (1982). Isolation of a fibronectin-binding protein from *Staphylococcus aureus*. *Infect Immun* **37**: 526–531.
Faassen, A.E., Mooradian, D.L., Tranquillo, R.T., Dickinson, R.B., Letourneau, P.C., Oegema, T.R. and McCarthy, J. B. (1993). Cell surface CD44-related chondroitin sulfate proteoglycan is required for transforming growth factor-β-stimulated mouse melanoma cell motility and invasive behavior on type I collagen. *J Cell Sci* **105**: 501–511.
Fath, K.R., Edgell, C.J. and Burridge, K. (1989). The distribution of distinct integrins in focal contacts is determined by the substratum composition. *J Cell Sci* **92**: 67–75.
Faull, R.J., Kovach, N.L., Harlan, J.M. and Ginsberg, M.H. (1993). Affinity modulation of integrin $\alpha5\beta1$: regulation of the functional response by soluble fibronectin. *J Cell Biol* **121**: 155–162.
Felding-Habermann, B., Mueller, B.M., Romerdahl, C.A. and Cheresh, D.A. (1992). Involvement of integrin αV gene expression in human melanoma tumorigenicity. *J Clin Invest* **89**: 2018–2022.

Ferns, G.A., Sprugel, K. H., Seifert, R.A., Bowen-Pope, D.F., Kelly, J.D., Murray, M., Raines, E.W. and Ross, R. (1990). Relative platelet-derived growth factor receptor subunit expression determines cell migration to different dimeric forms of PDGF. *Growth Factors* **3**: 315–324.

Fessler, J.H. and Fessler, L.I. (1989). *Drosophila* extracellular matrix. *Annu. Rev. Cell Biol.* **5**: 309–339.

Fogerty, F.J., Fessler, L.I., Bunch, T.A., Yaron, Y., Parker, C.G., Nelson, R.E., Brower, D.L., Gullberg, D. and Fessler, J. H. (1994). Tiggrin, a novel *Drosophila* extracellular matrix protein that functions as a ligand for *Drosophila* $\alpha_{PS2}\beta_{PS3}$. *Development* **120**: 1747–1758.

Fogerty, F. J. and Mosher, D.F. (1990). Mechanisms for organization of fibronectin matrix. *Cell Differ Dev* **32**: 439–450.

Folkman, J. and Moscona, A. (1978). Role of cell shape in growth control. *Nature* **273**: 345–349.

Forsberg, E., Ek, B., Engström, Å. and Johansson, S. (1994) Purification and characterization of integrin $\alpha_9\beta_1$. *Exp Cell Res* **213**: 183–190

Freed, E., Gailit, J., van der Geer, P., Ruoslahti, E. and Hunter, T. (1989). A novel integrin β subunit is associated with the vitronectin receptor a subunit (αv) in a human osteosarcoma cell line and is a substrate for protein kinase C. *EMBO J* **8**: 2955–2965.

Frelinger, A. 3., Cohen, I., Plow, E.F., Smith, M.A., Roberts, J., Lam, S.C. and Ginsberg, M.H. (1990). Selective inhibition of integrin function by antibodies specific for ligand- occupied receptor conformers. *J Biol Chem* **265**: 6346–6352.

Fristrom, D., Wilcox, M. and Fristrom, J. (1993). The distribution of PS integrins, laminin A and F-actin during key stages in *Drosophila* wing development. *Development* **117**: 509–523.

Fässler, R., Pfaff, M., Murphy, J., Noegel, A. A., Johansson, S., Timpl, R. and Albrecht, R. (In Press). Lack of β1 integrin gene in embryonic stem cells affects morphology, adhesion and migration but not integration into the inner cell mass of blastocysts. *J Cell Biol* **128**: 979–988.

Gailit, J. and Ruoslahti, E. (1988). Regulation of the fibronectin receptor affinity by divalent cations. *J Biol Chem* **263**: 12927–12932.

Geiger, B. and Ayalon, O. (1992). Cadherins. *Annu Rev Cell Biol* **8**: 307–332.

Geiger, B., Salomon, D., Takeichi, M. and Hynes, R.O. (1992). A chimeric N-cadherin/β 1-integrin receptor which localizes to both cell- cell and cell-matrix adhesions. *J Cell Sci* **103**: 943–951.

George, E.L., Georges-Labouesse, E.N., Patel-King, R.S., Rayburn, H. and Hynes, R. O. (1993). Defects in meso-derm, neural tube and vascular development in mouse embryos lacking fibronectin. *Development* **119**: 1079–1091.

Giancotti, F. G. and Ruoslahti, E. (1990). Elevated levels of the $\alpha_5\beta_1$ fibronectin receptor suppress the transformed phenotype of Chinese hamster ovary cells. *Cell* **60**: 849–859.

Gillery, P., Coustry, F., Pujol, J.P. and Borel, J.P. (1989). Inhibition of collagen synthesis by interleukin-1 in three-dimensional collagen lattice cultures of fibroblasts. *Experientia* **45**: 98–101.

Ginsberg, M.H., Loftus, J., Ryckwaert, J.J., Pierschbacher, M., Pytela, R., Ruoslahti, E. and Plow, E. F. (1987). Immunochemical and amino-terminal sequence comparison of two cytoadhesins indicates they contain similar or identical b subunits and distinct a subunits. *J Biol Chem* **262**: 5437–5440.

Ginsberg, M.H., Loftus, J.C. and Plow, E.F. (1988). Cytoadhesins, integrins, and platelets. *Thromb Haemost* **59**: 1–6.

Gipson, I.K., Spurr-Michaud, S., Tisdale, A., Elwell, J. and Stepp, M.A. (1993). Redistribution of the hemi-desmosome components $\alpha_6\beta_4$ integrin and bullous pemphigoid antigens during epithelial wound healing. *Exp Cell Res* **207**: 86–98.

Gladson, C.L. and Cheresh, D.A. (1991). Glioblastoma expression of vitronectin and the αvβ3 integrin. Adhesion mechanism for transformed glial cells. *J Clin Invest* **88**: 1924–1932.

Gotwals, P.J., Fessler, L.I., Wherli, M. and Hynes, R.O. (1994). *Drosophila* PS1 integrin is a laminin receptor and differs in ligand specificity from PS2. *Proc Natl Acad Sci (USA)* **91**: 11447–11451

Gotwals, P.J., Paine-Saunders, S.E., Stark, K.A. and Hynes, R.O. (1994). *Drosophila* integrins and their ligands. *Curr Op Cell Biol* **6**: 734–739.

Goward, C.R., Scawen, M.D., Murphy, J.P. and Atkinson, T. (1993). Molecular evolution of bacterial cell-surface proteins. *TIBS* **18**: 136–140.

Greve, H., Blumberg, P., Schmidt, G., Schlumberger, W., Rauterberg, J. and Kresse, H. (1990). Influence of colla-gen lattice on the metabolism of small proteoglycan II by cultured fibroblasts. *Biochem J* **269**: 149–155.

Greve, J.M. and Gottlieb, D.I. (1982). Monoclonal antibodies which alter the morphology of cultured chick myogenic cells. *J Cell Biochem* **18**: 221–229.

Grinnell, F. (1994). Fibroblasts, myofibroblasts, and wound contraction. *J Cell Biol* **124**: 401–404.

Grinnell, F., Hays, D.G. and Minter, D. (1977). Cell adhesion and spreading factor: partial purification and prop-erties. *Exp Cell Res* **110**: 175–190.

Grinnell, F. and Lamke, C. R. (1984). Reorganization of hydrated collagen lattices by human skin fibroblasts. *J Cell Sci* **66**: 51–63.

Grinnell, F. and Minter, D. (1978). Attachment and spreading of baby hamster kidney cells to collagen substrata: Effects of cold-insoluble globulin. *Proc Natl Acad Sci (USA)* **75**: 4409–4412.

Guan, J. L. and Shalloway, D. (1992). Regulation of focal adhesion-associated protein tyrosine kinase by both cellular adhesion and oncogenic transformation. *Nature* **358**: 690–692.

Guidry, C. (1993). Fibroblast contraction of collagen gels requires activation of protein kinase C. *J Cell Physiol* **155**: 358–367.

Guidry, C. and Höök, M. (1991). Endothelins produced by endothelial cells promote collagen gel contraction by fibroblasts. *J Cell Biol* **115**: 873–880.

Gullberg, D., Fessler, L.I. and Fessler, J.H. (1994). Differentiation, extracellular matrix synthesis, and integrin assembly by *Drosophila* embryo cells cultured on vitronectin and laminin substrates. *Dev. Dynamics* **199**: 116–128.

Gullberg, D., Gehlsen, K.R., Turner, D.C., Åhlén, K., Zijenah, L.S., Barnes, M. J. and Rubin, K. (1992). Analysis of $\alpha_1\beta_1$, $\alpha_2\beta_1$ and $\alpha_3\beta_1$ integrins in cell -collagen interactions: identification of conformation dependent $\alpha_1\beta_1$ binding sites in collagen type I. *EMBO J* **11**: 3865–3873.

Gullberg, D., Tingström, A., Thuresson, A.C., Olsson, L., Terracio, L., Borg, T.K. and Rubin, K. (1990). β_1 integrin-mediated collagen gel contraction is stimulated by PDGF. *Exp Cell Res* **186**: 264–272.

Gunthert, U., Hofman, M., Rudy, W., Reber, S., Zöller, M., Haussann, I., Matzku, S., Wenzel, A., Ponta, H. and Herrlich, P. (1991). A new variant of glycoprotein CD44 confers metastatic poential to rat carcinoma cells. *Cell* **65**: 13–24.

Hanks, S.K., Calalb, M.B., Harper, M.C. and Patel, S.K. (1992). Focal adhesion protein-tyrosine kinase phosphorylated in response to cell attachment to fibronectin. *Proc Natl Acad Sci U S A* **89**: 8487–8491.

Hausman, R.E. and Moscona, A.A. (1975). Purification and characterization of the retina-specific cell-aggregating factor. *Proc Natl Acad Sci (USA)* **72**: 916–920.

Haynes, B.F., Telen, M.J., Hale, L.P. and Denning, S.M. (1989). CD44: A molecule involved in leukocyte adherence and T-cell activation. *Immunol Today* **10**: 432–428.

Heino, J., Ignotz, R.A., Hemler, M.E., Crouse, C. and Massague, J. (1989). Regulation of cell adhesion receptors by transforming growth factor-b. Concomitant regulation of integrins that share a common β_1 subunit. *J Biol Chem* **264**: 380–388.

Heino, J. and Massague, J. (1989). Transforming growth factor-β switches the pattern of integrins expressed in MG-63 human osteosarcoma cells and causes a selective loss of cell adhesion to laminin. *J Biol Chem* **264**: 21806–21811.

Hemler, M.E. (1990). VLA proteins in the integrin family: structures, functions, and their role on leukocytes. *Annu Rev Immunol* **8**: 365–400.

Hemler, M.E., Jacobson, J.G., Brenner, M.B., Mann, D. and Strominger, J. L. (1985). VLA-1: a T cell surface antigen which defines a novel late stage of human T cell activation. *Eur J Immunol* **15**: 502–508.

Henchcliffe, C., Garcia-Alonso, L., Tang, J. and Goodman, C.S. (1993). Genetic analysis of laminin A reveals diverse functions during morphogenesis in *Drosophila*. *Development* **118**: 325–337.

Herman, B. and Pledger, W.J. (1985). Platelet-derived growth factor-induced alterations in vinculin and actin distribution in BALB/c-3T3 cells. *J Cell Biol* **100**: 1031–1040.

Hirano, S., Ui, K., Miyake, T., Uemura, T. and Takaeichi, M. (1991). *Drosophila* PS integrins recognize vertebrate vitronectin and function as cell-substratum adhesion receptors *in vitro*. *Development* **113**: 1007–1116.

Hirsch, E., Gullberg, D., Balzac, F., Altruda, F., Silengo, L. and Tarone, G. (1994). α_v integrin subunit is predominantly located in nervous tissue and skeletal muscle during mouse development. *Dev. Dynamics* **201**: 108–120.

Hoffman, E.P., Brown, R.H. and Kunkel, L.M. (1987). Dystrophin: the protein product of Duchenne muscular dystrophy locus. *Cell* **51**: 919–928.

Holderbaum, D., Hale, G.S. and Erhart, L.A. (1986). Collagen binding to *Staphylococcus aureus*. *Infect Immun* **54**: 359–364.

Horwitz, A., Duggan, K., Buck, C., Beckerle, M.C. and Burridge, K. (1986) Interaction of plasma membrane receptor with talin — a transmembrane linkage. *Nature* **320**: 531–533.

Hormia, M., Virtanen, I. and Quaranta, V. (1992). Immunolocalization of integrin $\alpha6\beta4$ in mouse junctional epithelium suggests an anchoring function to both the internal and the external basal lamina. *J Dent Res* **71**: 1503–1508.

Hoschutzky, H., Lottspeich, F. and Jann, K. (1989). Isolation and characterization of the α-galactosyl-1,4-β-galactosyl-specific adhesin (P adhesin) from fimbriated *Escherichia coli*. *Infect Immun* **57**: 76–81.

Hresko, M.C., Williams, B.D. and Waterston, R.H. (1994). Assembly of body wall muscle and muscle cell attachment structures in Caenorhabditis elegans. *J. Cell Biol.* **124**: 491–506.

Humphries, M.J., Mould, A.P. and Tuckwell, D.S. (1993). Dynamic aspects of adhesion receptor function-integrins both twist and shout. *Bioessays* **15**: 391–397.

Humphries, M.J., Yamada, K.M. and Olden, K. (1988). Investigation of the biological effects of anti-cell adhesive synthetic peptides that inhibit experimental metastasis of B16-F10 murine melanoma cells. *J Clin Invest* **81**: 782–790.

Hunt, R.C., Pakalnis, V.A., Choudhury, P. and Black, E.P. (1994). Cytokines and serum cause $\alpha_2\beta_1$ integrin-mediated contraction of collagen gels by cultured retinal pigment epithelial cells. *Invest Ophthalmol Vis Sci* **35**: 955–963.

Hynes, R.O. (1987). Integrins: a family of cell surface receptors. *Cell* **48**: 549–554.

Hynes, R.O. (1992). Integrins: versatility, modulation, and signaling in cell adhesion. *Cell* **69**: 11–25.

Hynes, R.O. and Yamada, K.M. (1982). Fibronectins: multifunctional modular glycoproteins. *J Cell Biol* **95**: 369 –377.

Höök, M., Rubin, K., Oldberg, Å., Öbrink, B. and Vaheri, A. (1977). Cold-insoluble globulin mediates the adhesion of rat liver cells to plastic petri dishes. *Biochem Biophys Res Commun* **79**: 726–733.

Ignotz, R.A., Heino, J. and Massague, J. (1989). Regulation of cell adhesion receptors by transforming growth factor-β. Regulation of vitronectin receptor and LFA-1. *J Biol Chem* **264**: 389–392.

Ignotz, R.A. and Massague, J. (1987). Cell adhesion protein receptors as targets for transforming growth factor-β action. *Cell* **51**: 189–197.

Ingber, D. (1991). Integrins as mechanochemical transducers. *Curr Opin Cell Biol* **3**: 841–848.

Isemura, M., Kazama, T., Takahashi, K. and Yamaguchi, Y. (1993). Immunochemical localization of integrin subunits in the human placenta. *Tohoku J Exp Med* **171**: 167–183.

Izzard, C.S. and Lochner, L.R. (1976). Cell-to-substrate contacts in living fibroblasts: an interference-reflexion study with an evaluation of the technique. *J Cell Sci* **21**: 129–159.

Jackson, D.G., Buckley, J. and Bell, J.I. (1992). Multiple variants of the human lymphocyte homing receptor CD44 generated by insertions at a single site in the extracellular domain. *J Biol Chem* **267**: 4732–4739.

Jaffredo, T., Horwitz, A.F., Buck, C.A., Rong, P.M. and Dieteren-Lievre, F. (1988). Myoblast migration specifically inhibited in the chick embryo by grafted CSAT hybridoma cells secreting an anti-integrin antibody. *Development* **103**: 431–446.

Jaken, S., Leach, K. and Klauck, T. (1989). Association of type 3 protein kinase C with focal contacts in rat embryo fibroblasts. *J Cell Biol* **109**: 697–704.

Jalkanen, M., Elenius, K. and Salmivirta, M. (1992). Syndecan—a cell surface proteoglycan that selectively binds extracellular effector molecules. *Adv Exp Med Biol* **313**: 79–85.

Janat, M.F., Argraves, W.S. and Liau, G. (1992). Regulation of vascular smooth muscle cell integrin expression by transforming growth factor $\beta 1$ and by platelet-derived growth factor-BB. *J Cell Physiol* **151**: 588–595.

Johansson, M.W., Larsson, E., Luning, B., Pasquale, E.B. and Ruoslahti, E. (1994). Altered localization and cytoplasmic domain-binding properties of tyrosine- phosphorylated β_1 integrin. *J Cell Biol* **126**: 1299–1309.

Johansson, S., Kjellén, L., Höök, M. and Timpl, R. (1981). Substrate adhesion of rat hepatocytes: a comparison of laminin and fibronectin as attachment proteins. *J Cell Biol* **90**: 260–264.

Johnson, B.A., Haines, G.K., Harlow, L.A. and Koch, A.E. (1993). Adhesion molecule expression in human synovial tissue. *Arthritis Rheum* **36**: 137–146.

Jones, J.C., Kurpakus, M.A., Cooper, H.M. and Quaranta, V. (1991). A function for the integrin $\alpha_6\beta_4$ in the hemidesmosome. *Cell Regul* **2**: 427–438.

Jones, J.C.R., Asmuth, J., Baker, S.E., Langhofer, M., Roth, S.I. and Hpkinson, S.B. (1994). Hemidesmosomes: extracellular matrix/intermediate filament connectors. *Exp Cell Res* **213**: 1–11.

Jonsson, P., Lindberg, M., Haraldsson, I. and Wadström, T. (1985). Virulence of *Staphylococcus aureus* in a mouse mastitis model: studies of α-hemolysin, coagulase and protein A as possible virulence determinants with protoplast fusion and gene cloning. *Infect Immun* **49**: 765–769.

Juliano, R.L. and Haskill, S. (1993). Signal transduction from the extracellular matrix. *J Cell Biol* **120**: 577–585.

Jönsson, K., Signäs, C., Muller, H.-P. and Lindberg, M. (1991). Two different genes encode fibronectin binding proteins in *Stahylococcus aureus*: The complete nucleotide sequence and characterization of the second gene. *Eur J Biochem* **202**: 1041–1048.

Kadoya, Y., Holmvall, K., Sorokin, L. and Ekblom, P. (1995). Distinct roles of domains E8 and E3 of laminin-1 during salivary branching morphogenesis. *J. Cell Biol.* **129**: 521–534.

Kelly, D.E. (1966). Fine structure of desmosomes, hemidesmosomes, and an adepidermal globular layer in developing newt epidermis. *J Cell Biol* **28**: 51–72.

Kennel, S.J., Godfrey, V., Ch'ang, L.Y., Lankford, T.K., Foote, L. J. and Makkinje, A. (1992). The β_4 subunit of the integrin family is displayed on a restricted subset of endothelium in mice. *J Cell Sci* **101**: 145–150.

Kern, A., Eble, J., Golbik, R. and Kuhn, K. (1993). Interaction of type IV collagen with the isolated integrins $\alpha_1\beta_1$ and $\alpha_2\beta_1$. *Eur J Biochem* **215**: 151–159.

Kieffer, N. and Phillips, D.R. (1990). Platelet membrane glycoproteins: functions in cellular interactions. *Annu Rev Cell Biol* **6**: 329–357.

Kishimoto, T.K., O'Connor, K., Lee, A., Roberts, T.M. and Springer, T.A. (1987) Cloning of the beta subunit of the leukocyte adhesion proteins: homology to an extracellular matrix receptor defines a novel supergene family. *Cell* **48:** 681–690.

Klein, C.E., Dressel, D., Steinmayer, T., Mauch, C., Eckes, B., Krieg, T., Bankert, R.B. and Weber, L. (1991). Integrin $\alpha_2\beta_1$ is upregulated in fibroblasts and highly aggressive melanoma cells in three-dimensional collagen lattices and mediates the reorganization of collagen I fibrils. *J Cell Biol* **115**: 1427–1436.

Klein, S., Giancotti, F.G., Presta, M., Albelda, S.M., Buck, C. A. and Rifkin, D.B. (1993). Basic fibroblast growth factor modulates integrin expression in microvascular endothelial cells. *Mol Biol Cell* **4**: 973–982.

Knudsen, K.A., Horwitz, A.F. and Buck, C.A. (1985). A monoclonal antibody identifies a glycoprotein complex involved in cell-substratum adhesion. *Exp Cell Res* **157**: 218–226.

Koretz, K., Schlag, P., Boumsell, L. and Moller, P. (1991). Expression of VLA-α2, VLA- α6, and VLA-β1 chains in normal mucosa and adenomas of the colon, and in colon carcinomas and their liver metastases. *Am J Pathol* **138**: 741–750.

Korhonen, M., Laitinen, L., Ylanne, J., Gould, V. E. and Virtanen, I. (1992). Integrins in developing, normal and malignant human kidney. *Kidney Int* **41**: 641–644.

Korhonen, M., Laitinen, L., Ylanne, J., Koukoulis, G. K., Quaranta, V., Juusela, H., Gould, V.E. and Virtanen, I. (1992). Integrin distributions in renal cell carcinomas of various grades of malignancy. *Am J Pathol* **141**: 1161–1171.

Korhonen, T.K., Väisenen-Rehn, V., Rehn, M., Pere, A., Parkkinen, J. and Finne, J. (1984). *Escherichia coli* fimbriae recognizing sialyl galactosides. *J Bacteriol* **159**: 762–766.

Kornberg, L., Earp, H. S., Parsons, J. T., Schaller, M. and Juliano, R. L. (1992). Cell adhesion or integrin clustering increases phosphorylation of a focal adhesion-associated tyrosine kinase. *J Biol Chem* **267**: 23439–23442.

Koukoulis, G.K., Howeedy, A.A., Korhonen, M., Virtanen, I. and Gould, V.E. (1993). Distribution of tenascin, cellular fibronectins and integrins in the normal, hyperplastic and neoplastic breast. *J Submicrosc Cytol Pathol* **25**: 285–295.

Koukoulis, G.K., Virtanen, I., Korhonen, M., Laitinen, L., Quaranta, V. and Gould, V.E. (1991). Immunohistochemical localization of integrins in the normal, hyperplastic, and neoplastic breast. Correlations with their functions as receptors and cell adhesion molecules. *Am J Pathol* **139**: 787–799.

Koukoulis, G.K., Virtanen, I., Moll, R., Quaranta, V. and Gould, V. E. (1993). Immunolocalization of integrins in the normal and neoplastic colonic epithelium. *Virchows Arch B Cell Pathol* **63**: 373–383.

Kovach, N.L., Carlos, T.M., Yee, E. and Harlan, J.M. (1992). A monoclonal antibody to β_1 integrin (CD29) stimulates VLA-dependent adherence of leukocytes to human umbilical vein endothelial cells and matrix components. *J Cell Biol* **116**: 499–509.

Kramer, R.H. and Marks, N. (1989). Identification of integrin collagen receptors on human melanoma cells. *J Biol Chem* **264**: 4684–4688.

Kunicki, T.J., Nugent, D.J., Staats, S.J., Orchekowski, R.P., Wayner, E.A. and Carter, W.G. (1988). The human fibroblast class II extracellular matrix receptor mediates platelet adhesion to collagen and is identical to the platelet glycoprotein Ia-IIa complex. *J Biol Chem* **263**: 4516–4519.

Kusche-Gullberg, M., Garrison, K., MacKrell, A.J., Fessler, L.I. and Fessler, J.H. (1992). Laminin A chain: expression during *Drosophila* development and genomic sequence. *EMBO J.* **11**: 4519–4527.

Kuusela, P. (1978). Fibronectin binds to *Staphylococcus aureus*. *Nature* **246**: 718–720.

Kuusela, P., Vartio, T., Vuento, M. and Myhre, E.B. (1984). Binding sites for streptococci and staphylococci in fibronectin. *Infect Immun* **45**: 433–436.

Kuypers, J.M. and Proctor, R.A. (1989). Reduced adherence to traumatized rat heart valves by a low-fibronectin-binding mutant of *Staphylococcus aureus*. *Infect Immun* **57**: 2306–2312.

LaFlamme, S.E., Akiyama, S.K. and Yamada, K.M. (1992). Regulation of fibronectin receptor distribution. *J Cell Biol* **117**: 437–447.

Languino, L.R., Gehlsen, K.R., Wayner, E., Carter, W.G., Engvall, E. and Ruoslahti, E. (1989). Endothelial cells use $\alpha_2\beta_1$ integrin as a laminin receptor. *J Cell Biol* **109**: 2455–2462.

Languino, L.R. and Ruoslahti, E. (1992). An alternative form of the integrin β_1 subunit with a variant cytoplasmic domain. *J Biol Chem* **267**: 7116–7120.

Larjava, H., Peltonen, J., Akiyama, S.K., Yamada, S.S., Gralnick, H.R., Uitto, J. and Yamada, K.M. (1990). Novel function for β_1 integrins in keratinocyte cell-cell interactions. *J Cell Biol* **110**: 803–815.

Leak, L.V. (1970). Electron microscopic observations on lymphatic capillaries and the structural components of the tissue-lymph interface. *Microvasc Res* **2**: 361–370.

Leavesley, D.I., Schwartz, M.A., Rosenfeld, M. and Cheresh, D.A. (1993). Integrin β1- and β3-mediated endothelial cell migration is triggered through distinct signaling mechanisms. *J Cell Biol* **121**: 163–170.

Leffler, H. and Svanborg-Edén, C. (1981). Glycolipid receptors for uropathogenic *Escherichia coli* on human erythrocytes and uroepithelial cells. *Infect Immun* **34**: 920–929.

Lenter, M. and Vestweber, D. (1994). The integrin chains β_1 and α_6 associate with the chaperone calnexin prior to integrin assembly. *J Biol Chem* **269**: 12263–12268.

Leptin, M., Grunewald, B. and Stein, D. (1992). No effect of RGDS peptides. *Nature* **355**: 777.

Liebert, M., Wedemeyer, G., Stein, J.A., Washington, R., Jr., Van Waes, C., Carey, T.E. and Grossman, H.B. (1993). The monoclonal antibody BQ16 identifies the $\alpha_6\beta_4$ integrin on bladder cancer. *Hybridoma* **12**: 67–80.

Lin, Y.C. and Grinnell, F. (1993). Decreased level of PDGF-stimulated receptor autophosphorylation by fibroblasts in mechanically relaxed collagen matrices. *J Cell Biol* **122**: 663–672.

Lindgren, P.-E., Speziale, P., McGavin, M., Monstein, H.-J., Höök, M., Visai, L., Kostiainen, T., Bozzini, S. and Lindberg, M. (1992). Cloning and expression of two different genes from *Streptococcus dysgalactiae* encoding fibronectin receptors. *J Biol Chem* **267**: 1924–1931.

Lindmark, G., Gerdin, B., Påhlman, L., Glimelius, B., Gehlsen, K. and Rubin, K. (1993). Interconnection of integrins α_2 and α_3 and structure of the basal membrane in colorectal cancer: relation to survival. *Eur J Surg Oncol* **19**: 50–60.

Loeber, C.P. and Runyan, R.B. (1990). A comparison of fibronectin, laminin and galactosyltransferase adhesion mechanisms during embryonic cardiac mesenchymal cell migration in vitro. *Developmental Biology* **140**: 410–412.

Loftus, J.C., Smith, J.W. and Ginsberg, M.H. (1994). Integrin-mediated cell adhesion: the extracellular face. *J Biol Chem* **269**: 25235–25238.

Lollo, B.A., Chan, K.W.H., Hanson, E.M., Moy, V.T. and Brian, A.A. (1993). Direct evidence for two affinity states for lymphocyte function-associated antigen 1 on activated T cells. *J Biol Chem* **268**: 21693–21700.

Lund, T., Onarheim, H., Wiig, H. and Reed, R.K. (1989). Mechanisms behind increased dermal inhibition pressure in acute burn edema. *Am J Physiol (Heart Circ Physiol)* **256**: H940–H948.

MacKrell, A.J., Blumberg, B., Haynes, S.R. and Fessler, J.H. (1988). The lethal myospheroid gene of *Drosophila* encodes a membrane protein homologous to vertebrate integrin β subunits. *Proc. Natl. Acad. Sci. USA* **85**: 2633–2637.

Masumoto, A. and Hemler, M.E. (1993). Multiple activation states of VLA-4. Mechanistic differences between adhesion to CS1/fibronectin and to vascular cell adhesion molecule-1. *J Biol Chem* **268**: 228–234.

Mauch, C., Hatamochi, A., Scharffetter, K. and Krieg, T. (1988). Regulation of collagen synthesis in fibroblasts within a three-dimensional collagen gel. *Exp Cell Res* **178**: 493–503.

Maxe, I., Rydén, C., Wadström, T. and Rubin, K. (1986). Specific attachment of *Staphylococcus aureus* to immobilized fibronectin. *Infect Immun* **54**: 695–704.

McNamee, H.P., Ingber, D.E. and Schwartz, M.A. (1993). Adhesion to fibronectin stimulates inositol lipid synthesis and enhances PDGF-induced inositol lipid breakdown. *J Cell Biol* **121**: 673–678.

McGavin Homonylo, M., Krajewska-Pietrasik, D., Rydén, C. and Höök, M. (1993). Identification of a Staphylococcus aureus extracellular matrix binding protein with broad specificity. *Infect Immun* **61**: 2479–2485.

Mecham, R.P. (1991). Receptors for laminin on mammalian cells. *FASEB J* **5**: 2538–2546.

Menko, S. and Boettiger, D. (1987). Occupation of the extracellular matrix receptor, integrin, is a control point for myogenic differentiation. *Cell* **51**: 51–57.

Meredith, J., Jr., Fazeli, B. and Schwartz, M.A. (1993). The extracellular matrix as a cell survival factor. *Mol Biol Cell* **4**: 953–961.

Mette, S.A., Pilewski, J., Buck, C.A. and Albelda, S.M. (1993). Distribution of integrin cell adhesion receptors on normal bronchial epithelial cells and lung cancer cells in vitro and in vivo. *Am J Respir Cell Mol Biol* **8**: 562 –572.

Meyer, F.A. (1983). Macromolecular basis of globular protein exclusion and of swelling pressure in loose connective tissue (umbilical cord). *Biochem Biophys Acta* **755**: 388–399.

Michishita, M., Videm, V. and Arnaout, M.A. (1993). A novel divalent cation-binding site in the A domain of the β_2 integrin CR3 (CD11b/CD18) is essential for ligand binding. *Cell* **72**: 857–867.

Miller, J. B. (1992). Myoblast diversity in skeletal myogenesis: how much and to what end. *Cell* **69**: 1–3.

Mimuro, J. and Loskutoff, D.J. (1989). Purification of a protein from bovine plasma that binds to type 1 plasminogen activator inhibitor and prevents its interaction with extracellular matrix. Evidence that the protein is vitronectin. *J Biol Chem* **264**: 936–939.

Mollenhauer, J. and von der Mark, K. (1983). Isolation and characterization of a collagen-binding glycoprotein from chondrocyte membranes. *EMBO J* **2**: 45–50.

Montesano, R. and Orci, L. (1988). Transforming growth factor β stimulates collagen-matrix contraction by fibroblasts: implications for wound healing. *Proc Natl Acad Sci U S A* **85**: 4894–4897.

Montgomery, A.M., Reisfeld, R.A. and Cheresh, D.A. (1994). Integrin $\alpha_v\beta_3$ rescues melanoma cells from apoptosis in three- dimensional dermal collagen. *Proc Natl Acad Sci USA* **91**: 8856–8860.

Mosher, D.F. and Proctor, R.A. (1980). Binding and factor XIIIa-mediated crosslinking of a 27-kilodalton fragment of fibronectin to *Staphylococcus aureus*. *Science* **209**: 927–929.

Muller, K. and Gerisch, G. (1978). A specific glycoprotein as the target site of adhesion blocking Fab in aggregating *Dictyostelium* cells. *Nature,* **274**: 445–449.

Murphy-Ullrich, J. E., Lightner, V. A., Aukhil, I., Yan, Y. Z., Erickson, H. P. and Höök, M. (1991). Focal adhesion integrity is downregulated by the alternatively spliced domain of human tenascin. *J Cell Biol* **115**: 1127–1136.

Muschler, J.L. and Horwitz, A.F. (1991). Down-regulation of the chicken $\alpha_5\beta_1$ integrin fibronectin receptor during development. *Development* **113**: 327–337.

Naidet, C., Semeriva, M., Yamada, K.M. and Thiery, J.P. (1987). Peptides containing the cell-attachment recognition sequence Arg-Gly-Asp prevent gastrulation in *Drosophila* embryos. *Nature* **325**: 348–350.

Nakagawa, S., Pawelek, P. and Grinnell, F. (1989). Extracellular matrix organization modulates fibroblast growth and growth factor responsiveness. *Exp Cell Res* **182**: 572–582.

Natali, P.G., Nicotra, M.R., Bartolazzi, A., Cavaliere, R. and Bigotti, A. (1993). Integrin expression in cutaneous malignant melanoma: association of the α_3/β_1 heterodimer with tumor progression. *Int J Cancer* **54**: 68–72.

Natali, P.G., Nicotra, M.R., Botti, C., Mottolese, M., Bigotti, A. and Segatto, O. (1992). Changes in expression of α_6/β_4 integrin heterodimer in primary and metastatic breast cancer. *Br J Cancer* **66**: 318–322.

Neff, N.T., Lowrey, C., Decker, C., Tovar, A., Damsky, C., Buck, C. and Horwitz, A.F. (1982). A monoclonal antibody detaches embryonic skeletal muscle from extracellular matrices. *J Cell Biol* **95**: 654–666.

Nelson, L., Flock, J.-I., Höök, M., Lindberg, M., Muller, H.P. and Wadström, T. (1991). Adhesins in staphylococcal mastitis as vaccine components. In *New Insights into the Pathogenesis of Mastitis*. Burvenich, Vandeputti, VanMessam, Hill Flem, Editors. *Vet J Suppl* **62**: 111–125.

Nelson, R.E., Fessler, L.I., Tagaki, Y., Blumberg, B., Kenne, D.R., Olson, P.F., Parker, C.G. and Fessler, J. H. (1994). Peroxidasin: a novel enzyme-matrix protein of *Drosophila* development. *EMBO J* **13**: 3438–3447.

Nermut, M.V., Green, N.M., Eason, P., Yamada, S.S. and Yamada, K.M. (1988). Electron microscopy and structural model of human fibronectin receptor. *EMBO J* **7**: 4093–4099.

Neugebauer, K.M. and Reichardt, L. F. (1991). Cell-surface regulation of β_1-integrin activity on developing retinal neurons. *Nature* **350**: 68–71.

Nip, J., Shibata, H., Loskutoff, D.J., Cheresh, D.A. and Brodt, P. (1992). Human melanoma cells derived from lymphatic metastases use integrin $\alpha_v\beta_3$ to adhere to lymph node vitronectin. *J Clin Invest* **90**: 1406–1413.

Nusgens, B., Merrill, C., Lapiere, C. and Bell, E. (1984). Collagen biosynthesis by cells in a tissue equivalent matrix *in vitro*. *Coll Relat Res* **4**: 351–363.

O'Toole, T.E., Katagiri, Y., Faull, R.J., Peter, K., Tamura, R., Quaranta, V., Loftus, J.C., Shattil, S.J. and Ginsberg, M.H. (1994). Integrin cytoplasmic domains mediate inside-out signal transduction. *J Cell Biol* **124**: 1047–1059.

O'Toole, T.E., Mandelman, D., Forsyth, J., Shattil, S.J., Plow, E.F. and Ginsberg, M.H. (1991). Modulation of the affinity of integrin $\alpha_{IIb}\beta_3$ (GPIIb-IIIa) by the cytoplasmic domain of α_{IIb}. *Science* **254**: 845–847.

Oldberg, A., Franzen, A. and Heinegård, D. (1986). Cloning and sequence analysis of rat bone sialoprote in (osteopontin) cDNA reveals an Arg-Gly-Asp cell-binding sequence. *Proc Natl Acad Sci USA* **83**: 8819– 8823.

Oldberg, A., Franzén, A., Heinegård, D., Pierschbacher, M. and Ruoslahti, E. (1988). Identification of a bone sialoprotein receptor in osteosarcoma cells. *J Biol Chem* **263**: 19433–19436.

Otey, C.A., Pavalko, F.M. and Burridge, K. (1990). An interaction between α-actinin and the β_1 integrin subunit in vitro. *J Cell Biol* **111**: 721–729.

Palmer, E.L., Ruegg, C., Ferrando, R., Pytela, R. and Sheppard, D. (1993). Sequence and tissue distribution of the integrin $\alpha 9$ subunit, a novel partner of $\beta 1$ that is widely distributed in epithelia and muscle. *J Cell Biol* **123**: 1289–1297.

Patti, J.M., Bremell, T., Krajewska-Pietrasik, D., Abdelnour, A., Tarkowski, A., Rydén, C. and Höök, M. (1994). The *Staphylococcus aureus* collagen adhesin is a virulence determinant in experimental septic arthritis. *Infect Immun* **62**: 152–161.

Patti, J.M., Jonsson, H., Guss, B., Switalski, L.M., Widberg, K., Lindberg, M. and Höök, M. (1992). Molecular characterization and expression of a gene encoding a *Staphylococcus aureus* collagen adhesin. *J Biol Chem* **267**: 4766–4772.

Paulus, W., Baur, I., Schuppan, D. and Roggendorf, W. (1993). Characterization of integrin receptors in normal and neoplastic human brain. *Am J Pathol* **143**: 154–163.

Pearlstein, E. (1976). Plasma membrane glycoprotein which mediates adhesion of fibroblasts to collagen. *Nature* **262**: 497–500.

Pfaff, M., Aumailley, M., Specks, U., Knolle, J., Zerwes, H.G. and Timpl, R. (1993). Integrin and Arg-Gly-Asp dependence of cell adhesion to the native and unfolded triple helix of collagen type VI. *Exp Cell Res* **206**: 167–176.

Pfäffle, M., Ruggiero, F., Hofmann, H., Fernández, M.P., Selmin, O., Yamada, Y., Garrone, R. and von der Mark, K. (1988). Biosynthesis, secretion and extracellular localization of anchorin CII, a collagen-binding protein of the calpactin family. *EMBO J* **7**: 2335–2342.

Pierschbacher, M., Hayman, E.G. and Ruoslahti, E. (1983). Synthetic peptide with cell attachment activity of fibronectin. *Proc Natl Acad Sci U S A* **80**: 1224–1227.

Pierschbacher, M.D. and Ruoslahti, E. (1984). Cell attachment activity of fibronectin can be duplicated by small synthetic fragments of the molecule. *Nature* **309**: 30–33.

Pignatelli, M., Cardillo, M.R., Hanby, A. and Stamp, G.W. (1992). Integrins and their accessory adhesion molecules in mammary carcinomas: loss of polarization in poorly differentiated tumors. *Hum Pathol* **23**: 1159–1166.

Pignatelli, M., Hanby, A.M. and Stamp, G.W. (1991). Low expression of β_1, α_2 and α_3 subunits of VLA integrins in malignant mammary tumours. *J Pathol* **165**: 25–32.
Pignatelli, M., Liu, D., Nasim, M.M., Stamp, G.W., Hirano, S. and Takeichi, M. (1992). Morphoregulatory activities of E-cadherin and beta-1 integrins in colorectal tumour cells. *Br J Cancer* **66**: 629–634.
Plantefaber, L.C. and Hynes, R.O. (1989). Changes in integrin receptors on oncogenically transformed cells. *Cell* **56**: 281–290.
Prieto, A.L., Edelman, G.M. and Crossin, K.L. (1993). Multiple integrins mediate cell attachment to cytotactin/tenascin. *Proc Natl Acad Sci USA* **90**: 10154–10158.
Proctor, R.A., Mosher, D.F. and Olbrantz, P.J. (1982). Fibronectin binding to *Staphylococcus aureus*. *J Biol Chem* **257**: 14788–14794.
Pytela, R., Pierschbacher, M.D. and Ruoslahti, E. (1985). A 125/115-kDa cell surface receptor specific for vitronectin interacts with the arginine-glycine-aspartic acid adhesion sequence derived from fibronectin. *Proc Natl Acad Sci USA* **82**: 5766–5770.
Pytela, R., Pierschbacher, M.D. and Ruoslahti, E. (1985). Identification and isolation of a 140 kd cell surface glycoprotein with properties expected of a fibronectin receptor. *Cell* **40**: 191–198.
Qian, F., Vaux, D.L. and Weissman, I.L. (1994). Expression of the integrin a_4b_1 on melanoma cells can inhibit the invasive stage of metastasis formation. *Cell* **77**: 335–347.
Rankin, S. and Rozengurt, E. (1994). Platelet-derived growth factor modulation of focal adhesion kinase (p125FAK) and paxillin tyrosine phosphorylation in Swiss 3T3 cells. Bell- shaped dose response and cross-talk with bombesin. *J Biol Chem* **269**: 704–710.
Rao, C.N., Castronovo, V., Schmitt, M.C., Wewer, U.M., Claysmith, A.P., Liotta, L.A. and Sobel, M.E. (1989). Evidence for a precursor of the high-affinity metastasis-associated murine laminin receptor. *Biochemistry* **28**: 7476–7486.
Reed, R.K. (1994). Regulation of intestitial fluid pressure and control of tissue fluid content. In: *Interstitium, Connective Tissue and Lymphatics*. Eds: R.K. Reed, N.G. McHale, J.L. Bert, C.P. Winlowe, G.A. Laine. Portland Press Proceedings. London pp. 85–100.
Reed, R.K., Rubin, K., Wiig, H. and Rodt, S.A. (1992). Blockade of β_1-integrins in skin causes edema through lowering of interstitial fluid pressure. *Circ Res* **71**: 978–983.
Reichardt, L.F. and Tomaselli (1991). Extracellular matrix molecules and their receptors: functions in neural development. *Annu. Rev. Neurosci.* **14**: 531–570.
Ridley, A.J. and Hall, A. (1992). The small GTP-binding protein rho regulates the assembly of focal adhesions and actin stress fibers in response to growth factors. *Cell* **70**: 389–399.
Ridley, A.J., Paterson, H.F., Johnston, C.L., Diekmann, D. and Hall, A. (1992). The small GTP-binding protein rac regulates growth factor-induced membrane ruffling. *Cell* **70**: 401–410.
Rodt, S. Å., Berg, A., Rubin, K., Åhlén, K. and Reed, R. K. (In Press). Platelet-derived growth factor (PDGF) normalizes increased negativity of interstitial pressure in skin (Pif) in dextran anaphylaxis (Abstract) (Microcirculatory society annual meeting 1995). *Microcirculation*
Rodt, S. Å., Reed, R.K., Ljungström, M., Gustafsson, T.O. and Rubin, K. (1994). The anti-inflammatory agent a-trinositol exerts its edema-preventing effect through modulation of β_1 integrin function. *Circ Res* **75**: 942–948.
Rohrschneider, L.R. (1980). Adhesion plaques of Rous sarcoma virus transformed cells contain the src gene product. *Proc. Natl. Acad. Sci. USA* **77**: 3514–3518.
Rosen, G.D., Barks, J.L., Iademarco, M.F., Fisher, R.J. and Dean, D.C. (1994). An intricate arrangement of binding sites for the Ets family of transcription factors regulates activity of the α4 integrin gene promoter. *J Biol Chem* **269**: 15652–15660.
Rosen, G.D., Sanes, J.R., LaChance, R., Cunningham, J.M., Roman, J. and Dean, D.C. (1992). Roles for the integrin VLA-4 and its counter receptor VCAM-1 in myogenesis. *Cell* **69**: 1107–1119.
Ross, R., Masuda, J., Raines, E.W., Gown, A.M., Katsuda, S., Sasahara, M., Malden, L.T., Masuko, H. and Sato, H. (1990). Localization of PDGF-B protein in macrophages in all phases of atherogenesis. *Science* **248**: 1009–1012.
Rossino, P., Defilippi, P., Silengo, L. and Tarone, G. (1991). Up-regulation of the integrin α_1/β_1 in human neuroblastoma cells differentiated by retinoic acid: correlation with increased neurite outgrowth response to laminin. *Cell Regul* **2**: 1021–1033.
Rubin, K., Höök, M., Öbrink, B. and Timpl, R. (1981). Substrate adhesion of rat hepatocytes: mechanism of attachment to collagen substrates. *Cell* **24**: 463–470.
Rubin, K., Johansson, S., Pettersson, I., Ocklind, C., Öbrink, B. and Höök, M. (1979). Attachment of rat hepatocytes to collagen and fibronectin: a study using antibodies directed against cell surface components. *Biochem Biophys Res Commun* **91**: 86–94.
Ruegg, C., Postigo, A.A., Sikorski, E.E., Butcher, E.C., Pytela, R. and Erle, D.J. (1992). Role of integrin $\alpha_4\beta_7/\alpha_4\beta_P$ in lymphocyte adherence to fibronectin and VCAM-1 and in homotypic cell clustering. *J Cell Biol* **117**: 179–189.

Ruoslahti, E. (1991). Integrins. *J Clin Invest* **87**: 1–5.
Ruoslahti, E. and Hayman, E. G. (1979). Two active sites with different characteristics in fibronectin. *FEBS Lett* **97**: 221–224.
Ruoslahti, E. and Pierschbacher, M.D. (1987). New perspectives in cell adhesion: RGD and integrins. *Science* **238**: 491–497.
Rydén, C., Maxe, I., Franzén, A., Ljungh, Å., Heinegård, D. and Rubin, K. (1987). Selective binding of bone matrix sialoprotein to *Staphylococcus aureus* in osteomyelitis. *Lancet* **ii**: 515.
Rydén, C., Yacoub, A., Maxe, I., Heinegård, D., Oldberg, Å., Franzén, A., Ljungh, Å. and Rubin, K. (1989). Specific binding of bone sialoprotein to Staphylococcus aureus from patients with osteomyelitis. *Eur J Biochem* **184**: 331–336.
Rydén, C., Rubin, K., Speziale, P., Höök, M., Lindberg, M. and Wadström, T. (1983). Fibronectin receptors from *Staphylococcus aureus*. *J Biol Chem* **258**: 3396–3401.
Sage, E.H. and Bornstein, P. (1991). Extracellular proteins that modulate cell-matrix interactions. SPARC, tenascin, and thrombospondin. *J Biol Chem* **266**: 14831–14834.
Sager, R., Anisowicz, A., Neveu, M., Liang, P. and Sotiropoulou, G. (1993). Identification by differential display of alpha6 integrin as a candidate tumor suppressor gene. *Faseb J* **7**: 964–970.
Sanes, J.R. (1989). Extracellular matrix molecules that influence neuronal development. *Ann. Rev. Neurosci.* **12**: 491–516.
Santala, P. and Heino, J. (1991). Regulation of integrin-type cell adhesion receptors by cytokines. *J Biol Chem* **266**: 23505–23509.
Sastry, S.K. and Horwitz, A.F. (1993). Integrin cytoplasmic domains: mediators of cytoskeletal linkages and extra- and intracellular initiated transmembrane signaling. *Curr Opin Cell Biol* **5**: 819–831.
Savoia, P., Trusolino, L., Pepino, E., Cremona, O. and Marchisio, P.C. (1993). Expression and topography of integrins and basement membrane proteins in epidermal carcinomas: basal but not squamous cell carcinomas display loss of $\alpha_6\beta_4$ and BM-600/nicein. *J Invest Dermatol* **101**: 352–358.
Schadendorf, D., Gawlik, C., Haney, U., Ostmeier, H., Suter, L. and Czarnetzki, B. M. (1993). Tumour progression and metastatic behaviour in vivo correlates with integrin expression on melanocytic tumours. *J Pathol* **170**: 429–434.
Schaller, J., Glander, H. J. and Dethloff, J. (1993). Evidence of β_1 integrins and fibronectin on spermatogenic cells in human testis. *Hum Reprod* **8**: 1873–1878.
Schaller, M.D., Hildebrand, J.D., Shannon, J.D., Fox, J.W., Vines, R.R. and Parsons, J.T. (1994). Autophosphorylation of the focal adhesion kinase, pp125FAK, directs SH2- dependent binding of pp60src. *Mol Cell Biol* **14**: 1680–1688.
Schaller, M.D. and Parsons, J. T. (1994). Focal adhesion kinase and associated proteins. *Curr Op Cell Biol* **6**: 705–710.
Schennings, T., Heimdahl, A., Coster, K. and Flock, J.-I. (1993). Immunization with fibronectin binding protein from *S. aureus* protects against experimental endocarditis in rats. *Microbial Pathogenesis* **15**: 227–236.
Schiro, J.A., Chan, B.M., Roswit, W.T., Kassner, P.D., Pentland, A.P., Hemler, M.E., Eisen, A.Z. and Kupper, T.S. (1991). Integrin $\alpha_2\beta_1$ (VLA-2) mediates reorganization and contraction of collagen matrices by human cells. *Cell* **67**: 403–410.
Schneewind, O., Model, P. and Fischetti, V. (1992). Sorting of protein A to the staphylococcal cell wall. *Cell* **70**: 267–281.
Schor, S. L. and Court, J. (1979). Different mechanisms in the attachment of cells to native and denatured collagen. *J Cell Sci* **38**: 267–281.
Schwartz, M.A. (1992). Transmembrane signalling by integrins. *Trends in Cell Biology* **2**: 304–308.
Seftor, R.E., Seftor, E.A., Gehlsen, K.R., Stetler-Stevenson, W.G., Brown, P.D., Ruoslahti, E. and Hendrix, M. J. (1992). Role of the $\alpha_v\beta_3$ integrin in human melanoma cell invasion. *Proc Natl Acad Sci USA* **89**: 1557–1561.
Seftor, R.E., Seftor, E.A., Stetler-Stevenson, W.G. and Hendrix, M.J. (1993). The 72 kDa type IV collagenase is modulated via differential expression of $\alpha_v\beta_3$ and $\alpha_5\beta_1$ integrins during human melanoma cell invasion. *Cancer Res* **53**: 3411–3415.
Shattil, S.J., Hoxie, J.A., Cunningham, M. and Brass, L.F. (1985). Changes in the platelet membrane glycoprotein IIb.IIIa complex during platelet activation. *J Biol Chem* **260**: 11107–11114.
Shur, B.D. (1993). Glycosyltransferases as cell adhesion molecules. *Curr Op Cell Biol* **8**: 854–863.
Siegbahn, A., Hammacher, A., Westermark, B. and Heldin, C.H. (1990). Differential effects of the various isoforms of platelet-derived growth factor on chemotaxis of fibroblasts, monocytes, and granulocytes. *J Clin Invest* **85**: 916–920.
Signäs, C., Raucci, G., Jönsson, K., Lindgren, P.-E., Anantharamaiah, G.M., Höök, M. and Lindberg, M. (1989). Nucleotide sequence of the gene for a fibronectin-binding protein from *Staphylococcus aureus*: use of this peptide sequence in the synthesis of biologically active peptides. *Proc Natl Acad Sci USA* **86**: 699–703.
Sims, P.J., Ginsberg, M.H., Plow, E.F. and Shattil, S.J. (1991). Effect of platelet activation on the conformation of the plasma membrane glycoprotein IIb-IIIa complex. *J Biol Chem* **266**: 7345–7352.

Singer, I.I. (1979). The fibronexus: a transmembrane association of fibronectin-containing fibers and bundles of 5 nm microfilaments in hamster and human fibroblasts. *Cell* **16**: 675–685.

Sjöbring, U. and Björck, L. (1992). Convergent evolution among immunoglobulin G-binding bacterial proteins. *Proc Natl Acad Sci (USA)* **89**: 8532–8536.

Smedsrød, B., Pertoft, H., Gustafson, S. and Laurent, T.C. (1990). Scavenger functions of the liver endothelial cell. *Biochem J* **266**: 313–327.

Song, W.K., Wang, W., Foster, R.F., Bielser, D.A. and Kaufman, S.J. (1992). H36-α_7 is a novel integrin a chain that is developmentally regulated during skeletal myogenesis. *J. Cell Biol.* **117**: 643–657.

Sorokin, L., Sonnenberg, A., Aumailley, M., Timpl, R. and Ekblom, P. (1990). Recognition of the laminin E8 cell-binding site by an integrin possessing the $\alpha6$ subunit is essential for epithelial polarization in developing kidney tubules. *J Cell Biol* **111**: 1265–1273.

Sorokin, L.M., Conzelmann, S., Ekblom, P., Battaglia, C., Aumailley, M. and Timpl, R. (1992). Monoclonal antibodies against laminin A chain of fragment E3 and their effects on binding to cells and proteoglycan and on kidney development. *Exp. Cell Res.* **201**: 137–144.

Speziale, P., Raucci, G., Visai, L., Switalski, L.M., Timpl, R. and Höök, M. (1986). Binding of collagen to *Staphylococcus aureus* Cowan I. *J Bacteriol* **167**: 77–81.

Spring, J., Paine-Saunders, S., Hynes, R.O. and Bernfield, M. (1994). *Drosophila* syndecan: conservation of a cell surface heparan sulfate proteoglycan. *Proc Natl Acad Sci (USA)* **91**: 3334–3338.

Springer, T.A. (1990). Adhesion receptors of the immune system. *Nature* **346**: 425–434.

Sriramarao, P., Mendler, M. and Bourdon, M.A. (1993). Endothelial cell attachment and spreading on human tenascin is mediated by $\alpha2\beta1$ and $\alpha v\beta3$ integrins. *J Cell Sci* **105**: 1001–1012.

Sriramarao, P., Steffner, P. and Gehlsen, K. R. (1993). Biochemical evidence for a homophilic interaction of the $\alpha3\beta1$ integrin. *J Biol Chem* **268**: 22036–22041.

Staatz, W.D., Rajpara, S.M., Wayner, E.A., Carter, W.G. and Santoro, S.A. (1989). The membrane glycoprotein Ia-IIa (VLA-2) complex mediates the Mg^{2+}- dependent adhesion of platelets to collagen. *J Cell Biol* **108**: 1917–1924.

Stallmach, A., von Lampe, B., Matthes, H., Bornhoft, G. and Riecken, E.O. (1992). Diminished expression of integrin adhesion molecules on human colonic epithelial cells during the benign to malign tumour transformation. *Gut* **33**: 342–346.

Stallmach, A., von Lampe, B., Orzechowski, H.D., Matthes, H. and Riecken, E.O. (1994). Increased fibronectin-receptor expression in colon carcinoma-derived HT 29 cells decreases tumorigenicity in nude mice. *Gastroenterology* **106**: 19–27.

Stephens, L.E., Sonne, J.E., Fitzgerald, M.L. and Damsky, C.H. (1993). Targeted deletion of ß1 integrins in F9 embryonal carcinoma cells affects morphological differentiation but not tissue-specific gene expression. *J. Cell Biol.* **123**: 1607–1620.

Stopak, D. and Harris, A.K. (1982). Connective tissue morphogenesis by fibroblast traction. I. Tissue culture observations. *Dev Biol* **90**: 383–398.

Svanborg-Edén, C., Freter, R., Hagberg, L., Hull, R., Hull, S., Leffler, H. and Schoolnik, G. (1982). Inhibition of experimental ascending urinary tract infection by an epithelial cell-surface receptor analogue. *Nature* **298**: 560–562.

Switalski, L.M., Patti, J.M., Butcher, W.G., Gristina, A.G., Speziale, P. and Höök, M. (1993). A collagen receptor on *Staphylococcus aureus* strains isolated from patients with septic arthritis mediates adhesion to cartilage. *Mol Microbiol* **7**: 99–107.

Switalski, L.M., Speziale, P. and Höök, M. (1989). Isolation and characterization of a putative collagen receptor from *Staphylococcus aureus*. *J Biol Chem* **264**: 21080–21086.

Sy, M.-S., Guo, Y.-J. and Stamenkovic, I. (1992). Inhibition of tumor growth in vivo with a soluble CD44-immunglobulin fusion protein. *J Exp Med* **176**: 623–627.

Symington, B.E., Takada, Y. and Carter, W.G. (1993). Interaction of integrins $\alpha_3\beta_1$ and $\alpha_2\beta_1$: potential role in keratinocyte intercellular adhesion. *J Cell Biol* **120**: 523–535.

Takada, Y., Wayner, E.A., Carter, W.G. and Hemler, M.E. (1988). Extracellular matrix receptors, ECMRII and ECMRI, for collagen and fibronectin correspond to VLA-2 and VLA-3 in the VLA family of heterodimers. *J Cell Biochem* **37**: 385–393.

Takeachi, M. (1988). The cadherins: cell-cell adhesion molecules controlling animal morphogenesis. *Development* **102**: 639–655.

Tamkun, J.W., DeSimone, D.W., Fonda, D., Patel, R.S., Buck, C., Horwitz, A.F. and Hynes, R.O. (1986). Structure of integrin, a glycoprotein involved in the transmembrane linkage between fibronectin and actin. *Cell* **46**: 271–282.

Tapley, P., Horwitz, A., Buck, C., Duggan, K. and Rohrschneider, L. (1989). Integrins isolated from Rous sarcoma virus-transformed chicken embryo fibroblasts. *Oncogene* **4**: 325–333.

Tarone, G., Cirillo, D., Giancotti, F.G., Comoglio, P.M. and Marchisio, P.C. (1985). Rous sarcoma virus-transformed fibroblasts adhere primarily at discrete protrusions of the ventral membrane called podosomes. *Exp Cell Res* **159**: 141–157.

Terraco, L., Rubin, K., Gullberg, D., Balog, E., Carver, W., Jyring, R. and Borg, T.K. (1991). Expression of collagen binding integrins during cardiac development and hypertrophy. *Circ Res* **68**: 734–744.

Terranova, V.P., Rohrbach, D.H. and Martin, G.R. (1980). Role of laminin in the attachment of PAM 212 (epithelial) cells to basement membrane collagen. *Cell* **2**: 719–722.

Thomas, L., Byers, H.R., Vink, J. and Stamenkovic, I. (1992). CD44H regulates tumor cell migration on hyaluronate-coated substrate. *J Cell Biol* **118**: 971–977.

Thiery, J.-P., Brackenbury, R., Rutishauser, U. and Edelman, G.M. (1977) Adhesion among neural cells of the chick embryo. II. Purification and characterization of a cell adhesion molecule from neural retina. *J Biol Chem* **252**: 6841–6845

Tidball, J.G. and Spencer, M.J. (1993). PDGF stimulation induces phosphorylation of talin and cytoskeletal reorganization in skeletal muscle. *J Cell Biol* **123**: 627–635.

Tingström, A., Heldin, C.H. and Rubin, K. (1992). Regulation of fibroblast-mediated collagen gel contraction by platelet- derived growth factor, interleukin-1 α and transforming growth factor-$\beta 1$. *J Cell Sci* **102**: 315–322.

Tomasek, J.J. and Hay, E.D. (1984). Analysis of the role of microfilaments and microtubules in acquisition of bipolarity and elongation of fibroblasts in hydrated collagen gels. *J Cell Biol* **99**: 536–549.

Turner, C.E. (1994). Paxillin: a cytoskeletal target for tyrosine kinases. *BioEssays* **16**: 47–52.

Turner, C.E. and Burridge, K. (1991). Transmembrane molecular assemblies in cell-extracellular matrix interactions. *Curr Opin Cell Biol* **3**: 849–853.

Uhlén, M., Guss, B., Nilsson, B., Gatenbeck, S., Philipsson, L. and Lindberg, M. (1984). Complete sequence of the staphylococcal gene encoding protein A. *J Biol Chem* **259**: 1695–1702.

Van Bockxmeer, F.M., Martin, C.E. and Constable, I.J. (1984). Effect of cyclic AMP on cellular contractility and DNA synthesis in chorioretinal fibroblasts maintained in collagen matrices. *Exp Cell Res* **155**: 413–421.

van de Wiel-van Kemenade, E., van Kooyk, Y., de Boer, A.J., Huijbens, R.J., Weder, P., van de Kasteele, W., Melief, C.J. and Figdor, C.G. (1992). Adhesion of T and B lymphocytes to extracellular matrix and endothelial cells can be regulated through the β subunit of VLA. *J Cell Biol* **117**: 461–470.

van den Berg, T.K., van der Ende, M., Dopp, E.A., Kraal, G. and Dijkstra, C.D. (1993). Localization of $\beta 1$ integrins and their extracellular ligands in human lymphoid tissues. *Am J Pathol* **143**: 1098–1110.

Van Waes, C. and Carey, T.E. (1992). Overexpression of the A9 antigen/$\alpha 6\beta 4$ integrin in head and neck cancer. *Otolaryngol Clin North Am* **25**: 1117–1139.

Vestweber, D. (1993). The selectins and their ligands. *Curr Top Microbiol Immunol* **184**: 65–75.

Vogel, B.E., Tarone, G., Giancotti, F.G., Gailit, J. and Ruoslahti, E. (1990). A novel fibronectin receptor with an unexpected subunit composition ($\alpha_v\beta_1$). *J Biol Chem* **265**: 5934–5937.

Volk, T., Fessler, L.I. and Fessler, J.H. (1991). A role for integrin in the formation of sarcomeric cytoarchitecture. *Cell* **63**: 525–536.

Voytek, A., Gristina, A.G., Barth, E., Myrvik, Q., Switalski, L.M., Höök, M. and Speziale, P. (1988). Staphylococcal adhesion to collagen in intra-articular sepsis. *Biomaterials* **9**: 107–110.

Vuori, K. and Ruoslahti, E. (1993). Activation of protein kinase C precedes $\alpha_5\beta_1$ integrin-mediated cell spreading on fibronectin. *J Biol Chem* **268**: 21459–21462.

Wayner, E.A., Carter, W.G., Piotrowicz, R.S. and Kunicki, T.J. (1988). The function of multiple extracellular matrix receptors in mediating cell adhesion to extracellular matrix: preparation of monoclonal antibodies to the fibronectin receptor that specifically inhibit cell adhesion to fibronectin and react with platelet glycoproteins Ic-IIa. *J Cell Biol* **107**: 1881–1891.

Wehrle-Haller, B. and Chiquet, M. (1993). Dual function of tenascin: simultaneous promotion of neurite growth and inhibition of glial migration. *J Cell Sci* **106**: 597–610.

Weinel, R. J., Rosendahl, A., Neumann, K., Chaloupka, B., Erβ, D., Rothmund, M. and Santoro, S. (1992). Expression and function of VLA -α_2, -α_3, -α_5 and -α_6— integrin receptors in pancreatic carcinoma. *Int J Cancer* **52**: 827–833.

Westerlund, B., Kuusela, P., Risteli, J., Vartio, T., Rauvala, H., Virkola, R. and Korhonen, T. (1989a). The O75X adhesin of uropathogenic *Escherichia coli* is a type IV collagen-binding protein. *Mol Microbiol* **3**: 329–337.

Westerlund, B., Kuusela, P., Vartio, T., I., v.D. and Korhonen, T.K. (1989b). A novel lectin-independent interaction of P fimbriae of *Escherichia coli* with immobilized fibronectin. *FEBS Lett* **243**: 199–204.

Williams, A.F. and Barclay, A.N. (1988). The immunoglobulin superfamily — domains for cell surface recognition. *Annu Rev Immunol* **6**: 381–405.

Williams, J.M., Hughes, P.E., O'Toole, T. E. and Ginsberg, M.H. (1994). The inner world of cell adhesion: integrin cytoplasmic domains. *Trends in Cell Biology* **4**: 109–112.

Woods, A. and Couchman, J.R. (1992). Protein kinase C involvement in focal adhesion formation. *J Cell Sci* **101**: 277–290.

Woods, A., Couchman, J.R., Johansson, S. and Höök, M. (1986). Adhesion and cytoskeletal organisation of fibroblasts in response to fibronectin fragments. *EMBO J* **5**: 665–670.

Xu, H., Wu, X.-R., Wewer, U.M. and Engvall, E. (1994). Murine muscular dystrophy caused by a mutation in the laminin α_2(Lama2) gene. *Nature Genetics* **8**: 297–302.

Yacoub, A., Lindahl, P., Rubin, K., Wendel, M., Heinegård, D. and Rydén, C. (1994). Purification of a bone sialoprotein-binding protein from *Staphylococcus aureus*. *Eur J Biochem* **222**: 919–925.

Yang, J.T., Rayburn, H. and Hynes, R.O. (1993). Embryonic mesodermal defects in α_5 integrin-deficient mice. *Development* **119**: 1093–1105

Yee, G. H. and Hynes, R.O. (1993). A novel, tissue specific integrin subunit, βv, expressed in the midgut of *Drosophila* melanogaster. *Development* **118**: 845–858.

Yokosaki, Y., Palmer, E.L., Prieto, A.L., Crossin, K.L., Bourdon, M.A., Pytela, R. and Sheppard, D. (1994) The integrin $\alpha_9\beta_1$ mediates cell attachment to a non-RGD site in the third fibronectin type III repeat of tenascin. *J Biol Chem* **269**: 26691–26693.

Yow, H.K., Wong, J. M., Chen, H.S., Lee, C.G., Davis, S., Steele, G., Jr. and Chen, L.B. (1988). Increased mRNA expression of a laminin-binding protein in human colon carcinoma: complete sequence of a full-length cDNA encoding the protein. *Proc Natl Acad Sci U S A* **85**: 6394–6398.

Zambonin-Zallone, A., Teti, A., Grano, M., Rubinacci, A., Abbadini, M., Gaboli, M. and Marchisio, P.C. (1989). Immunocytochemical distribution of extracellular matrix receptors in human osteoclasts: $\alpha_v\beta_3$ integrin is colocalized with vinculin and talin in the podosomes of osteoclastoma giant cells. *Exp Cell Res* **182**: 645–652.

Zhang, Z., Tarone, G. and Turner, D.C. (1993). Expression of integrin $\alpha_1\beta_1$ is regulated by nerve growth factor and dexamethasone in PC12 cells. Functional consequences for adhesion and neurite outgrowth. *J Biol Chem* **268**: 5557–5565.

Ziober, B. L., Vu, M.P., Waleh, N., Crawford, J., Lin, C.S. and Kramer, R.H. (1993). Alternative extracellular and cytoplasmic domains of the integrin α_7 subunit are differentially expressed during development. *J Biol Chem* **268**: 26773–26783.

Zusman, S., Grinblat, Y., Yee, G., Kafatos, F.C. and Hynes, R.O. (1993). Analyses of PS integrin functions during *Drosophila* development. *Development* **118**: 737–750.

Zutter, M.M., Krigman, H.R. and Santoro, S.A. (1993). Altered integrin expression in adenocarcinoma of the breast. Analysis by in situ hybridization. *Am J Pathol* **142**: 1439–448.

Zutter, M.M. and Santoro, S.A. (1990). Widespread histologic distribution of the $\alpha_2\beta_1$ integrin cell- surface collagen receptor. *Am J Pathol* **137**: 113–120.

Zutter, M.M., Santoro, S.A., Painter, A.S., Tsung, Y.L. and Gafford, A. (1994). The human α_2 integrin gene promoter. Identification of positive and negative regulatory elements important for cell-type and developmentally restricted gene expression. *J Biol Chem* **269**: 463–469.

10 The Role of Specific Macromolecules in Cell–Matrix Interactions and in Matrix Function:Physicochemical and Mechanical Mediators of Chondrocyte Biosynthesis

Alan J. Grodzinsky,[1] Eliot H. Frank,[1] Young-Jo Kim,[1] and Michael D. Buschmann[2]

[1]*Continuum Electromechanics Group, Department of Electrical Engineering and Computer Science, Department of Mechanical Engineering, Massachusetts Institute of Technology, Cambridge, MA, USA*

[2]*Institute of Biomedical Engineering, Ecole Polytechnique, University of Montreal, Montreal, Quebec, Canada*

INTRODUCTION

Articular cartilage provides a low-friction wear-resistant bearing surface which distributes and transmits stresses generated by body weight and muscle contraction to the underlying bone. Loss of cartilage in disease states such as osteoarthritis results in joint pain, immobility, and deformity causing great morbidity. The ability of cartilage to withstand compressive, tensile, and shear forces depends on the composition and structural integrity of its hydrated extracellular matrix (ECM), which consists of collagen fibrils, proteoglycans (PG), and other proteins and glycoproteins. While collagen fibrils are strong in tension, proteoglycans (aggrecan) resist compression primarily due to osmotic interactions (electrostatic repulsive interactions between glycosaminoglycan chains) (Kempson, 1979; Maroudas, 1979; Grodzinsky, 1983) and fluid/solid frictional interactions (Mow *et al.*, 1984). The maintenance of a functionally intact matrix requires the coordinated synthesis, assembly, degradation of proteoglycans, collagens, and other matrix molecules. The regulation of these metabolic processes *in vivo* appears to involve a combination of cell biological and physical mechanisms.

Clinical observations and studies *in vivo* suggest that joint loading and motion can induce a wide range of metabolic responses in cartilage. Immobilization or reduced loading can cause a decrease in proteoglycan synthesis and content (Akeson *et al.*, 1973; Caterson and Lowther, 1978; Kiviranta *et al.*, 1987; Olah and Kostenszky, 1972; Palmoski *et al.*, 1979) and a resultant softening of the tissue (Jurvelin *et al.*, 1989). In contrast, dynamic loading or remobilization of the joint can increase proteoglycan synthesis and content (Caterson and Lowther, 1978; Kiviranta *et al.*, 1987), and a restoration of biomechanical properties (Jurvelin *et al.*, 1989). More severe static (Gritzka *et al.*, 1973) or impact (Radin *et al.*, 1984) loading often causes cartilage deterioration and can lead to osteoarthritic changes

Extracellular Matrix, Volume 2, Molecular Components and Interactions, edited by Wayne D. Comper.

(Thompson *et al.*, 1991). Thus, while some degree of "normal" joint loading appears to promote structural adaptation, "abnormal" mechanical forces predispose cartilage to degeneration (Arokoski *et al.*, 1993). The physical and biological transduction mechanisms responsible for these alterations at the cellular level are not fully understood and are difficult to identify *in vivo*. As a result, cartilage explant systems have become increasingly important for studies aimed at understanding the mechanisms by which physical forces may regulate cartilage metabolism.

Many physical phenomena which occur in cartilage during loading *in vivo* have been identified and quantified *in vitro* (Figure 10.1). Cartilage compression can be considered to consist of a time-varying **dynamic** component superimposed on a slowly evolving time-averaged **static** component. Compression of cartilage results in deformation of cells and extracellular matrix (Poole *et al.*, 1984). The static component of compression results in fluid loss and compaction of the matrix. The resulting increase in the local proteoglycan negative fixed charge density caused by matrix compaction will attract and concentrate positively charged ions (including H^+, Na^+, and K^+), causing a decrease in intratissue pH and an increase in the local osmotic pressure of the intratissue fluid (Maroudas, 1979; Urban *et al.*, 1979; Grodzinsky, 1983; Schneiderman *et al.*, 1986; Gray *et al.*, 1988) (Figure 10.1). The dynamic component of cartilage loading can cause hydrostatic pressure gradients and a concomitant flow of interstitial fluid through the matrix (Mow *et al.*, 1984; Mak, 1986). In turn, fluid flow will induce electrical streaming potentials and currents within the matrix in the neighborhood of the cells (Grodzinsky *et al.*, 1978; Frank and Grodzinsky, 1987a) (Figure 10.1). Any of these physicochemical, mechanical, or electrical phenomena may modulate matrix metabolism. An understanding of the spatial distribution of these forces and flows within cartilage during compression has been aided by the development of theoretical models for the mechanical (Mak, 1986; Armstrong *et al.*, 1984; Setton *et al.*, 1993), physicochemical (Eisenberg and Grodzinsky, 1987; Lai *et al.*, 1991), and electromechanical (Frank and Grodzinsky, 1987b) behavior of cartilage. Such models can facilitate the study of transduction mechanisms by providing a framework for correlating the spatial distributions of biosynthesis and physical stimuli that occur within cartilage explants during compression (Kim *et al.*, 1994b)

In this chapter, we first review the effects of physicochemical and mechanical stimuli on chondrocyte biosynthetic response in *intact cartilage explants*. Cartilage provides an important model system for the study of physical regulation of cell metabolism in many respects. Specific molecular components in cartilage's dense ECM provide this tissue with its functional mechanical and physicochemical properties. In turn, static and dynamic compression have distinct effects on the synthesis and assembly of specific matrix macromolecules. An understanding of the cellular transduction mechanisms responsible for this feedback system could provide insight concerning the ability of chondrocytes adapt to their biophysical environment. Next, we describe the effects of compression on chondrocytes cultured in agarose gels. With this *in vitro* model system, we can highlight the importance of cell-matrix interactions on the transduction of mechanical/physicochemical stimuli by comparing chondrocyte/agarose cultures at early times (before deposition of pericellular/territorial matrix) and at later times (after deposition of dense matrix around and between the cells in agarose). Finally, we explore the relevance of the known molecular structure of aggrecan and its constituent GAG chains to the measurable physicochemical properties and mechanical stiffness of cartilage. These results highlight the importance of

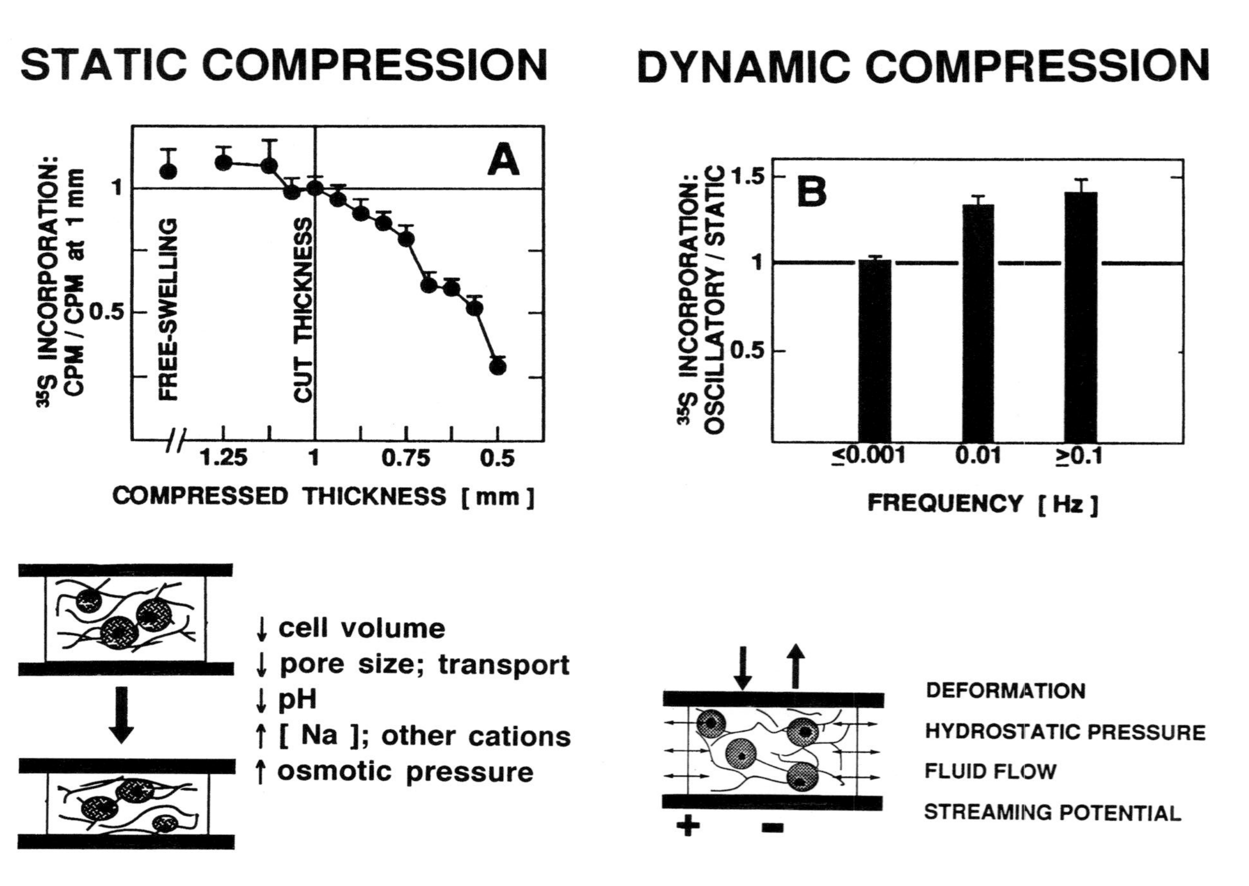

Figure 10.1 Effect of static (A) and dynamic (B) compression on proteoglycan synthesis in calf cartilage disks as assessed by ^{35}S-sulfate incorporation into compressed disks normalized to that in control disks statically held at 1 mm thickness for both cases (Data adapted from Sah *et al.*, 1989 and Sah *et al.*, 1990b)). Diagrams indicate possible physical stimuli associated with static and dynamic compression, respectively.

specific molecular-level structures to the macroscopic, functional material properties of an extracellular matrix; in addition, a rationale is envisioned for the sensitivity of chondrocyte biosynthetic response to mechanical loading, and the specific need for chondrocytes to synthesize such highly complex molecules as aggrecan in response to mechanical forces in their environment.

PHYSICAL REGULATION OF MATRIX METABOLISM IN CARTILAGE EXPLANTS

Differential Effects of Static and Dynamic Compression on Matrix Synthesis

Studies *in vitro* during the past decade have demonstrated several of the relationships between mechanical compression and matrix biosynthesis that had been observed previously *in vivo*. Most studies *in vitro* have focused on the effects of static compression and the release of static compression. Investigators have found consistently that static compression and the concomitant matrix compaction of cartilage explants, applied by mechanical or osmotic stresses, resulted in a decrease in synthesis of proteoglycans and proteins (Bayliss *et al.*, 1986; Gray *et al.*, 1988; Gray *et al.*, 1989; Jones *et al.*, 1982; Palmoski and Brandt, 1984; Sah *et al.*, 1989; Schneiderman *et al.*, 1986; Urban and Bayliss, 1989) as measured by radiolabel incorporation.

As an example, Sah *et al.* (1989) applied a 12-hour static compression to groups of 3 mm diameter disks of femoropatellar groove cartilage from 1–2 week old calves. In each polysulfone static compression chamber, 6–12 cartilage disks were sandwiched between two fluid-impermeable platens; by varying the distance between the platens, cartilage disks were compressed in a uniaxial, radially unconfined configuration (Sah *et al.*, 1989; Sah *et al.*, 1990b). Using ^{35}S-sulfate and L-[5-^{3}H]-proline to assess synthesis of proteoglycans and total protein, respectively, they observed a dose dependent decrease in proteoglycan synthesis (Figure 10.1A) with increasing static compression. Total protein synthesis decreased with increasing compression in a similar manner (Sah*et al.*, 1989). The 90% radiolabel diffusion time for 3 mm disks was measured to be ~0.5 hr (Sah *et al.*, 1989). In addition, mechanical measurements showed that stress relaxation of the 3 mm diameter disks was 90% complete within 15 mins. Thus, the period to reach mechanical equilibrium and biosynthetic steady state was short relative to the 12 hr of static compression. Kinetic studies indicated that radiolabel incorporation rates had decreased within the first hour after application of such static compressions (Gray *et al.*, 1989; Sah *et al.*, 1989). After *release* of static compression, proteoglycan synthesis returned to normal by 60 hr, even in disks that had been compressed to 0.5 mm (Kim *et al.*, 1992, 1994b).

In contrast to the effects of static compression, investigators have found that dynamic compression can stimulate matrix biosynthesis (Palmoski and Brandt, 1984; Sah *et al.*, 1989; Parkkinen *et al.*, 1992), though the results have been variable and depended on the specific loading protocol, specimen geometry, and compression frequency. In one study, dynamic compressive loads (60 sec on/60 sec off, and 4 sec on/11 sec off) for two hours led to a decrease and increase, respectively, in proteoglycan biosynthesis in the two hours following compression (Palmoski and Brandt, 1984). The compression that occurred during loading was not measured, and the peak applied stress was relatively low (0.001–0.011 MPa). In another study, cyclic tensile stretching (5.5%) of high density chick chondrocyte cultures at 0.2 Hz for 24 hours was accompanied by an increase in proteoglycan synthesis during the last 4 hours of stretching (DeWitt *et al.*, 1984).

Figure 10.1B shows the effects of dynamic compression on proteoglycan synthesis in calf cartilage explants similar to those of Figure 10.1A. A specialized dynamic compression chamber was constructed (Sah *et al.*, 1989; Kim *et al.*, 1994b) to enable oscillatory unconfined compression of experimental and control disks within the same incubator-like environment. This chamber was mounted in a mechanical spectrometer and feedback controlled compressive displacements (or loads) were applied to the experimental disks over a wide range of frequencies and amplitudes, while simultaneously measuring the resulting load (displacement). Experimental disks were compressed statically to 1.0 mm with a superimposed *low amplitude oscillatory compression* of 1–4%; control disks were held statically at 1.0 mm. Oscillatory compression at frequencies f = 0.01–1 Hz stimulated incorporation of ^{35}S-sulfate (Figure 10.1B) and ^{3}H-proline by ~30–40% but had no effect at $f \leq$ 0.001 Hz (Sah *et al.*, 1989; Kim *et al.*, 1994b). The low amplitude of such oscillations ensured that there was little change in tissue hydration and fixed charge density; thus, the biosynthetic effects of such oscillations could not have been induced by the higher strain physicochemical phenomena associated with the static compression tests of Figure 10.1A.

Effects of Compression on Aggregation of Newly Synthesized Proteoglycans

The effects of static and dynamic compression on the assembly (aggregation) of newly synthesized aggrecan molecules into functional proteoglycan aggregates was recently examined in calf cartilage explants (Sah *et al.*, 1990a). Pulse-chase experiments showed that conversion of newly synthesized aggrecan to a form that would bind with high affinity to hyaluronan occurred with a time constant ($t_{1/2}$) of about 5.7 hr in free-swelling disks at pH 7.45. *Static compression* (in pH 7.45 medium) slowed this conversion to high affinity binding. In parallel studies, incubation in acidic medium (without compression) also delayed conversion (Sah *et al.*, 1990a). Both effects were dose-dependent. Oscillatory compression of 2% amplitude at 0.001, 0.01, or 0.1 cycles/sec during chase did not, however, affect the conversion.

The similar change in intratissue pH induced by static compression or by incubation in acidic medium, and the similarly delayed kinetics of binding affinity with static compression and acidic medium suggested that this delayed aggregation may be elicited through a common *physicochemical mechanism of decreased intratissue pH* (Sah *et al.*, 1990a). (Though dynamic strains generated physiologic levels of dynamic stress, the low amplitude of dynamic strain would leave intratissue pH relatively unaltered.) Such a pH mechanism might have a physiologic role, promoting proteoglycan deposition in regions of low proteoglycan concentration during cartilage remodeling or repair.

Effects of Compression on Matrix Degradation

The effects of large amplitude compression and changes in tissue pH on the catabolism and loss to the medium of proteoglycans and proteins from cartilage explants were quantified in a series of pulse-chase experiments (Sah *et al.*, 1991). Slow, high amplitude cyclic unconfined compression (2 hr on/2 hr off) from 1.25 mm to 1.0, 0.75, or 0.5 mm for 24-hr led to sustained increases in loss of ^{35}S-sulfate and ^{3}H-proline labeled macromolecules. In addition, high amplitude cyclic compression led to a sustained increase in the rate of loss of ^{3}H-hydroxyproline residues and an increase in tissue swelling. Thus, such compression appeared to have caused disruption of the collagen network (Sah *et al.*, 1991). The ^{35}S-proteoglycans lost during such acute loading were of smaller size than those from controls, but contained a similarly low proportion (~15%) that could form aggregates with excess hyaluronan and link protein. The size distribution and aggregability of the remaining tissue aggrecan were not markedly affected. The loss of tissue aggrecan paralleled the loss of ^{35}S-labeled macromolecules. The loss rates from these radiolabeled disks were not greatly altered by incubation in acid-titrated medium (Sah *et al.*, 1991).

Taken together, these macromolecular release patterns suggested that a threshold in high amplitude loading was identified which could cause disruption the collagen matrix meshwork. These acute cyclic loads additionally appeared to affect matrix loss by altering diffusive and convective transport. This study provides a model for mechanically induced cartilage degeneration *in vitro*, and suggests a means for distinguishing between direct mechanical damage to the matrix versus cell-mediated (e.g., enzymatic) degradation. The possibility that excessive compression may alter the forms of the catabolic fragments of aggrecan and collagen associated with specific matrix proteinases appears to be an important issue for further studies.

Physical Mechanisms — Static Compression

It has been hypothesized that the inhibition of biosynthesis and changes in proteoglycan binding affinity (aggregation) during static compression may be associated, in part, with physicochemical changes in the environment of the cells (Bayliss *et al.*, 1986; Schneiderman *et al.*, 1986; Gray *et al.*, 1988; Sah *et al.*, 1989; 1990a; Urban and Bayliss, 1989; Urban *et al.*, 1993). With compression and fluid exudation, the density of fixed negative charges in the matrix is increased, resulting in attraction and concentration of positive counterions such as H^+ and Na^+. Thus, changes in intratissue pH and osmolarity have been identified as possible mediators of cellular response. A theoretical model based on Donnan equilibrium and electroneutrality has been used to estimate changes in intratissue ion concentrations caused by compression to test such a hypothesis (Gray *et al.*, 1988; Urban and Bayliss, 1989; Urban *et al.*, 1993). Indeed, proteoglycan biosynthesis was similarly inhibited by equivalent increases in the intra-tissue concentration of H^+ (i.e., decreased pH) (Gray *et al.*, 1988) or Na^+ (Urban and Bayliss, 1989) brought about by either (a) tissue compression or (b) incubation in bathing medium of low pH or high NaCl in the absence of compression. Interestingly, biosynthetic rates were found to be maximal in isolated chondrocytes and cartilage explants when medium osmolarity was adjusted to achieve physiologic levels of extracellular osmolarity (Urban *et al.*, 1993). Other recent studies have shown that altered medium pH or osmolarity can induce altered growth rates in neonatal cartilage over several weeks in culture (Garcia *et al.*, 1994). In addition, ongoing studies have addressed the effects of changes in extracellular pH on the intracellular pH of chondrocytes, and the range over which intracellular pH is buffered (Wilkins and Hall, 1992).

Static compression or deformation may cause changes in matrix and cell morphology that could also alter chondrocyte biosynthesis. Tissue compaction results in decreased matrix pore size, which could hinder *transport* of nutrients, waste products, growth factors, and newly synthesized matrix molecules. Previous studies have addressed certain issues regarding diffusion limitation and altered biosynthesis during static compression (Gray *et al.*, 1988) or osmotic dehydration (Schneiderman *et al.*, 1986). More recently, a series of experiments (Kim *et al.*, 1994b) in which the *surface area-to-volume ratio* of disks or the *concentration of labeling substrate* or *serum* were varied provided no evidence that limited diffusive transport of nutrients, waste products and/or precursors was responsible for the inhibition of biosynthesis by large displacement static compression, such as that shown in Fig 10.1A. (Recovery of biosynthesis from static compression, histological analyses of compressed tissue, and measured DNA content suggested that there was no significant cell damage even during 12-h of 50% static compression (Kim *et al.*, 1994b).) Cell shape changes have also been shown to alter chondrocyte phenotype (Benya *et al.*, 1988; Benya and Shaffer, 1982; Brown *et al.*, 1988) and PG synthesis (Watt, 1986); therefore, cell shape changes caused by matrix compression (Freeman *et al.*, 1990; Guilak *et al.*, 1991) or osmotic dehydration could conceivably alter chondrocyte metabolism.

Physical Mechanisms — Dynamic Compression

Dynamic compression can induce a variety of physical phenomena, including hydrostatic pressure gradients, fluid flow, streaming potentials and currents within the tissue (Mak, 1986; Mow *et al.*, 1984; Frank *et al.*, 1987) (Figure 10.1B), depending on the frequency of loading, specimen geometry, and boundary conditions. At higher frequencies (e.g., 1 Hz), there may be very little fluid exudation/imbibition during each loading cycle, with elastic-like deformation (Eberhardt *et al.*, 1990). In contrast, at lower frequencies (e.g., 0.001–0.01 Hz), fluid flow within the tissue could be more significant during each cycle.

To explore the effect of the dynamic component of compression on biosynthesis, small amplitude cyclic unconfined compressions were applied to cartilage disks superimposed on a *fixed* static offset compression (Figure 10.1B) (Sah *et al.*, 1989; Kim *et al.*, 1994b); experimental and control explants were subjected to the same level of static offset compression. Such low amplitude cyclic compressions do not significantly alter the time-averaged disk water content or fixed charge density and, therefore, do not result in physicochemical changes associated with large amplitude static compressions (e.g., Figure 10.1A) (Sah *et al.*, 1990b). In this manner, the effects of oscillatory deformation of cells and matrix, fluid flow, streaming potentials, and hydrostatic pressure could be explored separately from physicochemical stimuli. In the experiments of Figure 10.1B, the load and compression sustained by the cartilage disks were simultaneously measured throughout mechanical stimulation and radiolabeling; in this manner, the dynamic stiffness of the cartilage was measured (Sah *et al.*, 1989; Kim *et al.*, 1995). The dynamic stiffness amplitude began to increase sharply at a frequency of at ~0.001 Hz, and plateaued as the frequency was increased to ~0.1 Hz (Sah *et al.*, 1989; Kim *et al.*, 1995). Based on poroelastic theory, this increase in stiffness is attributed to fluid/matrix frictional interactions which become significant as the frequency is increased from zero. The theoretical frequency at which the increase in stiffness becomes steepest is of order $1/\tau_m$, where $\tau_m = a^2 / H_a k$ is the characteristic unconfined compression stress relaxation time constant (Armstrong *et al.*, 1984) (a is the disk radius, H_a is the equilibrium confined compression modulus, and k is the hydraulic permeability). For $a = 1.5$ mm, $H_a \sim 0.5$ MPa (Mow *et al.*, 1984; Frank and Grodzinsky, 1987b), and $k \sim 2 \times 10^{-15}$ m^2 /Pa.s (Mow *et al.*, 1984; Frank and Grodzinsky, 1987b), the characteristic frequency $\sim 1/\tau_m \sim 0.001$ Hz.

Poroelastic theory shows that at the higher frequencies, there is an increase in hydrostatic pressure in the central region of the cartilage disk; however, fluid velocity (i.e., pressure gradients) and streaming potential are highest near the radial periphery of the disk. Therefore, the relative importance of oscillatory *fluid flow, hydrostatic pressure, streaming potential, and cell deformation* in modulating chondrocyte metabolism during dynamic compression was explored by quantifying the *frequency dependence* and the *spatial (radial) distribution* of the biosynthetic response within the 3-mm diameter explant disks (Kim *et al.*, 1994b; 1995). After dynamic compression of 3-mm disks as in Figure 10.1B, the center 2-mm core of each disk was removed and the core and outer ring were analyzed separately for radiolabel incorporation. While compression at frequencies between 0.002 and 0.01 Hz caused a stimulation of biosynthesis that was distributed throughout the core and outer ring of the disk, compression at 0.1 Hz caused a stimulation that was confined mainly to the outer ring (Figure 10.2B) (Kim *et al.*, 1994b). These distributions in incorporation were compared to theoretical estimates of the radial distribution of physical forces and flows within the matrix (Kim *et al.*, 1994b; 1995) (Figure 10.2A). These biosynthetic patterns most closely matched the spatial profiles of fluid flow (which is coupled to streaming potential) and cell deformation (Kim *et al.*, 1995). Thus, the stimulation of chondrocyte biosynthesis by *dynamic* mechanical compression appears associated mainly with changes in fluid flow, streaming potential, and/or cell shape (which were greatest near the disk periphery at higher frequencies (Figure 10.2)), while hydrostatic pressure probably played a less important role under the conditions of our experiments. Nevertheless, the slight biosynthetic stimulation in the central region together with measured peak dynamic loads of ~0.5–1 MPa at the higher frequencies would be consistent with recent findings that hydrostatic pressure can stimulate biosynthesis (Hall *et al.*, 1991).

Although relative flow between the solid and fluid phases of the matrix may stimulate biosynthesis via many mechanisms, one possible effect of increased relative flow may be increased convective transport of nutrients and growth factors (O'Hara *et al.*, 1990). Preliminary data suggest that even when the serum concentration is reduced from 10% to 0.5%, dynamic compression at 0.1 Hz stimulates proline and sulfate incorporation into 3-mm diameter disks (Wu, 1992). Furthermore, dynamic compression of 3-mm diameter chondrocyte/agarose disks, which have 1/4 the GAG concentration of intact calf cartilage and, hence, less resistance to diffusive transport, showed increased biosynthesis at 0.1 Hz (Buschmann *et al.*, 1992) (see below). Therefore, if relative fluid flow was an operative mechanism during dynamic compression, it might act to directly stimulate chondrocytes (e.g., via membrane shear) or alter the pericellular concentrations of peptide factors or newly synthesized matrix macromolecules, and not just indirectly to increase convective transport of nutrients.

With increasing frequency (strain rates), dynamic compression of cartilage would induce increasing levels of hydrostatic pressure within the tissue (Armstrong *et al.*, 1984). The effect of hydrostatic pressure on cartilage metabolism could be studied via mechanical compression of cartilage (as above), or more directly by pressurizing the fluid in a vessel containing chondrocytes or cartilage explants. The effects of hydrostatic pressure on chondrocyte activity in embryonic chicken, fetal mouse, bovine and human cartilage, and in rat chondrosarcoma cells have been reported (Bourret and Rodan, 1976; Hall *et al.*, 1991; Kimura *et al.*, 1985; Klein-Nulend *et al.*, 1986; 1987; Lippiello *et al.*, 1985; van Kampen *et al.*, 1985; Veldhuijzen *et al.*, 1979; 1987). In general, static or low frequency hydrostatic pressure caused a decrease in GAG synthesis, while higher frequency hydrostatic loading

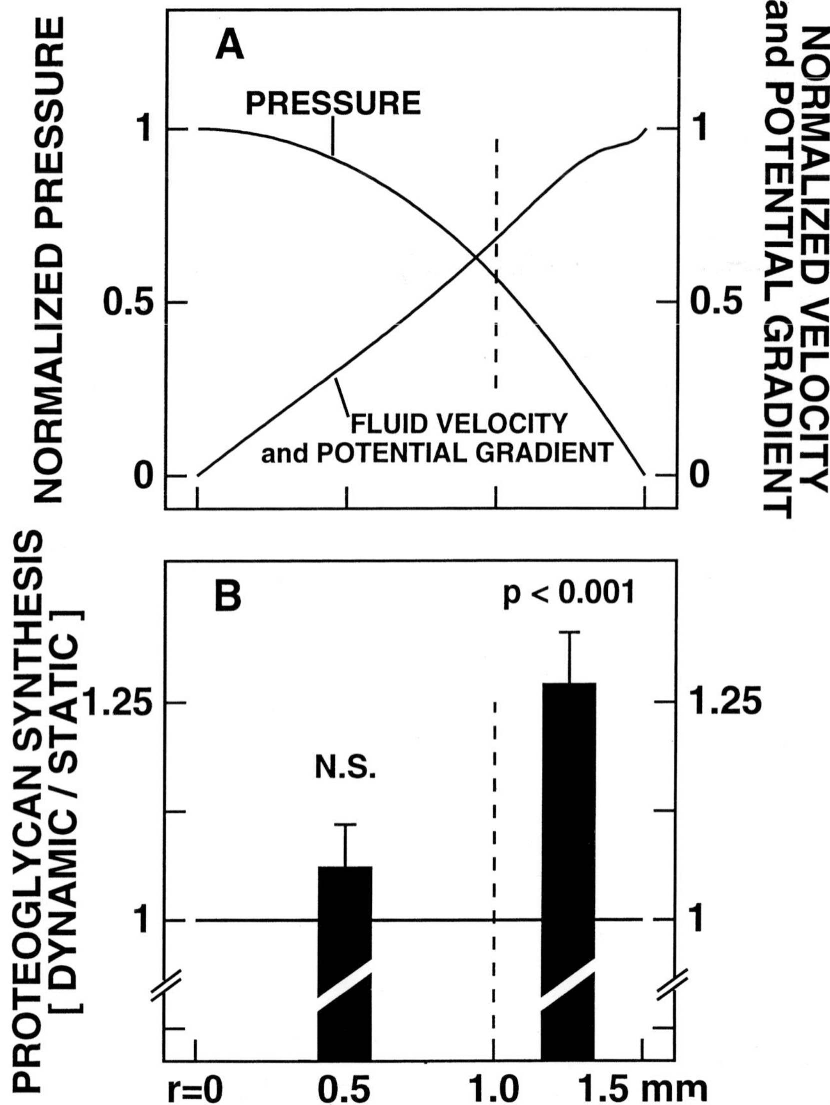

Figure 10.2A Normalized amplitudes of streaming potential gradient, interstitial fluid velocity (relative to the matrix), and hydrostatic pressure as a function of radial position at the mid-plane ($z = 0$) within a 3-mm diameter cartilage disk, at the frequency 0.1 Hz. The pressure amplitude is normalized to the value at $r = 0$; the relative fluid velocity/streaming potential gradient amplitude is normalized to the value at $r = 1$. These profiles represent the maximum amplitudes that would occur at each radial position within a 3 mm diameter disk during sinusoidal, uniaxial unconfined compression of a disk by solid (fluid-impermeable) platens (see Kim *et al.*, 1995 for details). **(B)** Bars represent proteoglycan synthesis measured as [^{35}S]-sulfate incorporation in the 2 mm diameter core and the outer ring of 3 mm diameter cartilage disks compressed at 0.1 Hz, relative to the incorporation rates in the statically held controls (data adapted from Kim *et al.*, 1994b). Dynamic compression increased proteoglycan synthesis substantially more in the outer ring, similar to the trends of the streaming potential gradient and fluid velocity profiles of **(A)**.

caused an increase in GAG synthesis. However, the effective threshold pressure for increased synthesis varied widely. In chick embryonic chondrocytes and fetal mouse cartilaginous long bone rudiments, cyclic pressures as low as 13 kPa (Klein-Nulend *et al.*, 1986; 1987; van Kampen *et al.*, 1985; Veldhuijzen *et al.*, 1987) elicited a stimulatory response, while pressures on the order of 5 MPa (Hall *et al.*, 1991; Kimura *et al.*, 1985; Lippiello *et al.*, 1985) were needed to induce similar increases in matrix synthesis in bovine and human articular cartilage.

Regarding changes in cell shape and volume, at very low frequency the chondrocytes are compressed axially in a radially uniform manner due to the inhibition of radial bulging by the collagen fibrils and minimal pressurization due to the slow rate of compression. Therefore, the chondrocytes throughout the disk can lose water, resulting in decreased cell volume (Freeman *et al.*, 1990; Guilak *et al.*, 1991). As the frequency of compression is increased, more fluid would be trapped within the matrix giving rise to increased radial bulging of the disks. The matrix would thus appear more incompressible (volume conserving) with increased frequency in the disk interior. Therefore, changes in cell volume could occur preferentially in the outer radial periphery, consistent with the radial dependence of the biosynthetic response observed. If the frequency is too low then the cell is unable to respond. Furthermore, preliminary experiments suggest that cartilage disks treated with 6 μM dihydrocytochalasin-B (DHCB, which disrupts actin polymerization in the cell) in media containing 0.5% FBS did not show any response to dynamic compression at 0.1 Hz (Wu, 1992). Previously, researchers have used cytochalasin to demonstrate the roles of the cytoskeleton and cell shape/volume changes in modulating collagen phenotype [19–21], sulfate incorporation by chondrocytes (Newman and Watt, 1988), and intracellular cAMP level (Watson, 1989; 1990). Therefore, it is a strong possibility that cell shape or volume change is part of the transduction mechanism operative during dynamic compression. The interaction of cell shape and cellular function during dynamic compression may be mediated by ECM receptors (Watt, 1986), stretch receptors on cell membranes (Guharay and Sachs, 1985), and mechanical forces transmitted through the cytoskeleton (Watson, 1991). DeWitt *et al.* (1984) demonstrated that dynamic mechanical stretching of 14-d old chondrocyte cultures caused a stimulation of PG and DNA synthesis and cAMP levels but no change in protein synthesis, though it was difficult to separate the effects of cell deformation from other physical stimuli. Lee *et al.* (1982) plated chondrocytes onto elastin membranes and observed a stimulation in GAG synthesis when the membranes were stretched at 1 Hz using 10% strain. However, simple agitation of the membrane also caused an increase in GAG synthesis, unlike the results of DeWitt *et al.* (1984).

DIFFERENTIAL EFFECTS OF COMPRESSION ON THE SYNTHESIS OF SPECIFIC MATRIX MACROMOLECULES

Effects of Static Compression on Aggrecan, Link Protein, Hyaluronan, Small Proteoglycans, and Collagen

The studies on the effects of mechanical loading on chondrocyte biosynthesis described above have focused on incorporation of ^{35}S-sulfate and ^{3}H-proline as a measure of the synthesis of total proteoglycans and proteins. Recently, experiments were initiated to quantify the effects of compression on the specific molecular components of the matrix, including the molecular components of the proteoglycan aggregate, the small proteoglycans, and collagen (Kim *et al.*, 1992; 1993). Groups of disks were radiolabeled during 12-hr of static compression and analyzed for structure of newly synthesized aggrecan (chromatographic profiles, sulfation pattern, and core protein processing kinetics using a cyclohexamide protocol (Kimura *et al.*, 1981)), link protein synthesis (immunoprecipitation, SDS-PAGE), hyaluronan synthesis (disaccharide isolation), decorin and biglycan synthesis (chromatography, SDS-PAGE), and collagen synthesis (^{3}H-hydroxyproline). During static compression, aggrecan synthesis was inhibited compared to uncompressed controls (Figure 10.3A, similar to Figure 10.1A); this inhibition was found to be associated specifically with a decreased core protein pool size and GAG synthesis capacity. Aggrecan core protein size, GAG size, and GAG sulfation were unchanged (Kim *et al.*, 1993). The synthesis of collagen and link protein (Figure 10.3A) and small proteoglycans (not shown) were also markedly inhibited by static compression (Kim, 1993). In marked contrast, hyaluronan synthesis was not affected by compression (Kim *et al.*, 1994a).

Kinetics of Recovery of Biosynthesis after Release of Compression

Groups of disks were compressed for 12 hours, released from compression, and cultured in the free swelling state for 3 days. After specified recovery times, disks were radiolabeled to assess matrix component biosynthetic rates. By 8 hours after release of compression, aggrecan synthesis was still markedly depressed (Figure 10.3B). Aggrecan synthesis recovered slowly to free swelling control values by 60 hr after release (Kim *et al.*, 1992; 1994a), and aggrecan was initially substituted with fewer and longer GAG chains. Consistent with this finding, the core protein processing time constant was almost 2-fold longer (Kim *et al.*, 1993). Collagen was also slow to recover, and the small proteoglycans essentially recovered by 3 days after release of compression. Surprisingly, however, synthesis of both forms of link protein completely recovered by 8 hr after release (Figure 10.3B), and there was no apparent change in size (Kim *et al.*, 1993). Hyaluronan was unaffected by release of compression (Figure 10.3B) (Kim *et al.*, 1994a).

The finding that hyaluronan synthesis was unaffected by static compression while aggrecan and link synthesis were markedly inhibited, suggests that static compression selectively affects protein synthesis. This inhibition is most likely a physiologically relevant response since aggrecan structure was unchanged and the processing rate of aggrecan core-protein was maintained during compression, whereas this rate was slowed during recovery. The fact that post-translational modification of link protein was unaffected immediately after release from compression suggests that alteration in aggrecan glycosylation is not due to nonspecific disruption of cellular machinery. Further, the response to static compression did not appear to be due to cell lysis since there was a complete recovery of link protein synthesis immediately after release from compression.

The finding that HA synthesis was unaffected by compression or release is exciting with respect to hypotheses on mechanisms, and has important implications regarding hypotheses linking biosynthetic pathways to the effects of mechanically generated cell deformation and to connections between ECM and cytoskeletal structures. Unlike aggrecan and link protein synthesis, which require extensive intracellular processing, synthesis of

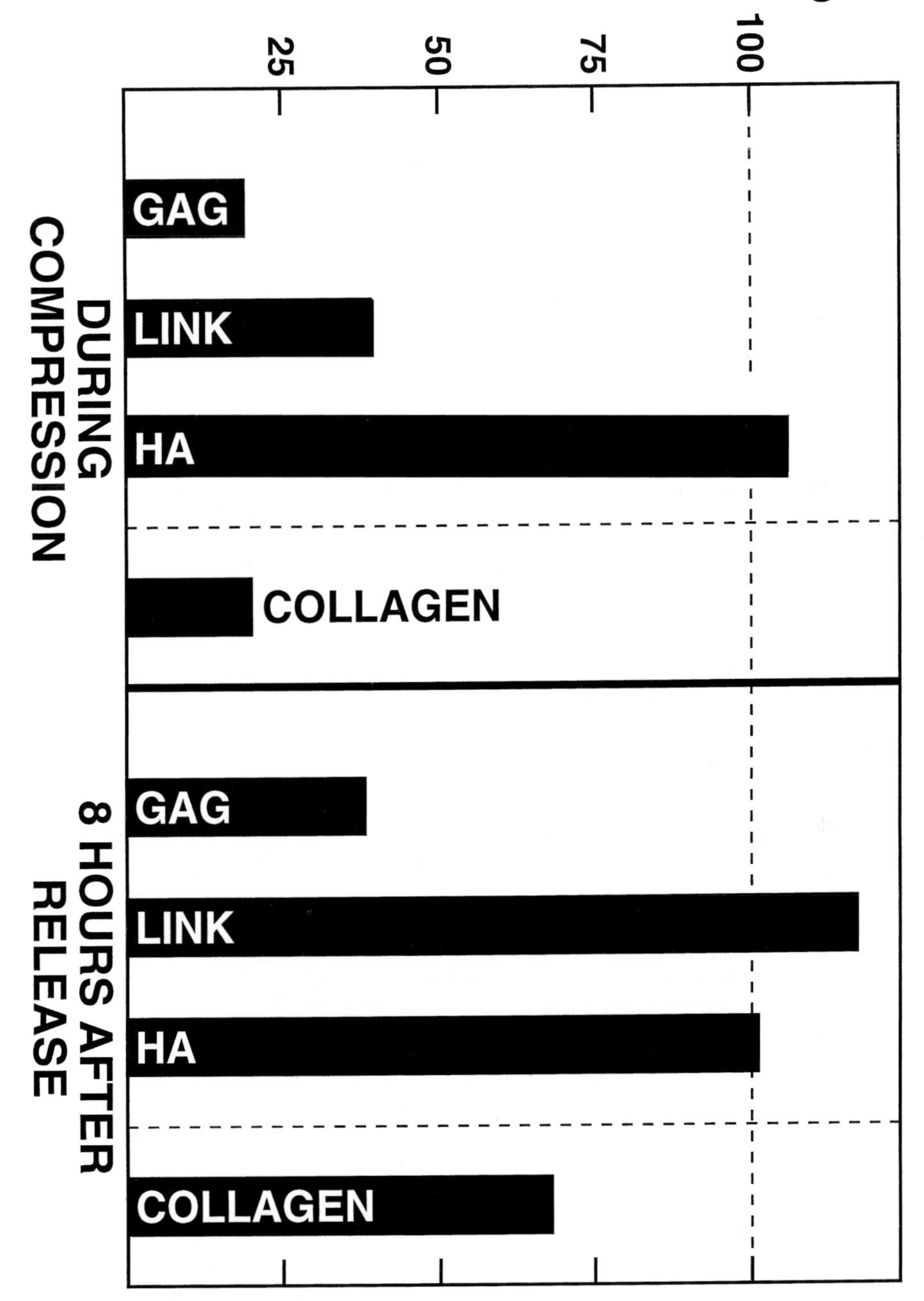

Figure 10.3 Biosynthetic rates of aggrecan, link protein, hyaluronan, and collagen during 12-hour static compression of calf cartilage disks to 50% of initial cut thickness (A), and 8-hours after release of compression (B), normalized to that in free swelling control disks (data from Kim *et al.*, 1992; 1993).

hyaluronan occurs at the plasma membrane catalyzed by the hyaluronan-synthetase complex (Spiro *et al.*, 1991; Ng *et al.*, 1992). In addition, the specific activity of the UDP-hexosamine pool increased with compression as measured by the relative amounts of [^{35}S]-sulfate and [^{3}H]-glucosamine incorporated into ΔDi6Sr (Kim *et al.*, 1994a). In this context, Morales and Hascall (Morales and Hascall, 1989) reported that incubation of calf cartilage explants with IL-1 also caused an inhibition of aggrecan synthesis accompanied by an increase in the specific activity of the UDP-hexosamine pool; however, HA synthesis during IL-1 treatment was not affected. The striking parallel in the results produced by IL-1 treatment (Morales and Hascall, 1989) and static mechanical compression suggests that the response of cartilage to static compression is not a general inhibition of cellular activity, but appears to be part of a transduction mechanism which results in alterations of specific pathways. Together, these results suggest that studies of the extent and kinetics of chondrocyte biosynthetic response to specific regimes of dynamic and static compression may give significant insights concerning the cellular targets and biosynthetic pathways that may underly the biosynthetic response to compression.

CELL-MATRIX INTERACTIONS: CHONDROCYTE BIOSYNTHESIS IN AGAROSE GEL CULTURE

Correlation of Matrix Synthesis with Matrix Content in Long Term Culture

The ability of chondrocytes from calf articular cartilage to synthesize and assemble a mechanically functional cartilage-like extracellular matrix was quantified in high cell density (~10^7 cells/ml) agarose gel culture (Buschmann *et al.*, 1992). The time evolution of chondrocyte proliferation, proteoglycan and protein synthesis and loss to the media, and total deposition of glycosaminoglycan — containing matrix within agarose gels was characterized during 10 weeks in culture. To assess whether the matrix deposited within the agarose gel was mechanically and electromechanically functional, the time evolution of dynamic mechanical stiffness and oscillatory streaming potential in uniaxial confined compression was measured in parallel cultures; from these data, the intrinsic equilibrium modulus, hydraulic permeability and electrokinetic coupling coefficient of the developing cultures (Frank and Grodzinsky, 1987a; 1987b) was calculated. At early times in culture before significant matrix had been deposited by the cells (e.g., low GAG content during the first few days, Figure 10.4A) the rate of proteoglycan synthesis was high (Figure 10.4B). By 1 month in culture, proteoglycan and protein synthesis had fallen (Figure 10.4B) to a level similar to that found in the parent calf cartilage explants from which the cells were extracted (Buschmann *et al.*, 1992). The majority of the newly synthesized proteoglycans remained in the gel, and their deposition appeared to have caused the disks to swell by more than 50% during this period. Histological sections showed that the morphological appearance of the cultured chondrocytes was similar to that of cells in native tissue (Buschmann *et al.*, 1992). A matrix coat rich in PG and collagen developed around each cell or cell group, and increased in thickness as the culture proceeded. The equilibrium modulus, dynamic stiffness, and compression-induced streaming potential rose to many times (>5×) their initial values at the start of the culture; the hydraulic permeability decreased to a fraction (~1/10) that of the cell-laden porous agarose at the beginning of the culture. By day 35 of culture, GAG concentration (Figure 10.4A), equilibrium modulus,

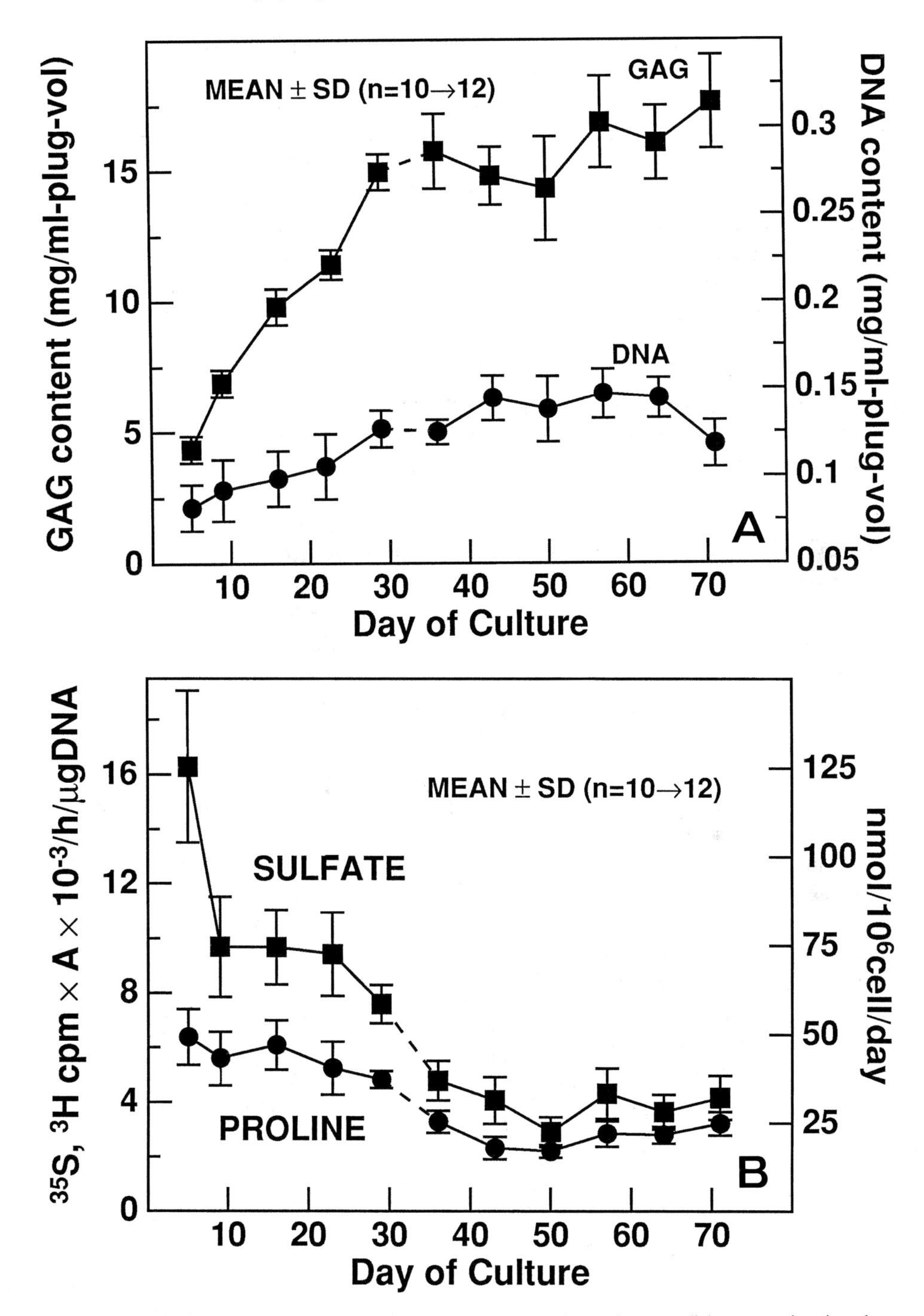

Figure 10.4 (A) GAG and DNA content in 3 mm diameter chondrocyte/agarose disks versus time in culture, in mg/ml of disk volume. (B) Sulfate and proline incorporation versus time in culture (A = 1.36 ± 0.04 for sulfate and 0.285 ± 0.008 for proline). (Data from Buschmann *et al.*, 1992).

and streaming potential had increased to ~ 25% that of calf articular cartilage. In particular, the non-destructive measurement of the streaming potential correlated well with changes in GAG concentration with time in culture (Buschmann *et al.*, 1992).

Together, these results suggested that chondrocytes in long-term agarose culture could synthesize a mechanically functional cartilage-like matrix. Interestingly, the matrix biosynthetic rates appeared to depend on the content of newly deposited pericellular and territorial matrix: synthesis was highest initially before significant matrix deposition had occurred (Figure 10.4). This observation is consistent with a physicochemical mechanism, since increasing levels of GAG deposition would lower the extracellular pH in the immediate vicinity of the cells. In this regard, the chondrocyte/agarose culture system provides a useful model system to study the role of matrix and cell-matrix interactions in the regulation of chondrocyte response to mechanical loads. Using selected cell populations and culture times (e.g., Aydelotte and Kuettner, 1988; Aydelotte *et al.*, 1988), it may be possible to perturb matrix assembly through the addition or removal of a particular matrix constituent and to subsequently determine the mechanical and biological significance of that constituent.

Effect of Static and Dynamic Compression on Chondrocyte Biosynthesis in Agarose Culture

The role of cell-matrix interactions and physicochemical mechanisms. In a series of experiments (Buschmann *et al.*, 1995), 3-mm diameter chondrocyte/agarose disks were placed in the static and dynamic compression chambers and subjected to compression protocols at different times in culture (2–41 days). [^{35}S]-sulfate and [^{3}H]-proline radiolabel incorporation rates were used as measures of proteoglycan and protein synthesis, respectively, as in the explant studies above. Graded levels of static compression (up to 50%) produced little or no change in biosynthesis at very early times, but resulted in a significant decrease in synthesis with increasing static compression at later times in culture (Buschmann *et al.*, 1995b) (Figure 10.5); the latter observation was qualitatively similar to that seen in cartilage explants (Sah *et al.*, 1989). (Control studies (Freeman *et al.*, 1994) utilizing imaging microscopy showed that compression of agarose caused a concomitant compression/deformation of chondrocytes even at early times before significant matrix deposition.) Dynamic compression (~3% dynamic strain amplitude at 0.01–1.0 Hz) stimulated radiolabel incorporation by an amount that increased with time in culture prior to loading as more matrix was deposited around and near the cells (Buschmann *et al.*, 1991; 1995). This stimulation was also similar to that observed in cartilage explants (Sah *et al.*, 1989; Kim *et al.*, 1994b). The presence of greater matrix content at later times in culture also created differences in biosynthetic response at the center versus near the periphery of the 3 mm chondrocyte/agarose disks, with trends similar to those observed in explant disks (i.e., Figure 10.2B).

The fact that chondrocyte response to static compression depended significantly on the presence or absence of a surrounding matrix, as did the physical properties of the disks, suggested that cell-matrix interactions (e.g., mechanical and/or receptor mediated) and possible physicochemical mechanisms (e.g., altered hydration and charge density, reduced pH) may be critical in determining the biosynthetic response to compression. For dynamic compression, fluid flow, streaming potentials, and cell-matrix interactions (deformation)

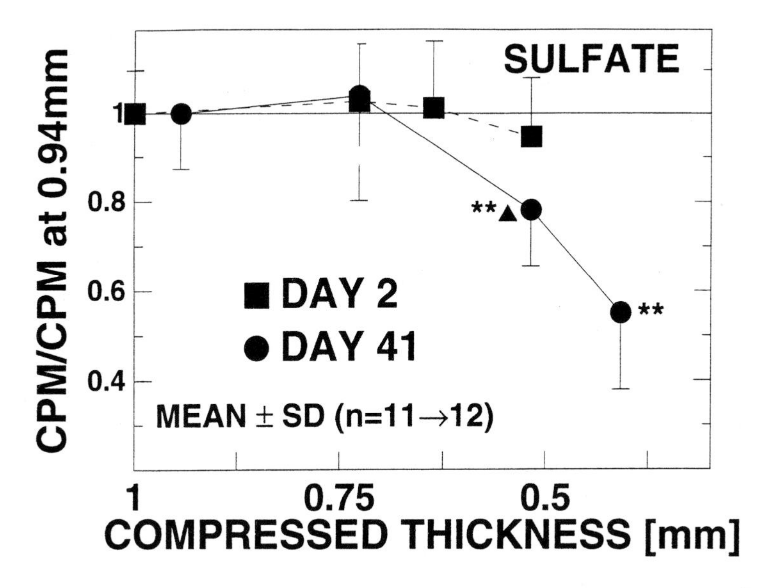

Figure 10.5 Effects of 12-hour static compression on proteoglycan synthesis in chondrocyte/agarose disks (such as those of Figure 10.4) at day 2 and day 41 of culture. (Data from Buschmann *et al.*, 1995b).

appeared to be more significant as stimuli than the small increase in fluid pressure, transport limitations, and matrix-independent cell deformation (Buschmann *et al.*, 1995b). The qualitative similarity in biosynthetic response to mechanical compression of chondrocytes cultured in agarose gel and chondrocytes in intact cartilage explants further indicates that the gel culture system preserves certain physiological features of chondrocyte behavior, and can be used to investigate chondrocyte response to physical and chemical stimuli in a controlled manner.

A MOLECULAR MODEL OF PROTEOGLYCAN-ASSOCIATED ELECTROSTATIC FORCES IN CARTILAGE MATRIX MECHANICS

Articular cartilage regions that are habitually subjected to high loads *in vivo* have been found to contain the highest concentrations of aggrecan (Slowman and Brandt, 1986; Roberts *et al.*, 1986). Studies *in vivo* and *in vitro* have shown that moderate dynamic compression of cartilage can stimulate proteoglycan synthesis by chondrocytes. Application of cyclic unconfined compression to fibrocartilaginous regions of bovine flexor tendon was found to induce and maintain the synthesis of large proteoglycans; in the absence of such loading, synthesis of small rather than large proteoglycans predominated (Koob *et al.*, 1992). From these studies and many others, it is clear that cell synthesis of large aggregating proteoglycans is very sensitively regulated by compressive loads. The resistance to compressive loading provided by aggrecan has been ascribed to the highly charged GAG chains substituted along the core protein. It is important to ask why such a complicated macromolecular structure is synthesized by the cells in order to provide the ECM with its necessary mechanical and osmotic properties. While aggregation of aggrecan along the hyaluronan backbone is thought to facilitate retention of proteoglycans within the tissue, the structure of the aggrecan monomer itself is not as easily motivated from a purely functional viewpoint. In particular, the localization of fixed charge groups along extended, rod-like GAG chains in a high density three dimensional array (corresponding to a high density of aggregates within the ECM) is only one means of achieving a high "fixed charge density" within the ECM. While the structure of aggrecan may be dictated in part by advantages associated with intracellular processing, we suggest here that localization of charge groups along GAG chains has functional advantages as well.

The importance of charged glycosaminoglycan chains in the functional properties of articular cartilage and other tissues has been long recognized (Maroudas, 1979). Swelling pressure and compressive stiffness has been shown to correlate with the concentration of matrix proteoglycans and their GAG constituents in a variety of cartilaginous tissues. To date, most models of cartilage swelling pressure have incorporated the Donnan theory directly (Maroudas, 1979; Urban *et al.*, 1979) or as part of a more general macroscopic continuum theory for cartilage behavior (Lai *et al.*, 1991). Molecular level models have not been commonly used to estimate explicitly the electrostatic contribution to proteoglycan swelling pressure and cartilage compressive stiffness.

The electrostatic contribution to compressive stiffness and swelling pressure arises from long range coulombic forces associated with the fixed charge groups of the GAG chains of the constituent aggrecan molecules. Macroscopically, this electrostatic contribution has often been viewed as a Donnan osmotic swelling pressure (Maroudas, 1979) since the presence of fixed charge increases the total concentration of mobile ions within the tissue and thereby increases the intratissue osmotic pressure. On the molecular level, the electrostatic contribution to stiffness and swelling pressure arises from repulsive electrostatic forces between GAG chains, which can be calculated from the fundamental laws of electrostatics and thermodynamics using the Poisson-Boltzmann (PB) equation in the context of a unit-cell model (Marcus, 1955).

It is widely accepted that the macroscopic Donnan model and molecular level PB model are two different views of the same fundamental electrostatic phenomenon (Overbeek, 1956; Sanfeld, 1968). The swelling pressure predicted by both models is the consequence of highly coupled electrostatic and osmotic phenomena, and could be called equivalently an osmotic swelling pressure or an electrostatic swelling pressure. In the Donnan approach, each "incremental volume" of extracellular matrix contains many macromolecules. There is no molecular level structure assumed; the electrostatic potential Φ is assumed to be constant, and thus the electric field is zero within the uniformly charged polyelectrolyte phase (Figure 10.6). In contrast, the length scale inherent to the Poisson-Boltzmann equation contains many ions but can be much smaller than an individual polyelectrolyte molecule. Because the electrical Debye length at physiological ionic strength (~0.8 nm) is on the order of or less than the spacing between neighboring GAG chains in tissues such as cartilage, the electrical potential will vary steeply between GAG chains (Figure 10.6) creating significant local electric fields in the vicinity of and between individual GAG chains. The inclusion of these molecular level electric fields in the quantitative prediction of swelling pressures and moduli is inherent to the Poisson-Boltzmann formulation and is absent from the Donnan approach. As a further comparison, it was recently shown (Basser and Grodzinsky, 1993) that the Donnan model can be derived directly from the micro PB model of a composite medium, using homogenization and scaling methods applied to a

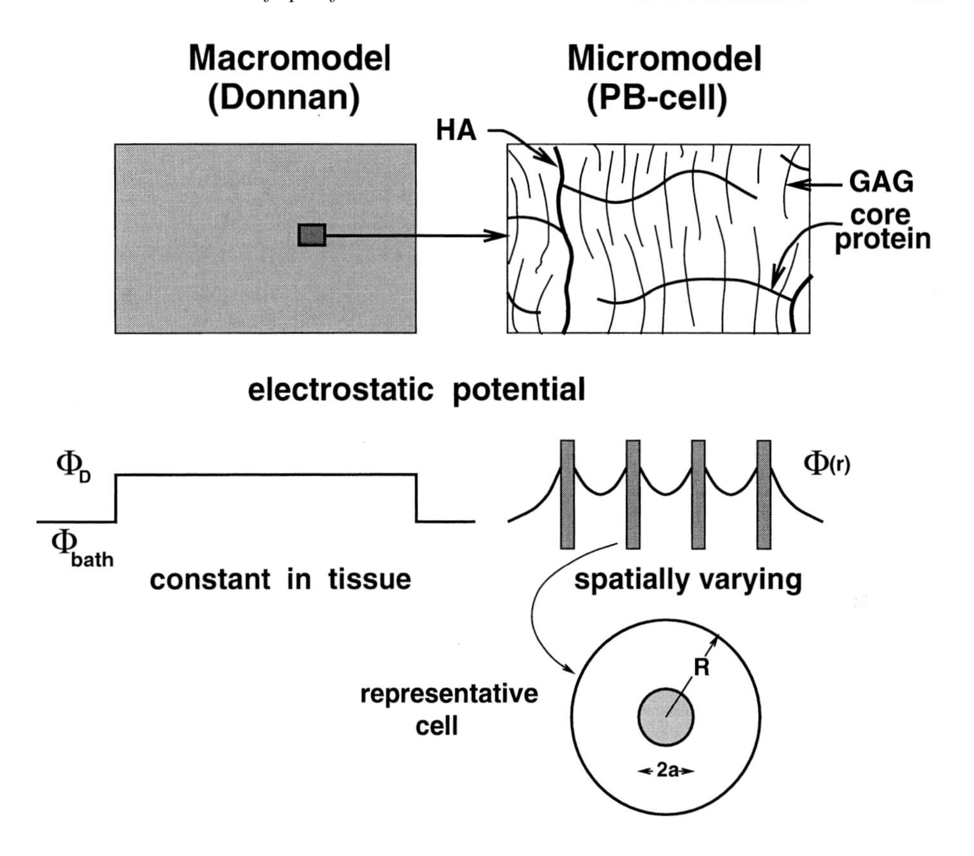

Figure 10.6 Schematic of ideal Donnan model and the microscopic Poisson-Boltzmann unit cell model. The Donnan model assumes a constant intratissue potential for a uniform matrix fixed charge density, while the PB model accounts for spatial variations in potential between GAG chains. The Donnan model is fully described by specifying the fixed charge density, while the PB model requires the radius of the GAG chain a, the intercharge spacing along the chain, and the radius of the unit cell (interchain spacing) R (from Buschmann and Grodzinsky, 1995).

periodic array of charged lamina in an ionic solution. The zeroth order solution of the homogenized PB equation was identical to the Donnan potential calculated from Donnan theory. This result is further evidence that the macro Donnan and micro Poisson-Boltzmann models describe the same electrostatic phenomena but at different length scales.

Buschmann and Grodzinsky (Buschmann and Grodzinsky, 1995) recently compared the ability of these two different theoretical models to predict previously published values of the swelling pressure of charged proteoglycans in solution (Williams and Comper, 1990) and the ionic strength dependence of the equilibrium modulus of PG-rich articular cartilage (Eisenberg and Grodzinsky, 1985). The microstructural model was based on the solution of the Poisson-Boltzmann equation within a unit cell containing a charged GAG molecule and its surrounding atmosphere of mobile counterions (Figure 10.6). In the prediction of PG swelling pressures at different PG concentrations, all model parameters were known from experiments — there were no adjustable parameters. In the prediction of the equilibrium modulus of cartilage at varying ionic strengths, there was one adjustable parameter in each

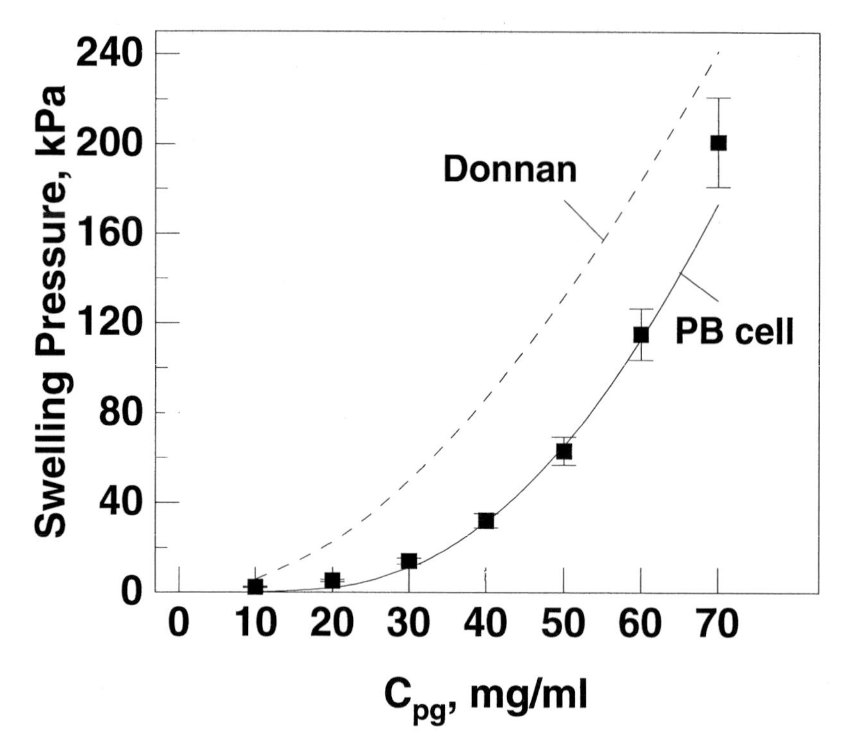

Figure 10.7 Donnan and PB model computations of swelling pressure of proteoglycans solutions as a function of PG concentration, C_{pg}. Data are from (Williams and Comper, 1990) (see Buschmann and Grodzinsky, 1995 for details).

model which was varied to obtain the best fit to the data. The resulting best fit parameter is compared to the range of literature values from other experiments.

Swelling Pressure of Proteoglycan Solutions

The predictions of the ideal Donnan model and the PB-cell model were compared to the data (Figure 10.7) of Williams and Comper (Williams and Comper, 1990), who used sedimentation-diffusion experiments to measure the osmotic swelling pressure of large, aggregating (chondroitin sulfate) proteoglycans from Swarm rat chondrosarcoma equilibrated in 0.15 M ionic strength phosphate-buffered saline PBS. These proteoglycans are of order 2×10^6 Da molecular weight, and are known to contain ~86% chondroitin sulfate (CS) as the primary GAG constituent with little other charged moieties. In the Donnan model, the fixed charge density ρ_m is proportional to the concentration of proteoglycans (C_{pg} in Figure 10.7). The proportionality constant is obtained from the structure of the chondroitin sulfate GAG assuming ~86% of the proteoglycans to be GAG (Buschmann and Grodzinsky, 1995). In the PB-cell model, the rod radius was set to $a = 0.55$ nm (Ogston *et al.*, 1973) and the intercharge distance b to 0.64 nm (Williams and Comper, 1990). The unit cell radius (inter-rod spacing), R, is determined from the PG concentration, by assigning a cylindrical volume of length $2b$ and radius R to each CS disaccharide (Buschmann and Grodzinsky, 1995). The concentrations of counter-ions and co-ions in the unit cell can be calculated from the Boltzmann distribution.

The range of PG concentration found in adult bovine articular cartilage is (20–80) mg/ml. Given the good agreement between the PB-cell model prediction, with no adjustable parameters, and the measured pressures (Figure 10.7) it would appear that within this concentration range, the swelling pressure of the PG is predominantly of electrostatic origin and is well described by the interactions included in the PB-cell model. The very steep rise in pressure predicted by the PB-cell model between 30 and 70 mg/ml, from 11 kPa to 173 kPa, is a consequence of decreasing the distance between these highly charged surfaces from ~5 Debye lengths at 30 mg/ml to ~2.8 Debye lengths at 70 mg/ml. Thus, using material parameter values from the literature, the PB-cell model agreed with the measured pressure of PG solutions to within experimental error (10%), whereas the ideal Donnan model overestimated the pressure by up to 3-fold (Figure 10.7).

Cartilage Equilibrium Modulus

The equilibrium modulus, H_A, of adult bovine articular cartilage has been measured in uniaxial confined compression for specimens equilibrated in NaCl baths ranging from 0.005 to 1.0 M (Eisenberg and Grodzinsky, 1985). The data of Figure 10.8 show that at neutral pH, H_A is relatively insensitive to NaCl concentration at the highest concentrations used. Thus, the modulus at 1.0 M NaCl approached an asymptote which is numerically equivalent to the nonelectrostatic contribution to the modulus. As the ionic strength is lowered, electrostatic repulsion forces between neighboring CS molecules come into

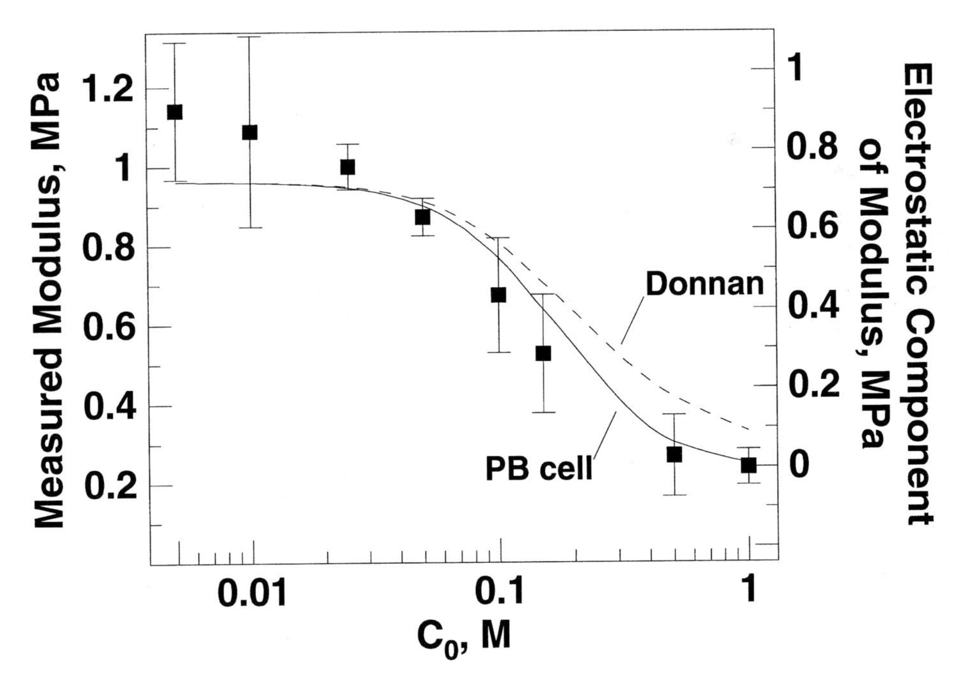

Figure 10.8 Electrostatic component of the equilibrium modulus of adult bovine articular cartilage measured in uniaxial confined compression at neutral pH, as a function of bath concentration of NaCl, C_o. Data are from (Eisenberg and Grodzinsky, 1985) (see Buschmann and Grodzinsky, 1995 for details).

play and add an electrostatic component, as indicated in Figure 10.8. This electrostatic component was estimated using the Donnan and PB-cell models. The measurements were performed at strains between 10% and 25%, and the stress-compression curves were seen to be essentially linear in this range. The electrostatic component was found from both models by dividing the difference in the swelling pressures at 10% and 20% compression by the change in strain (0.1) (Figure 10.8). The effect of strain on the charge density (Donnan) and the unit cell radius (PB) was included (Buschmann and Grodzinsky, 1995).

The PB model also quantitatively fit the ionic strength dependence of the confined compression equilibrium modulus of cartilage for a reasonable value of one adjustable parameter (Figure 10.8). When R is much less than the Debye length (low ionic strength in Figure 10.8), the two models could be expected to behave in similar fashions since the flat Donnan potential will be a closer approximation to the actual microscopic potential. At 0.15 M ionic strength, $1= \kappa \simeq 0.8$ nm; hence, a flat Donnan potential is not a reasonable approximation. Thus, in computations involving one adjustable parameter for each model, the PB-cell model predicted the ionic strength dependence of the equilibrium modulus of articular cartilage (Figure 10.8). Near physiologic ionic strength, the Donnan model overpredicted the modulus data by 2-fold, but the two models coincided for low ionic strengths ($C_0 < 0.025$ M) where the spatially invariant Donnan potential is a closer approximation to the PB potential distribution (Figure 10.8).

The ideal Donnan model was acceptable in determining small ion partitioning between polyelectrolyte and bath. At physiologic and higher ionic strengths, however, Donnan overestimated the potential and swelling pressure. Activity and osmotic coefficients can be computed from the PB model to use in a nonideal Donnan model, as has been described previously (Marcus, 1955).

In summary, the PB-cell model result (Buschmann and Grodzinsky, 1995) indicated that electrostatic forces between adjacent GAGs predominate in determining the swelling pressure of PG in the concentration range found in articular cartilage (20–80 mg/ml). The PB-cell model is also consistent with data (Eisenberg and Grodzinsky, 1985; Lai *et al.*, 1991) showing that these electrostatic forces account for ~1/2 (290 kPa) the equilibrium modulus of cartilage at physiological ionic strength while absolute swelling pressures may be as low as ~25–100 kPa. This important property of electrostatic repulsion between GAG chains that are highly charged but spaced a few Debye lengths apart allows cartilage to resist compression (high modulus) without generating excessive intratissue swelling pressures. It is for this reason that the complex microstructure of aggrecan is particularly appropriate for tissues that must sustain high compressive loads.

Acknowledgments

This research was supported by NIH Grant AR33236. The authors are indebted to Drs. Anna Plaas, John Sandy, Ernst Hunziker and James Kimura for critical discussions and ongoing collaborations related to these studies.

References

Akeson, W.H., Woo, S. J.Y., Amiel, D., Coutts, R.D., and Daniel, D. (1973) The connective tissue response to immobility: biochemical changes in periarticular connective tissue of the immobilized rabbit knee. *Clin. Orthop.*, **93**, 356–362.

Armstrong, C.G., Lai, W.M., and Mow, V.C. (1984) An analysis of the unconfined compression of articular cartilage. *J. Biomech. Eng.*, **106**, 165–173.

Arokoski, J., Kiviranta, I., Jurvelin, J., Tammi, M., and Helminen, H.J. (1993) Long-distance running causes site-dependent decrease of cartilage glycosaminoglycan content in the knee joints of beagle dogs. *Arthritis Rheum.*, **36**, 1451–1459.

Aydelotte, M.B. and Kuettner, K.E. (1988) Differences between sub-populations of cultured bovine articular chondrocytes. I. morphology and cartilage matrix production. *Connect. Tissue Res.*, **18**, 205–222.

Aydelotte, M.B., Greenhill, R.R., and Kuettner, K.E. (1988) Differences between sub-populations of cultured bovine articular chondrocytes. II. Proteoglycan metabolism. *Connect. Tissue Res.*, **18**, 223–234.

Basser, P. J. and Grodzinsky, A. J. (1993) The donnan model derived from microstructure. *Biophys. Chem.*, **46**, 57–68.

Bayliss, M.T., Urban, J.P.G., Johnstone, B., and Holm, S. (1986) In vitro method for measuring synthesis rates in the intervertebral disc. *J. Orthop. Res.*, **4**, 10–17.

Benya, P.D. and Shaffer, J.D. (1982) Dedifferentiated chondrocytes reexpress the differentiated collagen phenotype when cultured in agarose gels. *Cell*, **30**, 215–224.

Benya, P.D., Brown, P.D., and Padilla, S.R. (1988) Microfilament modification by dihydrocytochalasin B causes retinoic acid-modulated chondrocytes to reexpress the differentiated collagen phenotype without a change in shape. *J. Cell Biol.*, **106**, 161–170.

Bourret, L.A. and Rodan, G.A. (1976) Inhibition of cAMP accumulation in epiphyseal cartilage cells exposed to physiological pressure. *Calcif. Tissue Res.*, **21**, 431–436.

Brown, T.D., Anderson, D.D., Nepola, J.V., ., Singerman, R.J., Pedersen, D. R., and Brand, R. A. (1988) Contact stress aberrations following imprecise reduction of simple tibial plateau fractures. *J. Orthop. Res.*, **6**, 851–862.

Buschmann, M.D. and Grodzinsky, A.J. (1995) A molecular model of proteoglycan-associated electrostatic forces in cartilage mechanics. *J. Biomech. Eng.*, **117**, 170–192.

Buschmann, M.D., Gluzband, Y.A., Grodzinsky, A.J., Kimura, J.H., and Hunziker, E.B. (1991) Mechanical compression modulates matrix biosynthesis in chondrocyte/agarose gel culture. *Trans. Comb. Orthop. Res. Soc. USA, Japan, Canada*, **1**, 75.

Buschmann, M.D., Gluzband, Y.A., Grodzinsky, A.J., Kimura, J.H., and Hunziker, E.B. (1992) Chondrocytes in agarose culture synthesize a mechanically functional extracellular matrix. *J. Orthop. Res.*, **10**, 745–758.

Buschmann, M.D., Gluzband, Y.A., Grodzinsky, A.J., and Hunziker, E.B. (1995b) Mechanical compression modulates matrix biosynthesis in chondrocyte/agarose culture. *J. Cell Science*, **108**, 1497–1508.

Caterson, B. and Lowther, D.A. (1978) Changes in the metabolism of the proteoglycans from sheep articular cartilage in response to mechanical stress. *Biochim. Biophys. Acta*, **540**, 412–422.

DeWitt, M.T., Handley, C.J., Oakes, B.W., and Lowther, D.A. (1984) In vitro response of chondrocytes to mechanical loading: the effect of short term mechanical tension. *Connect. Tissue Res.*, **12**, 97–109.

Eberhardt, A.W., Keer, L.M., Lewis, J.L., and Vithoontien, V. (1990) An analytical model of joint contact. *J. Biomech. Eng.*, **112**, 407–413.

Eisenberg, S.R. and Grodzinsky, A.J. (1985) Swelling of articular cartilage and other connective tissues: Electromechanochemical forces. *J. Orthop. Res.*, **3**, 148–159.

Eisenberg, S.R. and Grodzinsky, A.J. (1987) The kinetics of chemically induced non-equilibrium swelling of articular cartilage and corneal stroma. *J. Biomech. Eng.*, **109**, 79–89.

Frank, E.H. and Grodzinsky, A.J. (1987a) Cartilage electromechanics-I. electrokinetic transduction and the effects of electrolyte pH and ionic strength. *J. Biomech.*, **20**, 615–627.

Frank, E.H. and Grodzinsky, A.J. (1987b) Cartilage electromechanics-II. a continuum model of cartilage electrokinetics and correlation with experiments. *J. Biomech.*, **20**, 629–639.

Frank, E.H., Grodzinsky, A.J., Koob, T.J., and Eyre, D.R. (1987) Streaming potentials: a sensitive index of enzymatic degradation in articular cartilage. *J. Orthop. Res.*, **5**, 497–508.

Freeman, P.M., Natarajan, R.N., Kimura, J.H. and Andriacchi, T.P. (1994) Chondrocyte cells respond mechanically to compressive loads. *J. Orthop. Res.*, **12**, 311–320.

Garcia, A.M., Black, A.C., and Gray, M.L. (1994) Effects of physiochemical factors on the growth of mandibular condyles *in vitro*. *Calcif. Tissue Int.*, **54**, 499–504.

Gray, M.L., Pizzanelli, A.M., Grodzinsky, A.J., and Lee, R.C. (1988) Mechanical and physicochemical determinants of the chondrocyte biosynthetic response. *J. Orthop. Res.*, **6**, 777–792.

Gray, M.L., Pizzanelli, A.M., Lee, R.C., Grodzinsky, A.J., and Swann, D.A. (1989) Kinetics of the chondrocyte biosynthetic response to compressive load and release. *Biochim. Biophys. Acta*, **991**, 415–425.

Gritzka, T.L., Fry, L.R., Cheesman, R.L., and Lavigne, A. (1973) Deterioration of articular cartilage caused by continuous compression in a moving rabbit joint. *J. Bone Joint Surg.*, **55A**, 1698–1720.

Grodzinsky, A.J. (1983) Electromechanical and physicochemical properties of connective tissue. *CRC Crit. Rev. Bioeng.*, **9**, 133–199.

Grodzinsky, A., Lipshitz, H., and Glimcher, M. (1978) Electromechanical properties of articular cartilage during compression and stress relaxation. *Nature*, **275**, 448–450.

Guharay, F. and Sachs, F. (1985) Mechanotransducer ion channels in chick skeletal muscle: the effects of extracellular pH. *J. Physiol.*, **363**, 119–134.

Guilak, F., Meyer, B.C., Ratcliffe, A., and Mow, V.C. (1991) The effect of static loading on proteoglycan biosynthesis and turnover in articular cartilage. *Trans. Orthop. Res. Soc.*, **16**, 50.

Hall, A.C., Urban, J.P.G., and Gehl, K.A. (1991) The effects of hydrostatic pressure on matrix synthesis in articular cartilage. *J. Orthop. Res.*, **9**, 1–10.

Jones, I.L., Klamfeldt, D.D.S., and Sandstrom, T. (1982) The effect of continuous mechanical pressure upon the turnover of articular cartilage proteoglycans *in vitro*. *Clin. Orthop.*, **165**, 283–289.

Jurvelin, J., Kiviranta, I., Saamanen, A.-M., Tammi, M., and Helminen, H.J. (1989) Partial restoration of immobilization-induced softening of canine articular cartilage after remobilization of the knee (stifle) joint. *J. Orthop. Res.*, **7**, 352–358.

Kempson, G.E. (1979) Mechanical properties of articular cartilage. In *Adult Articular Cartilage*, 2nd ed. (Freeman, M.A.R., Ed.), pp. 333-414, Pitman, Tunbridge Wells, England.

Kim, Y.-J. (1993) *Physical Regulation of Cartilage Metabolism: Effects of Compression on Specific Matrix Molecules and their Spatial Distribution.* PhD thesis, Massachusetts Institute of Technology, Cambridge, MA.

Kim, Y.-J., Grodzinsky, A.J., Plaas, A.H.K., and Sandy, J.D. (1992) The differential effects of static compression on the synthesis of specific cartilage matrix components. *Trans. Orthop. Res. Soc.*, **17**, 108.

Kim, Y.-J., Kung, S., Grodzinsky, A.J., Sandy, J.D., and Plaas, A.H.K. (1993) Effects of compression on cartilage link synthesis, aggrecan structure, and core-protein processing: Cellular mechanisms. *Trans. Orthop. Res. Soc.*, **18**, 13.

Kim, Y.-J., Grodzinsky, A.J., and Plaas, A.H.K. Differential effects of compression on synthesis and intracellular processing of aggrecan, link protein, and hyaluronan, (1994a). (submitted).

Kim, Y.-J., Sah, R. L.-Y., Grodzinsky, A.J., Plaas, A.H.K., and Sandy, J.D. (1994b) Mechanical regulation of cartilage biosynthetic behavior: Physical stimuli. *Arch. Biochem. Biophys.*, **311**, 1–12.

Kim, Y.-J., Bonassar, L.J., and Grodzinsky, A.J. (1995) The role of cartilage streaming potential, fluid flow and pressure in the stimulation of chondrocyte biosynthesis during dynamic compression. *J. Biomech.*, **28**, 1055–1066.

Kimura, J.H., Caputo, C.B., and Hascall, V.C. (1981) The effect of cycloheximide on synthesis of proteoglycans by cultured chondrocytes from the swarm rat chondrosarcoma. *J. Biol. Chem.*, **256**, 4368–4376.

Kimura, J.H., Schipplein, O.D., Kuettner, K.E., and Andriacchi, T.P. (1985) Effects of hydrostatic loading on extracellular matrix formation. *Trans. Orthop. Res. Soc.*, **10**, 365.

Kiviranta, I., Jurvelin, J., Tammi, M., Saamanen, A.-M., and Helminen, H.J. (1987) Weight bearing controls glycosaminoglycan concentration and articular cartilage thickness in the knee joints of young beagle dogs. *Arthritis Rheum.*, **30**, 801–809.

Klein-Nulend, J., Veldhuijzen, J.P., and Burger, E.H. (1986) Increased calcification of growth plate cartilage as a result of compressive force *in vitro*. *Arthritis Rheum.*, **29**, 1002–1009.

Klein-Nulend, J., Veldhuijzen, J.P., van de Stadt, R.J., van Kampen, G.P.J., Kuijer, R., and Burger, E.H. (1987) Influence of intermittent compressive force on proteoglycan content in calcifying growth plate cartilage *in vitro*. *J. Biol. Chem.*, **262**, 15490–15495.

Koob, T.J., Clark, P.E., Hernandez, D.J., Thurmond, F.A., and Vogel, K.G. (1992) Compression loading *in vitro* regulates proteoglycan synthesis by tendon fibrocyte. *Arch. Biochem. Biophys.*, **298**, 303–312.

Lai, W.M., Hou, J.S., and Mow, V.C. (1991) A triphasic theory for the swelling and deformation behaviors of articular cartilage. *J. Biomech. Eng.*, **113**, 245–258.

Lee, R. C., Rich, J. B., Kelley, K. M., Weiman, D. S., and Mathews, M. B. (1982) A comparison of *in vitro* cellular responses to mechanical and electrical stimulation. *American Surgeon* , **48**, 567–574.

Lippiello, L., Kaye, C., Neumata, T., and Mankin, H.J. (1985) In vitro metabolic response of articular cartilage segments to low levels of hydrostatic pressure. *Connect. Tissue Res.*, **13**, 99–107.

Mak, A.F. (1986) Unconfined compression of hydrated viscoelastic tissues: a biphasic poroviscoelastic analysis. *Biorheology* , **23**, 371–383.

Marcus, R.A. (1955) Calculation of thermodynamic properties of polyelectrolytes. *J. Chem.Phys.*, **23**, 1057–1068.

Maroudas, A. (1979) Physicochemical properties of articular cartilage. In *Adult Articular Cartilage*, 2nd ed. (Freeman, M.A.R., Ed.), pp. 215–290, Pitman, Tunbridge Wells, England.

Morales, T.I. and Hascall, V.C. (1989) Effects of interleukin-1 and lipopolysaccharides on protein and carbohydrate metabolism in bovine articular cartilage organ cultures. *Connect. Tissue Res.*, **10**, 255–275.

Mow, V.C., Holmes, M.H., and Lai, W.M. (1984) Fluid transport and mechanical properties of articular cartilage: a review. *J. Biomech.*, **17**, 377–394.

Newman, P. and Watt, F.M. (1988) Influence of cytochalasin d-induced changes in cell shape on proteoglycan synthesis by cultured articular chondrocytes. *Exp. Cell. Res.*, **178**, 199–210.

Ng, C.K., Handley, C.J., Preston, B.N., and Robinson, H.C. (1992) The extracellular processing and catabolism of hyaluronan in cultured adult articular cartilage explants. *Arch. Biochem. Biophys.*, **294**, 70–79.

Ogston, A.G., Preston, B.N., and Wells, J.D. (1973) On the transport of compact particles through solutions of chain-polymers. *Proc. R. Soc. London* , **333**, 297–316.

O'Hara, B.P., Urban, J.P.G., and Maroudas, A. (1990) Influence of cyclic loading on the nutrition of articular cartilage. *Ann. Rheum. Dis.*, **49**, 536–539.

Olah, E.H. and Kostenszky, K.S. (1972) Effect of altered functional demand on the glycosaminoglycan content of the articular cartilage of dogs. *Acta Biol. Hung.*, **23**, 195–200.

Overbeek, J.T.G. (1956) The Donnan equilibrium. *Prog. Biophys. Biophys. Chem.*, **6**, 58–84.

Palmoski, M.J. and Brandt, K.D. (1984) Effects of static and cyclic compressive loading on articular cartilage plugs *in vitro. Arthritis Rheum.*, **27**, 675–681.

Palmoski, M.J., Perricone, E., and Brandt, K.D. (1979) Development and reversal of a proteoglycan aggregation defect in normal canine knee cartilage after immobilization. *Arthritis Rheum.*, **22**, 508–517.

Parkkinen, J.J., Lammi, M.J., Helminen, H.J., and Tammi, M. (1992) Local stimulation of proteoglycan synthesis in articular cartilage explants by dynamic compression *in vitro. J. Orthop. Res.*, **10**, 610–620.

Poole, C.A., Flint, M.H., and Beaumont, B.W. (1984) Morphological and functional interrelationships of articular cartilage matrices. *J. Anat.*, **138**, 113–138.

Radin, E.L., Martin, R.B., Burr, D.B., Caterson, B., Boyd, R.D., and Goodwin, C. (1984) Effects of mechanical loading on the tissues of the rabbit knee. *J. Orthop. Res.*, **2**, 221–234.

Roberts, S., Weightman, B., Urban, J., and Chappell, D. (1986) Mechanical and biochemical properties of human articular cartilage in osteoarthritic femoral heads and in autopsy specimens. *J. Bone Joint Surg. [Br]*, **68-B**, 278–288.

Sah, R.L., Kim, Y.-J., Doong, J.H., Grodzinsky, A.J., Plaas, A.H.K., and Sandy, J.D. (1989) Biosynthetic response of cartilage explants to dynamic compression. *J. Orthop. Res.*, **7**, 619–636.

Sah, R.L., Grodzinsky, A.J., Plaas, A.H.K., and Sandy, J.D. (1990a) Effects of tissue compression on the hyaluronate binding properties of newly synthesized proteoglycans in cartilage explants. *Biochem. J.*, **267**, 803–808.

Sah, R.L., Kim, Y.-J., and Grodzinsky, A.J. (1990b) The effect of mechanical compression on cartilage metabolism. In *Methods for Cartilage Research* (Maroudas, A. and Kuettner, K.E., Eds.), pp. 116–119, Academic Press, New York.

Sah, R.L., Doong, J.Y.H., Grodzinsky, A.J., Plaas, A.H.K., and Sandy, J.D. (1991) Effects of compression on the loss of newly synthesized proteoglycans and proteins from cartilage explants. *Arch. Biochem. Biophys.*, **286**, 20–29.

Sanfeld, A. (1968) *Introduction to the Thermodynamics of Charged and Polarized Layers*, John Wiley and Sons, New York, NY.

Schneiderman, R., Kevet, D., and Maroudas, A. (1986) Effects of mechanical and osmotic pressure on the rate of glycosaminoglycan synthesis in the human adult femoral head cartilage: an *in vitro* study. *J. Orthop. Res.*, **4**, 393–408.

Setton, L.A., Zhu, W., and Mow, V.C. (1993) The biphasic poroviscoelastic behavior of articular cartilage: role of the surface zone in governing the compressive behavior. *J. Biomech.*, **26**, 581–592.

Slowman, S.D. and Brandt, K.D. (1986) Composition and glycosaminoglycan metabolism of articular cartilage from habitually loaded and habitually unloaded sites. *Arthritis Rheum.*, **29**, 88–94.

Spiro, R.C., Freeze, H.H., Sampath, D., and Garcia, J.A. (1991) Uncoupling of chondroitin sulfate glycosaminoglycan synthesis by brefeldin A. *J. Cell Biol.*, **115**, 1463–1473.

Thompson, R.C., Oegema, T.R., Lewis, J.L., and Wallace, L. (1991) Osteoarthrotic changes after acute transarticular load: an animal model. *J. Bone Joint Surg.*, **73A**, 990–1001.

Urban, J.P.G. and Bayliss, M.T. (1989) Regulation of proteoglycan synthesis rate in cartilage *in vitro*: influence of extracellular ionic composition. *Biochim. Biophys. Acta* , **992**, 59–65.

Urban, J.P.G., Maroudas, A., Bayliss, M.T., and Dillon, J. (1979) Swelling pressures of proteoglycans at the concentrations found in cartilaginous tissues. *Biorheology*, **16**, 447–464.

Urban, J.P.G., Hall, A.C., and Gehl, K.A. (1993) Regulation of matrix synthesis rates by the ionic and osmotic environment of articular chondrocytes. *J. Cell Physiol.*, **154**, 262–270.

van Kampen, G.P.J., Veldhuijzen, R., Kuijer, R., van de Stadt, R.J., and Schipper, C.A. (1985) Cartilage response to mechancial force in high-density chondrocyte cultures. *Arthritis Rheum.*, **28**, 419–424.

Veldhuijzen, J.P., Bourret, L.A., and Rodan, G.A. (1979) In vitro studies of the effect of intermittent compressive forces on cartilage cell proliferation. *J. Cell Physiol.*, **98**, 299–306.

Veldhuijzen, J.P., Huisman, A.H., Vermeiden, J.P.W., and Prahl-Andersen, B. (1987) The growth of cartilage cells *in vitro* and the effect of intermittent compressive force. a histological evaluation. *Connect. Tissue Res.*, **16**, 187–196.

Watson, P.A. (1989) Accumulation of camp and calcium in s49 mouse lymphoma cells following hyposmotic swelling. *J. Biol. Chem.*, **264**, 14735–14740.

Watson, P.A. (1990) Direct stimulation of adenylate cyclase by mechanical forces in s49 mouse lymphoma cells during hyposmotic swelling. *J. Biol. Chem.*, **265**, 6569–6575.

Watson, P.A. (1991) Function follows form: generation of intracellular signals by cell deformation. *FASEB* , **5**, 2013–2019.

Watt, F.M. (1986) The extracellular matrix and cell shape. *TIBS* , **11**, 482–485.

Wilkins, R.J. and Hall, A.C. (1992) Measurement of intracellular pH in isolated bovine articular chondrocytes. *Experimental Physiology* , **77**, 521–524.

Williams, R.P.W. and Comper, W.D. (1990) Osmotic flow caused by polyelectrolytes. *Biophys. Chem.*, **36**, 223–234.

Wu, H. (1992) *The Effect of Cytoskeletal Disruption on Cartilage Metabolic Response to Compression*. Master's thesis, Massachusetts Institute of Technology, Cambridge, MA.

11 Morphogenesis of Connective Tissues

Stuart A. Newman[1] **and James J. Tomasek**[2]

[1]*Department of Cell Biology and Anatomy, New York Medical College, Valhalla, NY, USA*

[2]*Department of Anatomical Sciences, University of Oklahoma Health Sciences Center, Oklahoma City, OK, USA*

INTRODUCTION

Morphogenesis is the molding of living tissues that occurs during development, regeneration, wound healing, and various pathological processes. During morphogenetic events tissue masses may disperse, lengthen or shorten, acquire lumens, form local condensations of cells, or develop one or more internal boundaries across which cell mixing is selective or prohibited. Distinct tissues, defined by a common boundary, can physically separate, or remain attached, where they may engulf, or become engulfed by one another. The outcomes of these processes are the various body plans and organ forms characteristic of metazoan organisms, as well as tumors, abnormal polyps and fibrotic lesions.

Morphogenesis clearly depends on a set of physical processes that act on tissues as materials. The material properties of mature tissues derive partly from the cells that make them up, and partly from the extracellular matrices they produce. Mature epithelioid tissues contain cells firmly attached to one another, and therefore have rheological properties akin to those of cytoplasm. For mature connective tissues the properties of their matrices, which can range from liquid, to gel-like, to solid, will predominate. But immature, morphogenetically active epithelioid tissues contain cells that readily rearrange and slip past one another, giving them many of the formal properties of elasticoviscous liquids (e.g., fluidity, surface tension), and the matrices of immature connective tissues, or mesenchymes, are devoid of mineral, and relatively poor in fibrous materials, giving them fluid properties as well.

The typical physical state of tissues during morphogenesis, then, is that of "soft matter", which is the designation given to semi-solid elasticoviscous materials like clays, putties, and polymer melts (de Gennes, 1992). The implication that morphogenetically active tissues will be subject to the class of physical effects characteristic to such materials (e.g.,

---

Extracellular Matrix, Volume 2, Molecular Components and Interactions, edited by Wayne D. Comper.

flow, resiliency, separation into immiscible phases, and the generation of interfacial tensions) raises the possibility that the branch of physics that deals with such materials can help us understand the molding of tissues into shapes and forms. However, unlike most nonliving soft materials, living tissues, by virtue of their cellular component, are mechanochemically active and capable of resisting, and even opposing, passive and externally imposed deformation. Despite tissues being soft matter, then, it is legitimate to ask whether the physics of non-living soft matter can tell us anything nontrivial about biological morphogenesis (Newman and Comper, 1990).

The active nature of morphogenesis clearly limits attempts to account for biological form on the basis of purely physical properties of tissues. But just as organisms are more susceptible to physical molding forces at earlier stages of their *development* than at later stages, it is probable that they were even more susceptible to such forces earlier in their *evolution* than at present (Newman, 1994). The earliest metazoan organisms undoubtedly lacked certain means of resisting physical influences which are found in their modern counterparts. A familiar example is that of hard body parts. It is clear that the calcified shells and chitinous exoskeletons that protect the soft tissues of many intervertebrate species against physical deformation evolved after the basic invertebrate body plans were established. Thus the evolution of all such forms occurred within living materials that were more plastic than those that currently prevail. Another example concerns the complex physiological integration of intracellular and extracellular events, coordinated by transmembrane proteins like cadherins and integrins in modern organisms. It is doubtful that such integration characterized the earliest multicellular organisms. Indeed, the relevant signalling pathways may have acquired their current degree of complexity well after the burst of morphological experimentation known as the "Cambrian explosion" that took place more than half a billion years ago (Gould, 1989; Newman, 1994).

The implication of such examples is that the advent of multicellularity, which led to more massive, more spatially extensive living forms than previously existed in the unicellular world, added a new set of physical determinants to the repertoire of living systems. Only with subsequent molecular evolution would organisms have acquired the ability to resist and counteract the direct effects of these physical forces (Newman, 1994).

Taking an evolutionary perspective on morphogenesis can thus provide a rational basis for accounting for organismal form. In particular, it leads to the hypothesis that forms assumed by the developing tissues of modern organisms exist partly by virtue of the fact that ancestral tissue forms were molded by a set of forces in which the physics of soft matter played a more important role than it does at present. We have suggested that such forces and their morphogenetic outcomes are "generic" to living tissues (Newman and Comper, 1990), in the same sense that waves and vortices are generic to liquids, and vibrations are generic to taut strings.

Although a morphogenetic event in any highly evolved organism can only be fully understood by analyzing all the physical and biochemical interactions that contribute to its realization, the view presented above suggests that it is useful to frame such an analysis in terms of the plausible evolutionary history of that morphogenetic event and the generic processes that may have contributed, or continue to contribute, to it (Newman, 1994; 1995). Let us assume for example, that the general outcome of a given morphogenetic process (e.g., multilayering, lumen formation, segmentation) had been established by tissue properties, including generic forces, at an early stage in the history of multicellular life, and that during the course of evolution the outcome became stabilized by a particular set of biochemical interactions. It would then be expected that the more recently integrated biochemical interactions might play a regulatory role in this morphogenetic process, but could be relatively dispensible to its broad outcome. The newly integrated interactions could modulate or fine-tune the original morphogenetic process, or buffer it against environmental perturbations or internal metabolic noise, causing a more reliable achievement of the previously evolved morphology. In effect, the original, physically guided process would provide a template for the accumulation, during subsequent evolution, of supporting molecular pathways. In any case, it would not be too surprising if organisms in which such "refining" components were eliminated exhibited relatively little morphological change. Alternatively, the recently acquired components may constitute a redundant pathway that allows some of the original morphogenetically important molecules to be dispensed with. The existence of redundancy in morphogenetic processes has been amply demonstrated by the lack of significant phenotypic effect of "knocking out" genes for proteins demonstrably involved in morphogenesis, such as the extracellular matrix protein tenascin (Saga *et al.*, 1992) or the integral membrane adhesion protein N-CAM (Cremer *et al.*, 1994).

The purpose of this review is to provide a basis for analyzing morphogenesis of connective tissues in terms of the physical mechanisms that are likely to underlie tissue transformations, either in the contemporary organism itself, or in an ancestral form. In this way the numerous molecular events that accompany morphological development can be analyzed as components of systems that have been progressively modified over the course of evolution. This approach represents a means of potentially identifying a "core" set of morphogenetic processes (Newman, 1993b), and of dealing with the obvious but often neglected fact that morphogenesis in contemporary organisms is "overdetermined" — the product of mechanisms more complex than those which originally defined its general character.

As noted by Ettinger and Doljanski (1992), connective tissues play a particularly prominent role in the morphogenesis of multicellular organisms by virtue of the great capacity for mesenchymal-fibroblast-like cells to remodel the extracellular matrix and sculpt it into specific shapes. While taking this viewpoint and the related phenomena into account, we propose here to broaden the scope of connective tissue morphogenesis by also considering those morphogenetic processes common to epithelial and connective tissues.

The perspective outlined above, that the material properties of tissues and the plausible ancestral origins of biological forms must be considered central in accounting for the mechanisms of morphogenesis, has led to the following order of presentation of topics in this review: First, a description is given of the physical properties and processes most likely to have been generic to the earliest multicellular organisms, since it is those processes which would have originally defined the basic rules of tissue morphogenesis, providing a set of templates upon which modern morphogenesis has evolved. Next, an additional set of morphogenetic mechanisms that pertain specifically to morphogenesis of mesenchymal and connective tissues are described. Finally a series of examples of mesenchymal morphogenesis and connective tissue remodeling will be discussed in relation to the various roles potentially played by the basic mechanisms described earlier. The examples discussed are drawn from vertebrate systems, but the principles that emerge are proposed to be generally applicable to the morphogenesis of invertebrate connective tissues as well.

GENERIC PHYSICAL MECHANISMS OF MORPHOGENESIS

Mechanisms Pertaining to Both Epithelial and Connective Tissues

Aggregation and Multicellularity

The ability of cells to attach directly to one another, or to the extracellular matrices that they produce, is the defining condition of tissues. Detailed knowledge of the specific molecules involved in cell-cell and cell-matrix adhesion will always be secondary to the fact that, regardless of the molecules involved, the result is always a multicellular aggregate.

There is little question that the evolutionarily earliest multicellular organisms were direct descendants of free-living unicellular organisms that remained attached for one reason or another after undergoing cytokinesis. The chemical and structural bases for such attachment can be quite varied. In modern organisms, epithelial tissues are composed of cells bound directly to one another by links between integral membrane proteins, by protein-rich plaques known as adherens junctions and desmosomes, or by the interleaving of their plasma membranes. In contrast, connective tissues are formed by the embedment of cells in macromolecular matrices of varying composition. Given the numerous possible routes to multicellularity, it is likely that aggregated forms emerged independently many times over the long history of life.

The existence of any mechanism by which cells may attach to one another, or to materials they produce, makes possible the simplest morphogenetic event, becoming multicellular. This demonstrates how intimately related are the evolutionary and developmental aspects of the generation of form. The molecular bases of the evolutionary origin of multicellularity is currently a matter of speculation. One hypothesis is that a change in the relative concentrations of sodium and calcium in the oceans could have caused certain cell surface proteins on what were previously unicellular organisms to take on an adhesive role (Kazmierczak and Degens, 1986). Regardless of how it was first accomplished (and, as indicated above, it could have occurred numerous times by disparate molecular means), the *evolution* of multicellularity brought into being single cells that had the capacity to *develop* into multicellular forms.

The size of such forms would have been variable when they first appeared, with the fully developed multicellular organism attaining a mean number of cells that was a direct function of the strength of the intercellular bonds, and an inverse function of the average shearing force that it experienced from its fluid environment. Later, perhaps driven by differential survival of better adapted forms, greater accuracy in size regulation would have evolved. Similarly for the regulation of shape: the earliest multicellular forms to appear would likely have been roughly spherical, albeit malleable. If one or another viable morphology in the primitive organism's physical repertoire proved more suitable for the survival of its progeny, subsequent evolution could have provided ancillary mechanisms that made their attainment more reproducible.

Contemporary organisms achieve multicellularity by numerous molecular mechanisms involving both cell-cell and cell matrix attachment. Descriptions and discussions of the various classes and functions of cell adhesion and extracellular matrix molecules will be found in other chapters of this series. The molecular interactions relevant to a number of specific examples of connective tissue morphogenesis will be discussed below.

Finally, the reversal of adhesive interactions, by the breakdown of adhesion molecules or extracellular matrices (Birkedale-Hansen *et al.*, 1994), or by the expression of anti-adhesive molecules such as tenascin (Chiquet-Ehrismann, 1991), would have the morphogenetic consequence of cell disaggregation or dispersal, a phenomenon utilized in a variety of developmental processes. Indeed, a number of morphogenetic processes involve epithelial-mesenchymal and mesenchymal-epithelial transitions (see, for example, Hay, 1993, and below).

Compartmentalization and Multilayering

At that point during the history of metazoan organisms when the strength or specificity of intercellular bonds became subject to modulation, a new class of morphogenetic processes was established: compartment formation. It has been demonstrated on both theoretical and experimental grounds that differential adhesion within a tissue mass can lead to the establishment of boundaries across which cells fail to mix (Steinberg, 1978; Armstrong, 1989; Steinberg and Takeichi, 1994; Graner, 1993). The phenomena observed are similar to what happens when two immiscible liquids, such as oil and water, are poured into the same container. As long as the molecules that make up one of the liquids have a greater binding affinity for one another than they do for the molecules of the other liquid, "phase separation" will take place. This is exhibited in an interface within the common fluid mass that neither type of molecule will cross, though they may move within their respective liquids by random Brownian motion. Eukaryotic cells are too large to move around by Brownian motion, but they can randomly perambulate through tissues by means of motile forces generated within their cytoplasm. When sufficient differential adhesion exists between two cell populations, not only will each type of cell keep to its own side of the interface, but when dissociated and randomly mixed, the two populations will "sort out", much like a shaken mixture of oil and vinegar, and for the same thermodynamic reasons.

The morphogenetic consequences of compartment formation may be subtle or dramatic. If the adhesive differential is small, the boundary between the two compartments will be relatively straight. If the differential is much greater, one tissue will thereby be more cohesive than the other. It will tend to minimize its surface area, while the less cohesive tissue spreads around, or engulfs, it. Any intermediate degree of differential adhesion will lead to interfaces with various degrees of curvature.

No matter how straight or curved the interface, the morphogenetic outcome of differential adhesion will be most striking if the cell types that abide by the common boundary are recognizably different from one another. When compartments form in the course of normal development the boundary may be covert to begin with, with the cells in the different compartments becoming structurally or functionally distinguishable only later. But even when the cell types of the cohesively different tissues cannot readily be distinguished, the generation of internal interfacial tensions can produce a multilayered or involuted structure where only a single layer or simple mass existed before, or cause more subtle alterations in the shape of a tissue primordium.

These considerations of differential adhesion pertain primarily to epithelioid tissues, in which the cells are in direct contact with one another. In mesenchymal tissues a different physical principle may be involved in the formation of regions of immiscibility. The manner in which any set of objects which can contact and interconnect with one another, and therefore may form themselves into networks with distinct physical properties, can be analyzed by a branch of mathematics known as "percolation theory" (Stauffer, 1985; Forgacs and Newman, 1994). This theory considers the properties of systems consisting of "bonds" and "sites" that can be connected by the bonds. If we consider a system in which the average number of bonds relative to sites is low, it stands to reason that only small clusters of objects will be interconnected. As the ratio increases, a threshold is reached beyond which a "macroscopic cluster" will form, and a single network can be considered to pervade the entire spatial domain in question. The system is said to "percolate" under these conditions. This change of state constitutes a phase transition, with the phases generally having distinguishable physical properties.

The theory of percolation has been used to study phenomena such as the formation of continents and galaxies, and the spread of forest fires. Since mesenchymal tissues contain macromolecular fibers such as type I collagen and hyaluronan, which may form networks that are subject to disruption by other macromolecules and by cells, the potential exists for forming distinct phases in the extracellular matrix. The relative arrangements of the tissue parcels defined by these phases will be subject to tensions at their mutual interfaces, in a manner analogous to the engulfment and other rearrangements possible with epithelial phases (Forgacs *et al.*, 1989, 1991; Forgacs and Newman, 1994).

An experimental system which demonstrates the possibility of phase transitions in connective tissue matrices is shown schematically in Figure 11.1 (Newman *et al.*, 1985). Here a droplet of soluble type I collagen is deposited adjacent to a second droplet, which is also populated with a small, but critical number of living cells, or cell-sized polystyrene beads. Surprisingly, an interface forms between the two droplets, indicating the presence of an interfacial tension, despite the fact that the composition of the two droplets is the same, with the exception of the presence of particles (constituting only a fraction of a percent of the volume ratio) in one of them. Over the next few minutes the droplet containing the particles spreads over and partially engulfs the droplet lacking particles (Forgacs *et al.*, 1989). At higher collagen concentrations the relative movement of the two phases only occurs when the extracellular matrix protein fibronectin, or its amino-terminal heparin binding domain, is present in the droplet lacking particles (Newman *et al.*, 1985; 1987).

An interpretation of this set of phenomena, referred to as "matrix-driven translocation", is that when the assembling collagen fibrils in the collagen solution reach a critical length, they can percolate, forming a pervasive network. But when their contacts are disrupted by the presence of cells or polystyrene beads, the network fails to form, hence the two droplets constitute separate "phases", with the bead-lacking drop more cohesive than the bead-containing drop (Forgacs *et al.*, 1991; Forgacs and Newman, 1994). The relative configurations of the droplets when fused with one another are dictated by the principles of thermodynamic equilibrium: the less cohesive phase envelops the more cohesive one. At the higher collagen concentrations the interface may be "metastable", only able to attain its equilibrium configuration when the fibronectin-particle surface interaction across the interface lowers an energy of activation.

The formation of compartments in tissues thus appears to be analogous to phase separation in liquids. As we have seen, generic physical properties of both epithelial and mesenchymal tissues can be mobilized to produce such effects. This implies that compartment formation could have arisen by various molecular routes relatively soon after the origin of multicellularity, simply by the reliable modulation of cell-cell or cell-matrix adhesivities. Indeed, there is evidence that an important function of homeobox proteins,

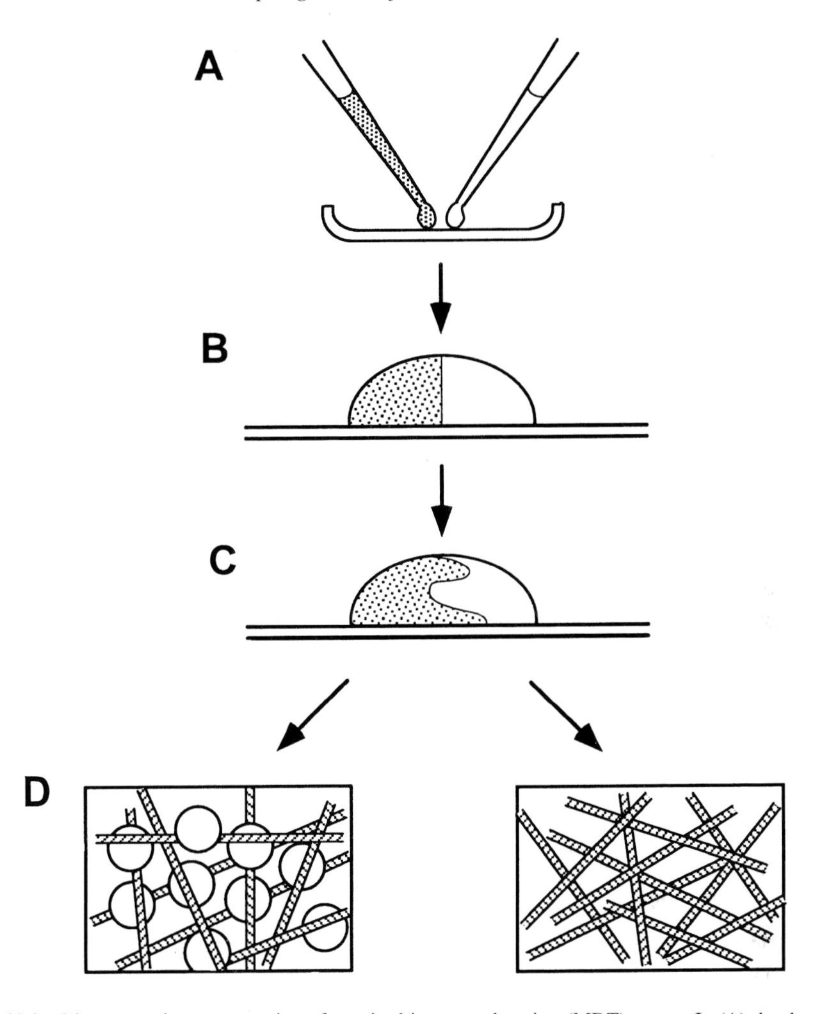

Figure 11.1 Diagrammatic representation of matrix-driven translocation (MDT) assay. In (A) the deposition on the surface of a petri dish of two droplets of soluble type I collagen, one containing cells or cell-sized polystyrene latex beads, and one lacking particles, but containing fibronectin, is shown. A higher magnification representation of a side view of the two droplets shortly after their fusion is shown in (B). The translocation effect — a reconfiguration of the apparent interface between the two droplets during the collagen assembly process — is shown in (C). The panels in (D) represent a higher magnification view of the growing collagen fibers in (C) disrupted by beads or cells on the left, and forming a pervasive network on the right. MDT is thought to result from a combination of differences between the material properties of the two droplets of collagen and the adhesive interactions between fibronectin and the cells or beads at the interface.

which are transcription factors frequently involved in tissue pattern formation and generation of regional identity, is to modulate the production of cell adhesion or extracellular matrix molecules (Chuong *et al.*, 1990; Jones *et al.*, 1992; Yokouchi *et al.*, 1995). Like the origin of multicellularity itself, the origin of compartmentalization must immediately have been accompanied by a whole new set of morphological possibilities, many of which are retained in modern organisms, albeit highly integrated with more recently evolved molecular machinery that would ensure reliability of developmental outcome.

Lumen Formation

Both epithelioid cells, which adhere directly to their neighbors, and mesenchymal cells, embedded in extracellular matrices, can in principle have uniform adhesive properties around their entire surfaces. Indeed, there is no reason to believe that the cells of the earliest metazoan forms had sophisticated means for regulating the distribution of their surface molecules. Many of the cell types in modern organisms, however, are polarized in the expression of several functions, notably adhesion (Rodriguez-Boulan and Nelson, 1993).

The targeting of adhesive molecules, or anti-adhesive molecules, to specific regions of the cell suface can have dramatic consequences. An epithelioid tissue mass consisting of cells that are non-adhesive over portions of their surfaces would quickly develop cavities or lumens (Tsarfaty *et al.*, 1992) (Figure 11.2). If such spaces were to come to adjoin one another, as a result of random cell movement, they would readily fuse. For mesenchymal tissues, in which cells are not initially in contact, the formation of cavities must proceed by a somewhat different route, but one which is also based on cell polarization and the acquisition of differential adhesivity over individual cell surfaces. It has been suggested, for example, that an initial step in the formation of kidney tubules is mesenchymal cell polarization and the partial expression of epithelial characteristics (Ekblom, 1990; see Section 3C below). The view that lumen formation originated as a simple consequence of differential adhesion in conjunction with cell polarity suggests that these properties may be central to various cavitation processes in the tissues of modern organisms, in which the epithelial transition would have come under more stringent control. In particular, the Met proto-oncogene product (Met), a member of the family of tyrosine kinase growth factor receptors, and its ligand hepatocyte growth factor/scatter factor (HGS/SF), may mediate conversion of mesenchyme to epithelium. When 3T3 mouse fibroblasts are transfected with human SF and Met cDNAs and induced to express high levels of HGF/SF and Met protein, these cells form duct-like structures, express cytokeratins, and form epithelial-like intercellular junctions (Tsarfaty *et al.*, 1994).

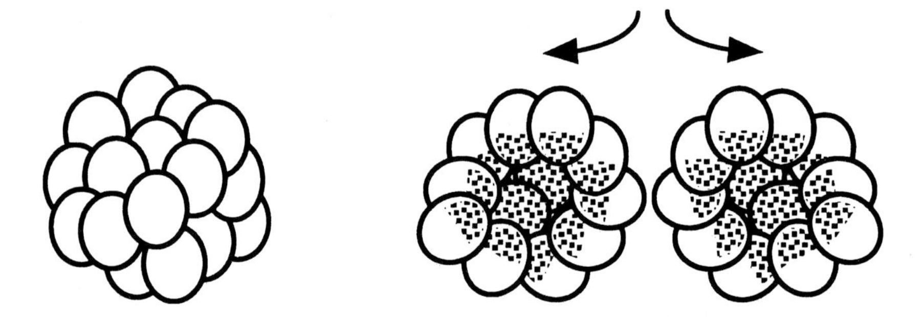

Figure 11.2 Schematic diagram showing the formation of a lumen in a tissue mass by the targeting of an anti-adhesive protein to a region of the cell surface. On the left cells are uniformly adhesive over their surfaces. On the right, an anti-adhesive protein has been targeted to a particular region of each cell. Random cell movement has maximized adhesive contacts over the remaining surface, and the nonadherent regions of each cell come to form the continuous lining of a lumen within the tissue mass. A break-away view of the resulting mass is shown.

Segmentation

Segmentation is the demarcation of tissue primordia into a linear arrangement of structurally similar domains. This is seen for example, in the establishment of body segments in insects, of the blocks of bone- and muscle-forming mesoderm, termed somites, along the embryonic axis of vertebrates, and of the periodic swellings in the vertebrate hindbrain, termed rhombomeres. The development of skeletal elements in the embryonic vertebrate limb exemplifies this process of tissue subdivision in two dimensions: the digits, for example, form within the limb bud mesenchyme as a set of parallel, elongated cell condensations which subsequently become subdivided into chains of cartilaginous rods.

In most instances, segmental organization appears to be based on the inability of otherwise similar tissues to exchange cells at their common boundaries or interfaces. This suggests a mechanism based on compartment formation, but in which the properties

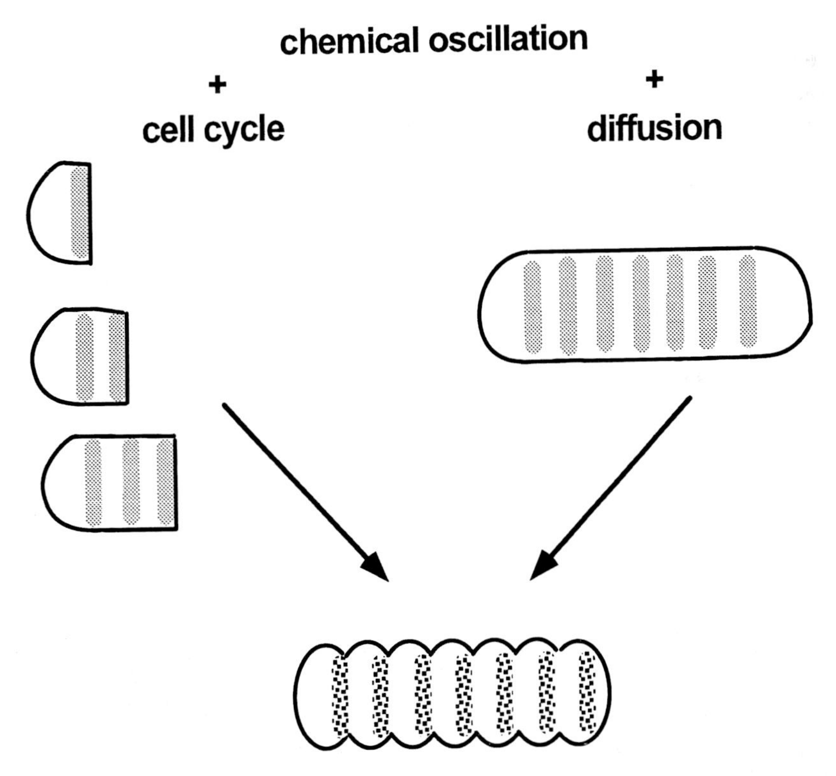

Figure 11.3 Schematic representation of possible relationship between two modes of tissue segmentation. Both modes have in common a biochemical circuit that admits of a chemical oscillation. One of the oscillating species directly or indirectly regulates the strength or specificity of cell adhesivity. In the mechanism shown on the left, the periodic change in cell adhesivity occurs in a growth zone in which the cell cycle has a different period from the regulatory oscillator; as a result, bands of tissue are sequentially generated with alternating cohesive properties. In the mechanism shown on the right, one or more of the biochemical species can diffuse, leading to a set of standing waves of concentration of the regulatory molecule by a reaction-diffusion mechanism. This leads to the simultaneous formation of bands of tissue with alternating cohesive properties. See Newman (1993a) for additional details.

defining the compartments (e.g., differential adhesion, network formation) are regulated in a periodic fashion (Newman, 1993a).

Several possible mechanisms exist for the establishment of both temporal and spatial biochemical periodicities in tissue masses. These mechanisms usually involve positive autoregulation of a molecule that is diffusible or otherwise transmissible from cell to cell (Turing, 1952; Meinhardt, 1982; Newman, 1993a, b; Boissonade *et al.*, 1994). The linking of such periodicities to the regulation of cell-cell or cell-matrix adhesion during phylogeny is a plausible basis for modern segmented body plans (Figure 11.3), which are believed to have emerged independently several times during evolution. The segmental or quasi-segmental organization of various organs, such as the vertebrate limb, the liver, etc., may have a similar physical basis.

Mechanisms that Pertain Specifically to Connective Tissues

In contrast to epithelioid tissues, in which cells are directly adherent to one another over a substantial portion of their surfaces, connective tissues consist of cells suspended in an extracellular matrix (ECM). There thus exist a set of additional morphogenetic mechanisms which depend on changes in the distance between cells, the effects of cells on the organization of the ECM, and the effects of the ECM on the shape and cytoskeletal organization of cells, that typically occur in connective, but not epithelioid tissue types. These will be described in this section.

Condensation

In this process, which is often a transient effect in development, mesenchymal cells, initially dispersed in a matrix, move closer to one another (Figure 11.4). Condensations generally progress to other structures, such as feather germs (Chuong and Edelman, 1985), cartilage or bone (Hall and Miyake, 1992; see below) or epithelial kidney tubules (Ekblom, 1992; see below).

A variety of cellular mechanisms for condensation formation have been suggested, including local loss of matrix materials, centripetal chemotactically-driven movement through the matrix, cell traction, and absence of cell movement away from a center (see Ettinger and Doljanski, 1992; Hall and Miyake, 1992, for further references). Mechanisms based on matrix loss, or on traction, which is usually associated with the collapse of ECM (see below), imply that the volume of the mesenchymal tissues within which condensations form should be reduced. Such volume changes have not been noted in the developing vertebrate limb, in which precartilage condensation is a prominent feature. Furthermore, the traction model for mesenchymal condensation implies that cells seeded upon or within collagen lattices should distort the lattices and align the collagen fibers (Harris *et al.*, 1984; Oster *et al.*, 1984, 1985; see below), but these effects are not seen with limb precartilage mesenchymal cells (Markwald *et al.*, 1990; Zanetti *et al.*, 1990).

In contrast, the major features of mesenchymal condensation can be accounted for by differential adhesion accompanied by random cell movements. Indeed, computer modeling has shown that changes in cell-matrix adhesive interactions are sufficient to bring about condensation (Graner and Sawada, 1993; Glazier and Graner, 1993). Experimentally, it has been shown that sites of local elevation of fibronectin in a mesenchymal mass will serve as

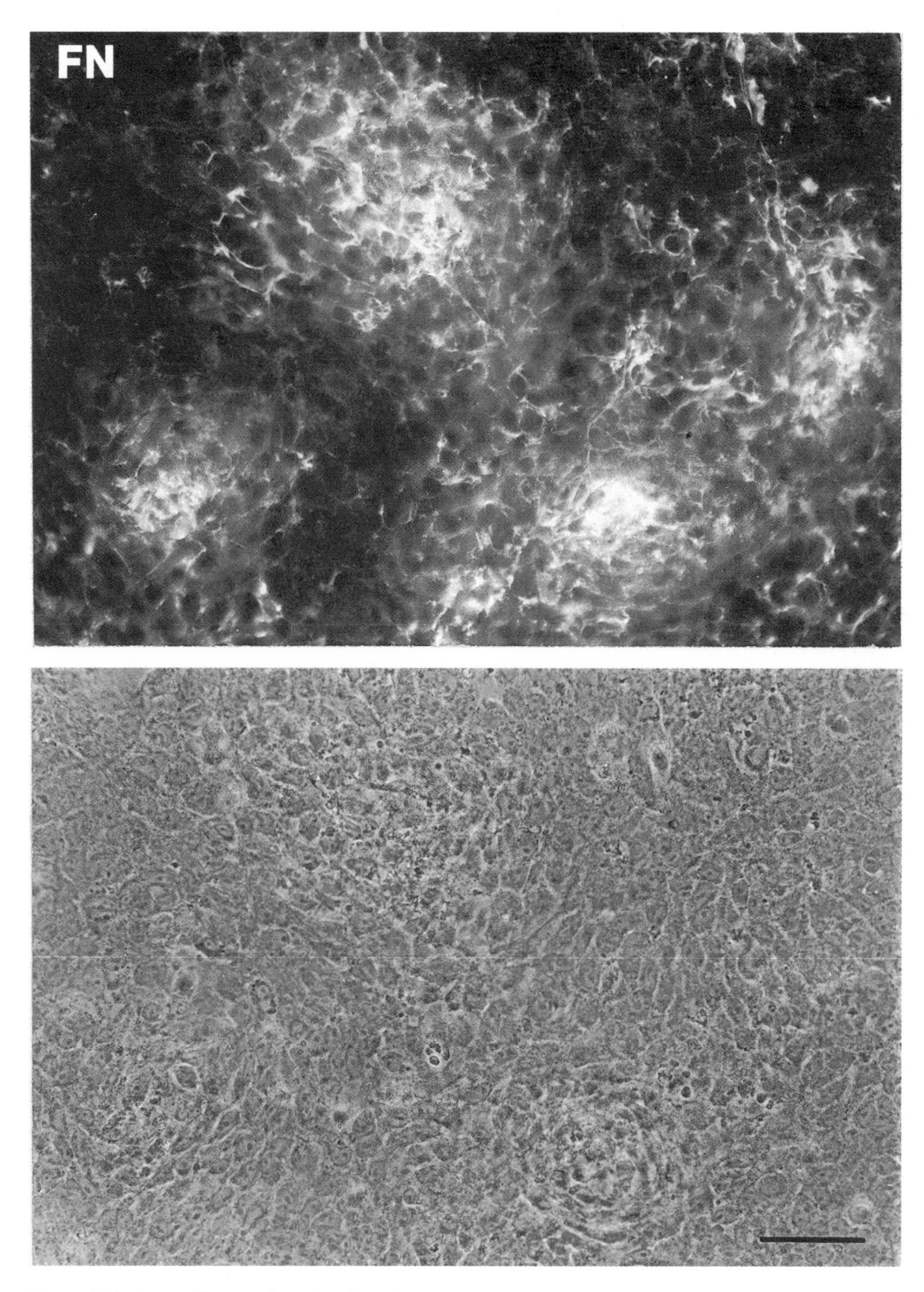

Figure 11.4 Precartilage condensations in a wing bud mesenchyme culture. Culture was fixed and immunostained for fibronectin during day 2 after plating. (Top panel) Fibronectin localization. (Bottom panel) Phase contrast microscopy. Note coincidence of fibronectin accumulation and sites of condensation. Scale bar represents 50 μm. (From Frenz et al., (1989b) with permission from Academic Press.).

foci of accumulation of passively conveyed latex particles, provided they have appropriate surface characteristics (Frenz *et al.*, 1989a); cells themselves may behave similarly (Frenz *et al.*, 1989b). Once the cells have moved close together, the condensations are consolidated by changes in cell surface adhesivity as the result of acquisition of epithelial-type surface molecules such as N-cadherin and N-CAM (Oberlender and Tuan, 1994; Widelitz *et al.*, 1993). The initiating event in this process is a local change in ECM composition, such as production of fibronectin under the influence of TGF-β (Newman, 1988; Leonard *et al.*, 1991; see Section 3B, below).

Contraction and Other Effects of Mechanical Stress

The formation of cell-cell and cell-ECM adhesive structures in tissues permit physical forces originating within cells to contribute to tissue morphogenesis. In particular, intracellular forces necessary for cell shape changes and migration can be imparted onto the surrounding cells and ECM, resulting in mechanical stress (Beloussov *et al.*, 1975; Grinnell, 1994; Ingber, 1994). Such cell-generated stresses have been demonstrated to act within mesenchymal tissues where they may lead to contraction or orientation of extracellular fibers or cytoskeletal filaments (Harris *et al.*, 1981; Stopak and Harris, 1982; Stopak *et al.*, 1985; Nogawa and Nakanishi, 1987; Sumida *et al.*, 1989).

The importance of mechanical stress in morphogenesis is also seen in the role of exogenous mechanical loads in determining tissue pattern. It has long been known, for instance, that the organization of the cardiovascular system is influenced by mechanical forces arising from blood pressure and flow (Russell, 1916). In tendons the fibroblasts and the extracellular matrix they produce appear to be modulated according to their mechanical requirements (Vogel and Koob, 1989). For example, in regions of tendons which wrap around bone, compressive forces lead to the expression of the proteoglycans aggrecan and bigycan and the formation of fibrocartilage (Evanko and Vogel, 1993). Similarly, in bone the ECM is deposited along lines of eventual tension and compression (Koch, 1917). This latter case is of interest in that the ECM organization first arises during embryonic development, before substantial mechanical stresses are placed on the bones. While some stress undoubtedly is generated during embryonic growth (Herring, 1993), it is also possible that this is an example of the phenomenon alluded to earlier, in which biochemical circuitry evolved that stabilized or reinforced an outcome that was originally dependent on external forces to bring it about. Such pathways could potentially come to be triggered earlier in the life history of the organism than the stage at which the external forces originally acted. Scenarios of this sort were discussed by Waddington (1957), who referred to them as "genetic assimilation".

An account of how mechanical stress may regulate morphogenetic processes requires an understanding of how forces are generated intracellularly and transmitted to the extracellular matrix, as well as how cells themselves respond to such mechanical stresses. Collagen matrix model systems have been developed which facilitate the analysis of the manner in which mechanical stress is generated in tissues and how it influences cell behavior. Three model systems with distinctly different mechanical properties are illustrated in Figure 11.5. In the floating collagen gel or lattice model (Figure 11.5A) fibroblasts spread and migrate on collagen fibers. Tractional forces generated by cells during spreading and migration are transmitted to the fibers, resulting in compaction of the collagen gel (Harris *et al.*, 1981; Tomasek and Akiyama, 1992). In this model, the collagen fibers are

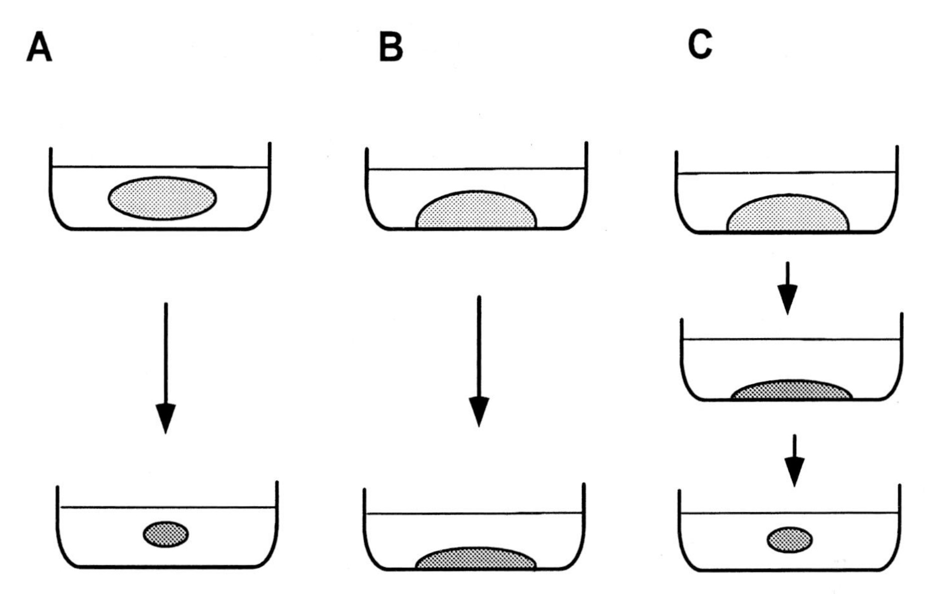

Figure 11.5 Diagrammatic representation of the three in vitro collagen gel models. In (A), the floating collagen gel model, the collagen matrix is freely-floating within the culture media. Fibroblast-generated tractional forces, resulting in gel contraction, are distributed equally throughout the matrix. No tension develops in this system. In (B), the stabilized collagen gel model, the matrix is attached to the underlying substratum. Cell-generated forces are distributed unequally along the lines of greatest stress resulting in a state of isometric tension. In (C), the stress-relaxed collagen gel model, the stabilized collagen gel is released from the underlying substratum, resulting in rapid collagen lattice contraction. Tension rapidly dissipates as the cells actively contract the matrix. (Adapted from Grinnell (1994)).

free to move in all directions; tension is therefore distributed isotropically, and the matrix remains mechanically relaxed (Grinnell, 1994).

In the stabilized collagen gel model (Figure 11.5B) tractional forces result in reduction of matrix height; however because the collagen is attached to the underlying plastic substratum tension is distributed anistropically along lines of stress (Mochitate *et al.*, 1991; Tomasek *et al.*, 1992). The final result is a mechanically stressed tissue (Grinnell, 1994). To evaluate the development and maintenance of mechanical stress, a third model has been utilized, in which the stabilized collagen gel is released from its points of attachment (Figure 11.5C). This results in a rapid contraction of the collagen gel (Mochitate *et al.*, 1991; Tomasek *et al.*, 1992) as a result of a contraction by the individual cells, involving an actin and myosin interaction similar to that which occurs in smooth muscle (Tomasek *et al.*, 1992).

These results demonstrate that fibroblasts can generate intracellular forces which are transmitted to the surrounding ECM and other cells. The physical state of the tissue can determine the types of mechanical stresses which develop. In the collagen gel model isotropic stress will develop in the floating gel, while anisotropic stress will develop in the stabilized gel.

The phenotype displayed by a fibroblast depends on the mechanical stress present in the matrix. Fibroblasts within a floating collagen gel develop a stellate shape with long, slender processes, and an actin cytoskeleton arranged into a fine meshwork throughout the cytoplasm (Bellows *et al.*, 1981; Tomasek *et al.*, 1982). In contrast, fibroblasts within a stabilized collagen gel develop a slender bipolar shape, oriented along the lines of stress (Stopak and Harris, 1982; Bellows *et al.*, 1982), prominent bundles of actin microfilaments and the transmembrane cytoskeleton-ECM attachment referred to as the "fibronexus" (Mochitate *et al.*, 1991; Tomasek *et al.*, 1992). Fibroblasts within floating collagen gels also have a low proliferative capacity, which appears to be the result of decreased responsiveness of the cells to growth factors (Nakagawa *et al.*, 1989; Lin and Grinnell, 1993). In addition, fibroblasts in floating collagen gels show a decreased collagen synthesis (Mauch *et al.*, 1988), increased release of collagenase (Unemori and Werb, 1986), and increased activation of 72 kD gelatinase (Halliday *et al.*, submitted for publication) compared to fibroblasts in stabilized collagen gels.

Apart from its role in changing connective tissue size and shape by contraction, mechanical stress plays an important role in the organization of the ECM and cells during morphogenesis. Intracellular forces transmitted to the ECM can align collagen fibers and cells both *in vitro* (Harris *et al.*, 1981; Mochitate *et al.*, 1991; Tomasek *et al.*, 1992) and *in vivo* (Stopak *et al.*, 1985), and may regulate the formation of fibronectin fibrils (Halliday and Tomasek, 1995). In addition, mechanical stress may also play an important role in morphogenesis by regulating cell behavior. Cells recognize and respond to mechanical stresses by changing their shape, growth, expression of specific gene products, and cytoskeletal organization (Grinnell, 1994; Ingber, 1994), as well as by remodeling their extracellular matrix (Unemori and Werb, 1986; Lambert *et al.*, 1992; Halliday *et al.*, submitted for publication).

Differences in mechanical stress may regulate cell behavior through the intermediary of the cytoskeleton. Recently it has been proposed that the cytoskeleton can be viewed as being built following the rules of tensional integrity or "tensegrity" in which cytoskeletal elements form an interconnected "prestressed" geodesic dome-like structure (Ingber, 1993; Ingber, 1994). An alternative to this idea is that mechanical continuity is established throughout the cytoskeleton by the formation of random networks according to the mathematical principle of percolation (Forgacs, 1995; see also above). From an evolutionary point of view it is plausible that percolation, which is a random phenomenon, was the primitive, "generic", basis for mechanical continuity in the cytoskeleton. Tensegrity, which depends on a precisely articulated set of connections among structural elements, could have evolved later, first locally within domains of the cytoskeletal network, and then perhaps more globally.

In either case, mechanical interactions between cells and their surrounding matrix may be transduced by integrins across the cell surface and throughout the cytoskeleton, possibly into the nucleus (Wang *et al.*, 1993). In addition, mechanical forces may be transduced into chemical signals within the focal adhesion complex (Kornberg *et al.*, 1992; McNamee *et al.*, 1993; Schaller and Parsons, 1994). In this manner differences in mechanical stress may play an important role in regulating cell shape, cytoskeletal organization, growth, and differentiation during morphogenesis.

MORPHOGENESIS OF CONNECTIVE TISSUES: EXAMPLES

The following examples represent only a small subset of the experimental systems currently under study. They have been chosen to illustrate the recurrent utilization of generic physical mechanisms in the morphogenesis of connective tissues as well as the wide range of molecular interactions by which these mechanisms are realized.

Connective Tissue Development from the Neural Crest

The neural crest consists of populations of cells that detach and migrate away from the dorsal ridges of the neural tube just before, or shortly after it closes (Langille and Hall, 1993). The detachment of neural crest cells from the neural tube must involve loss or modification of cell-cell adhesion, but which molecules and mechanisms may be involved is not yet clear. At least two Ca^{2+}-dependent cell-cell adhesion molecules, N-cadherin and E-cadherin, are lost from the neural crest cells prior to, or shortly after detachment from the neural tube (Duband *et al.*, 1988; Akitaya and Bronner-Fraser, 1992; Bronner-Fraser *et al.*, 1992). Although the Ca^{2+}-independent cell adhesion molecule N-CAM is partially lost from the surfaces of neural crest cells as they initiate migration (Thiery *et al.*, 1982) N-CAM is probably not a key regulator of the detachment process since N-CAM-deficient mice are phenotypically normal (Cremer *et al.*, 1994).

Alternative mechanisms for detachment of neural crest cells from the neural tube are proteolytic degradation of attachment molecules, and generation of tractional forces (see above) that would mechanically rip cells away from their neighbors (Erickson and Perris, 1993). Although neural crest cells are known to produce proteases that are associated with increased motility in culture (Valinsky and Le Douarin, 1985; Erickson and Isseroff, 1989), and there is electron microscopic evidence for cell rupture in the vicinity of detached neural crest cells (Nichols, 1986; Bilozur and Hay, 1989), there is no experimental evidence that indicates that either of these detachment mechanisms operates *in situ*. Some form of epithelial to mesenchymal transformation, based on the acquisition of new cell surface properties, remains the most probable mode of detachment of these cells.

The connective tissues of the vertebrate head are largely the product of the cephalic (cranial) neural crest. These cells arise from the dorsal-most regions of the rhombomeres that constitute the future brain, and migrate dorsolaterally along several pathways to form the branchial apparatus (Noden, 1984). Cephalic neural crest migration follows a "micro-finger" morphology consisting of several parallel streams (Horstadius and Sellman, 1946). While the invading sheet of neural crest cells may be divided into streams by anatomical obstructions encountered in their path (Noden, 1988), it is significant that ectopic neural crest cells transplanted into ventral regions of axolotl embryos break into microfingers of similar dimension as they migrate dorsal through a region of the embryo distant from the endogenous streams (Horstadius and Sellman, 1946). This has led Comper *et al.* (1987) to suggest that such microfingering flows through the ECM might be based on a generic physical transport mechanism they observed in non-uniform polymer solutions (Preston *et al.*, 1980) including some composed of ECM macromolecules such as proteoglycans (Harper *et al.*, 1984) and collagen (Ghosh and Comper, 1988).

Although cephalic neural crest cells, like those of the trunk, can form peripheral ganglia and melanocytes (Le Douarin *et al.*, 1992), unlike the trunk neural crest cells they also contain a subpopulation that can form cartilage and bone. These cells participate in the formation of skeletal structures of the head and neck. In mammals, for example, the incus, malleus and stapes bones of the middle ear, and portions of the mandible, maxilla, temporal and hyoid bones, are all derived from cranial neural crest (Larsen, 1992).

If the concept of skeletogenesis is expanded to include both the endoskeleton (i.e., cartilage and bone) and the dermal, or exoskeleton (e.g., teeth, the mineralized terminal portions of fin rays in modern fish, and the armor plates of many fossil fish), then it appears that the trunk neural crest shares skeletogenic potential with the cephalic neural crest. Smith and Hall (1993) adduce evidence that trunk neural crest can give rise to portions of the exoskeleton in lower vertebrates, and suggest that it had even more extensive skeletogenic potential earlier in evolution. Interestingly, Lumsden (1987) recombined mouse mandibular epithelium with premigratory trunk neural crest from the cervical region and found that the neural crest cells gave rise to teeth and surrounding bone (i.e., part of the endoskelton), but not cartilage.

Having detached from the neural tube and migrated to the sites of the future branchial arches, the various populations of skeletogenic neural crest cells aggregate and begin to undergo morphogenesis. The formation of the mandible in the mouse is an instructive example (Atchley and Hall, 1991; Figure 11.6). Four separate cell populations can be distinguished in the formation of this structure: one chondrogenic, two osteogenic, and one odontogenic (Atchley and Hall, 1991). In the alveolar region of the developing mandible one population of ectomesenchyme cells produces osteoblasts, which generate the alveolar bone for attachment of the teeth, and odontoblasts, which produce the dentine of the teeth. Another ectomesenchymal population differentiates into osteoblasts which produce the ramal bone. A separate subpopulation of cells produces osteoblasts and precartilage condensations leading to chondrocytes, which generate the coronoid, condyloid and angular processes by a combination of intramembranous and endochondral routes. A fourth, chondrogenic, population produces Meckel's cartilage which, in mammals, subsequently disappears.

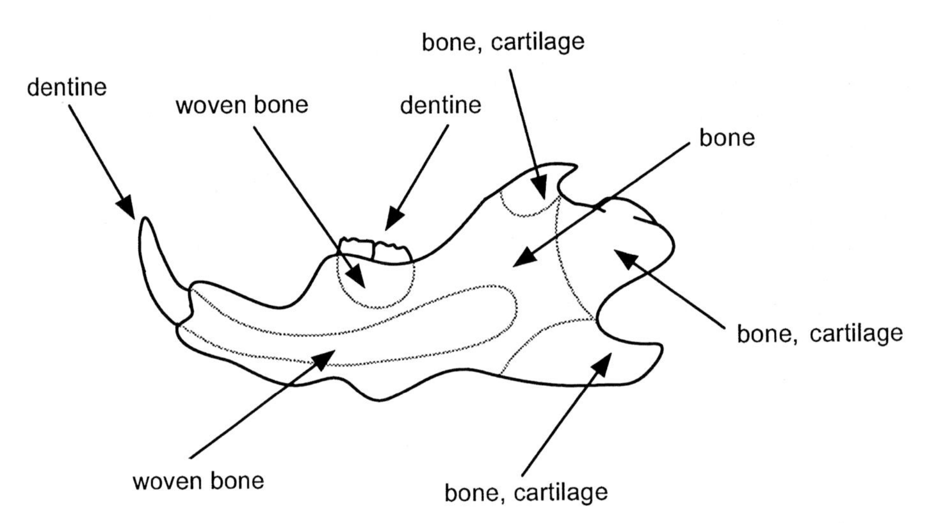

Figure 11.6 A drawing of the mouse mandible showing the distribution of the four major neural crest-derived connective tissue types that comprise this complex structure. The "bone" of the central ramus and the three projections or "processes" on the right are lamellar bone, in contrast to the woven bone of the tooth sockets. These bone types differ in the organization of their collagen fibers. (Adapted, with modifications, from Atchley and Hall (1991)).

At what point in development these four neural crest cell-derived subpopulations diversify, and how they interact with one another and adjacent epithelia to generate a unified structure, the mandible, remain obscure, but recent work on the patterns of expression of the *Msx1* and *Msx2* genes provide some insight into molecular mechanisms that may be involved. McKenzie *et al.* (1991, 1992) found both these genes to be expressed in tooth mesenchyme during early morphogenesis, with *Msx1* reaching its maximal level of expression in the mesenchymal condensations. Interestingly, mice in which this gene has been inactivated lack teeth (Satokata and Maas, 1994). Interaction with epithelium is required in order for mandibular mesenchyme to undergo proper bone differentiation and morphogenesis in avian species (Hall, 1983; Takahashi *et al.*, 1991). It is therefore significant that *Msx2* expression in the mandibular mesenchyme of the embryonic quail is dependent on the presence of mandibular epithelium (Takahashi *et al.*, 1991). (See also discussion of the *Msx* genes in epithelial-mesenchymal interactions in the developing limb in next section.)

Concerning the potential of cephalic neural crest cells to form specific structures, Noden (1983) showed by heterotopic grafts in chicken embryos that these cells acquire regional specification prior to their emigration to and residence in the branchial apparatus. For example, when regions of the neural crest that would normally contribute to the second branchial arch were replaced with cells that would have gone to the first arch, the grafted cells produced a second mandible. Although the original regional specification was thus sufficient to cause the grafted tissue to produce first arch rather than second arch derivatives, it is significant that it did not determine the actual structure made by the tissue. Thus, grafted *mesencephalic* crest cells which would normally participate in the formation of maxillary, periocular, and frontonasal skeletal structures, formed the same mandibular-type structures in their new location as did grafted *metencephalic* crest cells, for which the mandibular arch is the normal fate.

The determinants of regional identity of the body axis, central nervous system, and neural crest cells in the vertebrate embryo are thought to be the Antennapedia-class *Hox* gene products (McGinnis and Krumlauf, 1992). However, there is also ample evidence that mechanical loading, including that arising from growth dynamics, muscle contraction, and other function-related activities, plays a substantial role in the formation of the cranium and face (Herring, 1993). Based on the earlier discussion of tissues as physical materials, we can speculate that the *Hox* gene products are acting as regulators of cell surface and extracellular matrix proteins, endowing the tissue at each level with a particular quantitative mix of these components, thus defining its material properties. The particular structures that any such material might form will depend on their characteristic responses to extrinsically applied molding forces, just as clay and jelly can be made into numerous objects that both reflect their identity as materials and the sculpting forces to which they have been subjected.

Morphogenesis of Connective Tissues in the Vertebrate Limb

The skeleton and musculature of the vertebrate limb arise from mesoblasts — masses of mesodermally-derived mesenchymal tissue — that form at four positions beneath the ectoderm covering the body wall. The formation of the limb "fields" containing the mesoblast precursor cells, shortly after the body axis is established, is a rare example of compartment formation (see above) in a purely mesenchymal cell population. (The early stages of somitogenesis (Stern *et al.*, 1988; Wood and Thorogood, 1994) is another.) Heintzelman *et al.* (1978) studied the behavior of chicken limb bud and flank mesenchyme in "tissue fusion" experiments. They found that these tissues acted like immiscible liquid phases, a result analogous to those obtained in tissue fusion experiments in which one or both of the fragments were epithelioid and the differential cohesiveness observed could be attributed to relative strengths of direct cell-cell adhesion (Steinberg and Poole, 1982).

The source of the tissue cohesivity differences in the mesenchyme-mesenchyme fusions was unclear, but an indirect implication of these studies was that apart from their both being more cohesive than flank mesenchyme, fore and hind limb mesenchyme were also different in this respect from one another, with hind limb tissue being the more cohesive (Heintzelman *et al.*, 1978). Recently this was tested directly (Downie and Newman, 1994), and the prediction was borne out: contiguously plated high density cultures of trypsin-dissociated wing and leg bud mesenchymal cells formed a sharp interface across which there was no mixing of cells. This interface was convex with respect to the leg mesenchymal culture, as predicted if the leg tissue was more cohesive than the wing (Figure 11.7). Interestingly, the two types of cultures differed markedly in the organization of fibronectin. As detected by immunofluorescence microscopy, fibronectin in the wing cultures was diffusely organized around the cells and concentrated in the central regions of forming condensations. In the leg cultures fibronectin was even more abundant in the condensing regions, but most striking is the presence of a network of long fibronectin-rich fibers not seen in the wing cultures (Downie and Newman, 1994; 1995) (Figure 11.7). This raises the possibility that the immiscibility of fore and hind limb mesenchyme (and perhaps that of other mesenchymal tissues that normally form contiguously, such as limb and flank, and adjacent somites) may be due to organizational differences of the ECM. (Recall that ECM organizational differences were proposed to be the basis of interface formation in the matrix-driven translocation model system discussed above.) It is perhaps significant in this regard that mouse embryos deficient in fibronectin fail to develop limb buds or somites (George *et al.*, 1993).

Although it is difficult to distinguish morphologically between mesenchymal cells in the early limb bud that will give rise to muscle and those that will give rise to skeletal and other connective tissues, those subpopulations of cells (in avian species, at least) abide by an early, covert determination during normal development (Newman, 1977; 1988). Whereas the precursor cells for cartilage and connective tissue arise from the somatopleure, or body wall, the myogenic precursors originate in the somites and migrate into the limb bud just before and during the time of its emergence (Chevallier *et al.*, 1977; Christ *et al.*, 1977). While limb development therefore does not involve myogenic versus chondrogenic decisions within the mesenchyme, skeletal pattern formation occurs in a population of mesenchymal cells with options to undergo chondrogenesis, apoptotic death (Saunders and Fallon, 1967; Zakeri *et al.*, 1994; Toné *et al.*, 1994), or differentiation into non-cartilage connective tissue (Newman, 1980).

Chondrogenic differentiation in the limb is non-uniform in time and space, occurring in a proximodistal sequence in all species examined (Hinchliffe and Johnson, 1980) (Figure 11.8). As in the cranifacial system discussed above (Section 3A), overt chondrogenesis is preceded by mesenchymal condensation. The entire limb skeleton is laid down as a cartilage model which is replaced by bone later in development in those species with bony skeletons (Hinchliffe and Johnson, 1980).

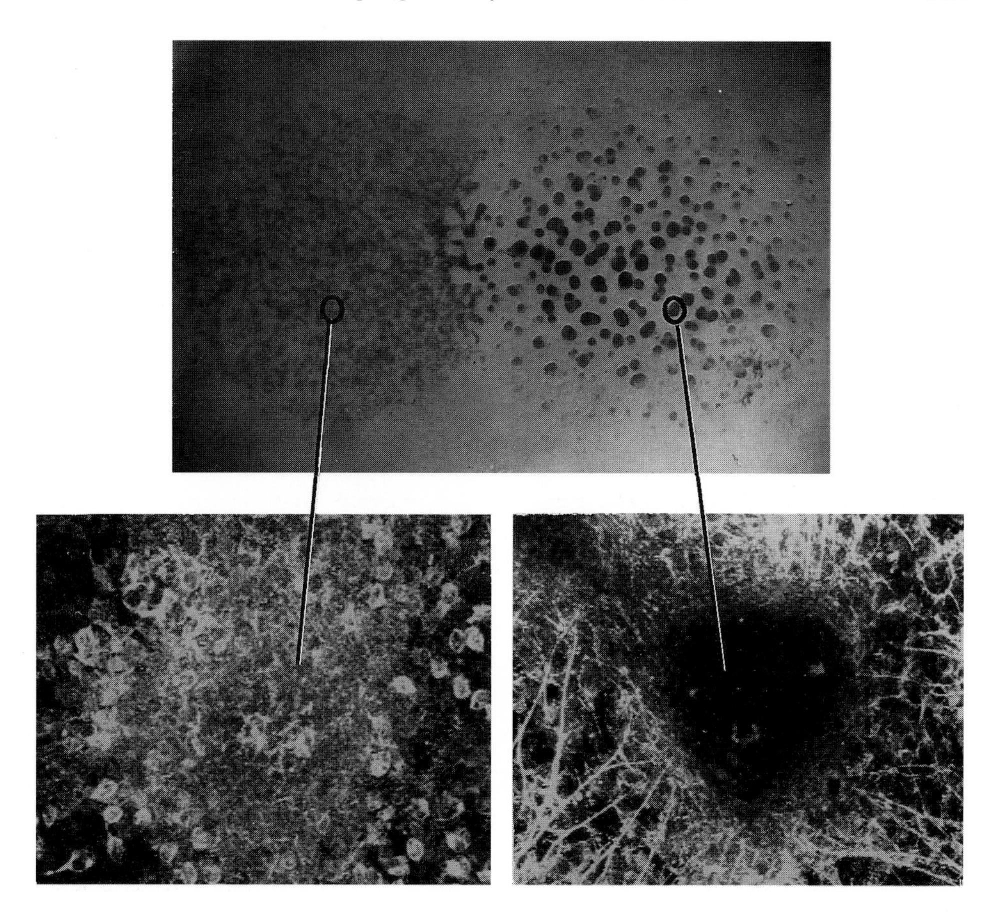

Figure 11.7 Upper panel, contiguous high density cocultures of wing bud (left) and leg bud (right) precartilage mesenchymal cells after 6 days in culture. Cartilage, which is stained with Alcian blue, has formed in both cultures, in a relatively diffuse pattern in the wing cell mass and in a nodular pattern in the leg bud cell mass. The internodular regions in the leg cultures contain nonstaining cells. The sharp interface which formed between the two cell masses was always convex with respect to the leg culture. The diameters of the two cell masses were approximately 5 mm. Lower panel, scanning confocal micrographs of areas of precartilage condensation from similar wing and leg mesenchyme cultures fixed at 2½ days after plating (e.g., 3½ days before the stage shown in upper panel) stained with a fluorescent antibody against fibronectin. Area of 6 day cultures corresponding to single 2½ day mesenchymal condensations are indicated by circles in upper panel. Note the extensive network of fibronectin-rich fibrils in the leg cell condensation, in contrast to the diffuse distribution of fibronectin staining in the wing cell condensation (also compare to lower magnification view of wing mesenchymal condensations in Figure 11.4). The dark area at the center of the leg condensation represents a region of the optical section in which the antibody label had not penetrated. (Adapted, with modifications, from Downie and Newman (1994)).

The formation of precartilage condensations in both the developing limb and in high density cultures derived from limb precartilage mesenchyme is, as noted above, accompanied by the accumulation of the extracellular glycoprotein fibronectin between the condensing cells (Tomasek *et al.*, 1982; Frenz *et al.*, 1989b). Although fibronectin's best-studied cell adhesive functions are mediated by the binding of its centrally located arg-gly-asp-ser (RGD) motif to members of the integrin family of integral membrane pro-

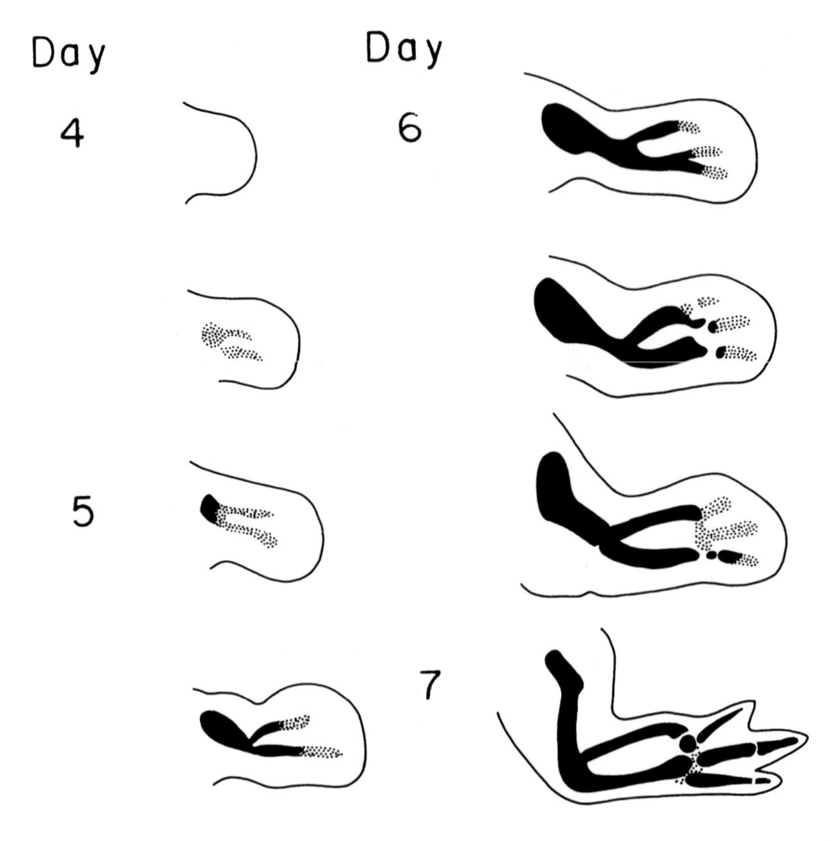

Figure 11.8 Development of the chick wing bud, represented in cross-section, between 4 and 7 days of incubation. Solid black regions represent definitive cartilage; stippled area represent regions of mesenchymal condensation.

teins (Hynes, 1990), this interaction does not appear to be important in the precartilage condensation process (Frenz *et al.*, 1989a,b). Instead, the amino-terminal heparin-binding domain of fibronectin, possibly interacting with heparan sulfate-containing syndecan proteoglycans on the precartilage cell surface (Gould *et al.*, 1992), appears to mediate early events in the condensation pathway (Frenz *et al.*, 1989a,b). Later, there is a transient epithelialization of the condensing cells, evidenced by the functional appearance of N-CAM (Widelitz *et al.*, 1993) and N-cadherin (Oberlender and Tuan, 1994). It is likely, however, that the condensation process is molecularly overdetermined: limb development proceeds normally in N-CAM-deficient mice (Cremer *et al.*, 1994).

A symmetrical limb skeleton, i.e., one in which there is no anteroposterior polarity, and all the digits, for example, are similar to each other, will form from randomized limb bud mesenchymal cells that have been packed into an epithelial hull (Zwilling, 1964; Ros *et al.*, 1994). This shows that gradients and other nonuniform distributions of molecules in the unperturbed mesoblast, such as retinoic acid (Thaller and Eichele, 1987) and its receptors (Dollé *et al.*, 1990), the *Hox* gene products (Duboule, 1992), and the *Sonic hedgehog* gene product (Riddle *et al.*, 1993), are not necessary for the establishment of the basic limb plan of proximodistally increasing numbers of parallel skeletal elements. However, grafting experiments using limb primordia suggest that some kind of chemical prepattern which specifies the positions of future skeletal elements is present prior to overt condensation formation (Wolpert and Hornbruch, 1990), contrary to the predictions of pattern forming mechanisms based on mechanochemical instabilities in the mesenchyme (Oster *et al.*, 1984; 1985; see Section 2B1, above). Rather than being the result of preexisting gradients in the mesoblast then, this chemical prepattern may instead be the outcome of a process of "self-organization" involving diffusion and cross regulatory effects of various differentiation factors, such as TGF-β, on the mesenchymal cells (Newman and Frisch, 1979; Newman, 1988). The intrinsic pattern forming properties of such systems were first demonstrated mathematically by Turing (1952) and have since been realized in a number of nonliving model chemical systems (Castets *et al.*, 1990; Ouyang and Swinney, 1991; Boissonade, *et al.*, 1992). These systems exhibit periodic arrays of spots and bars of chemical concentration on a spatial scale similar to that of the incipient skeletal elements of the limb.

It is perhaps significant in this regard that members of the TGF-β family of growth factors, which induce the production of fibronectin in mesenchymal tissues (Massagué, 1990; Leonard *et al.*, 1991) are also positively autoregulatory (van Obberghen-Schilling *et al.*, 1988), a requirement that must be met by at least one of the components in a Turing-type pattern forming system (Meinhardt, 1982). This generic mechanism may have determined the form of the limb at its evolutionary origin (Newman, 1984; Shubin and Alberch, 1986) and may continue to play a part in mesenchymal morphogenesis in the limbs of modern vertebrates (Leonard *et al.*, 1991; Downie and Newman, 1994). Over the course of evolution, regional differences in expression of numbers of the *Hox* gene family, possibly by regulating the degree of expression of ECM and cell adhesion molecules (Newman, 1993b; Yokouchi *et al.*, 1995), may have led to the distinctions among skeletal elements seen in modern limbs.

As in the case of craniofacial skeletogenesis (previous section) skeletal development in the limb is dependent on epithelial-mesenchymal interactions. Rimming the distal margin of the paddle-shaped limb bud during the chondrogenic pattern-forming phase is a ridge of ectodermal cells morphologically distinct from the rest of the cells of the ectodermal sheath. These taller apical ectodermal ridge (AER) cells are absolutely required for limb outgrowth and the consequent generation of progressively more distal parts during development (Saunders, 1948). Recent work suggests that AER activity can be accounted for by one or more of the FGF family of growth factors produced by these specialized epithelial cells (Niswander *et al.*, 1993; Fallon *et al.*, 1994). As in the craniofacial regions, the production of *Msx1* and *Msx2* gene products correlate with epithelial-mesenchymal interactions (Coelho *et al.*, 1993). Although *Msx1* is normally expressed by mesenchyme directly beneath the AER in chickens (Coelho *et al.*, 1993), mice that lack this gene develop normal limbs (Satokota and Maas, 1994). If gene products of this class are essential for epithelial-mesenchymal interaction during limb development they may act in a redundant fashion. Indeed, *Msx2* is expressed ectopically, in a pattern identical to that of *Msx1*, in limb buds formed by recombining randomized mesoblast cells with ectoderm (Ros *et al.*, 1994).

Conversion of Mesenchyme to Epithelium in Kidney Development

The metanephric kidney develops from two distinct but interacting populations of cells, the epithelial ureteric bud and the mesenchymal metanephric mesoderm. The metanephric mesenchyme induces the ureteric bud to branch and form the collecting system, while the ureteric bud induces the metanephric mesenchyme to form the epithelium of the glomeruli and the kidney tubules. This conversion of mesenchyme to epithelium has been well studied in the kidney and involves three basic steps:

(1) condensation of mesenchymal cells;
(2) acquisition of epithelial cell characteristics including cell polarity, basement membrane formation and adhesion of lateral cell surfaces;
(3) differentiation of tubule segments.

The conversion of metanephric mesenchyme to kidney tubule epithelium is induced by the ureteric bud *in vivo* (Gruenwald, 1943). Other embryonic tissues can induce nephrogenic mesenchyme to convert into epithelium, neuronal tissues being the best known of these (Unsworth and Grobstein, 1970). No tissue other than metanephric mesenchyme has been shown to produce nephrons in response to inducer tissues (Saxén, 1970), demonstrating that the interaction is permissive rather than directive.

So far, it has not been possible to induce epithelial tubule development with cell-free extracts, but in recent studies using a metanephric mesenchymal cell clone, hepatocyte growth factor/scatter factor caused the appearance of uvomorulin and cytokeratin, consistent with a conversion toward epithelium (Karp *et al.*, 1994). In addition, in the developing murine kidney HGF/SF and Met are highly expressed at the time corresponding to the onset of tubulogenesis and branching morphogenesis and an antiserum against HGF/SF inhibits kidney development in organ culture (Santos *et al.*, 1994). Thus, the HGF/SF-Met receptor pathway may regulate conversion of mesenchyme to epithelium during kidney development. Misfunctioning of this pathway may contribute to cyst-forming aberrations of epithelia, such as polycystic kidney disease (Rosen *et al.*, 1994).

The conversion of metanephric mesenchyme to epithelium appears to be driven by a gradual increase in intercellular adhesion. This is not surprising as mesenchymal cells are separated from each other by extracellular matrix and must first come together in order to form an epithelium. Morphologically, the mesenchyme is observed to condense into multicellular aggregates. Later the cells comprising the aggregate become polarized, acquiring a basement membrane and a lumen.

The conversion of metanephric mesenchyme to epithelium appears to involve alteration in the expression of a number of adhesive proteins. These include the extracellular matrix macromolecules types I and III collagen (Ekblom *et al.*, 1981), fibronectin (Ekblom, 1981), tenascin (Aufderheide *et al.*, 1987), and laminin (Klein *et al.*, 1988b; Ekblom *et al.*, 1990), as well as the cell-surface adhesion molecules N-CAM (Klein *et al.*, 1988a), uvomorulin (E-cadherin) (Vestweber *et al.*, 1985) and the integrin receptor $\alpha_6\beta_1$ (Sorokin *et al.*, 1990) (see Figure 11.9).

The mechanism by which alterations in these adhesive proteins result in mesenchymal condensations is not well understood, but may involve differential adhesion accompanied by random cell movements, as described previously. Fibronectin, found in the undifferentiated and uninduced mesenchyme (Wartiovaara *et al.*, 1976), may increase in amount at sites of incipient condensation, providing an early matrix for mesenchymal cell accumulation, as it does in developing limb mesenchyme (Frenz *et al.*, 1989 a,b). N-CAM, expressed in both mesenchymal and condensation stages (Klein *et al.*, 1988a), may assist in consolidation when cells come into contact with each other, as suggested for other types of

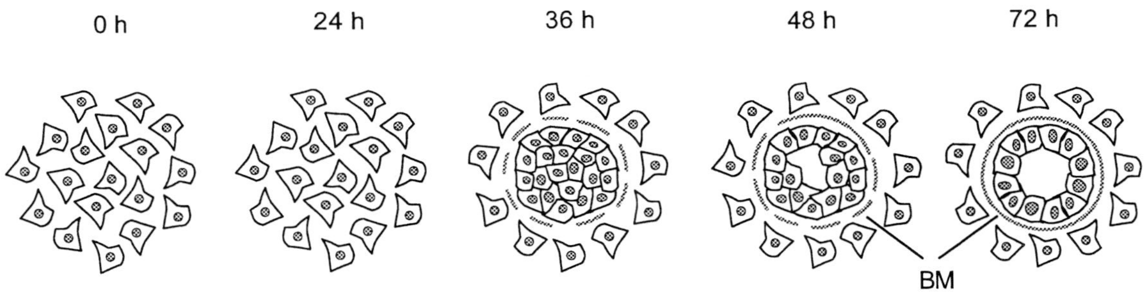

	0 h	24 h	36 h	48 h	72 h
Collagen I, III	+	-	-	-	-
Fibronectin	+	+/-	+/-	+	+
Tenascin	-	-	-	-	-
Laminin B	+	+	+	+	+
Laminin A	-	-	+	+	+
$\alpha6\beta1$ Integrin	-	-	+	+	+
E-Cadherin	-	-	+	+	+
N-CAM	+	+	+	-	-

BM

Figure 11.9 Summary of expression of some adhesion proteins during conversion of metanephric mesenchyme to epithelium. Marks refer to expression in the stem cells and in developing epithelium. Note that the E-cadherin and laminin A chains appear at the condensation stage. Tenascin also appears on day 2 of in vitro development, but in the matrix of the surrounding mesenchyme not undergoing the epithelial transformation. (Adapted, with slight modifications, from Ekblom, 1992).

condensations (Widelitz *et al.*, 1994).

After condensation the metanephric mesenchyme develops into kidney tubule epithelium. The distinguishing characteristic of epithelial cells is their polarity — their basal surfaces are attached to an underlying basement membrane, while their apical surfaces are free. The first evidence of polarity in the condensed kidney mesenchyme is the formation of a basement membrane (Ekblom *et al.*, 1981). This structure is composed of laminin, type IV collagen and basement membrane proteoglycan (Ekblom *et al.*, 1981; Lash *et al.*, 1983; Klein *et al.*, 1988b). The importance of the basement membrane in development of epithelial polarity has been demonstrated in studies on the role of laminin in the conversion of metanephric mesenchyme to kidney tubule epithelium (Ekblom, 1989; Ekblom, 1992).

Laminin is a large extracellular glycoprotein composed of several chains which interact in a variety of ways to form a large family of complexes (Timpl, 1989). The 220-kD B1 and B2 chains of laminin are expressed by metanephric mesenchyme prior to and during epithelial development (Klein *et al.*, 1988b; Ekblom *et al.*, 1990). Once epithelial cell development begins A-chain mRNA levels increase dramatically in those areas where new tubules will form (Klein *et al.*, 1988b; Ekblom *et al.*, 1990). Presumably the newly synthesized A chain associates with the B1 and B2 chains to form a large laminin complex. Deposition of laminin A-chain is seen prominently only in the extracellular matrix around condensates and correlates with the onset of basement membrane formation (Klein *et al.*, 1988b). At the time the cells begin expressing laminin A-chain they are not morphologically polarized, suggesting that a polarized deposition of the A-chain is an early stage of cell polarity development (Ekblom, 1992).

Functional studies using antiserum known to interfere with binding of laminin to cells have provided more direct evidence for the importance of laminin in the development of kidney epithelium. Antiserum against the E8 and E3 fragments of laminin inhibited epithelial development, while antisera against the central parts of laminin did not affect development (Klein *et al.*, 1988b). The E8 fragment is composed of adjacent domains from the carboxy termini of B1 and B2 chains and parts of the A-chain. The interaction of these domains could thus be rate-limiting for the appearance of functionally active laminin molecules which promote the development of a polarized kidney epithelium (Ekblom, 1992).

Binding of the A-B1-B2 laminin to the basal cell surface to promote development of a polarized epithelium may be accomplished through the integrin receptor $\alpha_6\beta_1$, which has been demonstrated to bind the E8 region of laminin (Sonnenberg *et al.*, 1990). In addition, the α_6 subunit is co-expressed with the A chain of laminin during conversion of mesenchyme to epithelium (Sorokin *et al.*, 1990). A-chain laminin may be instrumental in localizing the $\alpha_6\beta_1$ integrin to the basal surface. Initially, the receptor subunit is present all over the cell surface; however, when cell polarization begins it becomes enriched at the basal surface (Sorokin *et al.*, 1990). The receptor is important in development of epithelial polarity: antibodies against the α_6 subunit can interfere with kidney tubule development *in vitro* (Sorokin *et al.*, 1990).

As polarity develops epithelial cells become highly adherent along their lateral borders and produce specialized cell-cell junctions in this location. Uvomorulin, also referred to as E-cadherin, is a member of the cadherin family of adhesive proteins (Takeichi, 1988). E-cadherin begins to be expressed at the time of conversion of mesenchyme to epithelium (Klein *et al.*, 1988a). Initially, E-cadherin is expressed on all sides of the cells and only later becomes enriched at the lateral surface (Klein *et al.*, 1988a), presumably reflecting the formation of adherens junctions (Ekblom, 1992). Other adhesion proteins also appear to be involved in the development of kidney epithelial cells since E-cadherin never appears at any developmental stage during the formation of the glomerulus (Vestweber *et al.*, 1985).

Once the early kidney tubule has been formed it is possible to distinguish between distal and proximal tubules and between different epithelial cells of the glomerulus. It may be that development of specific characteristics of the free apical surface which characterize the different segments of the kidney tubule and glomerulus are late differentiation events in consequence of the earlier adhesive cell-matrix and cell-cell interactions (Ekblom, 1992).

Connective Tissue Morphogenesis During Wound Healing

Wound repair in response to injury is characterized by acute inflammation followed by the deposition of a transitional granulation tissue consisting of fibroblasts surrounded by an ECM containing types I and III collagen and fibronectin. Granulation tissue acts to retract the wound space, leading to closure of the defect and subsequent scar formation (Peacock, 1984). In addition to the orderly repair of a wound, errant tissue contraction may lead to pathological contractures (Rudolph *et al.*, 1992). Fibroblasts present in granulation tissue are capable of exerting mechanical force on surrounding ECM leading to tissue contraction. The resulting mechanical stresses induce dramatic changes in the fibroblasts themselves during tissue contraction. Some of the cellular mechanisms by which fibroblasts exert forces on their surroundings and the means by which mechanical stress regulates fibroblast phenotype, gene expression, and proliferation may represent relics of the single-celled stage of eukaryotic evolution. The machinery underlying these processes in contemporary multicellular tissues is just beginning to be characterized (see above).

During the initial stages of wound healing, fibroblasts in the granulation tissue are migratory. These cells have a dispersed actin cytoskeleton with cortical actin microfilaments (Ehrlich, 1988; Rudolph *et al.*, 1992). Collagen lattice studies have suggested that migratory fibroblasts can exert tractional forces on the collagen matrix resulting in the reorganization of the collagen fibers (Harris *et al.*, 1981; Grinnell, 1994). In the floating collagen lattice model tractional forces generated by migratory fibroblasts result in collagen lattice contraction (Bell *et al.*, 1979; Figure 11.5A). The migratory wound fibroblasts in granulation tissue presumably generate sufficient force to initiate wound contraction (Ehrlich, 1988; Rudolph *et al.*, 1992; Grinnell, 1994).

During tissue contraction fibroblasts switch from a migratory to a contractile phenotype (Gabbiani *et al.*, 1972). The fibroblasts acquire large bundles of actin microfilaments (Gabbiani *et al.*, 1972) and begin to express the α-smooth muscle isoform of actin (Darby and Gabbiani, 1990). In addition, they assemble fibronectin into fibrils at their surfaces and form a specialized transmembrane connection, the fibronexus, linking the intracellular actin microfilaments with extracellular fibronectin fibrils (Singer *et al.*, 1984; Welch *et al.*, 1990; Tomasek and Haaksma, 1991). These modified fibroblasts have been termed "myofibroblasts" because of their contractile phenotype (Gabbiani *et al.*, 1972).

Studies using collagen lattice model systems suggest that mechanical stress is responsible for the acquisition of a contractile phenotype by fibroblasts. In the stabilized collagen gel model (Figure 11.5B), tractional forces result in a reduction of matrix height, but because the collagen is stabilized a mechanically stressed tissue develops (Mochitate *et al.*,

1991; Tomasek *et al.*, 1992; Grinnell, 1994). In this environment fibroblasts acquire a contractile phenotype and appear similar to the myofibroblasts seen in contracting tissues (Mochitate *et al.*, 1991; Tomasek *et al.*, 1992) (Figure 11.10).

In contracting tissues mechanical stress may develop in a similar manner (Rudolph *et al.*, 1992; Grinnell, 1994). Migrating fibroblasts generate sufficient tractional force to initiate contraction. As contraction proceeds there is continued outward tension exerted by the surrounding tissues, resulting in increased mechanical stress. In response to this stress migrating fibroblasts differentiate into myofibroblasts. Earlier studies demonstrated that contracting granulation tissue develops isometric tension (Higton and James, 1964). When exogenous mechanical stress is applied to skin by stretching it myofibroblasts appear (Squier, 1981).

Stabilized collagen gels, when released from their points of attachment, undergo a rapid contraction (Figure 11.5C), which, like the contraction seen in granulation tissue, involves an actin-myosin interaction within the resident cells (Tomasek *et al.*, 1992). Such cells can also generate isometric force. Indeed, when stabilized collagen gels are attached to a force transducer (Delvoye *et al.*, 1991; Kolodney and Wysolmerski, 1992; Pilcher *et al.*, 1995), the amount of isometric force generated is comparable to that generated in contracting granulation tissue (Higton and James, 1964).

The contraction of myofibroblasts is promoted by specific exogenous factors. Thrombin can promote both isotonic (Pilcher *et al.*, 1994) and isometric (Kolodney and Elson, 1993; Pilcher *et al.*, 1995) contraction of collagen lattices populated with cells. Lysophosphatidic acid, a potent mitogenic component of serum, can also promote collagen lattice contraction (Kolodney and Elson, 1993).

Mechanical stress may also regulate the assembly of fibronectin fibrils observed at the surfaces of myofibroblasts in contracting tissues (Welch *et al.*, 1990; Tomasek and Haaksma, 1991). Fibroblasts in floating collagen gels synthesize and secrete fibronectin, which is not assembled into fibrils (Halliday *et al.*, 1995). In contrast, fibroblasts in stabilized collagen gels assemble fibronectin into fibrils once mechanical stress has developed

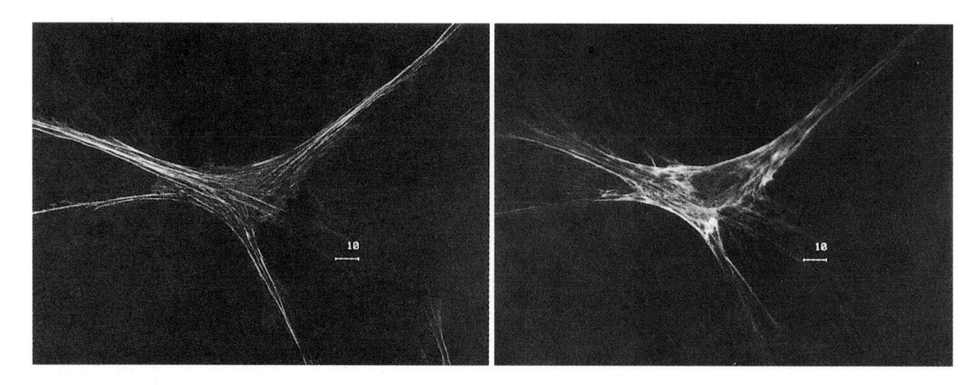

Figure 11.10 Confocal laser scanning fluorescence images of a human fibroblast cultured for 5 days in a stabilized collagen gel and double-stained for actin and fibronectin. In this environment the fibroblast has acquired a contractile phenotype similar to that of myofibroblasts seen in contracting tissues. Large bundles of actin microfilaments are seen to traverse the cell (A; stained with bodipy phallacidin). In addition, fibronectin has been assembled into fibrils present at the cell surface (B; stained with an anti-fibronectin antibody).

in the collagen matrix (Mochitate *et al.*, 1991; Tomasek *et al.*, 1992; Halliday *et al.*, 1995).

While it is clear that fibroblasts are capable of generating forces that can exert stress on and move extracellular matrices, recent studies suggest that there are additional mechanisms that promote wound closure. In particular, daily full-thickness excisions of the central granulation tissue of back and flank wounds, in either the guinea pig or domestic pig, did not alter the rate or completeness of wound closure or final pattern of the scar (Gross *et al.*, 1995). These results suggest that the polarized coordinated migration of a rim of densely packed, freshly proliferated, fibroblasts underlying and pulling inward the dermal edges, may promote wound closure. The ability of both contractile forces generated by granulation tissue myofibroblasts and the directed migration of peripheral fibroblasts to contribute to wound closure may represent another example of redundancy in morphogenetic processes. One may speculate that the mechanically-based contraction mechanism may have provided a "generic template" for the evolution of the directed cell movements, but a more reciprocal evolutionary pathway is also possible.

Lastly, during normal wound healing myofibroblasts disappear, leaving a relatively acellular scar tissue. Mechanical stress also plays an important role in this process. The rapid contraction that occurs upon release of stabilized collagen gels (Figure 11.5C) results in an inactivation of PDGF receptors, which decreases cell proliferation and responsiveness to growth factors (Lin and Grinnell, 1993). In addition, loss of mechanical stress may trigger a form of programmed cell death, or apoptosis (Rudolph *et al.*, 1992; Grinnell, 1994). Consequently, as mechanical stress is relieved during wound repair by wound contraction and ECM deposition, myofibroblasts would die off. If, in contrast, mechanical stress were not relieved, cell proliferation, ECM deposition and contraction would continue in a vicious circle, resulting in a pathological contracture. This can be seen in Dupuytren's disease where contraction of the palmar aponeurosis results in irreversible flexion of the digits because of the continued presence of myofibroblasts in the palmar aponeurosis (Schultz and Tomasek, 1990).

CONCLUSIONS

Given the variety of gene products and other molecules which have been identified as participating in changes in tissue form and pattern, it is remarkable how limited in kind are the outcomes of these morphogenetic processes. The studies reviewed here suggest that connective tissues become organized into stereotypical forms during the course of development and pathogenesis not primarily because specific genes and their products have been conserved over the course of evolution, but because the generic physical properties of multicellular aggregates and their ECMs, which would have come into play with the advent of the earliest multicellular forms, are capable of producing only a limited spectrum of tissue configurations. The effects of genetic change on the evolution of morphogenesis are hypothesized to have been indirect — the modulation of adhesivity, cohesivity, elasticity, viscosity — that is, material properties which are shared by all tissues but which also provide the basis for their morphogenetic differences.

We reiterate that the morphology of a connective tissue structure in any contemporary organism need not be produced during the course of development by the same generic physical processes that may have been responsible for its evolutionary origin. There are numerous examples of apparent co-optation of developmental events by new gene activities during the course of evolution, leading to functional redundancy, and in principle, to the complete mechanistic supercession of morphogenetic processes that originally brought these events about (Newman, 1994). Since these more recently evolved developmental mechanisms are, in our interpretation, built upon "generic templates", the stereotypical form remains even after the process that was originally responsible for it has become less important.

On this basis it has been proposed that a causal account of why organisms have the forms that they do can be accomplished by extrapolation back through evolutionary time — the "biological fifth dimension" — to those periods when bodies and organs were less "molecularly overdetermined" than they are today, and thus more subject to physical molding forces (Newman, 1995). The molecular processes that drive morphogenetic changes in contemporary organisms can then be considered in terms of the order in which they may have been acquired during the course of evolution, and how their relative efficacies as determinative events may have changed.

The particular morphogenetic events which have been discussed here — aggregation and disaggregation, condensation, segmentation, epithelialization and lumen formation, contraction — all involve the consequences of variations in adhesive and tensile forces, and therefore can be accommodated within a view that seeks to account for tissue changes by the generic properties of elasticoviscous materials. It must be noted, however, that certain unusual connective tissue reorganization processes are more difficult to accommodate into this picture. One such example is the "convergent extension" of the archenteron during amphibian gastrulation, in which cells intercalate to narrow the tissue and at the same time to move it forward (Keller *et al.*, 1985). This movement, which can occur in the absence of the substratum upon which the tissue normally moves (Keller and Jansa, 1992) and even in isolated explants (Shih and Keller, 1992), violates the expectation that a tissue will "round up" as a result of its generic elasticoviscous properties (Steinberg and Poole, 1982; Newman and Comper, 1990). It is possible that such movements, although they appear unitary, evolved as composites of the more basic generic processes discussed in this review.

Acknowledgements

This work was supported, in part, by grants from the National Science Foundation and National Institutes of Health to S.A.N., and from the Oklahoma Center for the Advancement of Science and Technology to J.J.T.

References

Akitaya, T. and Bronner-Fraser, M (1992) Expression of cell adhesion molecules during initiation and cessation of neural crest cell migration. *Dev. Dyn.* **194**:12–20.

Armstrong, P.B. (1989) Cell sorting out: The self-assembly of tissues in vitro. *Crit. Rev. Biochem. and Mol. Biol.* **24**:119–149.

Atchley, W.R. and Hall, B.K. (1991) A model for development and evolution of complex morphological structures. *Biol. Rev. Camb. Philos. Soc.* **66**:101–157.

Aufderheide, E., Chiquet-Ehrismann, R. and Ekblom, P. (1987) Epithelial-mesenchymal interactions in the developing kidney lead to expression of tenascin in the mesenchyme. *J. Cell.* **105**:599–608.

Bell, E., Ivarsson, B. and Merrill, C. (1979) Production of a tissue-like structure by contraction of collagen lattices by human fibroblasts of different proliferative potential in vitro. *Proc. Natl. Acad. Sci. U.S.A.* **76**:1274–1278.

Bellows, C.G., Melcher, A.H. and Aubin, J.E. (1981) Contraction and organization of collagen gels by cells cultured from periodontal ligament, gingiva and bone suggest functional differences between cell types. *J. Cell Sci.* **50**:299–314.

Bellows, C.G., Melcher, A.H. and Aubin, J.E. (1982) Association between tension and orientation of periodontal ligament fibroblasts and exogenous collagen fibres in collagen gels in vitro. *J. Cell Sci.* **58**:125–138.

Beloussov, L.V., Dorfman, J.G. and Cherdantzev, V.G. (1975) Mechanical stresses and morphological patterns in amphibian embryos. *J. Embryol. Exp. Morphol.* **34**:559–574.

Bilozur, M.E. and Hay, E.D. (1989) Cell migration into neural tube lumen provides evidence for the "fixed cortex" theory of cell motility. *Cell. Motil. Cytoskeleton* **14**:469–484.

Birkedale-Hansen, H., Moore, W.G., Bodden, M.K., Windsor, L.J., Birkedale-Hansen, B., DeCarlo, A. and Engler, J.A. (1994) Matrix metalloproteinases: A review. *Crit. Rev. Oral Biol. Med.* **4**:197–250.

Boissonade, J., Dulos, E. and DeKepper, P. (1994) Turing patterns: From myth to reality. *In:* Chemical Waves and Patterns. (Eds: Kapral, R. and Showalter, K.) Kluwer, Boston (in press).

Bronner-Fraser, M., Wolf, J.J. and Murray, B.A. (1992) Effects of antibodies against N-cadherin and N-CAM on the cranial neural crest and neural tube. *Dev. Biol.* **153**:291–301.

Castets, V., Dulos, E., Boissonade, J. and DeKepper, P. (1990) Experimental evidence of a sustained standing Turing-type nonequilibrium chemical pattern. *Phys. Rev. Lett.* **64**:2953–2956.

Chevallier, A., Kieny, M., Mauger, A. and Sengel, P. (1977) Developmental fate of the somitic mesoderm in the chick embryo. In: *Vertebrate Limb and Somite Morphogenesis.* (Eds: Ede, D.A., Hinchliffe, J.R. and Balls, M.) Cambridge University Press, Cambridge, 421–432.

Chiquet-Ehrismann, R. (1991) Anti-adhesive molecules of the extracellular matrix. *Curr. Opin. Cell. Biol.* **3**:800–804.

Christ, B., Jacob, H.J. and Jacob, M. (1977) Experimental analysis of the origin of the wing musculature in avian embryos. *Anat. Embryol. Berl.* **150**:171–186.

Chuong, C.M. and Edelman, G.M. (1985) Expression of cell-adhesion molecules in embryonic induction. I. Morphogenesis of nestling feathers. *J. Cell. Biol.* **101**:1009–1026.

Chuong, C.M., Oliver, G., Ting, S.A., Jegalian, B.G., Chen, H.M. and De Robertis, E.M. (1990) Gradients of homeoproteins in developing feather buds. *Development* **110**:1021–1030.

Coelho, C.N., Upholt, W.B. and Kosher, R.A. (1993) Ectoderm from various regions of the developing chick limb bud differentially regulates the expression of the chicken homeobox-containing genes GHox-7 and GHox-8 by limb mesenchymal cells. *Dev. Biol.* **156**:303–306.

Comper, W.D., Pratt, L., Handley, C.J. and Harper, G.S. (1987) Cell transport in model extracellular matrices. *Arch. Biochem. Biophys.* **252**:60–70.

Cremer, H., Lange, R., Christoph, A., Plomann, M., Vopper, G., Roes, J., Brown, R., Baldwin, S., Kraemer, P., Scheff, S., *et al.* (1994) Inactivation of the N-CAM gene in mice results in size reduction of the olfactory bulb and deficits in spatial learning. *Nature* **367**:455–459.

Darby, I., Skalli, O., Gabbiani, G. (1990) Alpha-smooth muscle actin is transiently expressed by myofibroblasts during experimental wound healing. *Lab. Invest.* **63**:21–29.

de Gennes, P.G. (1992) Soft matter. *Science* **256**:495–497.

Delvoye, P., Wiliquet, P., Leveque, J.L., Nusgens, B.V. and Lapiere, C.M. (1991) Measurement of mechanical forces generated by skin fibroblasts embedded in a three-dimensional collagen gel. *J. Invest. Dermatol.* **97**:898–902.

Dollé, P., Ruberte, E., Kastner, P., Petkovich, M., Stoner, C.M., Gudas, L.J., and Chambon, P (1989) Differential expression of genes encoding, alpha, beta and retinoic acid receptors and CRABP in the developing limbs of the mouse. *Nature* **342**:702–705.

Downie, S.A. and Newman, S.A. (1994) Morphogenetic differences between fore and hind limb precartilage mesenchyme: Relation to mechanisms of skeletal pattern formation. *Dev. Biol.* **162**:195–208.

Downie, S.A. and Newman, S.A. (1995) Different roles for fibronectin in the generation of fore and hind limb precartilage condensations. *Develop. Biol.* **172**: 519–530.

Duband, J.L., Volberg, T., Sabanay, I., Thiery, J.P. and Geiger, B. (1988) Spatial and temporal distribution of the adherens-junction-associated adhesion molecule A-CAM during avian embryogenesis. *Development* **103**: 325–344.

Duboule, D. (1992) The vertebrate limb: A model system to study the Hox/Hom gene network during development and evolution. *Bioessays* **14**:375–384.

Ehrlich, H.P. (1988) Wound closure: evidence of cooperation between fibroblasts and collagen matrix. *Eye* **2**:149–157.

Ekblom, M., Klein, G., Mugrauer, G., Fecker, L., Deutzmann, R., Timpl, R. and Ekblom, P. (1990) Transient and locally restricted expression of laminin A chain mRNA by developing epithelial cells during kidney organogenesis. *Cell* **60**:337–346.

Ekblom, P. (1981) Formation of basement membranes in the embryonic kidney: An immunohistological study. *J. Cell. Biol.* **91**:1–10.

Ekblom, P. (1992) Renal development. In: *The Kidney: Physiology and Pathophysiology.* 2nd ed (Eds: Seldin, D.W. and Giebish, G.) Raven Press, New York, 475–500.

Ekblom, P., Lehtonen, E., Saxen, L. and Timpl, R. (1981) Shift in collagen type as an early response to induction of the metanephric mesenchyme. *J. Cell. Biol.* **89**:276–283.

Erickson, C.A. and Isseroff, R.R. (1989) Plasminogen activator activity is associated with neural crest cell motility in tissue culture. *J. Exp. Zool.* **251**:123–133.

Erickson, C.A. and Perris, R. (1993) The role of cell-cell and cell-matrix interactions in the morphogenesis of the neural crest. *Dev. Biol.* **159**:60–74.

Ettinger, L. and Doljanski, F. (1992) On the generation of form by the continuous interactions between cells and their extracellular matrix. *Biol. Rev. Camb. Philos. Soc.* **67**:459–489.

Evanko, S.P. and Vogel, K.G. (1993) Proteoglycan synthesis in fetal tendon is differentially regulated by cyclic compression in vitro. *Arch. Biochem. Biophys.* **307**:153–164.

Fallon, J.F., Lopez, A., Ros, M.A., Savage, M.P., Olwin, B.B., and Simandl, B.K. (1994). FGF-2: apical ectodermal ridge growth signal for chick limb development. *Science.* **2641**:104–107.

Forgacs, G. (1995) On the possible role of cytoskeletal filamentous networks in intracellular signalling: an approach based on percolation. *J. Cell Sci.,* **108**: 2131–2144.

Forgacs, G., Jaikaria, N.S., Frisch, H.L. and Newman, S.A. (1989) Wetting, percolation and morphogenesis in a model tissue system. *J. Theor. Biol.* **140**:417–430.

Forgacs, G. and Newman, S.A. (1994) Phase transitions, interfaces, and morphogenesis in a network of protein fibers. *Int. Rev. Cytol.* **150**:139–148.

Forgacs, G., Newman, S.A., Obukhov, S.P. and Birk, D.E. (1991) Phase transition and morphogenesis in a model biological system. *Phys. Rev. Lett.* **67**:2399–2402.

Frenz, D.A., Akiyama, S.K., Paulsen, D.F. and Newman, S.A. (1989) Latex beads as probes of cell surface-extracellular matrix interactions during chondrogenesis: Evidence for a role for amino-terminal heparin-binding domain of fibronectin. *Dev. Biol.* **136**:87–96.

Frenz, D.A., Jaikaria, N.S. and Newman, S.A. (1989) The mechanism of precartilage mesenchymal condensation: A major role for interaction of the cell surface with the amino-terminal heparin-binding domain of fibronectin. *Dev. Biol.* **136**:97–103.

Gabbiani, G., Hirschel, B.J., Ryan, G.B., Statkov, P.R. and Majno, G. (1972) Granulation tissue as a contractile organ. A study of structure and function. *J. Exp. Med.* **135**:719–734.

George, E.L., Georges-Labouesse, E.N., Patel-King, R.S., Rayburn, H. and Hynes, R.O. (1993) Defects in mesoderm, neural tube and vascular development in mouse embryos lacking fibronectin. *Development* **119**:1079–1091.

Ghosh, S. and Comper, W.D. (1988) Oriented fibrillogenesis of collagen in vitro by ordered convection. *Connect. Tissue. Res.* **17**:33–41.

Glazier, J.A. and Graner, F. (1993) A simulation of the differential adhesion driven rearrangement of biological cells. *Phys. Rev.* **E47**:2128–2154.

Gould, S.E., Upholt, W.B. and Kosher, R.A. (1992) Syndecan 3: A member of the syndecan family of membrane-intercalated proteoglycans that is expressed in high amounts at the onset of chicken limb cartilage differentiation. *Proc. Natl. Acad. Sci. U.S.A.* **89**:3271–3275.

Gould, S.J. (1989) *Wonderful Life.* W.W. Norton, New York.

Graner, F. (1993) Can surface adhesion drive cell-rearrangement? Part I: Biological cell-sorting. *J. Theor. Biol.* **164**:455–476.

Graner, F. and Sawada, Y. (1993) Can cell surface adhesion drive cell rearrangement? Part II: A geometrical model. *J. Theor. Biol.* **164**:477–506.

Grinnell, F. (1994) Fibroblasts, myofibroblasts, and wound contraction. *J. Cell. Biol.* **124**:401–404.

Gross, J., Farinelli, W., Sadow, P., Anderson, R., and Bruns, R. (1995). On the mechanism of skin wound "contraction": A granulation tissue "knockout" with a normal phenotype. *Proc. Nat. Acad. Sci. USA* **92**: 5982–5986.

Gruenwald, P. (1943) Stimulation of nephrogenic tissues by normal and abnormal inducers. *Anat. Rec.* **86**:321–335.

Hall, B.K. (1983) Epithelial-mesenchymal interactions in cartilage and bone development. In: *Epithelial-Mesenchymal Interactions in Development.* (Eds: Sawyer, R.H. and Fallon, J.F.) Praeger, New York, 189–214.

Hall, B.K. and Miyake, T. (1992) The membranous skeleton: The role of cell condensations in vertebrate skeletogenesis. *Anat. Embryol. Berl.* **186**:107–124.

Halliday, N.L. and Tomasek, J.J. (1995) Mechanical properties of the extracellular matrix influence fibronectin fibril assembly in vitro. *Exp. Cell. Res.* **217**:109–117.

Halliday, N.L., Howard, E., Updike, D., and Tomasek, J.J. Mechanical properties of the extracellular matrix influence activation of 72 kD gelatinase. (submitted for publication).

Harper, G.S., Comper, W.D. and Preston, B.N. (1984) Dissipative structures in proteoglycan solutions. *J. Biol. Chem.* **259**:10582–10589.

Harris, A.K., Stopak, D. and Warner, P. (1984) Generation of spatially periodic patterns by a mechanical instability: A mechanical alternative to the Turing model. *J. Embryol. Exp. Morphol.* **80**:1–20.

Harris, A.K., Stopak, D. and Wild, P. (1981) Fibroblast traction as a mechanism for collagen morphogenesis. *Nature* **290**:249–251.

Hay, E.D. (Ed.) (1991) *Cell Biology of Extracellular Matrix.* 2nd ed. Plenum Press, New York.

Hay, E.D. (1993) Extracellular matrix alters epithelial differentiation. *Curr. Opin. Cell. Biol.* **5**:1029–1035.

Heintzelman, K.F., Phillips, H.M. and Davis, G.S. (1978) Liquid-tissue behavior and differential cohesiveness during chick limb budding. *J. Embryol. Exp. Morphol.* **47**:1–15.

Herring, S.W. (1993) Formation of the vertebrate face: Epigenetic and functional influences. *Amer. Zool.* **33**:472–483.

Higton, D.I.R. and James, D.W. (1964) The force of contraction of full-thickness wounds of rabbit skin. *Br. J. Surg.* **51**:462–466.

Hinchliffe, J.R. and Johnson, D.R. (1980) *The Development of the Vertebrate Limb.* Oxford University Press, Oxford.

Horstadius, S. and Sellman, S. (1946) Experimentelle untersuchungen uber die determination des knorpeligen kopfskelettes bei urodelen. *Nova Acta R. Soc. Scient. Upsal. Ser.* **4(13)**:1–170.

Hynes, R.O. (1990) *Fibronectins.* Springer-Verlag, New York.

Ingber, D.E. (1993) Cellular tensegrity: defining new rules of biological design that govern the cytoskeleton. *J. Cell Sci.* **104**:613–627.

Ingber, D.E., Dike, L., Hansen, L., Karp, S., Liley, H., Maniotis, A., McNamee, H., Mooney, D., Plopper, G., Sims, J., and Wang, N. (1994) Cellular tensegrity: exploring how mechanical changes in the cytoskeleton regulate cell growth, migration, and tissue pattern during morphogenesis. *Int. Rev. Cytol.* **150**:173–224.

Jones, F.S., Prediger, E.A. Bittner, D.A., DeRobertis, E.M. and Edelman, G.M. (1992) Cell adhesion molecules as targets for Hox genes: Neural cell adhesion molecule promoter activity is modulated by cotransfection with Hox-2.5 and -2.4. *Proc. Nat. Acad. Sci. U.S.A.* **89**:2086–2090.

Kazmierczak, J. and Degens, E.T. (1986) Calcium and the early eukaryotes. *Mitt. Geol.-Palaeont. Inst. Univ. Hamburg* **61**:1–20.

Karp, S.L., Ortiz-Arduan, A., Li, S. and Neilson, E.G. (1994) Epithelial differentiation of metanephric mesenchymal cells after stimulation with hepatocyte growth factor or embryonic spinal cord. *Proc. Natl. Acad. Sci. U.S.A.* **91**:5286–5290.

Keller, R.E., Danilchik, M., Gimlich, R. and Shih, J. (1985) Convergent extension by cell intercalation during gastrulation of *Xenopus laevis.* In: *Molecular Determinants of Animal Form.* (Ed: Edelman, G.M.) Alan R. Liss, New York, 111–141.

Keller, R. and Jansa, S. (1992) Xenopus gastrulation without a blastocoel roof. *Dev. Dyn.* **195**:162–176.

Klein, G., Langegger, M., Goridis, C. and Ekblom, P. (1988) Neural cell adhesion molecules during embryonic induction and development of the kidney. *Development* **102**:749–761.

Klein, G., Langegger, M., Timpl, R. and Ekblom, P. (1988) Role of laminin A chain in the development of epithelial cell polarity. *Cell* **55**:331–341.

Koch, J. (1917). The laws of bone architecture. *Am. J. Anat.* **21**:177–298.

Kolodney, M.S. and Elson, E.L. (1993) Correlation of myosin light chain phosphorylation with isometric contraction of fibroblasts. *J. Biol. Chem.* **268**:23850–23855.

Kolodney, M.S. and Wysolmerski, R.B. (1992) Isometric contraction by fibroblasts and endothelial cells in tissue culture: a quantitative study. *J. Cell. Biol.***117**:73–82.

Kornberg, L., Earp, H.S., Parsons, J.T., Schaller, M. and Juliano, R.L. (1992) Cell adhesion or integrin clustering increases phosphorylation of a focal adhesion–associated tyrosine kinase. *J. Biol. Chem.* **267**:23439–23442.

Kreis, T. and Vale, R. (Eds.) (1993) *Guidebook to the Extracellular Matrix and Adhesion Proteins.* Oxford University Press, Oxford.

Lambert, C.A., Soudant, E.P., Nusgens, B.V. and Lapiere, C.M. (1992) Pretranslational regulation of extracellular matrix macromolecules and collagenase expression in fibroblasts by mechanical forces. *Lab. Invest.* **66**:444–451.

Langille, R.M. and Hall, B.K. (1993) Pattern formation and the neural crest. In: *The Vertebrate Skull.* Vol. 1: Development. (Eds: Hanken, J. and Hall, B.K.) University of Chicago Press, Chicago, 77–111.

Larsen, W.J. (1992) Human Embryology. Churchill Livingstone, New York.

Lash, J.W., Saxén, L. and Ekblom, P. (1983) Biosynthesis of proteoglycans in organ cultures of developing kidney mesenchyme. *Exp. Cell. Res.* **147**:85–93.

Le Douarin, N.M., Dupin, E., Baroffio, A. and Dulac, C. (1992) New insights into the development of neural crest derivatives. *Int. Rev. Cytol.* **138**:269–314.

Leonard, C.M., Fuld, H.M., Frenz, D.A., Downie, S.A., Massagué, J. and Newman, S.A. (1991) Role of transforming growth factor-beta in chondrogenic pattern formation in the embryonic limb: stimulation of mesenchymal condensation and fibronectin gene expression by exogenous TGF-beta and evidence for endogenous TGF-beta-like activity. *Dev. Biol.* **145**:99–109.

Lin, Y.C. and Grinnell, F. (1993) Decreased level of PDGF-stimulated receptor autophosphorylation by fibroblasts in mechanically relaxed collagen matrices. *J. Cell. Biol.* **122**:663–672.

Lumsden, A.S.G. (1987) The neural crest contribution to tooth development in the mammalian embryo. In: *Developmental and Evolutionary Aspects of the Neural Crest.* (Ed: Maderson, P.F.A.) John Wiley and Sons, New York, 261–300.

MacKenzie, A., Ferguson, M.W. and Sharpe, P.T. (1992) Expression patterns of the homeobox gene, *Hox-8*, in the mouse embryo suggest a role in specifying tooth initiation and shape. *Development* **115**:403–420.

Mackenzie, A., Leeming, G.L., Jowett, A.K., Ferguson, M.W. and Sharpe, P.T. (1991) The homeobox gene *Hox 7.1* has specific regional and temporal expression patterns during early murine craniofacial embryogenesis, especially tooth development *in vivo* and *in vitro*. *Development* **111**:269–285.

Markwald, R.R., Bolender, D.L., Krug, E.L. and Lepera, R. (1990) Morphogenesis of precursor subpopulations of chicken limb mesenchyme in three dimensional collagen gel culture. *Anat. Rec.* **226**:91–107.

Massagué, J. (1990) The transforming growth factor-beta family. *Ann. Rev. Cell Biol.* **6**:597–641.

Mauch, C., Hatamochi, A., Scharffetter, K. and Krieg, T.(1988) Regulation of collagen synthesis in fibroblasts within a three-dimensional collagen gel. *Exp. Cell. Res.* **178**:493–503.

McGinnis, W. and Krumlauf, R. (1992) Homeobox genes and axial patterning. *Cell* **68**:283–302.

McNamee, H.P., Ingber, D.E. and Schwartz, M.A. (1993) Adhesion to fibronectin stimulates inositol lipid synthesis and enhances PDGF-induced inositol lipid breakdown. *J. Cell. Biol.***121**:673–678.

Meinhardt, H. (1982) *Models of Biological Pattern Formation.* Academic, New York.

Mochitate, K., Pawelek, P. and Grinnell, F. (1991) Stress relaxation of contracted collagen gels: Disruption of actin filament bundles, release of cell surface fibronectin, and down-regulation of DNA and protein synthesis. *Exp. Cell. Res.* **193**:198–207.

Nakagawa, S., Pawelek, P. and Grinnell, F. (1989) Extracellular matrix organization modulates fibroblast growth and growth factor responsiveness. *Exp. Cell. Res.* **182**:572–582.

Newman, S.A. (1977) Lineage and pattern in the developing wing bud. In: *Vertebrate Limb and Somite Morphogenesis.* (Eds: Ede, D.A., Hinchliffe, J.R. and Balls, M.) Cambridge University Press, Cambridge, 181–200.

Newman, S.A. (1980) Fibroblast progenitor cells of the embryonic chick limb. *J. Embryol. Exp. Morphol.* **56**: 191–200.

Newman, S.A. (1984) Vertebrate bones and violin tones: Music and the making of limbs. *The Sciences (NY Acad of Sciences)* **24**:38–43.

Newman, S.A. (1988) Lineage and pattern in the developing vertebrate limb. *Trends. Genet.* **4**:329–332.

Newman, S.A. (1993a) Is segmentation generic? *Bioessays* **15**:277–283.

Newman, S.A. (1993b) Why does a limb look like a limb? *Prog. Clin. Biol. Res.* **383A**:89–98.

Newman, S.A. (1994) Generic physical mechanisms of tissue morphogenesis: A common basis for development and evolution. *J. Evol. Biol.* **7**:467–488.

Newman, S.A. (1995) Interplay of genetics and physical processes of tissue morphogenesis in development and evolution: The biological fifth dimension. In: *Interplay of Genetic and Physical Processes in the Development of Biological Form.* (Eds: Beysens, D., Forgacs, G. and Gaill, F.) World Scientific, Singapore, 3–12.

Newman, S.A. and Comper, W.D. (1990) 'Generic' physical mechanisms of morphogenesis and pattern formation. *Development* **110**:1–18.

Newman, S.A., Frenz, D.A., Hasegawa, E. and Akiyama, S.K. (1987) Matrix-driven translocation: Dependence on interaction of amino-terminal domain of fibronectin with heparin-like surface components of cells or particles. *Proc. Natl. Acad. Sci. U.S.A.* **84**:4791–4795.

Newman, S.A., Frenz, D.A., Tomasek, J.J. and Rabuzzi, D.D. (1985) Matrix-driven translocation of cells and nonliving particles. *Science* **228**:885–889.

Newman, S.A. and Frisch, H.L. (1979) Dynamics of skeletal pattern formation in developing chick limb. *Science* **205**:662–668.

Nichols, D.H. (1986) Mesenchyme formation from the trigeminal placodes of the mouse embryo. *Am. J. Anat* **176**:19–31.

Niswander, L., Tickle, C., Vogel, A., Booth, I., and Martin G.R. (1993) FGF-4 replaces the apical ectodermal ridge and directs outgrowth and patterning of the limb. *Cell.* **75**, 579–587.

Noden, D.M. (1983) The role of the neural crest in patterning of avian cranial skeletal, connective, and muscle tissues. *Dev. Biol.* **96**:144–165.

Noden, D.M. (1984) Craniofacial development: New views on old problems. *Anat. Rec.* **208**:1–13.

Noden, D.M. (1988) Interactions and fates of avian craniofacial mesenchyme. *Development* **103 (Suppl)**:121–140.

Nogawa, H. and Nakanishi, Y. (1987) Mechanical aspects of the mesenchymal influence on epithelial branching morphogenesis of mouse salivary gland. *Development* **101**:491–500.

Oberlender, S.A. and Tuan, R.S. (1994) Expression and functional involvement of N-cadherin in embryonic limb chondrogenesis. *Development* **120**:177–187.

Oster, G.F., Murray, J.D. and Harris, A.K. (1983) Mechanical aspects of mesenchymal morphogenesis. *J. Embryol. Exp. Morphol.* **78**:83–125.

Ouyang, Q. and Swinney, H. (1991) Transition from a uniform state to hexagonal and striped Turing patterns. *Nature* **352**:610–612.

Peacock, E.E. (1984) *Wound Repair*. 3rd ed. WB Saunders Co, Philadelphia.

Pilcher, B.K., Kim, D.W., Carney, D.H. and Tomasek, J.J. (1994) Thrombin stimulates fibroblast-mediated collagen lattice contraction by its proteolytic activated receptor. *Exp. Cell. Res.* **211**:368–373.

Pilcher, B.K., Levine, N.S. and Tomasek, J.J. (1995) Thrombin promotion of isometric contraction in fibroblasts: Its extracellular action. *Plastic Reconstr. Surg.* **96**:1188–1195.

Preston, B.N., Laurent, T.C., Comper, W.D. and Checkley, G.J. (1980) Rapid polymer transport in concentrated solutions through the formation of ordered structures. *Nature* **287**:499–503.

Riddle, R.D., Johnson, R.L., Laufer, E. and Tabin, C. (1993) Sonic hedgehog mediates the polarizing activity of the ZPA. *Cell* **75**:1401–1416.

Rodriguez-Boulan, E. and Nelson, W.J. (Eds.) (1993) *Epithelial and Neuronal Cell Polarity*. Company of Biologists, Cambridge (J. Cell. Sci. Suppl. 17).

Ros, M.A., Lyons, G., Kosher, R.A., Upholt, W.B., Coelho, C.N. and Fallon, J.F. (1992) Apical ridge dependent and independent mesodermal domains of GHox-7 and GHox-8 expression in chick limb buds. *Development* **116**:811–818.

Ros, M.A., Lyons, G.E., Mackem, S. and Fallon, J.F. (1994) Recombinant limbs as a model to study homeobox gene regulation during limb development. *Develop. Biol.* **166**:59–72.

Rosen, E.M., Nigam, S.K. and Goldberg, I.D. (1994) Scatter factor and the c-met receptor: A paradigm for mesenchymal/ epithelial interaction. *J. Cell Biol.* **127**:1783–1787.

Rudolph, R., Vande Berg, J. and Ehrlich, H.P. (1992) *Wound Contraction and Scar Contracture*. WB Sanders, Philadelphia.

Russell, E.S. (1982). *Form and Function*. Univ. of Chicago Press, Chicago.

Saga, Y., Yagi, T., Ikawa, Y., Sakakura, T. and Aizawa, S. (1992) Mice develop normally without tenascin. *Genes. Dev.* **6**:1821–1831.

Santos, O.F.P., Barros, E.J.G., Yang, X.-M., Matsumoto, K., Nakamura, T., Park, M. and Nigam, S.K. (1994) Involvement of hepatocyte growth factor in kidney development. *Develop. Biol.* **163**:525–529.

Satokata, I. and Maas, R. (1994) Msx1 deficient mice exhibit cleft palate and abnormalities of craniofacial and tooth development [see comments]. *Nat. Genet.* **6**:348–356.

Saunders, J.W. (1948) The proximo-distal sequence of origin of the parts of the chick wing and the role of the ectoderm. *J. Exp. Zool.* **108**:363–402.

Saunders, J.W. and Fallon, J.F. (1967) Cell death in embryonic morphogenesis. In: *Major Problems in Developmental Biology*. (Ed: Locke, M.) Academic Press, New York, 289–314.

Saxén, L. (1970) Failure to demonstrate tubule induction in a heterologous mesenchyme. *Dev. Biol.* **23**:511–523.

Schaller, M.D. and Parsons, J.T. (1994) Focal adhesion kinase and associated proteins. *Curr. Opin. Cell. Biol.* **6**:705–710.

Schaller, M.D. and Parsons, J.T. (1994) Focal adhesion kinase and associated proteins. *Curr. Opin. Cell Biol.* **6**:705–710.

Schultz, R.J. and Tomasek, J.J. (1990) Cellular structure and interactions. In: *Dupuytren's Disease* (Eds: McFarlane, R.M., McGrouther, D.A., Flint, M.H.) Churchill Livingston, Edinburgh, 86–98.

Shih, J. and Keller, R. (1992) Cell motility driving mediolateral intercalation in explants of Xenopus laevis. *Development* **116**:901–914.

Shubin, N.H. and Alberch, P. (1986) A morphogenetic approach to the origin and basic organization of the tetrapod limb. *Evol. Biol.* **20**:319–387.

Singer, I.I., Kawka, D.W., Kazazis, D.M. and Clark, R.A. (1984) In vivo co-distribution of fibronectin and actin fibers in granulation tissue: Immunofluorescence and electron microscope studies of the fibronexus at the myofibroblast surface. *J. Cell. Biol.* **98**:2091–2106.

Smith, M.M. and Hall, B.K. (1993) A developmental model for evolution of the vertebrate exoskeleton and teeth: The role of cranial and trunk neural crest. *Evol. Biol.* **27**:387–448.

Sonnenberg, A., Linders, C.J., Modderman, P.W., Damsky, C.H., Aumailley, M. and Timpl, R. (1990) Integrin recognition of different cell-binding fragments of laminin (P1, E3, E8) and evidence that alpha 6 beta 1 but not alpha 6 beta 4 functions as a major receptor for fragment E8. *J. Cell. Biol.* **110**:2145–2155.

Sorokin, L., Sonnenberg, A., Aumailley, M., Timpl, R. and Ekblom, P. (1990) Recognition of the laminin E8 cell-binding site by an integrin possessing the alpha 6 subunit is essential for epithelial polarization in developing kidney tubules. *J. Cell. Biol.* **111**:1265–1273.

Squier, C.A. (1981) The effect of stretching on formation of myofibroblasts in mouse skin. *Cell Tissue Res.* **220(2)**:325–335.

Stauffer, D. (1985) *Introduction to Percolation Theory*. Taylor and Francis, London.

Steinberg, M.S. (1978) Specific cell ligands and the differential adhesion hypothesis: How do they fit together? In: *Specificity of Embryological Interactions*. (Ed: Garrod, D.R.) London: Chapman and Hall, 97–130.

Steinberg, M.S. and Poole, T.J. (1982) Liquid behavior of embryonic tissues. In: *Cell Behavior*. (Eds: Bellairs, R. and Curtis, A.S.G.) Cambridge: Cambridge University Press, 583–607.

Steinberg, M.S. and Takeichi, M. (1994) Experimental specification of cell sorting, tissue spreading, and specific spatial patterning by quantitative differences in cadherin expression. *Proc. Natl. Acad. Sci. U.S.A.* **91**:206–209.

Stern, C.D., Fraser, S.E., Keynes, R.J. and Primmett, D.R. (1988) A cell lineage analysis of segmentation in the chick embryo. *Development* **104(Suppl)**:231–244.

Stopak, D. and Harris, A.K. (1982) Connective tissue morphogenesis by fibroblast traction. I. Tissue culture observations. *Dev. Biol.* **90**:383–398.

Stopak, D., Wessells, N.K. and Harris, A.K. (1985) Morphogenetic rearrangement of injected collagen in developing chicken limb buds. *Proc. Natl. Acad. Sci. U.S.A.* **82**:2804–2808.

Sumida, H., Ashcraft, R.A. Jr, and Thompson, R.P. (1989) Cytoplasmic stress fibers in the developing heart. *Anat. Rec.* **223**:82–89.

Takahashi, Y., Bontoux, M. and Le Douarin, N.M. (1991) Epithelio-mesenchymal interactions are critical for Quox 7 expression and membrane bone differentiation in the neural crest derived mandibular mesenchyme. *Embo. J.* **10**:2387–2393.

Takeichi, M. (1988) The cadherins: Cell-cell adhesion molecules controlling animal morphogenesis. *Development* **102**:639–655.

Takeichi, M. (1991) Cadherin cell adhesion receptors as a morphogenetic regulator. *Science* **251**:1451–1455.

Thaller, C. and Eichele, G. (1987) Identification and spatial distribution of retinoids in the developing chick limb bud. *Nature* **327**:625–628.

Thiery, J.P., Duband, J.L., Rutishauser, U. and Edelman, G.M. (1982) Cell adhesion molecules in early chicken embryogenesis. *Proc. Natl. Acad. Sci. U.S.A.* **79**:6737–6741.

Timpl, R. (1989) Structure and biological activity of basement membrane proteins. *Eur. J. Biochem.* **180**:487–502.

Tomasek, J.J. and Akiyama, S.K. (1992) Fibroblast-mediated collagen gel contraction does not require fibronectin-alpha 5 beta 1 integrin interaction. *Anat. Rec.* **234**:153–160.

Tomasek, J.J. and Haaksma, C.J. (1991) Fibronectin filaments and actin microfilaments are organized into a fibronexus in Dupuytren's diseased tissue. *Anat. Rec.* **230**:175–182.

Tomasek, J.J., Haaksma, C.J., Eddy, R.J. and Vaughan, M.B. (1992) Fibroblast contraction occurs on release of tension in attached collagen lattices: dependency on an organized actin cytoskeleton and serum. *Anat. Rec.* **232**:359–368.

Tomasek, J.J., Hay, E.D. and Fujiwara, K. (1982) Collagen modulates cell shape and cytoskeleton of embryonic corneal and fibroma fibroblasts: Distribution of actin, alpha-actinin, and myosin. *Dev. Biol.* **92**:107–122.

Tomasek, J.J., Mazurkiewicz, J.E. and Newman, S.A. (1982) Nonuniform distribution of fibronectin during avian limb development. *Dev. Biol.* **90**:118–126.

Toné, S., Tanaka, S., Minatogawa, Y. and Kido, R. (1994) DNA fragmentation during the programmed cell death in the chick limb buds. *Exp. Cell. Res.* **215**:234–236.

Tsarfaty, I., Resau, J.H., Rulong, S., Keydar, I., Faletto, D.L. and Vande Woude, G.F. (1992) The met proto-oncogene receptor and lumen formation. *Science* **257**:1258–1261.

Tsarfaty, I., Rong, S., Resau, J.H., Rulong, S., da Salva, P.O. and Vande Woude, G.F. (1994) The met proto-oncogene mesenchymal to epithelial cell conversion. *Science* **263**:98–101.

Turing, A. (1952) The chemical basis of morphogenesis. *Phil. Trans. Roy. Soc. Lond. B* **237**:37–72.

Turner, M.L. (1992) Cell adhesion molecules: a unifying approach to topographic biology. *Biol. Rev. Camb. Philos. Soc.* **67**:359–377.

Unemori, E.N. and Werb, Z. (1986) Reorganization of polymerized actin: a possible trigger for induction of procollagenase in fibroblasts cultured in and on collagen gels. *J. Cell. Biol.* **103**:1021–1031.

Unsworth, B. and Grobstein, C. (1970) Induction of kidney tubules in mouse metanephrogenic mesenchyme by various embryonic mesenchymal tissues. *Dev. Biol.* **21**:547–556.

Valinsky, J.E. and Le Douarin, N.M. (1985) Production of plasminogen activator by migrating cephalic neural crest cells. *Embo. J.* **4**:1403–1406.

Van Obberghen-Schilling, E., Roche, N.S., Flanders, K.C., Sporn, M.B. and Roberts, A. (1988) Transforming growth factor beta-1 positively regulates its own expression in normal and transformed cells. *J. Biol. Chem.* **263**:7741–7746.

Vestweber, D., Kemler, R. and Ekblom, P (1985) Cell-adhesion molecule uvomorulin during kidney development. *Dev. Biol.* **112**:213–221.

Vogel, K.G. and Koob, T.J. (1989) Structural specialization in tendons under compression. *Int. Rev. Cytol.* **115**:267–293.

Waddington, C.H. (1957) *The Strategy of the Genes.* Allen and Unwin, London.

Wang, N., Butler, J.P. and Ingber, D.E. (1993) Mechanotransduction across the cell surface and through the cytoskeleton [see comments]. *Science* **260**:1124–1127.

Wartiovaara, J., Stenman, S. and Vaheri, A. (1976) Changes in expression of fibroblast surface antigen (SFA) during cytodifferentiation and heterokaryon formation. *Differentiation* **5**:85–89.

Welch, M.P., Odland, G.F. and Clark, R.A. (1990) Temporal relationships of F-actin bundle formation, collagen and fibronectin matrix assembly, and fibronectin receptor expression to wound contraction. *J. Cell. Biol.* **110**:133–145.

Widelitz, R.B., Jiang, T.X., Murray, B.A. and Chuong, C.M. (1993) Adhesion molecules in skeletogenesis: II. Neural cell adhesion molecules mediate precartilaginous mesenchymal condensations and enhance chondrogenesis. *J. Cell Physiol.* **156**:399–411.

Wolpert, L. and Hornbruch, A. (1990) Double anterior chick limb buds and models for cartilage rudiment specification. *Development* **109**:961–966.

Wood, A. and Thorogood, P. (1994) Patterns of cell behaviour underlying somitogenesis and notochord formation in intact vertebrate embryos. *Develop. Dynamics* **201**:151–167.

Yokouchi, Y., Nakazato, S., Yamamoto, M., Goto, Y., Kameda, T., Iba, H., and Kuroiwa, A. (1995). Misexpression of *Hoxa-13* induces cartilage homeotic transformation and changes cell adhesiveness in chick limb buds. *Genes Devel.* **9**:2509–2522.

Zakeri, Z., Quaglino, D. and Ahuja, H.S. (1994) Apoptotic cell death in the mouse limb and its suppression in the hammertoe mutant. *Dev. Biol.* **165**:294–297.

Zanetti, N.C., Dress, V.M., and Solursh, M. (1990). Comparison between ectoderm-conditioned medium and fibronectin in their effects on chondrogenesis by limb bud mesenchymal cells. *Dev. Biol.* **139**:383–395.

Zwilling, E. (1964) Development of fragmented and dissociated limb bud mesoderm. *Develop. Biol.* **9**:20–37.

Index